IUTAM SYMPOSIUM ON MECHANICS OF PASSIVE AND ACTIVE FLOW CONTROL

FLUID MECHANICS AND ITS APPLICATIONS
Volume 53

Series Editor: R. MOREAU
MADYLAM
Ecole Nationale Supérieure d'Hydraulique de Grenoble
Boîte Postale 95
38402 Saint Martin d'Hères Cedex, France

Aims and Scope of the Series

The purpose of this series is to focus on subjects in which fluid mechanics plays a fundamental role.

As well as the more traditional applications of aeronautics, hydraulics, heat and mass transfer etc., books will be published dealing with topics which are currently in a state of rapid development, such as turbulence, suspensions and multiphase fluids, super and hypersonic flows and numerical modelling techniques.

It is a widely held view that it is the interdisciplinary subjects that will receive intense scientific attention, bringing them to the forefront of technological advancement. Fluids have the ability to transport matter and its properties as well as transmit force, therefore fluid mechanics is a subject that is particulary open to cross fertilisation with other sciences and disciplines of engineering. The subject of fluid mechanics will be highly relevant in domains such as chemical, metallurgical, biological and ecological engineering. This series is particularly open to such new multidisciplinary domains.

The median level of presentation is the first year graduate student. Some texts are monographs defining the current state of a field; others are accessible to final year undergraduates; but essentially the emphasis is on readability and clarity.

For a list of related mechanics titles, see final pages.

IUTAM Symposium on

Mechanics of Passive and Active Flow Control

Proceedings of the IUTAM Symposium
held in Göttingen, Germany,
7–11 September 1998

Edited by

G.E.A. MEIER

DLR, Göttingen, Germany

and

P.R. VISWANATH

NAL, Bangalore, India

SPRINGER SCIENCE+BUSINESS MEDIA, B.V.

A C.I.P. Catalogue record for this book is available from the Library of Congress.

ISBN 978-94-010-5826-1 ISBN 978-94-011-4199-4 (eBook)
DOI 10.1007/978-94-011-4199-4

Printed on acid-free paper

Chairmen of the IUTAM Symposium

G.E.A. Meier (DLR) P.R. Viswanath (NAL)

Scientific Committee

J. Cousteix (France)
A. Gyr (Switzerland)
V.V. Kozlov (Russia)
G.E.A. Meier (Germany), Chairman
P. Moin (USA)
P. A. Monkewitz, (Switzerland)
R. Narasimha, (India)
T. Tatsumi, (Japan)
P.R. Viswanath, (India), Chairman
I. Wygnanski, (USA)

Local Organizing Committee

A. Dinkelacker
B. Gölling
H.-P. Kreplin
G.E.A. Meier
H. Rosemann
E. Winkels (Conference Secretary)

Symposium Sponsors

Deutsches Zentrum für Luft- und Raumfahrt e.V., Köln, Germany
International Union of Theoretical and Applied Mechanics (IUTAM)
Offset- und Dissertationsdruck Jürgen Kinzel, Göttingen, Germany
Sartorius AG, Göttingen, Germany
Sparkasse Göttingen, Germany
Volksbank Göttingen e.G., Germany

Contents

Session 4

Session 5

II. ACTIVE CONTROL METHODS

Session 1

Session 2

Session 3

Session 4

Session 5

Session 6

Preface

The call for papers for the IUTAM-Symposium on *Mechanics of Passive and Active Flow Control* brought an overwhelming response of applications for contributions. Finally 12 invited lectures, 48 papers and 23 posters were selected by the Scientific Committee to be presented in the conference. 58 papers are published in this volume. Due to the limited number of pages available, poster presentations could not be considered for publication. The editors would like to thank all the members of the Scientific Committee for their very valuable assistance.

The papers presented at the IUTAM Symposium were classified under three groups devoted to

- Passive Control Methods,
- Active Control Methods and
- Control Concepts.

This was done to contrast at first between the passive techniques where the control power is mainly supplied by the flow itself and the active techniques where the power is provided by external sources; the third group was devoted to control concepts for presenting methods of control theory and new techniques of flow control.

The opening lecture was delivered by Prof. T. Tatsumi on Fluid Mechanics in the turn of the century, which was followed by a lecture on turbulent drag reduction by Prof. Fazle Hussain

In the first group dealing with *Passive Control Methods* problems of transonic airfoil flow have been presented in several papers. Especially the problem of shock control was in the center of interest because this is a practical problem of civil transport aircraft right now. Among the different control concepts the methods of passive venting and the use of bumps in the vicinity of the shock root have been presented in several papers. Also the related problem of buffet excitation a wing at high incidence and control by passive venting was addressed.

The next topic in sessions 2 and 3 of group 1 was the use of mechanical means to influence the flow especially the phenomenon of flow separation. Dynamic stall of helicopter airfoils can be influenced by changes of the shape of the airfoil with the angle of attack, micro-vortex generators can influence the turbulent flow separation and also flaps which are models of bird feathers which are bent up in back-flow areas can be used in technical applications. In session 4 the influence of riblets on the turbulent boundary layer was discussed elaborately in three lectures. Session 5 was devoted to investigations of the influence of venting on the drag of bluff bodies and the influence of vortices and jets on turbulent flows. All these papers showed that passive flow control is already a practically used tool in drag reduction and separation control.

In the second group devoted to *Active Control Methods,* there are several papers considering receptivity problems and the influence of suction on different types of boundary layer instabilities. The second session in this group dealt with active flow control by micro systems especially with respect to the influence on boundary layers and separation. Session 3 and 4 dealt with the influence of heat, sound and surface oscillations on boundary layers. Also the wake problem and the separation control with active mechanical means have been treated in this session. The session number 5 was devoted to the wake control problem especially in case of blunt bodies using jet actuators, rotating cylinders and moving surfaces. Session number 6 dealt with papers on control by acoustic excitation from outer and inner acoustic sources and the related method of periodic blowing and suction.

In the third group on *Control Concepts,* several papers addressed the problem of active control of turbulent shear flows for drag reduction, separation control and evaluation of different control schemes; also the theoretical background of instabilities and the possible strategies to interfere have been discussed elaborately.

The symposium brought together the different groups working in the field of flow control from Europe, Asia and USA. The meeting was just in time because some methods of flow control have recently been introduced into technical applications successfully. At least some promising results for most of the described control techniques are already available.

The symposium was organized in cooperation between the DLR Institute of Fluid Mechanics in Göttingen, Germany and the National Aerospace Laboratories (NAL) in Bangalore, India. In the preparation of the conference a lot of valuable contributions have been made by the Indian colleagues. The German side put a lot of effort in the practical local arrangements for the conference. Especially Dr. Kreplin, Mrs. Sander and Mrs. Winkels have to be mentioned here for arranging the meeting and preparing also the social events. The Local Organizing Committee was also supported by S. Alberti, M. Bodmann, D. Brennecke, T. Dardart, G. Elias, H. Frahnert, P. Gaab, M. Jelinski, E. Keppler, S. Koch, M. Rütten, N. Schumann, K. Zysk. The social events have been highlighted by a reception by the Lord mayor of the town of Göttingen in the ancient city hall of Göttingen and a common dinner party in the restaurant facilities of the DLR.

The Organizers of this Symposium are very grateful to IUTAM for the opportunity to have this conference on such an interesting topic. All the participants have recommended to repeat such a meeting in the future after a time of about five years. The financial support of the IUTAM and of the listed sponsors and institutions are gratefully acknowledged. We also appreciate the offer of Kluwer Academic Publishers to publish the proceedings of the symposium.

Göttingen, February 1999

The Editors
Gerd E.A. Meier
P.R. Viswanath

OPENING ADDRESS

TOMOMASA TATSUMI
International Institute for Advanced Studies
9-3 Kizugawadai, Kizu-cho, Kyoto 619-0225 Japan

Professor Meier, Professor Viswanath, and Ladies and Gentlemen!
It is my great pleasure to give an official IUTAM address in the Opening Session of this Symposium on behalf of the President of IUTAM, Professor Werner Schiehlen.

The purpose of the IUTAM address is principally to present the long history of IUTAM and its recent activities and goals to the Symposium participants, but here I would like to note the significance of the fact that the present Symposium is held in Goettingen or the fatherland of "modern fluid mechanics".

1. IUTAM, ICTAM and IUTAM Symposia

IUTAM or International Union of Theoretical and Applied Mechanics was settled in 1947, shortly after the end of the Second World War as an international scientific organization under ICSU or International Council of Scientific Unions. Now it has a history of about half a century.

It should be noted that IUTAM has its prehistory of a quarter of a century as ICAM or "International Congress of Applied Mechanics" , which dates back to 1924. Still two years before, an international conference on "hydro- and aerodynamics" was held in 1922 in Innsbruck, Austria, under the joint chairmanship of *Theodore von Karman*, then the Director of Aerodynamical Institute of Aachen Polytechnic and *Tullio Levi-Civita*, an Italian mathematician in Rome (see [1]).

Karman's idea of having such an international meeting was his recognition of a large gap between the rapid development in the field of "hydro- and aerodynamics" with interesting new results being obtained on one hand, and only a little space devoted to such problems at scientific meetings and scarce personal contacts between scientists in the field on the other. He proposed to break the dependence of "hydro- and aerodynamics" from its mother disciplines as mathematics, physics and engineerings and call an informal meeting of people interested in the field.

His proposal was warmly supported by Levi-Civita and they succeeded in organizing the above-mentioned conference in September 1922. Thirty three scientists gathered from Germany, Austria, Holland, Scandinavia and Italy. It may be interesting to note two big names of attendants, *Ludwig Prandtl*, Karman's former teacher from Goettingen, and *Werner Heisenberg* from Muenchen who was by *Arnold Sommerfeld* to report on his research on turbulences.

Heisenberg's work was on the stability of the plane Poiseuille flow and later published as a fifty pages dissertation [2]. He was able to derive in this paper the lower

G.E.A. Meier and P.R. Viswanath (eds.),
IUTAM Symposium on Mechanics of Passive and Active Flow Control, 1–4.

branch of the neutral stability curve but not to obtain the critical Reynolds number according to the insufficient accuracy of the eigen functions employed. This line of research was succeeded by *W. Tollmien* and *H. Schlichting* who completed the stability theory of the laminar boundary layer, which is now known by the name of the "Tollmien-Schlichting wave".

In addition to the success of the Innsbruck Conference, Karman found a new enthusiastic partner in young Dutch scientists, *Jan Burgers*. They planned to have a wider conference, no longer restricted to "hydro- and aerodynamics" but embracing the whole fields of "applied mechanics". Their devoted efforts resulted in organizing the First International Congress of Applied Mechanics in April 1924, Delft, The Netherlands. This Congress turned out to be a great success with over two hundred participants and seventy six papers presented.

It was decided that the Congress would assemble every fourth year in principle, and this decision has been kept up to the Fifth Congress held in 1938, Cambridge, USA. After eight years intermission due to the Second World War, the Sixth Congress was held in 1946, Paris, France, under the auspices of IUTAM with the new title of the International Congress of Theoretical and Applied Mechanics or ICTAM.

The most recent Congress is the Nineteenth ICTAM which was held in 1996, Kyoto, Japan, and the next Congress is being held as the Twentieth ICTAM in the year of 2000, Chicago, USA.

Another Scientific activity of IUTAM is to support yearly eight to ten international symposia on various subjects of Applied Mechanics. These "IUTAM Symposia" are selected from numerous proposals through screening by the IUTAM Symposia Panels and finally accepted by the General Assembly of IUTAM. The Proceedings of the IUTAM Symposia are usually published by the Kluwer in Fluid and Solid Mechanics series. We are pleased to hear that IUTAM Symposia enjoy a high esteem among others. The present Symposium on "Mechanics of Passive and Active Flow Control" or "FLOWCON" is nothing but one of such IUTAM Symposia.

2. **The Century of Fluid Mechanics**

In the beginning of my Address, I called Goettingen as the fatherland of "modern fluid mechanics". In this context, I referred to the fact that the era of "modern fluid mechanics" was really opened by *Ludwig Prandtl* with the start of this century.

Until the end of the last century, the theoretical contents of fluid mechanics essentially consisted of the potential flows and wave motions in an ideal fluid and slow motions in a viscous fluid (see [3]). Though the Navier-Stokes equation was already known as the fundamental equation for the viscous fluid by the middle of the last century, no general mathematical method was available for dealing with this nonlinear dissipative differential equation.

Such a stalemate was broken by Prandtl [4], who introduced the notion of the "boundary layer" existing between an inviscid flow and its solid boundary. In this way, all potential flows of an ideal fluid have been made to satisfy the non-slip condition at their solid boundaries.

The boundary layer theory has proved to be a very powerful means having universal applicability to the viscous flows at large Reynolds numbers. Probably, the merit of the boundary layer theory exists in decomposing the flow field of a viscous fluid at large Reynolds numbers into two parts, a global flow region and thin layers along the solid boundaries, and apply different mathematical approximation to different flow domains. Such a "composite method" is not only necessary but useful for solving a non-linear dissipative differential equation such as the Navier-Stokes equation at large Reynolds numbers.

Actually, almost all fluid flows of theoretical and practical importance have been solved using this method during this century (see [5]). In view of such achievements in the real fluid mechanics and remarkable progress in experimental and numerical fluid mechanics, the present century may be called the Century of Fluid Mechanics.

3. The Century of Complex Fluid Mechanics

It should be noted that the Century of Fluid Mechanics by no means implies that the study of Fluid Mechanics is completed in this century. Instead, the Century of Fluid Mechanics will be succeeded by the Century of Complex Fluid Mechanics.

The realm of Fluid Mechanics is always expanding with increasing "complexity". Even in this century, turbulent and chaotic flows have a "complex" property that they are deterministic and random. Besides turbulence, there are several non-classical fluids such as quantum fluids, relativistic fluids and biological fluids, and various flows associated with thermal convection, combustion, phase-change, chemical reactions, nuclear and thermo-nuclear reactions, and other fluid phenomena in bio-science, space science and environmental and energy engineerings, all of which have some kind of "complexity" beyond the conventional and simple fluids. Some of the above subjects already constitute the applied fields of the contemporary Fluid Mechanics, but I believe that they will become the hard core of the Complex Fluid Mechanics in the coming century.

4. Turbulence Research in the Next Century

Coming back to the turbulence research, I do not forget the strong impression I received from the two-volume book edited by *Sydney Goldstein:* "Modern Developments in Fluid Dynamics" [6]. This book brought me to an entirely new world of fluid mechanics, that is "boundary layers and turbulence".

Since that time, great progress has been made in turbulence research and our knowledge of turbulence has been enormously enriched compared with the days of 1938. Although its fundamental framework has still to be constructed, the turbulence research which has been developed on a large variety of "turbulent phenomena" has provided great contributions and benefits to science, technology and our human life.

The present IUTAM Symposium on Mechanics of Passive and Active Flow Control of FLOWCON is actually concerned with the most advanced front of such "turbulence research".

We are looking forward to enjoying interesting presentations and exciting discussions in this Symposium with its great success.

Thank you!

References

1. Battimelli, G.: The Early Internation Congresses of Applied Mechanics. In S. Juhas ed. *IUTAM-Short History,* Springer, 1988, pp. 9-18.
2. Heisenberg, W.: *Ueber Stabilitaet und Turbulenz von Fluessigkeitsstroemen.* Ann. Phys. Leipzig (4), 74 (1924), pp. 577-627.
3. Lamb, H.: *Hydrodynamics,* 6th ed. Cambridge Univ. Press, 1932.
4. Prandtl, L.: *Ueber Fluessigkeitsbewegung bei sehr kleiner Reibung.* Verh. III Intern. Math. Kongr. Heidelberg, 1904, pp. 484-491.
5. Rosenhead, L. ed.: *Laminar Boundary Layers.* Clarendon, Press, 1966.
6. Goldstein, S. ed.: *Modern Developments in Fluid Dynamics,* Vols. I, II. Clarendon Press, 1938; Dover edition, 1965.

DYNAMICS AND CONTROL OF NEAR-WALL COHERENT STRUCTURES

WADE SCHOPPA & FAZLE HUSSAIN
University of Houston
Department of Mechanical Engineering, Houston, TX 77204-4792

Abstract

Using direct numerical simulations of turbulent channel flow, we present a new method for skin friction reduction by prevention of streamwise vortex formation near the wall. Instability of lifted, vortex-free low-speed streaks is shown to generate new streamwise vortices, which dominate near-wall turbulence phenomena. Significantly, the 3D features of the (instantaneous) instability-generated vortices agree well with the coherent structures educed (i.e. ensemble-averaged) from fully turbulent flow. Based on this crucial role of streak instability in vortex generation, we develop a new technique for drag reduction, enabling large-scale flow forcing without requiring instantaneous flow information. As proof-of-principle, x-independent forcing, with a wavelength of 400 wall units and an amplitude of only 6% of the centerline velocity, produces a significant sustained drag reduction: 20% for imposed counterrotating streamwise vortices and 50% for colliding, z-directed wall jets. The drag reduction results from weakened longitudinal vortices near the wall, due to forcing-induced suppression of the underlying streak instability. In particular, the forcing significantly weakens the wall-normal vorticity flanking lifted low-speed streaks, thereby arresting the streaks' instability responsible for vortex generation. These results suggest promising new drag reduction strategies, *e.g.* passive vortex generators or colliding spanwise jets from *x*-aligned slots, involving large-scale (hence more durable) actuation and requiring no wall sensors or control logic.

1. Introduction

The boundary layers of practical interest are invariably turbulent, with drastically increased drag due to near-wall vortical coherent structures (CS). Viable control of near-wall turbulence, as yet largely unrealized in practice, has the potential to save millions of dollars in energy costs for engineering applications. Although a barrage of drag reduction strategies have been studied extensively – *e.g.* compliant walls, polymer addition, riblets, microbubbles, electromagnetic forces, active walls with MEMS, among many others – their engineering application has remained scarce. A lack of successful implementation of boundary layer control can generally be traced to two key difficulties: (i) tiny spatial scales of near-wall streamwise CS (~0.1 mm), and (ii) poor understanding of the dynamics of CS initiation and evolution.

To address these inherent obstacles, we propose here to develop new control approaches which explicitly utilize recent advances in the understanding of near-wall turbulence physics. The prominence of streamwise vortical coherent structures (CS) in

G.E.A. Meier and P.R. Viswanath (eds.),
IUTAM Symposium on Mechanics of Passive and Active Flow Control, 5–14.

near-wall turbulence is now well accepted (*e.g.* see [1]), as is their critical role in the elevated drag in turbulent boundary layers. The transport enhancing effect of near-wall CS is well understood. Due to their streamwise orientation, these CS sweep near-wall fluid toward the wall on one CS flank and eject it away from the wall on the other. Drag and heat transfer are enhanced by the wallward motion, which steepens the wall gradients of streamwise velocity U. Note that the gradient reduction on the outward motion side of vortices is relatively smaller, resulting in mean transport enhancement due to near-wall vortices.

The most logical approach to CS-based reduction of drag and heat transfer is to simply prevent vortex regeneration in the first place (in contrast to many approaches which counteract the wall interaction of fully developed CS). It has long been hypothesized that a major source of turbulence production near the wall is the instability of inflectional low-speed streaks (*e.g.* [2-4]), although the issue still remains unresolved. In particular, the following key issues have yet to be addressed in detail: (i) the relationship between streak instability and the formation mechanism of longitudinal vortices, (ii) physical space (3D) vortex dynamics arising from streak instability, and (iii) streak instability control strategies aimed at drag reduction.

Recently, we have discovered[5] that the CS[6,7] extracted from fully developed near-wall turbulence are directly created by 3D inviscid instability of lifted streaks near a single wall (created by previous "parent" vortices, no longer present), the generation mechanism being akin to that of streamwise vortices in free shear layers by oblique mode instability. This new-found association of near-wall CS formation with instability mechanisms opens up promising avenues for explaining and especially controlling near-wall turbulence, noting the documented success of experimental instability control in both free- and wall-bounded shear flows (*e.g.* see [8]).

To suppress CS via control of primary streak instability (responsible for CS formation), there are two possibilities: either (i) counteract existing perturbations which would otherwise excite instability, or (ii) stabilize the base flow streaks. Pursuit of (i) would necessitate small-scale detection and control, which would suffer from the durability problems faced by microscale active walls. Approach (ii) is very attractive from the standpoint of large-scale (hence more robust) control, wherein numerous (perhaps thousands of) streaks may be stabilized together – hence suppressing new CS formation over an extended spatial domain – with a single robust actuator.

As an alternative to microscale control approaches, our objective here is to investigate a new large-scale control approach explicitly designed to disrupt the naturally occurring vortex regeneration mechanism. We first demonstrate the underlying mechanism of CS formation, driven by nonlinear evolution of an inviscid, 3D instability of lifted low-speed streaks. We then investigate CS suppression through large-scale manipulation of streak instability, and explain the observed control effect using instability and vortex dynamics concepts.

2. Computational Approach

In the following, we address vortex regeneration and its control using direct numerical simulations of the Navier-Stokes equations. Periodic boundary conditions are used in x and z, and the no-slip condition is applied on the two walls normal to y; see *Kim et al.*[9] for the simulation algorithm details. The spatial discretization and *Re* are chosen so that all

Figure 1. Lifted low-speed streaks (black) illustrated by $u'<0$ at $y^+=20$ and streamwise vortices (grey) indicated by Jeong vortex definition[11] in the region $0<y^+<60$.

dynamically significant lengthscales are resolved (*i.e.* a finer computational grid does not markedly affect the solution); thus, no subgrid-scale turbulence model is necessary. Code validation and accuracy checks were performed by comparing the growth rates for simulated 2D and 3D Orr-Sommerfeld modes of the laminar (parabolic profile) flow with independent stability analysis results (agrees within 1%).

To better isolate instability and the subsequent vortex formation, we use the minimum outer Reynolds number $Re=U_c h/\nu=2000$ (U_c is the centerline velocity of the $2h$ wide channel) and the minimum domain sizes in x and z for sustained channel flow turbulence – the so-called "minimal flow unit" of Jimenez & Moin[10]. For the simulations of isolated vortex regeneration, a constant volume flux is maintained, and 32x129x32 grid points are used in x, y, and z respectively. The control simulations are initialized with full-domain channel flow turbulence at Re=1800 and 3200[19], with 48x65x48 and 192x129x192 dealiased Fourier modes respectively. Actuation is represented by an applied control flow, either maintained at a constant amplitude or allowed to freely evolve, superimposed onto the turbulence.

3. Vortex Formation Mechanism

The two most prominent structural features of near-wall turbulence are illustrated in Fig. 1: (i) "streaks" of low momentum fluid which has been lifted into the buffer region, and (ii) elongated longitudinal vortices, illustrated by the Jeong vortex definition[11]. It is now well-accepted that the streaks are generated by the lifting of low-speed fluid near the wall by the vertical velocity induced by streamwise vortices; this is consistent with the close proximity of streaks to streamwise vortices in Fig. 1. Note also that many streaks are devoid of nearby streamwise vortices (instantaneously), indicating that the characteristic elongation of streaks is due to the advection of streamwise vortices, which leave lifted low-speed fluid underneath them in their wake. Here, we reveal a more subtle and dynamically significant role of streaks, as a breeding ground for new streamwise CS via streak instability.

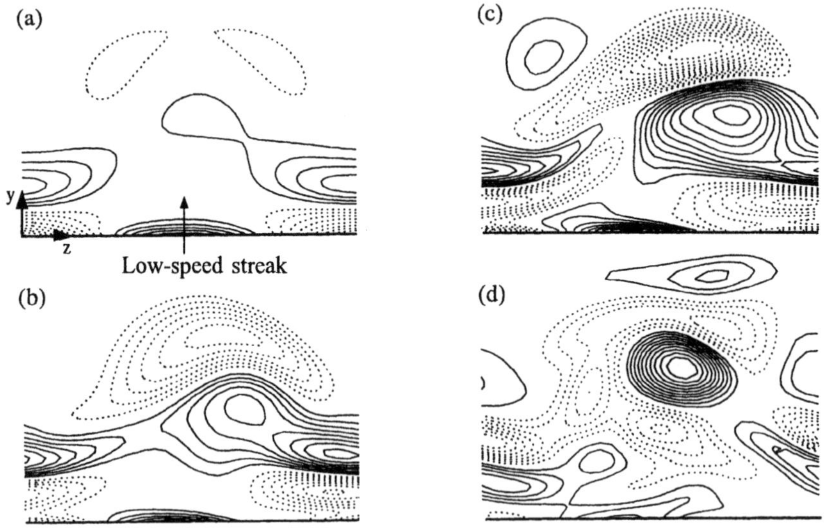

Figure 2. Streamwise vortex formation due to finite-amplitude streak instability, illustrated by cross-stream distributions of ω_x at (a) $t^+=17$, (b) $t^+=51$, (c) $t^+=103$, (d) $t^+=928$. Planes in (b) and (c) are tracked with the instability phase speed of approximately $0.6U_c$.

Having confirmed in our prior studies[5] that (one-walled) streaks with sufficient y circulation are indeed linearly unstable, we now illustrate the subsequent nonlinear evolution using DNS. Results clearly illustrate the genesis of streamwise CS, suggesting that streak instability is the dominant mechanism of vortex generation and thus turbulence production. Most significantly, as the mode grows to a nonlinear amplitude (initially $w'/U_c=1\%$ at $y^+=30$), new collapsed streamwise vortices are directly created (Fig. 2a-c). At early times, instability growth is characterized by increased circulation of flattened ω_x sheets, with the spanwise symmetry of the linear eigenmode approximately maintained. Subsequently, as nonlinear effects (described below) become prominent, $+\omega_x$ begins to concentrate on the $+z$ flank of the low-speed streak (Fig. 2b). By symmetry, the ω_x distribution at a half wavelength in x away is obtained by z reflection and sign inversion; thus, $-\omega_x$ is generated on the $-z$ flank here. As this ω_x amplification continues, collapsed (*i.e.* with compact cross-section) streamwise vortices quickly emerge (Fig. 2c). This genesis of new vortices from ω_x layers is strikingly similar to that frequently observed in minimal channel flow. Previous studies (*e.g.* [12]) presumed that the layer simply rolls up due to its own (2D) advection. Our results (discussed below) indicate that the vortex formation is not in reality a (Kelvin-Helmholtz type) roll up process; the formation is inherently 3D, dominated by intense ω_x stretching. Even well past their initial formation, streamwise vortices and hence turbulence continue to be sustained (*e.g.* Fig. 2d), indicating the importance of streak instability to turbulence sustenance.

The 3D geometry of the instability-generated vortices (Figs. 3a,b) (say, the x-overlapping of tilted, opposite-signed streamwise vortices on either side of a low-speed

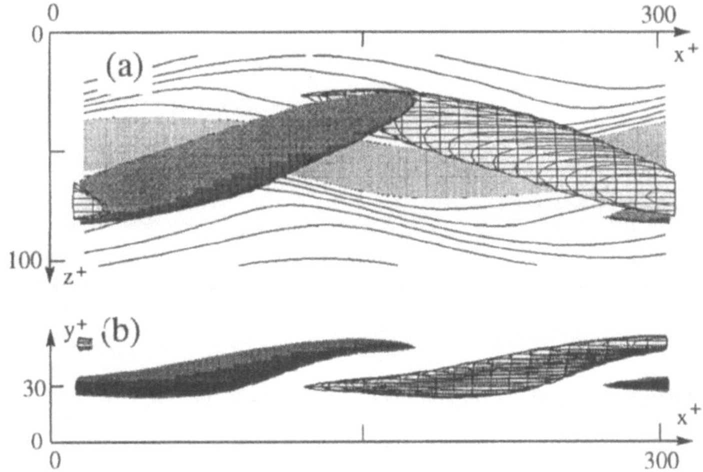

Figure 3. Streamwise vortices' (x,z) plane tilting, x-overlapping, and location relative to a low-speed streak in (a) top view, (b) side view. The 80% isosurfaces of $+\omega_x$ and $-\omega_x$ at $t^+=103$ are (dark) shaded and hatched respectively; contours of u at $y^+=20$ are overlaid in (a), with low levels of u light-shaded to demarcate the low-speed streak.

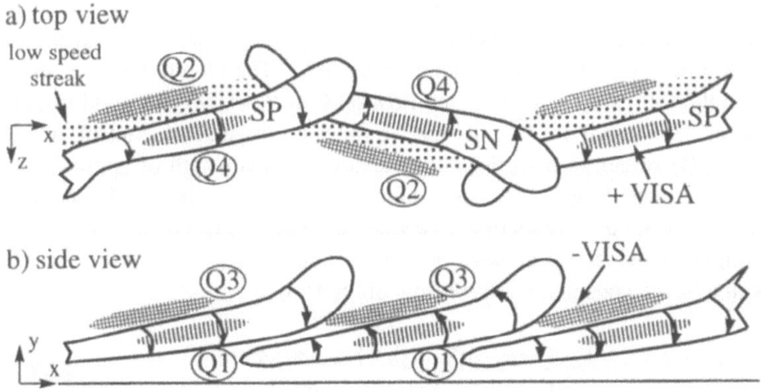

Figure 4. Near-wall educed CS and associated coherent events (adapted from [7]); including ±VISA events ($\pm\partial u/\partial x$); quadrant *Re* stresses Q1, Q2 (ejection), Q3, and Q4 (sweep); and a kinked low-speed streak.

streak) agrees well with the typical flow structure during the active phase of minimal channel regeneration. Most significantly, this vortex geometry (maintained upon evolution except for increasing overlap) is strikingly similar to that of 3D CS educed (from more than 100 vortex realizations) in full-domain turbulence (Fig. 4), which has been shown to capture all important near-wall events[7]. Irregularities (*e.g.* kinks) of the base flow streaks and finite-amplitude incoherent turbulence will surely occur, causing variations in vortices from one realization to another. If an underlying instability mechanism is present, it should be revealed by ensemble averaging over a large number of base flow/perturbation combinations, *i.e.* by CS eduction. The close correspondence of Figs. 3 and 4 indicates that this is in fact the case, serving as strong evidence that this

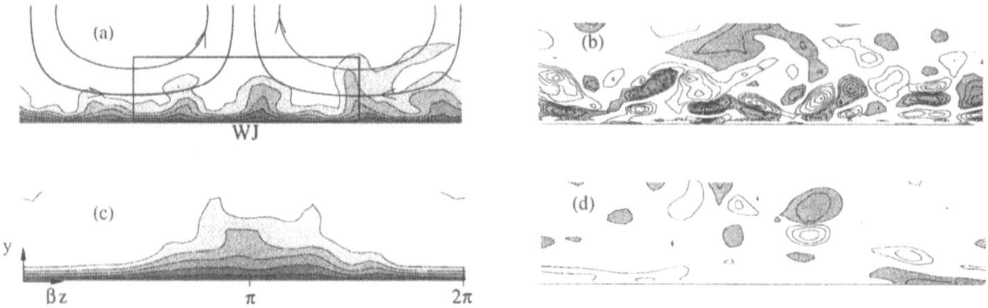

Figure 5. Distributions of $u(y,z)$ in (a,c) and $\omega_x(y,z)$ in (b,d) near one wall of turbulent channel flow at Re=1800, without (a,b) and with (c,d) an imposed large-scale control flow. The controlled flows, shown at t_0^+=500 (after control starts), have a frozen forcing amplitude of 6%. Note the disruption of streaks and the attenuation of streamwise vortices near the wall by control.

vortex formation process is a dominant mechanism in fully developed near-wall turbulence.

4. Large-scale Strategy for Drag Reduction

Having shown that lifted low-speed streaks are a necessary precondition for near-wall vortex formation and hence turbulence production, we now consider a new large-scale control technique aimed at streak stabilization. In particular, we investigate drag reduction by a spanwise row of counter-rotating, *x*-independent streamwise vortices, centered in the outer region (at the channel centerline). The effectiveness of this control is evaluated in the following using selected flowfields at Re=1800 and 3200 from the Kim *et al.*[9] database. It is important to note that we focus on large-scale control, in which the forced row of (2D) streamwise vortices has a spanwise wavelength of approximately 400 wall units, four times larger than the characteristic near-wall streak spacing. For additional details of our drag reduction approach, the reader is referred to Ref. [13].

As a simple model of streamwise vortices produced, say, by vortex generators or spanwise wall jets, we consider a control flow of the form

$$U_{con}=0$$
$$V_{con}(y,z)=-A\beta\cos(\beta z)(1+\cos\pi(y/h-1))$$
$$W_{con}(y,z)=-A\pi\sin(\beta z)\sin\pi(y/h-1) \qquad (1)$$

which satisfies the continuity equation and the no-slip condition on the channel walls (at y=0,2h), where A is the control amplitude and $2\pi/\beta^+\cong 400$. To demonstrate proof-of-principle for large-scale forcing, the z wavelength of the control flow is four times the characteristic streak spacing of approximately 100 wall units; even much larger-scale control (although intractable via computation) may be possible in practice. As illustrated in Fig. 5(a), the control flow (1) has a much larger scale than local minima of $u(y,z)$ near the wall, representing lifted low-speed streaks. For simplicity, we focus on the lower half of the channel (i.e. $y\in[0,h]$) in this and other figures; the upper half yields similar results. For a full period in z, (1) represents an array of counter-rotating 2D streamwise vortices (Fig. 5a), termed *vortex control*. Over the half period $\beta z\in[\pi/2,3\pi/2]$, (1) resembles colliding, spanwise-directed 2D wall jets (region *WJ* in Fig. 5a), referred to as *wall jet control*. Thus, we actually simulate a single control flow, distinguishing vortex and wall jet control by the region of z considered. In practice, the relative extents of diverging (outside *WJ*) and converging (inside *WJ*) wall jets can be adjusted to reduce the former.

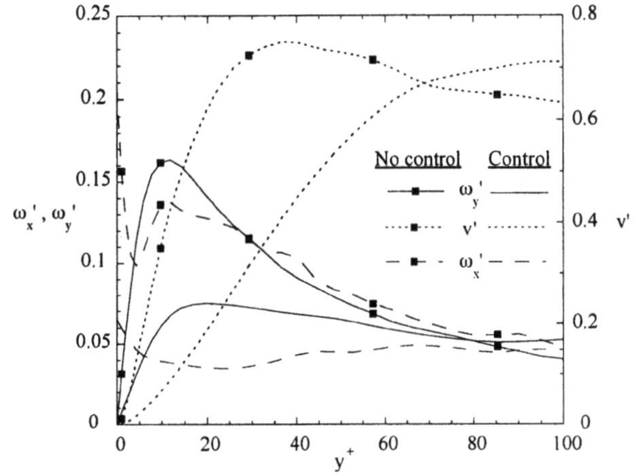

Figure 6. Suppression of v', ω_x', and ω_y' by large-scale control at t_0^+=500, with Re=1800 and frozen 6% forcing.

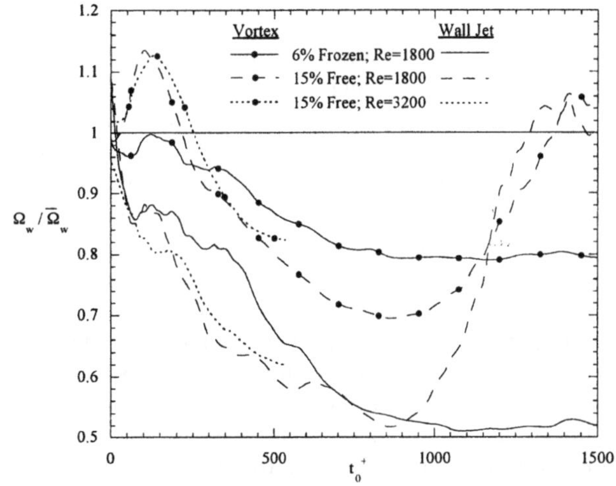

Figure 7. Time evolution of wall-integrated shear stress (normalized by the time-mean of the uncontrolled flow), illustrating significant drag reduction by large-scale control due to the suppressed vortex formation mechanism by control-induced streak ω_y reduction.

In essence, the objective of this forcing is to break the near-wall vortex regeneration cycle by disrupting the naturally occurring (unstable) distributions of streaks generated by previous or preexisting streamwise vortices. The effect of the control on preexisting streaks is best illustrated in Figs. 5(a,c), for an extremely weak control amplitude, in which the maximum control velocity is only 6% of the freestream centerline velocity. The numerous preexisting lifted streaks (Fig. 5a) are flattened by splatting where V_{con} is directed toward the wall and W_{con} diverges in z (outside region *WJ*). Within the wall jet control region *WJ*, V_{con} is directed away from the wall and W_{con} converges in z, causing cross-diffusion of compressed streaks and thereby weakening the streaks' ω_y. Clearly, even this very weak control drastically alters the near-wall streaks (*cf.* Figs. 5a,c); the lifted regions now have a much larger scale, and are separated by large regions free from any significant streaks.

Recalling the necessity of sufficient streak vorticity ω_y for instability and hence vortex formation, we expect this control to provide significant attenuation of near-wall streamwise vortices. A comparison of the near-wall ω_x with and without control (Figs. 5b,d) indicates that this is indeed the case. Without control (Fig. 5b), numerous compact, drag-producing vortices with strong ω_x are present immediately near the wall. In contrast, due to the disruption of streaks by control, ω_x in the controlled flow is significantly weaker, with no compact vortices present near the wall (Fig. 5d). Statistics of ω_y' confirm a strong reduction of local ω_y maxima by control, accompanied by large suppression of ω_x' and drag-producing v' (Fig. 6). The latter occurs only after existing vortices, which eventually weaken by annihilation due to cross-diffusion and dissipate, are not replaced by equally strong and numerous vortices,

Due to the streak and streamwise vortex attenuation illustrated in Fig. 5, a significant drag reduction is attained (Fig. 7). Significantly, *sustained* drag reduction occurs when the x-mean Fourier modes corresponding to the control (1) (*i.e.* (x,z) modes (0,n); n≠0) are frozen in time (after the initial inviscid transient of the control vortices due to their nonlinear self-evolution). With the control amplitude equal to 6% of U_c, a 20% drag reduction is attained for vortex control (integration of wall vorticity over a full period), while nearly 50% reduction occurs for wall jet control (integration over region *WJ*), compared to the time mean of the uncontrolled case. The control amplitude required for this significant drag reduction is surprisingly low, in general desirable from the practical standpoint of low power consumption (active control) or parasitic drag (passive control) of actuators. The dependence of drag on the (frozen) forcing amplitude (not shown) indicates a cusp-like effect of the control amplitude; the control effect is insignificant for 2% and weaker forcing, while 15% and stronger forcing leads to increased drag for both vortex and wall jet control. The increased drag at higher amplitudes reflect direct generation of drag by the control flow itself, occurring even in the absence of background turbulence. The optimum control is attained when the control is strong enough to stabilize near-wall streaks, yet weak enough not to induce significant additional drag. Significant drag reduction is also observed for free forcing, in which the control flow (1) is superimposed on the turbulence and allowed to freely evolve, at both Re=1800 and 3200 (Fig. 7). Although the control effect is temporary for free forcing, due to eventual dissipation of the control flow, significant drag reduction is observed for O(1000) wall time units. During this time, the control flow advects $(U_c^+)(\Delta t^+)$~16,000 wall units downstream, thus suggesting the practical feasibility of large-scale, effective control in both x and z (i.e. simultaneously many streaks, covering numerous wall vortices).

As shown schematically in Fig. 8, our drag reduction approach consists of an array of large-scale control devices, *i.e.* S_x and S_z in Fig. 8 much larger than the characteristic near-wall flow lengthscale of O(0.1 mm), say O(10 cm) if feasible. No sensors for flow measurement or electronic control hardware are necessary. Two possible embodiments for the control device are: (1) wall-mounted vortex generators, and (2) spanwise-directed, colliding wall jets. Embodiment (1) involves wall-mounted tabs at a slight angle θ to the flow direction, designed to produce long, streamwise-aligned control vortices. The direction of control vortex circulation is determined by the tab angle orientation (Fig. 8), which alternates in the spanwise direction z. The tab height h controls the height of the control vortex above the wall, which along with θ will require adjustment for specific applications. Computational results for proof-of-principle indicate that a very small

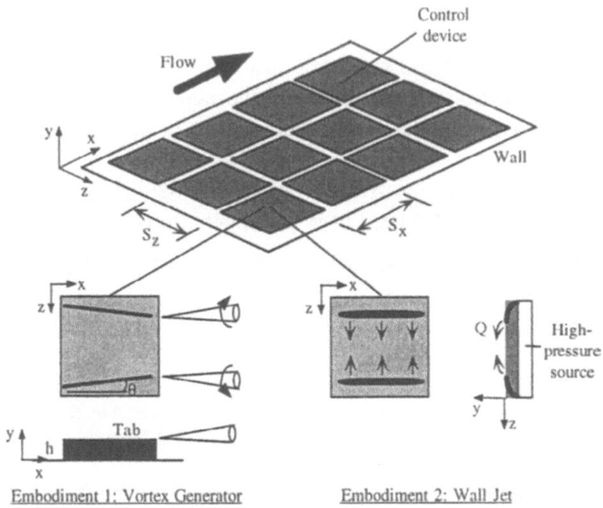

Figure 8. Schematic of large-scale drag reduction strategies, illustrating vortex and spanwise wall jet embodiments.

control vortex circulation is optimum; thus a small tab angle θ and hence little additional parasitic drag will be involved.

Embodiment (2) involves wall jets issued from narrow, streamwise slots in the wall, driven by an external high pressure source (Fig. 8). The jet volume flowrate Q is constant in time and will require adjustment for specific applications. The slots will be built into the wall surface so that an attached, spanwise-directed wall jet is produced. Slots neighboring in z are oriented so that adjacent wall jets collide to produce an upwelling (*i.e.* normal to the wall) control flow (Fig. 8), as modeled by region *WJ* in our computations (Fig. 5a). In practice, the relative extents of diverging (outside *WJ*) and converging (inside *WJ*) wall jets can be adjusted to reduce the former, noting that the converging wall jet region produces the largest drag reduction.

5. Concluding Remarks

To summarize, we have shown that (nonlinear) instability of ejected low-speed streaks, initially without any vortices whatsoever, directly generates new streamwise vortices near the wall. The resulting 3D vortex geometry is identical to that of the dominant CS, educed from fully developed near-wall turbulence, which in turn capture all important, extensively reported near-wall events. This serves as strong evidence that vortex-less streaks are the main breeding ground for new streamwise vortices, commonly accepted as dominant in turbulence production. In turn, the geometry of the newly generated vortices constitutes a built-in mechanism which sustains ejected streaks against their otherwise rapid self-annihilation due to cross-diffusion. Vortex-less streaks, the vehicle for instability-based vortex formation, are expected to arise inherently due to the differential advection of vortices and the streaks they generate.

Since streamwise vortex formation and the associated enhanced drag appear to be reliant on lifted low-speed streaks with strong ω_y, large-scale (relative to the natural streak spacing) control of streaks is a potentially effective approach to drag reduction. We demonstrate here the feasibility of drag reduction via bulk forcing using either counterrotating vortex generators or colliding spanwise wall jets, requiring no instantaneous flow information (otherwise necessary for adaptive control). For implementation at very high *Re*, the physical scale of our control will likely decrease, but being significantly larger than the near-wall structures, will surely alleviate the micro-scale requirement for controllers and eliminate the need for sensors. Experiments are necessary at higher *Re* to realize the drag reducing effect of the control considered here.

This research was supported by AFOSR grant F49620-97-1-0131 and the NASA Graduate Fellowship grant NGT-51022 of W.S. Supercomputer time was provided by the NASA Ames Research Center.

6. References

1. Robinson, S.K. 1991 Coherent motions in the turbulent boundary layer. *Ann. Rev. Fluid Mech*. 23, 601.
2. Kline, S.J., Reynolds, W.C., Schraub, F.A. & Rundstadler, P.W. 1967 The structure of turbulent boundary layers. *J. Fluid Mech*. 30, 741.
3. Swearingen, J.D. & Blackwelder, R.F. 1987 The growth and breakdown of streamwise vortices in the presence of a wall. *J. Fluid Mech*. 182, 225.
4. Hamilton, J., Kim, J. & Waleffe, F. 1995 Regeneration mechanisms of near-wall turbulence structures. *J. Fluid Mech*. 287, 317.
5. Schoppa, W. & Hussain, F. 1997 Genesis and dynamics of coherent structures in near-wall turbulence. In *Self-sustaining Mechanisms of Wall Turbulence* (ed. R. Panton). Computational Mechanics Publications, p. 385.
6. Jeong, J. & Hussain, F. 1992 Coherent structures near the wall in a turbulent channel flow. In *Proc. of Fifth Asian Congress of Fluid Mech*., Taejon, Korea (eds. K.S. Chang & H. Choi), p. 1262.
7. Jeong, J., Hussain, F., Schoppa, W. & Kim, J. 1997 Coherent structures near the wall in a turbulent channel flow. *J. Fluid Mech*. 332, 185.
8. Hussain, F. 1986 Coherent structures and turbulence. *J. Fluid Mech*. 173, 303.
9. Kim, J., Moin, P. & Moser, R.D. 1987 Turbulence statistics in fully developed channel flow at low Reynolds number. *J. Fluid Mech*. 177, 133.
10. Jimenez, J. & Moin, P. 1991 The minimal flow unit in near-wall turbulence. *J. Fluid Mech*. 225, 213.
11. Jeong, J. & Hussain, F. 1995 On the identification of a vortex. *J. Fluid Mech*. 285, 69
12. Jimenez, J. & Orlandi, P. 1993 The rollup of a vortex layer near a wall *J. Fluid Mech*. 248, 297.
13. Schoppa, W. & Hussain, F. 1998 A large-scale control strategy for drag reduction in turbulent boundary layers. *Phys. Fluids* 10, 1049.

FLOW PHYSICS INVOLVED IN SHOCK WAVE/BOUNDARY LAYER INTERACTION CONTROL

J. M. DÉLERY
ONERA
92190 MEUDON, France

1. INTRODUCTION

Shock waves almost inevitably occur when a flow becomes supersonic or transonic. The presence of shock waves entails the existence of discontinuities and regions of high gradients which are the shocks themselves and the shear layers resulting from the interaction with the boundary layers developing on the vehicle surface. These gradients "activate" the viscous terms producing entropy, which makes shock waves an important source of drag :

(1) directly by entropy generation in the thickness of the shock : this contribution is the *wave drag ;*

(2) indirectly, be enhancing dissipation in the boundary layers : hence an amplification of the *viscous drag.*

In addition, strong interactions with the boundary layers may lead to catastrophic separation with possible occurrence of large scale unsteadiness (buffeting, air intake buzz).

The idea to control shock wave/boundary layer interaction in order to avoid, or minimize, its detrimental effects is nearly contemporaneous with the advent of studies on high-speed flows. Thus, as early as 1941, Regenscheit[1] studied the effect of suction through a slit at the surface of an airfoil tested at transonic velocity. Since that time, a great deal of research has been devoted to this important practical problem[2].

2. BASIC PROPERTIES OF SHOCK INDUCED PHENOMENA

When a boundary layer, most often turbulent, is submitted to the strong retardation imparted by a shock wave, complex phenomena occur within its structure which is basically a parallel rotational flow with the Mach number varying from the outer supersonic value to zero at the wall. The process has been analyzed by Lighthill in the framework of its famous triple-deck theory[3].

The interaction resulting from the reflection of an oblique shock (C_1) is sketched in Fig. 1. The presence of the shock is felt upstream of its impact point in the perfect fluid model. This upstream influence phenomenon is in great part an inviscid mechanism, the pressure rise caused by the shock being transmitted upstream through the subsonic part of the boundary layer. The thickness of the subsonic layer depending of the velocity distribution, a fuller profile involves a thinner subsonic channel, hence a shorter upstream influence length. At the same time, a boundary layer profile with a low velocity deficit, has a higher momentum, hence a greater resistance to the retardation imparted by an adverse pressure gradient. The dilatation of the boundary layer subsonic region is felt by the outer supersonic flow - which includes the major part of the

G.E.A. Meier and P.R. Viswanath (eds.),
IUTAM Symposium on Mechanics of Passive and Active Flow Control, 15–22.

boundary layer - as a ramp effect inducing compression waves whose coalescence forms the reflected shock (C_2). This mechanism, which is amplified by a general retardation of the boundary layer flow, can be interpreted as a *viscous ramp effect* resulting in a spreading of the shock near the surface.

Since the retardation effect is larger in the boundary layer inner part, a situation can be reached where the flow is pushed in the upstream direction by the adverse pressure gradient so that a separated region forms. Then, the flow adopts the structure sketched in Fig. 2. Downstream of the separation point exists a bubble made of a recirculating flow bounded by a *discriminating streamline* (S) originating at the separation point S and ending at the reattachment point R. Due to the action of the strong mixing taking place in the detached shear layer which emanates from S, an energy transfer is operated from the outer high speed flow toward the separated region. As a consequence, the velocity U_s on the discriminating streamline (S) steadily increases, until the reattachement process begins.

The transmitted shock (C_3) penetrates in the separated viscous flow where it is reflected as an expansion wave. There results a deflection of the shear layer toward the wall on which it reattaches. At reattachment, the separated bubble vanishes, the flow on (S) being decelerated until it stagnates at R. This process is accompanied by a compression wave ending into a reattachment shock in the outer stream.

The extent of the separated region is dictated by the ability of the shear layer issuing from the separation point S to "negociate" the pressure rise at reattachment. This ability is function of the momentum available on (S) – or maximum velocity $(U_S)_{max}$ - at the starting of the reattachment process. When there is separation, the interaction of the shock wave with the boundary layer has deep repercussions on the contiguous inviscid flow, as shown in Fig. 2, the simple inviscid shock pattern made of an incident plus reflected shock being replaced by a pattern involving 5 shock waves :

- the incident shock (C_1),
- the separation shock (C_2),
- the transmitted shock (C_3) emanating from the intersection point I of (C_1) and (C_2),
- the second transmitted shock (C_4),
- the reattachment shock (C_5).

Transonic and ramp induced interactions lead to a similar organization for the separated flow, the boundary layer reacting to a pressure rise, no matter the cause of this pressure rise. What changes is the shock pattern associated with the interaction. In transonic flow, the interaction leads to the formation of a *lambda pattern.*

3. MECHANISMS FOR CONTROL ACTION

The upstream influence of the shock and the resistance of a turbulent boundary layer depending mainly of its momentum, a means to restrict the effect of the shock is to increase the boundary layer momentum prior to its interaction with the shock. This can be done by proper *boundary layer manipulation :*

- One can perform an injection through one or several slots located upstream of the shock origin, this technique being called *boundary layer blowing.*

- A distributed suction applied over a certain boundary layer run before the interaction reduces its shape parameter, thus producing a fuller velocity profile. On the contrary, distributed injection - or transpiration - increases the shape parameter, which renders the boundary layer more sensible to the shock. Vortex generators can be classified in this category of control actions, since the vortical structures that they create operate a momentum transfer from the outer high speed flow toward the boundary layer.

- It is also possible to eliminate the low speed part of the boundary layer by making a strong suction through a slot located at a well chosen location upstream of the interaction.

- The interaction being in great part dictated by the boundary layer properties, one can envisage to modify these properties *by changing the temperature of the wall.* An increase of the wall temperature lowers the Mach number level by increasing the sound speed ; a lowering of the wall temperature increases the Mach number by diminishing the sound speed, upstream propagation of the shock influence will be affected by this effect, the thickness of the boundary layer subsonic channel being augmented on a heated wall and reduced on a cooled wall. In addition, the temperature level in the boundary layer affects the density, lower temperature leading to higher density and higher temperature to lower density. There results a change of the boundary layer momentum, favorable in the first case, unfavorable in the second. The effect of the wall temperature has been mainly studied for the cold wall situation[4]. In this case, a strong contraction of the interaction domain is observed, with in some circumstances a suppression of the separation bubble. The situation of a heated wall has been investigated for an oblique shock reflection at a Mach number equal to 2.4, for a ratio of the wall temperature T_w to the recovery temperature T_r of two[5]. The iso-Mach contours plotted in Fig. 3 show the extension of the interaction occurring when the the wall is heated.

A key role being played by the velocity - or momentum - level which must be reached on the discriminating streamline to allow shear layer reattachment, any action modifying this level will influence the interaction. Thus, if some fluid is sucked through the wall, topological considerations lead to the flow structure sketched in Fig. 4b. The streamline (S_2) stagnating at the reattachment point R being located at a greater distance from the wall, the velocity $(U_S)_{max}$ on (S_2) is greater than in the basic case of Fig. 4a : hence, an increase of the ability of the flow to withstand a more important compression and a subsequent contraction of the interaction domain. The case of a fluid injection at low velocity is sketched in Fig. 4c. Now, the velocity on (S_2) is smaller, since (S_2) is located at a lower altitude on the velocity profile. The resulting effect is a lengthening of the separation bubble. However, if the injected mass flow is increased, there will be a reversal of the effect, since the velocities on the bottom part of the profile - and in particular $(U_S)_{max}$ - will increase if the mass flow fed into the separated region goes beyond a certain limit.

4. OBJECTIVES AND APPLICATION OF SOME CONTROL ACTIONS

When considering shock wave/boundary layer interaction control, it is essential to state the objective which is looked for :

- A control action can be considered to prevent separation and/or to stabilize the shock in a duct or a nozzle. Boundary layer blowing, suction, wall cooling can be very efficient for these purposes.

- If the objective is to decrease the drag of a profile, or limit the loss of efficiency in an air intake, the situation is more subtle since the drag has its origin both in the shock and in the boundary layer.

Any action "strengthening" the boundary layer tends also to strengthen the shock since the spreading caused by the interaction is reduced. This effect is illustrated in Fig. 5 which shows a comparison between a basic transonic interaction, without control, and the flow resulting from the suction of the boundary layer. In the second case, the entropy production through the shock is more important and the wave drag higher. On the other hand, since the downstream boundary layer profile is more filled, the momentum loss in the boundary layer is much reduced.

When separation occurs, we have seen that the simple inviscid shock system is replaced by a pattern made of continuous compression waves and multiple shocks through which the entropy production is less compared to the inviscid solution (for identical upstream and downstream conditions). Thus, the smearing of the shock system and splitting of the compression achieved by the interaction reduce the wave drag - or the efficiency loss - due to the shock. On the other hand, the momentum loss in a separated boundary layer is far more important than through an attached boundary layer, so that separation most often results in a sharp increase of drag or efficiency loss (not speaking of unsteadiness).

Since separation has a favorable effect on the wave drag, one can envisage to replace a strong, but not separated interaction, by a separated or separated-like flow organization. The *passive control* concept suggested in the early 80s, aims at combining the two effects by spreading the shock system while reducing the boundary layer thickening[6]. The principle of passive control consists in replacing a part of the surface by a perforated plate installed over with a closed cavity. The plate being implemented in the shock region, there occurs a natural flow circulation - via the cavity - from the downstream high pressure part of the interaction to the upstream low pressure part. The resulting effect on the flow is sketched in Fig. 6. The upstream transpiration provokes a thickening of the boundary layer, hence a rapid growth of its displacement thickness which is felt by the outer supersonic flow as a viscous ramp effect inducing an oblique shock (C_1). The meeting of (C_1) with the main normal shock (C_2) at the point I leads to a lambda shock pattern with the trailing shock (C_3). The situation is similar to the case of a natural shock-induced separation, the "strong" normal shock (C_2) being replaced by a two shock system in the vicinity of the surface : hence a reduction of the wave drag. The negative effect on the boundary layer, which would increase the friction losses, is limited by the suction operated in the downstream part of the plate. The effect of passive control is illustrated by the schlieren photographs in Fig. 7 which show the splitting of the original shock into a two shock system and the thickening of the boundary layer due to the transpiration effect[7].

The obvious merit of the concept is the absence of an energy source, the system being self operating. However, the effectiveness of passive control to reduce the total drag produced by the shock is impaired by the high friction drag along the perforated plate, the gain being in general modest when not negative. The concept could be improved by placing a suction slot downstream of the passive control cavity, as sketched in Fig. 8. This *hybrid control* device combine the advantage of passive control to decrease the wave drag and the high effectiveness of suction to reduce friction losses.

Since friction drag production in passive control is frequently unacceptable, one may consider the possibility to reproduce the separated flow structure by a *local deformation of the surface* having a double wedge like shape which materializes the viscous separated fluid of the original interaction. Such a bump induces a first shock at its origin and a second shock at its trailing edge. This system affects only slightly the boundary layer, while reducing substantially the wave drag.

5. CONCLUDING REMARKS

The signification of shock wave/boundary layer interaction control is ambiguous, in the sense that control can aim at minimizing the effect of the shock on the boundary layer properties or reducing the overall losses caused by the interaction, the two objectives being partly contradictory. In the first case, the main objective is most often to avoid boundary layer separation which can be achieved either by manipulation of the boundary layer prior to its interaction with the shock (upstream blowing, suction through a slot, cooling of the wall) or by suction or bleeding in the interaction region. In the second case, one has to reduce also the wave drag which is achieved by a splitting or spreading of the shock near the wall, which can be done by creation of a separated-like flow structure by means of a passive control cavity or a bump.

The energetic and economic aspects of interaction control, as also the problem of the installation of a control device in an airplane wing, have not been examined in this paper, although their consideration is essential. Neither we have discussed the problems raised by the modelling of an interaction under control conditions, considering the incidence of flow manipulation on turbulence behavior and the definition of a law to represent the suction/transpiration velocity in the control region.

6. REFERENCES

[1] Regenscheit, B. (1941) Versuche zur Widerstandsverringerung eines Flügels bei hoher Machscher - Zahl durch Absaugung der hinter dem Gebiet unstetiger Verdichtung abgelösten Grenzschicht. ZWB, Forschungsbericht N° 1424; English translation : NACA TM N° 1168

[2] Délery, J. (1985) Shock wave/turbulent boundary layer interaction and its control. *Prog. Aerospace Sci.*, Vol. 22, pp. 209-280

[3] Lighthill, M. J. (1953) On boundary-layer upstream influence. II Supersonic flows without separation. *Proc. R. Soc.*, A 217, pp. 478-507
Aeronautical Research Institute of Sweeden, FFA Report N° 115

[4] Kilburg, R.F. and Kotansky, D.R. (1969) Experimental investigation of the interaction of a plane oblique incident-reflecting shock wave with a turbulent boundary layer on a cooled surface. NACA CR-66-841

[5] Délery, J. (1992) Etude expérimentale de la réflexion d'une onde de choc sur une paroi chauffée en présence d'une couche limite turbulente. *La Recherche Aérospatiale*, N° 1992-1, pp. 1-23

[6] Savu, G.; Trifu, O. and Dumitrescu, L. Z. (1983) Suppression of shocks on transonic airfoil. *14th Int. Symp. On Shock Tubes and Waves*, Sydney, Australia

[7] Bur, R. ; Délery, J. and Corbel, B. (1997) Basic study of passive control applied to a two-dimensional transonic interaction. In E. Stanewsky, J. Délery, J. Fulker, W. Geisller (Eds), *Notes on Numerical Fluid Mechanics*, Vol. 56, Vieweg

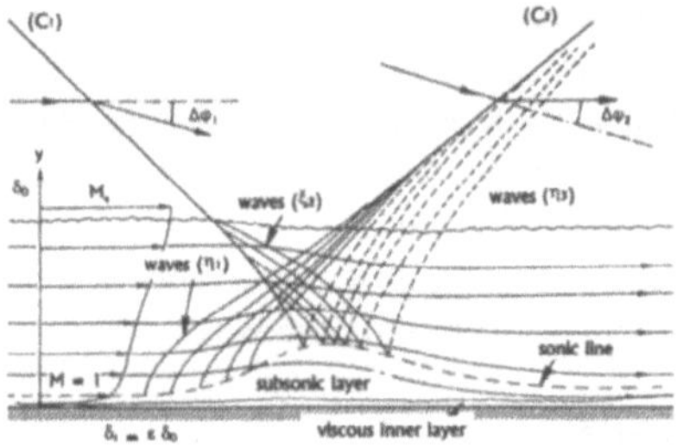

Fig. 1 – Reflection of an oblique shock wave without boundary layer separation. Upstream propagation mechanism and interaction spreading.

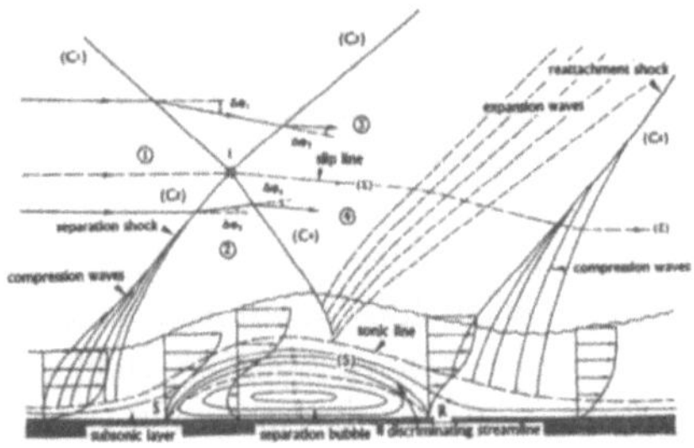

Fig. 2 – Reflection of an oblique shock wave with extended boundary layer separation. Separated flow topology and associated shock pattern.

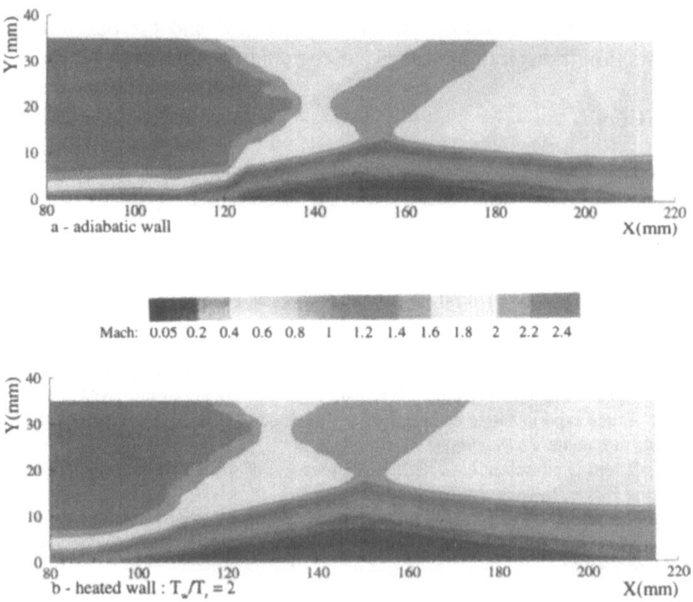

Fig. 3 – Wall temperature effect on an oblique shock reflection at Mach number 2.4 Mach number contours.

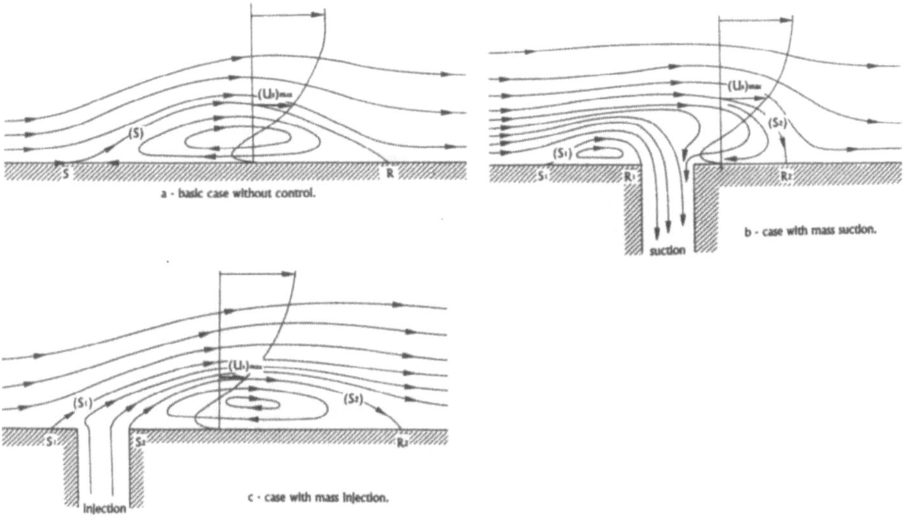

Fig. 4 – Control by mass suction or injection in the separated bubble. Schematic flow organization.

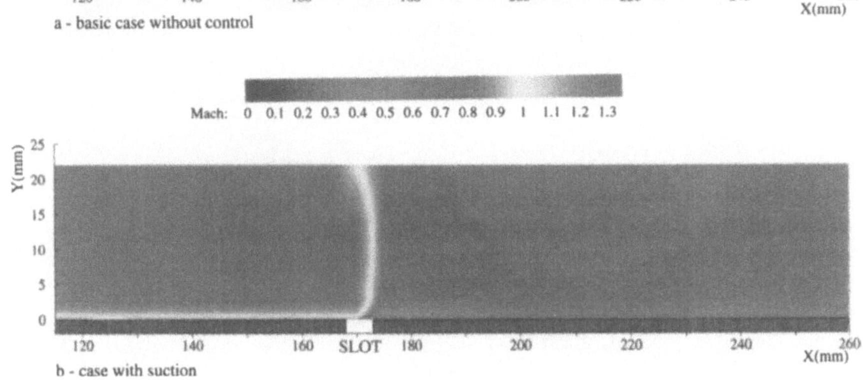

Fig. 5 – Transonic interaction with suction of the boundary layer through a slot. Mach number contours.

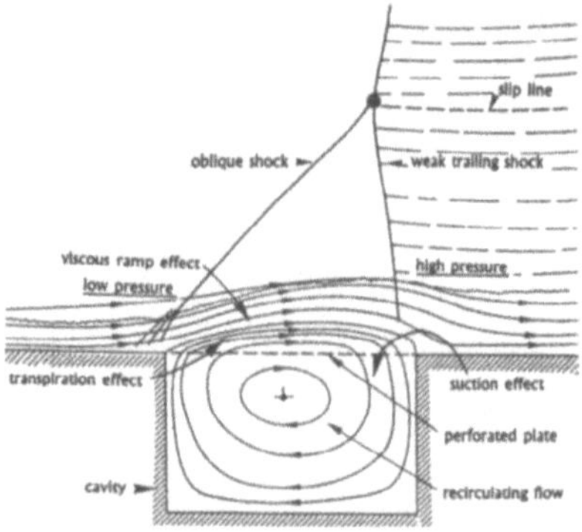

Fig. 6 – Principle of passive control applied to a transonic interaction.

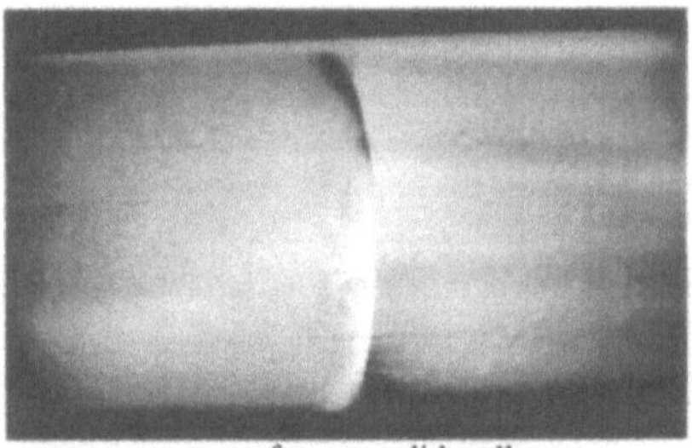

a – reference solid wall case

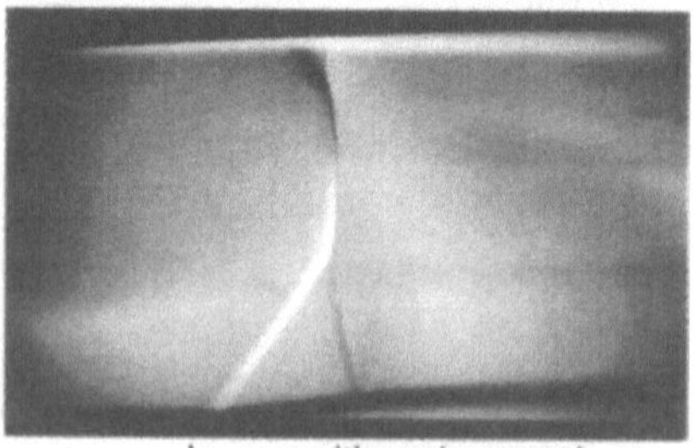

b – case with passive control

Fig. 7 – Schlieren visualization of transonic flow under passive control conditions.

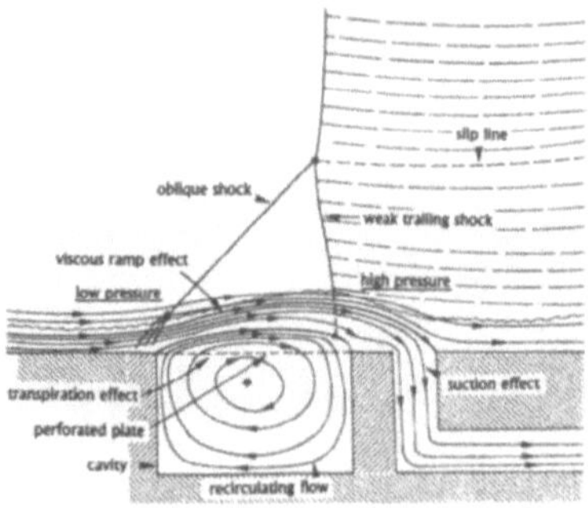

Fig. 8 – Principle of passive/active – or hybrid control – applied to a transonic interaction.

SHOCK/BOUNDARY LAYER CONTROL BY VARIOUS CONCEPTS

R. BOHNING AND D. STERN
Institut für Strömungslehre, Universität Karlsruhe (TH)
Kaiserstraße 12, 76128 Karlsruhe

1. Introduction

The shock/boundary layer interaction has a strong influence on the aerodynamic behaviour of transonic wings. Therefore many research programs, such as EUROSHOCK or RaWid focus on this topic. The results presented here were obtained through RaWid in cooperation with DASA (Bremen). The objective of this numerical investigation is to critically study several control methods (fig. 1) which aspire to the following aims:
drag reduction of the airfoil or wing, prevention of the shock-induced separation, fixation of the shock and postponement of the buffet onset.

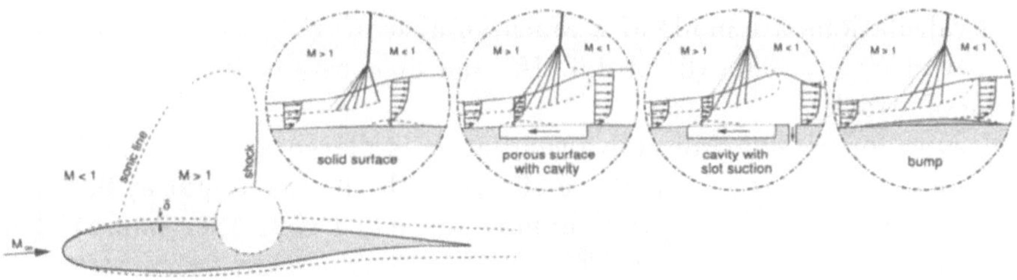

Figure 1. control of shock/boundary layer interaction

These methods include:

- passive ventilation through a cavity covered by a porous plate
- additional active suction out of the cavity
- suction slots located in front of or behind the shock
- contour bumps
- combined concepts (hybrid methods).

G.E.A. Meier and P.R. Viswanath (eds.),
IUTAM Symposium on Mechanics of Passive and Active Flow Control, 23–28.

2. 2D Numerical Investigations

For the testing of the different control methods the laminar airfoil LVA-1a is used with a fixed transition point at 0.5 x/c.
Verification of the 3D-Navier-Stokes-Solver KAPPA (KArlsruhe Parallel Program for Aerodynamics) is performed with the RAE 2822 and the LVA-1a with control through a passive cavity and two bumps with different heights as well as without control. For the passive ventilation the experimental value of the cavity-pressure is used.
In addition, the ventilation is verified separately by calculating the exercises suggested in the EUROSHOCK II project.
Turbulence modelling is implemented according to the extended formulation of the Baldwin-Lomax model by Cebeci and Smith [3], which takes into account mass flow through the wall and strong pressure gradients. Furthermore, it shows much better convergence and a smoother profile of the ventilation velocity.
The ventilation through the porous plate is modelled according to laws found by Poll [2] and Bohning, Doerffer [1], the latter take into account the influence of an outer tangential flow on the mass flux through a porous surface.
The wavedrag induced by the shock is calculated by the postprocessor EULWID, which has been made available by the DASA. The EULWID-KAPPA-interface, respecting the logic concatenations and boundary conditions of the grid-blocks, has been elaborated in this project.
Most calculations are made at a Mach number of $M_\infty = 0.76$, a Reynolds number of $Re_\infty = 6.0 \cdot 10^6$ or $4.6 \cdot 10^6$ and at constant lift c_l or α-sweeps (angle of attack).
The cavity for passive and active ventilation is placed at 0.58 – 0.695 x/c, the bumps are located at 0.59 – 0.79 x/c with the maximum at 0.73 x/c and a height of 0.002 C_b/c (Sc1 bump) and 0.004 C_b/c (Sc2 bump). The porosity of the porous plate is 5%.

2.1. RESULTS

The passive ventilation confirms the positive effect expected for the shock. The wave drag is reduced by nearly 50% (fig. 3, table 1), and the shock shows a λ-structure due to the enlarged compression region. But because of the strong disturbance of the boundary layer caused by the blowing part of the ventilated flow the viscous drag increases rapidly and the total drag increases up to 6% in these cases (fig. 5).
Additional suction out of the cavity shifts the shock downstream (fig. 3),

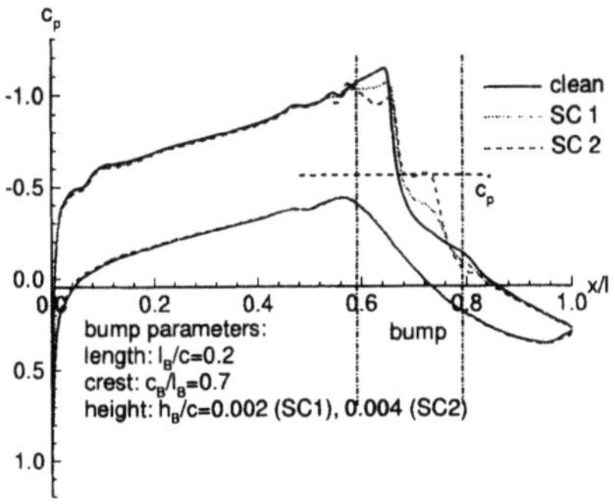

Figure 2. c_p, LVA-1a with bump

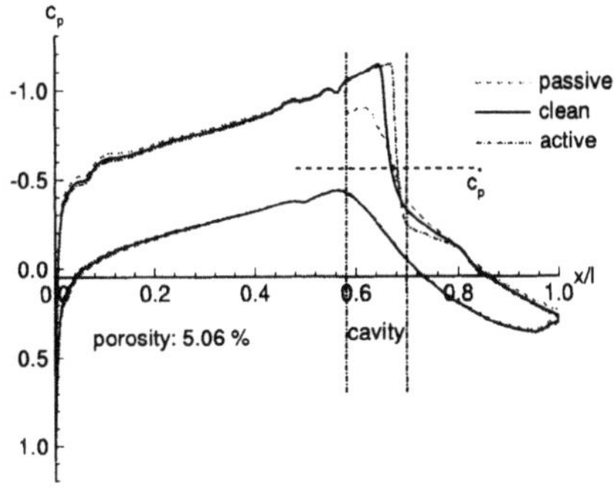

Figure 3. c_p, LVA-1a with ventilation

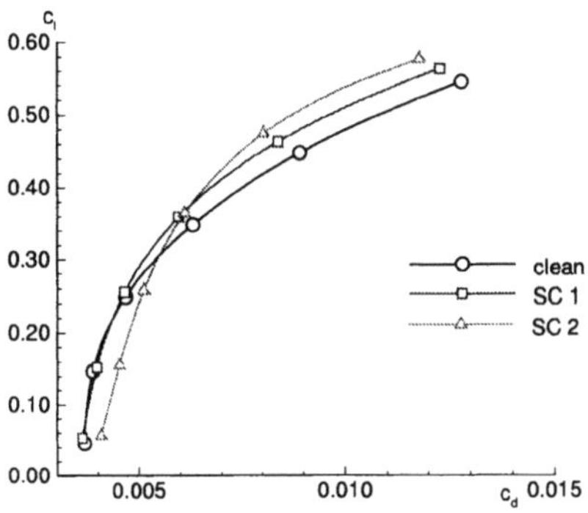

Figure 4. α-sweep, LVA-1a with bump

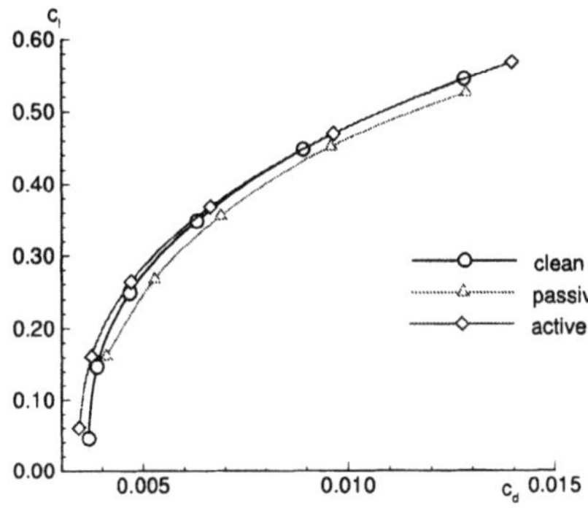

Figure 5. α-sweep, LVA-1a with ventilation

the wave drag increases according to the case without control. But now the viscous and the total drag are reduced, without considering the drag induced by the suction.

Suction slots placed in front of the shock cause better results than if they are placed behind. But only at low angles of attack α can significantly reduced total drags be observed, however, without the induced pump drag c_{qd}. In general these slight improvements are not justifying the technical effort for realising suction on this airfoil.

A two-cavity-system with a passive in front of an active cavity offers no improvement, neither does a passive cavity with a following suction slot.

Local modifying of the shape (bump) results indeed in a decrease of the total drag. Similar to the passive cavity there is a λ-structure of the shock, especially with the higher bump. The shocks themselves are weaker and located more downstream (fig. 2). The viscous drag does not increase, as compared to passive ventilation. The lower bump improves the drag in the whole range of the α-sweep, whereas the higher one is better at high angles of attack, but worse at low α (fig. 4). An additional suction slot at the beginning of the bump enhances these improvements.

TABLE 1. control methods at constant lift ($c_l = 0.452$)

control method	α	c_d	c_{wd}	c_q	c_{qd}
clean	1.025	0.00908	0.00316	-	-
passive	1.11	0.00932	0.00186	0.0	0.0
given p_c	1.23	0.00959	0.00176	0.0	0.0
active	0.9	0.00895	0.00347	-0.0005	0.00114
slot (0.58 – 0.60)	0.933	0.00902	0.00520	-0.00055	0.00133
slot (0.75 – 0.77)	0.86	0.00923	0.00319	-0.00095	0.00128
cavity + cavity	0.94	0.00915	0.00322	-0.0005	0.00087
cavity + slot	0.9	0.00928	0.00247	-0.00089	0.00166
SC1 Bump	0.948	0.00806	0.00245	-	-
SC2 Bump	0.897	0.00748	0.00260	-	-
SC1 Bump + slot	0.858	0.00788	0.00465	-0.00053	0.00126
SC2 Bump + slot	0.803	0.00715	0.00423	-0.00053	0.00126

2.2. CONCLUSIONS 2D

Obviously, for the laminar airfoil LVA-1a the bump is the most effective type of control method if drag reduction is of main interest. These results are in agreement with experimental and numerical results found in EUROSHOCK for the DRA airfoil.
Discrete suction, passive and active ventilation and combined concepts do not effect the drag significantly at the investigated conditions.
The ventilation is the best method to get a fixed shock on the airfoil or wing because of its self-regulation, which can be important in case of laminarisation or when reducing the buffeting.

3. 3D Numerical Investigations

Because of the results of the 2D calculations, the bump, considered to be the best method reducing the drag, is chosen to be tested on swept wings. Therefore the PHLF-1 and the ADIF airfoil are used with a sweeping angle of $\Phi = 26°$ and periodic boundary conditions in the third dimension.

3.1. PHLF-1

This airfoil is investigated at a Mach number of $M_\infty = 0.85$ and a variation of the angle of attack from $\alpha = -0.5°$ to $\alpha = 1.5°$. Analogically to the 2D cases two bumps with different heights are studied.
Similar to the results obtained for the LVA-1a airfoil, there is an improvement over the whole α-sweep (fig. 6), when using a small modification of the shape. Doubling the bump height worsens the results at low α, however, there are further enhancements at higher angles of attack.
Comparing the c_p-plots (fig. 7) at a constant Mach number $M_\infty = 0.85$ and a constant angle of attack $\alpha = 1.0°$ the difference in the shock region is obvious. The shock is weaker, the pressure rises in front of the shock and there is a bigger expansion area after the shock.

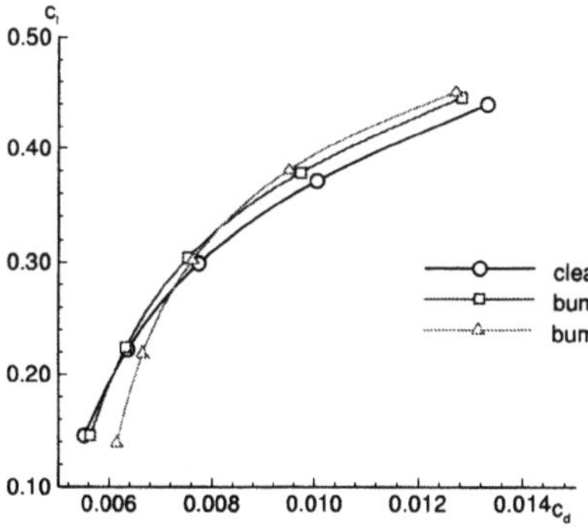

Figure 6. c_a/c_w, α-sweep

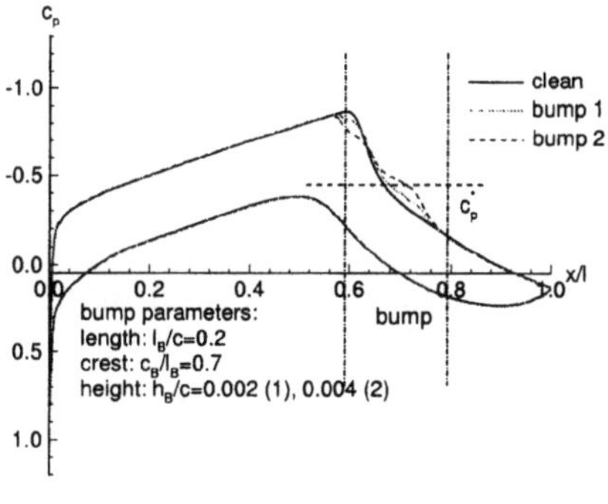

Figure 7. c_p-plot, swept PHLF-1

3.2. ADIF

The second swept wing bases on the ADIF airfoil, which is also calculated 2D. The 2D Mach number is $M_\infty = 0.755$ with a Reynolds number of $Re_\infty = 8.0 \cdot 10^6$, the angle of attack was chosen to meet the lift-coefficient of $c_l = 0.7481$. The higher bump reduces the total drag from $c_d = 0.01716$ to $c_d = 0.01456$, a decrease of 15.1%.

For the 3D calculations the corresponding Mach number is $M_\infty = 0.84$ with a Reynolds number of $Re_\infty = 9.9 \cdot 10^6$. The uncontrolled wing is calculated at $\alpha = 0.42°$, the wing with the higher bump at $\alpha = 0.32°$ to get the same lift ($c_l = 0.6152$).
The smaller bump shows slight improvements in the lift/drag-coefficient. The higher bump reduces the total drag from $c_d = 0.01259$ to $c_d = 0.01143$, a decrease of 9.2%. The effect of the bump is very impressing regarding iso-mach contours in the shock region (fig. 8,9). The λ-structure is clearly visible with a strongly enlarged expansion area after the shock, which is caused by the higher curvature of the shape.

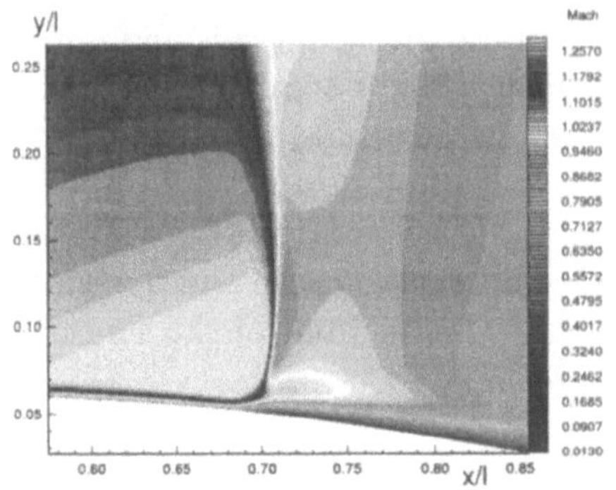

Figure 8. iso-mach contours, no control

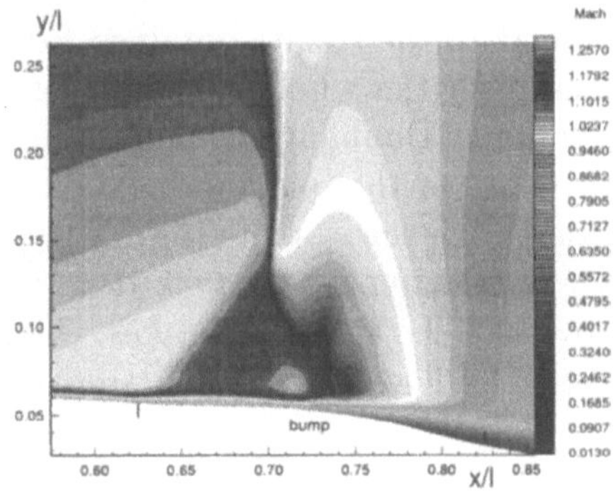

Figure 9. iso-mach contours, bump control

3.3. CONCLUSIONS 3D

The reduction of the total drag in the 3D calculations is about 1/3 smaller than in calculations with 2D cases. Nevertheless, the bump has proven to be a method to reduce the drag not only on airfoils but on wings, as well. The 3D phenomena do not significantly change the effect of the bump in the shock region.

4. References

1. Bohning, R.; Doerffer, P.: Effect of a Tangential Flow along the Porous Plate on the Transpiration Flow, Universität Karlsruhe 1997
2. Dargel, G.: Prediction of Ventilation Velocity Using Poll's Formula for Porous Sheets (Euroshock - Task 2.1), Daimler–Benz Aerospace Airbus GmbH 1996
3. Cebeci, T.; Smith, A.M.O.: Analysis of Turbulent Boundary Layers, Academic Press, New York - San Francisco - London 1974

EXPANSION SHOULDER BUMP FOR WING SECTION VISCOUS / WAVE DRAG CONTROL

H. SOBIECZKY, W. GEISSLER, M. HANNEMANN
DLR German Aerospace Research Center
Inst. Fluid Mechanics, Bunsenstr. 10, D-37073 Göttingen

1. Introduction

This is an attempt to use a systematic, theory-based method to design supercritical airfoils for the development of variable geometry wing sections with practically feasible mechanical surface deformations. The goal is, especially in this effort, to present results for airfoils with only local deformations which can be obtained by mechanical devices and which still should result in attractive aerodynamic performance improvements in varying transonic operating conditions.
During the past years, progress in aerodynamics included systematic work to add local surface modifications on transonic airfoils to influence flow quality, by reducing the negative effects of interaction between a shock wave and the boundary layer, see the related presentations in this book. Surface deformations, therefore, usually are located in the shock foot point domain, to spread recompression over a certain interval of the airfoil chord.
In this contribution surface modifications are proposed to be located far upstream of the shock, in the airfoil nose area. The concept derives from earlier developed systematic design methods to remove recompression shocks altogether: the knowledge base of transonic flow physics has given us an idea how to create typical 'transonic' airfoils and wings, and also theoretical concepts how to 'adapt' airfoils to maintain shock-free conditions if operating conditions change within a limited domain of flight Mach number and lift, see Sobieczky (1979). In the following, the route from an arbitrary airfoil, via a 'shock-free' modification, toward simply an 'improved' airfoil, is illustrated. Finding the associated shape modifications to be spread over a large portion of the upper (suction) surface of the modified airfoil, we subsequently will ask about a trade-off between more local changes and reduced improvements.

2. Transonic knowledge base for shock-free design

Thirty years ago, airfoil design methodology prior to the advent of large scale computers heavily relied on experiment and mathematical modeling of the physical background. The reduction of shocks and the related viscous losses through refined shaping of the airfoil became a 'knowledge base' before optimization strategies were proposed to attack such

G.E.A. Meier and P.R. Viswanath (eds.),
IUTAM Symposium on Mechanics of Passive and Active Flow Control, 29–34.

tasks with numerical methods. Today such knowledge may contribute to a reduced effort in selecting the essential parameters of mathematically defined geometry models of aerodynamically favorable configurations.
A method used for systematic design of shock-free configurations has become known as 'Fictitious Gas (FG) Method' referring to a helpful physical interpretation of the mathematical manipulation of the governing set of flow differential equations. For a recent review of this method see Sobieczky (1997a). A part of this method consists of the inverse method of characteristics, a linear method based on potential theory and applied to the supersonic part of transonic flow past a 2D configuration. Characteristics (Mach waves) are essential for understanding the propagation of perturbations in the flow, see the illustration in Fig. 1: For a focussing on the structure of supersonic flow patterns we present a typical result of the F.G. method, discussing the shape changes between the baseline and resulting airfoil, and the characteristics pattern for shock-free flow.

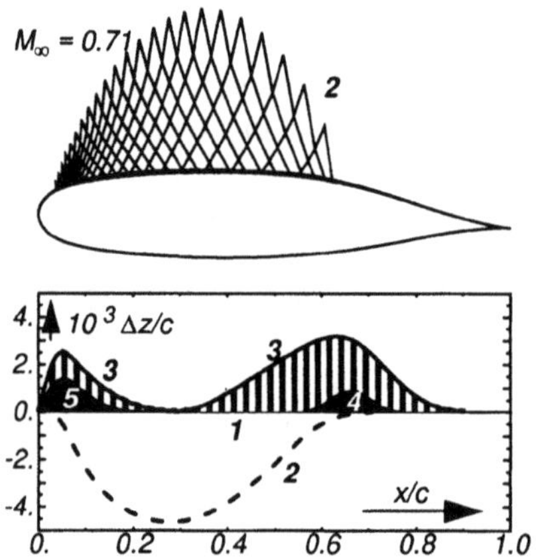

Figure 1:
Airfoil in inviscid transonic flow, (M_∞= 0.71). Shape changes of an initial airfoil (1) to become shock-free (2), and a new airfoil (3) with surface bumps to model design (2) curvature. Characteristics pattern of shock-free flow (2).
Reduced size bumps (4) and (5) to pick up part of the aerodynamic performance improvements.

The examples (1, 2, 3) illustrated here were computed using our Navier/Stokes code (see Geissler (1992)) which has been extended to serve as a design tool using the F. G. method, presented by Sobieczky, Geissler & Hannemann (1996). Fast analysis of steady airfoil flow is carried out with Drela's Euler code (MSES) for 2D airfoils, which is coupled with an efficient boundary layer method. We use this method in the software frame of a comfortable expert system, see Zores (1995).
The essential details of the shape changes resulting in shock-free flow are the curvature changes within the supersonic domain, leading to the desired balance of expansion and subsequent recompression along the characteristics in the adjacent flow field. We learn that increased curvature near the sonic expansion and recompression ('sonic shoulders') areas goes with reduced airfoil curvature within this domain, necessarily consistent with a somewhat flattened airfoil crest. Learning from this, we try to avoid the unpractical thickness ***reduction*** by obtaining similar curvatures by ***addition*** of two suitable bumps (3), see the diagram in Fig. 1, one in the expansion region and one in the recompression region. Immediately the question arises whether such an approximation of a systematic design by the simple geometry manipulation (3) leads to improvements in the lift-to-drag ratio comparable to those observed from the exact shock-free redesign (2). In Fig. 2 we

compare therefore the drag polar and the drag rise curve for off-design conditions: We see that favorable properties of the redesigned airfoil (2) are not completely, but partly reached.

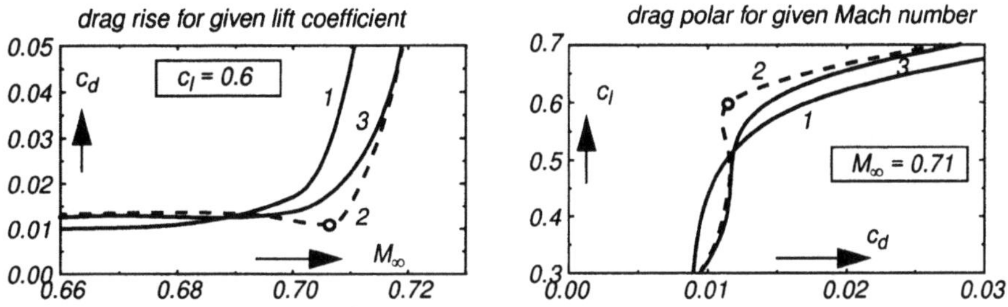

Figure 2:
Aerodynamic performance of original (1), shock-free (2) and modified (3) airfoil, Re = 20 Mill: drag rise for given lift coefficient (a) and drag polar for given Mach number (b).

3. Expansion and Recompression Shoulder Bumps

We return to Fig. 1 and observe that the dual bump (3) modeled for this example to create two curvature peaks adds up 0.3% to local airfoil ordinate z/c. Compared to the large negative bump (2) of -4.5% height, prescribed by the design method, this is desirable not only because of keeping the original airfoil thickness, but also because the substantial surface changes are required in the front and in the rear portion of the wing section, off the main wing box.

We may think of a mechanical realization of such surface modification by an adaptive device and see that surface deformation is still proposed to happen along a large part along the upper surface. This may be practically too difficult to install by some elastic, pneumatic or other device. We ask for a trade-off: If the bumps occur locally, with reduced amplitude, (4) and (5) in diagram Fig. 1, it is still possible to obtain the front and rear curvature peak, but we loose the crest curvature reduction.

At this point we realize that the idea of a rear bump (4) may be close to the concept of shock - boundary layer interaction (SBLI) control which has been investigated by various authors as reviewed elsewhere in this meeting. If this is correct, we should propose to extend this SBLI control concept by applying not only one bump (4) but another one (5) in the nose area, too. Before we further investigate such possibilities (with a necessarily doubled mechanical effort because of two bumps to be put into reality), we follow the more modest idea to install only the front bump (5) and neglect the rear bump (4).

In the following a first example of airfoil modification by such a front bump is presented with some results obtained with the above mentioned Euler and Navier/Stokes codes. Because of this bump situated amidst the above mentioned sonic expansion region, causing a concentration of wedge-shoulder type expansion waves, we call the surface modification an "Expansion Shoulder Bump" (ESB), its rear counterpart would be a "Recompression Shoulder Bump" (RSB). From the exact and approximated design case studies we are encouraged to assume that an only partial application of the surface deformation (3) may lead also to some improvements in efficiency, relative to the small and local surface changes of a baseline airfoil (1). We set the ambitious goal to start from an already aerodynamically well-designed airfoil, with the goal to improve it for some higher Mach number and higher load.

4. Improving a given transonic airfoil

Choosing a given airfoil shape and adding some of the discussed bumps gives an idea about the potential of increasing aerodynamic efficiency. As a baseline we use the transonic airfoil "OA15T", designed at the French aerospace research establishment, see Rodde & Archambaud (1994).
The airfoil was designed for a Mach number $M_\infty = 0.73$ and a lift coefficient $c_l = 0.65$. First we used our analysis methods and verified the published experimental data to a reasonable accuracy. These methods should then suffice to show the trends toward possible improvements, before further investigations in the transonic wind tunnel will confirm the concept. Here we show only the computational results.
We set the goal to allow for a local surface modification within the first 15% of airfoil chord on the upper surface in order to improve aerodynamic efficiency, i. e. the ratio of lift to drag, for the higher Mach number $M_\infty = 0.75$ and the higher lift of $c_l = 0.7$ at a Reynolds number of Re = 6 millions. Analysis of the baseline airfoil shows a moderate shock at these operating conditions.
Choice of the surface bump function should ultimately be determined by the means of a mechanical realization: use of an elastic structure leads to modelling by spline functions but the use of materials with variable flexibility by the means of varying thickness presently suggests to study more general functions: Identifying the bump shape parameters of importance for obtaining efficient changes in aerodynamic performance will require parametric computational studies prior to an experiment. Here we still use a set of mathematical functions which can also model whole airfoil shapes, see Sobieczky (1998). Figure 3 shows the front portion of the upper airfoil surface in the nose area with the contour changed by the bump geometry. A number of bump parameters can be varied, their numerical data determine the bump geometry very precisely for any model function which observes these data:

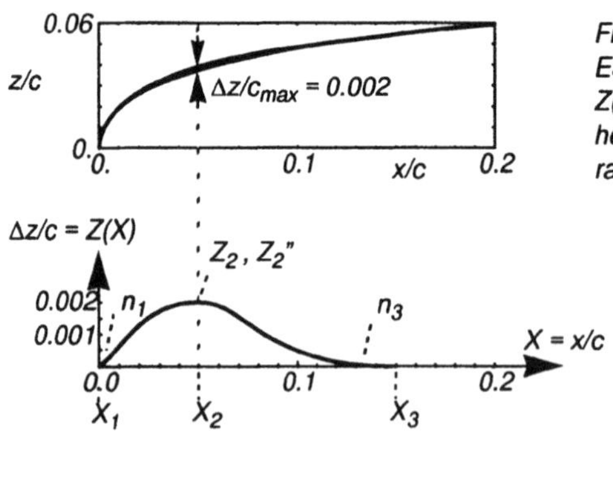

Figure 3:
ESB modeled by a parameterized function Z(X). Basic parameters defining range and height, (anti-)symmetry, crest curvature and ramp exponents.

Here we start the bump at the nose, $x/c = X_1 = 0$ and place its maximum at $X_2 = 5\%$ chord. Starting the bump at the parabolic nose allows a ramp exponent $n_1 = 1$ (wedge) there without curvature discontinuity around the nose, but for any $X_1 > 0$, n_1 would be cubic, like the ramp ending with a cubic ($n_3 = 3$) ramp exponent ensures smooth connection curvature at $X_3 = 15\%$ chord. Crest curvature Z_2" is an important parameter, the ratio Z_2"/Z_2 determines local flow quality substantially. A refined analysis of the role of these parameters seems worthwhile.
Flow analysis using both the MSES (Euler + boundary layer) code and the Navier/Stokes code confirms the concept nicely: Both numerical methods show good agreement in local surface pressure and give improved ratios of lift over drag. The improvement resulting from the MSES code is $\Delta(L/D) = 15\%$, from the N/S code it's 8.3%. The discrepancies are attributed to the still imperfect computation of drag, refined analysis with comparable grid resolution in both codes will follow these first investigations.
Of main interest here are the relative changes obtainable from such surface modifications, from the preliminary CFD results we find a promising confirmation of the concept as we see here.
Next we need to know the changes in off-design performance, both codes are therefore used to vary the angle of attack at the design Mach number of $M_\infty = 0.75$. Figure 4 shows drag polars for both the baseline airfoil and the ESB-modification. We see optimal improvements at design conditions while both at lower and at higher load the gains become marginal. At this point we may compare such improvements of those which are obtainable from SBLI control using what we termed earlier as 'RSB': a bump added at the recompression area of the local supersonic flow field, see the related contributions in this book. It is too early to oversee the advantages of an ESB over the use of RSB, parametric studies of both concepts, and also of a use of both an ESB and a RSB seems worthwhile.

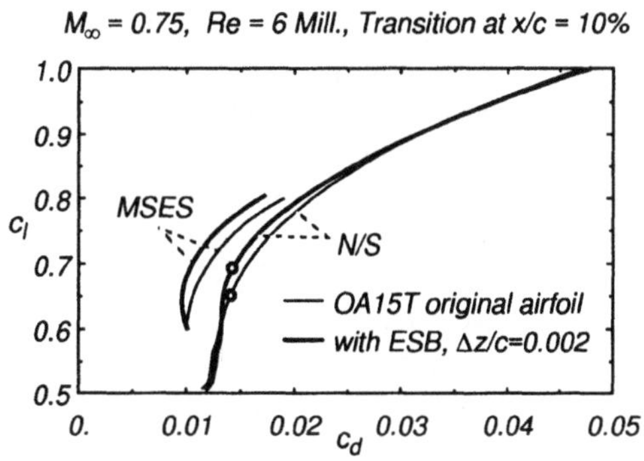

Figure 4:
Drag polars for the baseline OA15T airfoil and a modification with an ESB of 0.2% height: results of Euler + b.l. and N/S codes.

5. Toward 3D and unsteady flow applications

Our computational tools creating surfaces and their modifications are currently being used for airfoils, wings and complete configurations, see Sobieczky (1997b). Bump shapes like those introduced above, if found to be practically useful for 2D airfoils, will be tuned along span of aircraft wings and they may be used for unsteady flow control by changing bump shapes with time. Some first applications of such geometry tools are presently used for the definition of unsteady boundary conditions modeling a periodic droop of the airfoil nose, see Sobieczky, Geissler & Hannemann (1996), these are used in the low speed retreating phase of a helicopter rotor blade. Combination of this variable geometry with the ESB concept may be found useful also in the transonic, advancing phase of the rotor blade, thus prescribing periodic shape changes of rotor blades for the full cycle.

6. Conclusion

We have shown a theory-based shape modification concept for transonic wing sections which might have a potential for application in flow control in the domain of cruise conditions: A local bump in the upper surface nose region can be adapted to increase aerodynamic efficiency in varying cruise Mach numbers and aerodynamic loads and thus widen the region of optimum lift-over-drag ratio. The presented procedure to create calibrated bumps for curvature control is a 'manual' approach, which can be replaced later by a computer program resting on the same knowledge base. Future use of mechanical devices cre ating such bumps may include active control of transonic flow along wings and unstead phenomena on rotorcraft.

7. References

Geissler, W. (1992) Instationäres Navier-Stokes-Verfahren für beschleunigt beweg Profile mit Ablösung. DLR-FB-92-03.

Rodde, A.M., Archambaud, J.P. (1994) OAT15A Airfoil Data, AGARD FDP AR 303

Sobieczky, H. (1979) Computational Methods for the Design of Adaptive Airfoils an Wings, E. H. Hirschel (Ed.): Notes on Numerical Fluid Mechanics, Vol. 2, Braunsch weig: Vieweg Verlag

Sobieczky, H., Geissler, W., Hannemann, M. (1996) Numerical Tools for Unsteady Vis cous Flow Control. Proc. 15th Int. Conf. on Num. Meth. in Fluid Dynamics. Lectur Notes in Physics, K. W. Morton (Ed.), Berlin, Heidelberg, New York: Springer Ve lag

Sobieczky, H. (1997a) Gasdynamic Knowledge Base for High Speed Flow Modelin New Design Concepts for High Speed Air Transport. H. Sobieczky (Ed.): CISM Courses and Lectures Vol. 366, Wien, New York: Springer Verlag. pp 105-120

Sobieczky, H. (1997b) Geometry Generator for CFD and Applied Aerodynamics. Ne Design Concepts for High Speed Air Transport. H. Sobieczky, (Ed.): CISM Course and Lectures Vol. 366, Wien, New York: Springer Verlag. pp 137-158

Sobieczky, H. (1998) Parametric Airfoils and Wings. Recent Developments in Aerody namic Design Methodologies, K. Fuji and G. S. Dulikravich (Eds.): Notes on Nume ical Fluid Mechanics, Vol: (to appear), Braunschweig: Vieweg Verlag

Zores, R. (1995) Transonic Airfoil Design Using Expert Systems. AIAA-95-1818

EUROSHOCK I AND II: A SURVEY

Drag Reduction by Shock and Boundary Layer Control

E. Stanewsky
DLR, Institute of Fluid Mechanics
Bunsenstrasse 10, D-37073 Göttingen

1. Introduction

The development of the boundary layer in general and shock boundary layer interaction in particular essentially establish the flight performance of transonic transport aircraft at cruise and high-speed off-design conditions. Consequently, employing shock and boundary layer control can be assumed to have a large potential for improving flight performance in terms of cruise drag, hence speed and / or fuel consumption, and with regard to the drag-rise and buffet boundaries.

The potential of shock control was confirmed by the results of early experiments for a turbulent airfoil which have shown that passive shock control by ventilation via a perforated surface with a cavity underneath can not only shift the drag-rise and buffet boundaries to higher Mach numbers and / or lift coefficients, but also reduce airfoil drag over a fair range of the transonic flight envelope, Figure 1 [1]. These encouraging results and the advance of laminar airfoil technology, led to the EC Research Project EUROSHOCK (I) with the objective of investigating all aspects of passive shock control by ventilation for laminar-type airfoils. The expectations concerning drag reduction for these airfoils were, however, as will be outlined below, not met although a wealth of experience and physical understanding was gained during this exercise [2]. Based on the latter, a new proposal was issued concerned with active shock and **boundary layer** control; the resulting EC Research Project EUROSHOCK II is presently being performed and some quite encouraging results have already been achieved.

2. Objectives, Approach and Partners of EUROSHOPCK I and II

The specific objective of the research discussed here is to investigate and devise methods of shock and boundary layer control aiming at the reduction of aircraft drag and the improvement of aircraft off-design performance, and to provide the tools needed for the design of transonic wings with control. In order to meet these objectives, the projects were essentially divided into four major research tasks: ① Modeling of Control Phenomena, ② Prediction of Transonic Airfoil / Wing Flow with Control, ③ Wind Tunnel Experiments on Airfoils and Wings with Control and, only within EUROSHOCK II, ④ Control Application Aspects. All tasks are, of course, strongly interrelated with all

G.E.A. Meier and P.R. Viswanath (eds.),
IUTAM Symposium on Mechanics of Passive and Active Flow Control, 35–42.

information in EUROSHOCK II essentially going to Task 4 for the purpose of applying the results of potential devices to real aircraft.

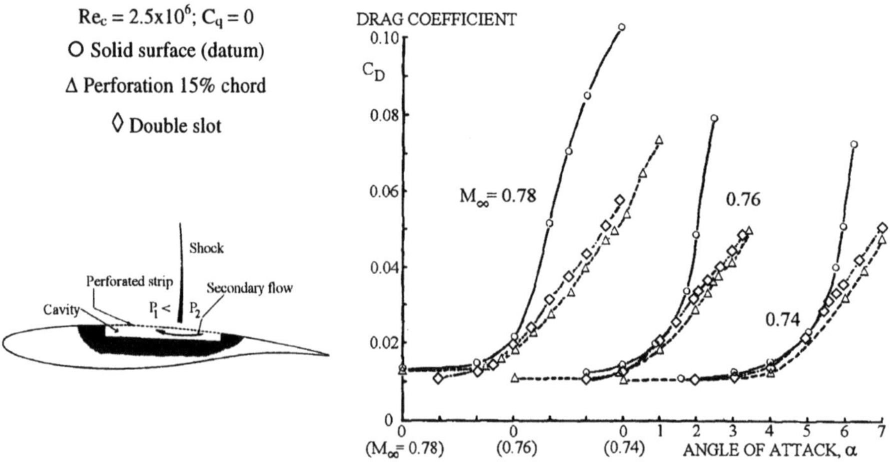

Figure 1. Mechanism and effect of passive shock control on drag, Airfoil VFW VA-2

The control mechanisms considered were in EUROSHOCK (I) passive shock control via a cavity covered by a perforated surface, Figure 1, left, and in EUROSHOCK II the active control devices shown in Figure 2. Partners in the projects were in EUROSHOCK (I): DLR, CIRA, DERA, NLR and ONERA, the Universities of Athens, Cambridge, Karlsruhe and Naples and DASA-Airbus, in EUROSHOCK II: DLR, CIRA, DERA, INTA and ONERA, the Universities of Cambridge, Karlsruhe and Naples and Alenia, BAe-Airbus, DASA-Airbus and Dassault Aviation.

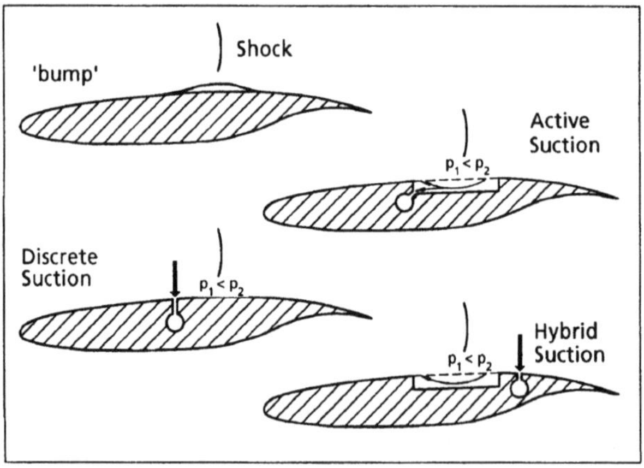

Figure 2. Active control mechanisms investigated in EUROSHOCK II

3. Basic Experiments and Physical Modeling

The specific objectives of the work performed here are ① to improve the physical understanding of the phenomena involved in a transonic shock wave / turbulent boundary layer interaction and its control, ② to assess and improve the physical laws associated with shock and boundary layer control and ③ to evaluate the efficiency of control systems. The partners involved here are ONERA and the Universities of Karlsruhe and Cambridge, the former investigating two-dimensional and Cambridge investigating three-dimensional aspects of control. All experiments were supplemented by Navier-Stokes computations employing various turbulence models.

Passive shock control by a porous plate and a cavity underneath exerts its influence on the flow by a mechanism that causes inflow into the cavity in the high-pressure area downstream of the shock and outflow from the cavity into the low-pressure area upstream of the shock. This generally spreads the steep pressure rise due to the shock over the forward part of the cavity region, Figure 3, which reduces wave drag and constitutes the main mechanism of drag reduction associated with passive control.

Of course, to assess the overall effect of passive control on drag, one must consider the development of the boundary layer displacement and momentum thicknesses, respectively, over the control region, Figure 4. It is obvious that, in comparison with the reference case, these characteristic boundary layer parameters are considerably increased due to control suggesting a substantial increase in viscous drag which, as will be outlined below, in the case of airfoil flow may outbalance the wave drag reduction.

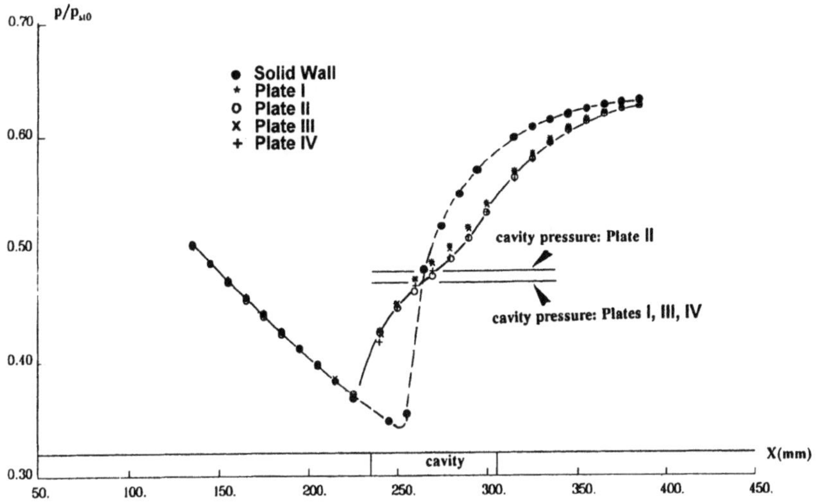

Figure 3 Surface pressure distributions for the ONERA channel-flow experiment

In **active shock control**, investigated up to now were different active and hybrid control schemes, including part-suction from a single cavity, hybrid control by a passive cavity followed by an active cavity with suction downstream, and discrete slot suction.

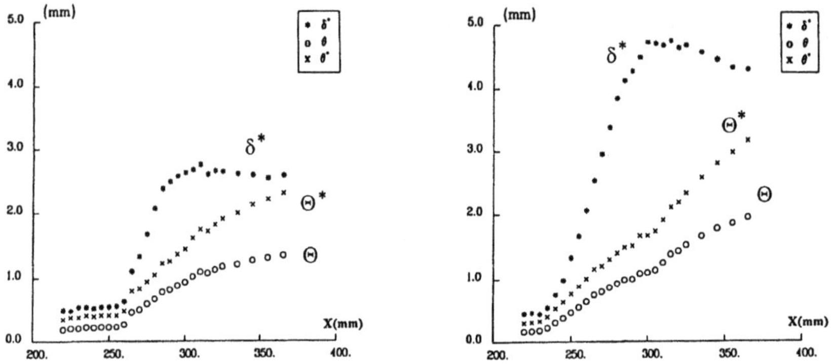

a. Solid surface reference case

b. Passive shock control

Figure 4. Development of boundary layer thickness parameters over the cavity region

Only one characteristic result, obtained by the University of Karlsruhe, shall be presented here, Figure 5: One observes in the case without control the "normal" boundary layer development due to the interaction with a shock of medium strength ($M_W = 1.30$). Introducing passive control leads, as already seen in Figure 4, to a significant thickening of the boundary layer downstream of the cavity, of course, coupled with a reduction in wave drag, while the addition of active control improves the state of the boundary layer, all dependent on the amount of suction applied. Note that the results for two- and three-dimensional flow are quite similar.

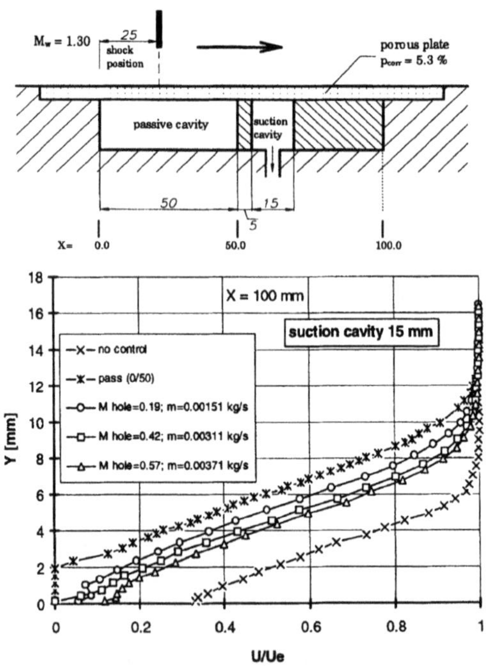

Figure 5 Hybrid control test set-up and boundary layer development with and w/o control

4. Extension of Numerical Methods and Preliminary Control Assessment

The main objective of the work carried out here is to extend existing steady and unsteady methods to treat **airfoil and wing** flows with shock and boundary layer control, to validate these methods with existing and new data and to perform a preliminary assessment of the merits of shock and boundary layer control. The computational methods and partners involved in this part of the project effort are presented in Figure 6.

Inviscid / Viscous	Transonic Small Perturbation (TSP)	Full Potential (FP)	Euler	Navier - Stokes (NS)
Integral Boundary Layer (IBL)	NLR* ONERA 2	ONERA 1	CIRA	
Triple Deck at Shock		DASA-Airbus		
Finite Difference (FD)		DASA-Airbus	Naples University	DLR Athens-Univ.* Alenia** INTA**
Navier - Stokes (NS)	* EUROSHOCK I only ** EUROSHOCK II only			

Steady Methods / 2D and Quasi 3D

Inviscid / Viscous	Transonic Small Perturbation (TSP)	Full Potential (FP)	Euler	Navier - Stokes (NS)
Integral Boundary Layer (IBL)	NLR* ONERA 2			
Triple Deck at Shock				
Finite Difference (FD)			Naples Univ.	
Navier - Stokes (NS)				DLR Athens-Univ.*

Unsteady Methods / 2D

Figure 6. Solution techniques employed by the partners

In passive shock control characteristic results were obtained by CIRA, and others, the former employing the EUBL2D VII-code, to the DRA-2303 laminar-airfoil test case $M\infty = 0.68$, $Re_c = 19x10^6$ at a fixed lift coefficient of $C_L = 0.81$, Figure 7: typically, passive control reduces the shock strength, hence wave drag, by spreading the steep pressure rise due to the shock over part of the control region, in close agreement with the experiment performed by DERA. Considering the corresponding development of the displacement thickness on the airfoil upper surface, one observes, however, that its growth downstream of the shock is stronger in the case of control since the initial increase in displacement thickness caused by the perforation and the re-circulating flow in the cavity region, also see Figure 4, is amplified by the rear adverse pressure gradients. This results in a considerable increase in viscous drag which overcompensates the reduction in wave drag. In the present case, the total drag coefficient increased by 8.4% due to control.

Note that all participating codes predicted, in agreement with experiment, an increase in drag and a positive effect on the buffet boundary due to passive control.

In **active shock control** the mechanisms for which computations were carried out here are, Figure 2, a contour bump in the shock region, discrete suction (plus these in combination), and partial suction via a single cavity with perforation. The computational results — as the experiments — have indicated that substantial drag reductions can only be achieved by bumps, placed "correctly" relative to the shock, and by discrete slot suction upstream of the shock. Note that a positive effect on buffet was predicted for all devices.

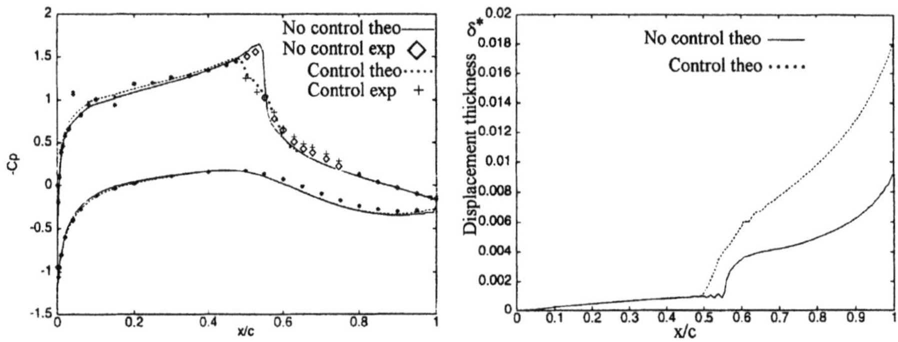

a. Surface pressure distributions b. Development of the displacement thickness

Figure 7. Experimental and computational results for the laminar-type airfoil DRA-2303

The effect of a bump on the flow development is demonstrated by computations performed by DASA-Airbus for the Airbus A340-type airfoil ADIF, Figure 8. The bump essentially spreads the pressure rise due to the shock, thus reducing wave drag, as does passive control by ventilation, however, this happens without increasing the viscous drag, as indicated by the displacement thickness in Figure 8. The predicted total drag reduction is for the freestream conditions shown in the figure 11.9%; the corresponding experimentally determined reduction is 21%.

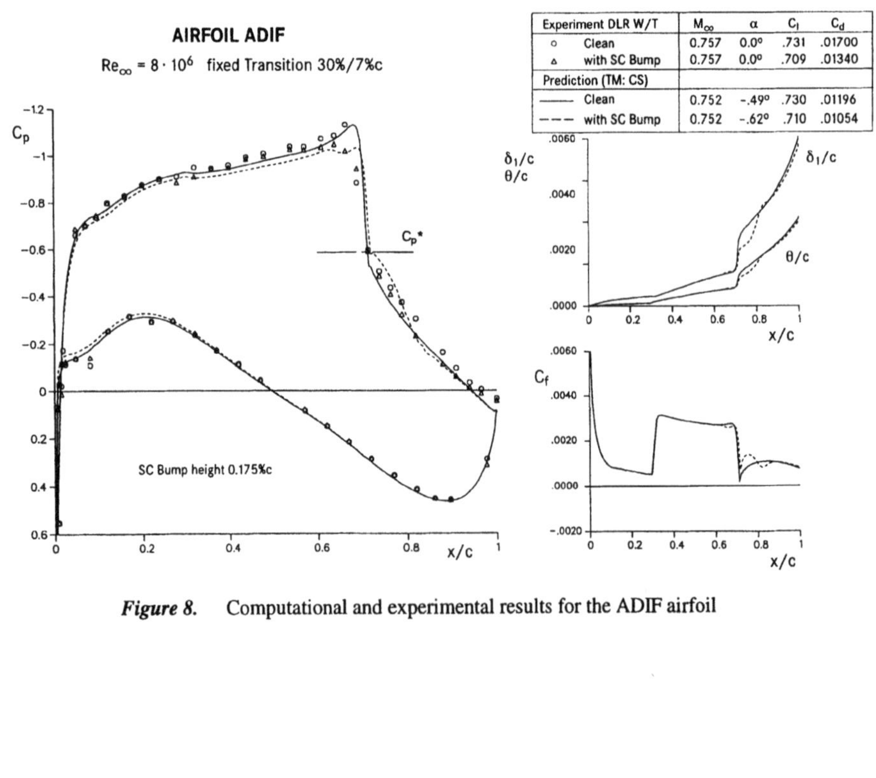

Figure 8. Computational and experimental results for the ADIF airfoil

5. Airfoil and Sheared-wing Experiments with and without Control

Within this effort detailed measurements on airfoils and wings up to flight Reynolds numbers are being performed in order to assess the improvements to be gained by shock and boundary layer control and to provide results for computer code validation, especially under control conditions. The partners are here DLR with airfoil and sheared-wing tests, ONERA with similar measurements with emphasis, however, on flow field surveys and DERA with passive and active control studies on the DRA-2303 airfoil. Again, only representative results can be given below.

In **passive control studies**, the laminar-type airfoil DA LVA-1A, designed by DASA-Airbus, was investigated in the DLR KRG and in the ONERA/CERT T2, both cryogenic adaptive-wall wind tunnels, without and with passive control by a perforated surface / cavity. The results of these measurements, Figure 9, clearly demonstrate the spreading of the shock due to passive control and also allow a conclusion concerning the development leading to the drag increase: the total pressure losses in the airfoil wake on the upper side of the wake show a reduction in these losses, i.e., a reduction in wave drag, caused by the near-surface spreading of the shock due to control, but also a dominating increase in pressure losses closer to the surface, i.e., a widening of the wake, due to increased viscous effects.

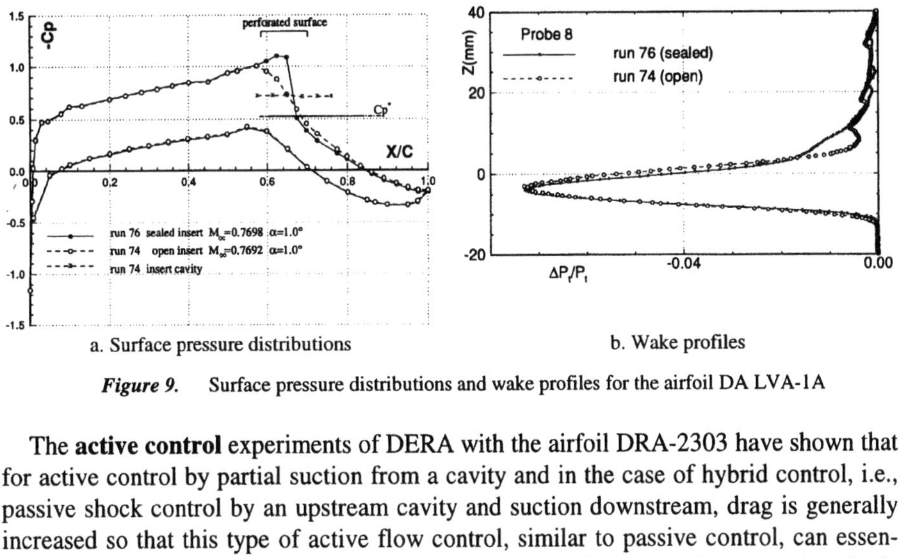

a. Surface pressure distributions

b. Wake profiles

Figure 9. Surface pressure distributions and wake profiles for the airfoil DA LVA-1A

The **active control** experiments of DERA with the airfoil DRA-2303 have shown that for active control by partial suction from a cavity and in the case of hybrid control, i.e., passive shock control by an upstream cavity and suction downstream, drag is generally increased so that this type of active flow control, similar to passive control, can essentially be ruled out if drag reduction is of main concern. Discrete suction through a single slot was found to reduce drag, even when accounting for pump drag.

Considering all control devices investigated, it seems that the contour bump has the largest potential for drag reduction. This is demonstrated by the results obtained with the ADIF airfoil, already cited, in the DLR KRG wind tunnel for the airfoil with two bumps, respectively, of different height, Figure 10. Drag reductions of up to 17 % at lift coefficients of $C_L \leq 0.60$ and up to 23 % at higher lift coefficients are achieved. It is, furthermore, interesting to note that the lower bump shows a much better performance at the lower lift coefficients, where the higher bump actually causes an increase in drag due to a strong double-shock system, while the higher bump is more effective at higher lift where stronger shocks are present. It seems obvious that for exploring the full potential of this device an "adaptive bump" is required.

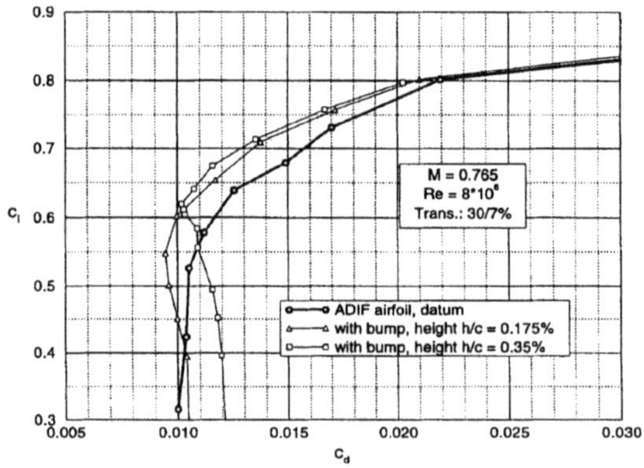

Figure 10. Measured drag polars for the ADIF airfoil without and with bumps of different height

6. Control Application Aspects

This topic is only being considered within EUROSHPCK II with main emphasis now being placed on contour bumps and discrete suction since these seem the only devices providing the potential for an overall performance improvement. The partners and specific objectives are here: BAe-Airbus and DASA-Airbus investigating benefits and penalties associated with the incorporation of these control devices into an A340-type turbulent and laminar aircraft wing, respectively, and Alenia and Dassault Aviation performing a similar task for a business-jet aircraft. To stay within the length given the paper, it shall only be mentioned here that in pursuing these objectives, DASA-Airbus in introducing a bump into an A340-type hybrid laminar flow wing and assuming a typical mission for such an aircraft of 600 flights per year over a range of 3500 NM, has determined that fuel savings of up to 2.11 % at a cruise Mach number of $M_\infty = 0.84$ can be achieved, an encouraging result which makes pursuing this concept further worthwhile.

7. Literature

1. Stanewsky, E. and Krogmann, P.: *Transonic Drag Rise and Drag Reduction by Active / Passive Boundary Layer Control*, AGARD Report No. 723, Special Course on Aircraft Drag Prediction and Reduction, July 1995.
2. Stanewsky, E., Délery, J., Fulker, J., Geissler, W. (Edit.): *EUROSHOCK-Drag Reduction by Passive Shock Control*, Notes on Numerical Fluid Mechanic, Volume 56, Verlag Vieweg & Sohn, 1997.

A REVIEW OF FLOW CONTROL RESEARCH AT DERA

P R ASHILL* AND J L FULKER
Defence Evaluation and Research Agency (DERA)
Bedford MK41 6AE, England

Abstract

Experimental and Computational Fluid Dynamic (CFD) studies at DERA of flow control are described. The investigations include wind-tunnel tests at a) low speed on the use of Sub Boundary-layer Vortex Generators (SBVGs) to control leading and trailing edge separations on highly swept wings and b) high subsonic speeds and high Reynolds number on an aerofoil model with and without shock control. The former tests revealed that SBVGs can reduce the lift-dependent drag of swept wings by up to about 16%. The latter tests confirmed predictions in showing that control by localised surface shaping offers significant reductions in drag and that passive control, by the use of ventilated surfaces with holes drilled normal to the surface, increases the drag. The benefits of a novel form of shock control for delaying buffet onset and increasing maximum lift are described and the possible advantages of streamwise slots over holes for passive ventilation are discussed.

1. Introduction

Increasingly, multi-disciplinary issues are playing an important part in aircraft design. Competitive pressures in the civil-transport aircraft industry are driving aircraft designers towards low-cost solutions, while combat aircraft have to operate efficiently over a wide range of conditions. This means that compromises have to be made in the aerodynamic design: thus, for example, geometries have to be considered featuring wing flows that may be less satisfactory than those over a wing that is aerodynamically optimum in some sense.

A way forward lies in the use of flow control devices. These devices may either be passive (e.g. Sub Boundary layer Vortex Generators or SBVGs), or active, being deployed only when required. Such devices provide a further degree of freedom in the design optimisation process.

This paper reviews a programme of research on flow control that has been proceeding at DERA for the past 6 years and describes work that has already been published, as well as work that is not yet in the public domain. The flow-control devices considered are:

---

* Consultant, employed by Cranfield University to work on DERA projects

G.E.A. Meier and P.R. Viswanath (eds.),
IUTAM Symposium on Mechanics of Passive and Active Flow Control, 43–56.

1) SBVGs of low parasitic drag, to control leading and trailing-edge flows on highly-swept wings (i.e. with sweep-backs greater than about 40°).

2) Devices for controlling the surface boundary condition in the region of shock waves to weaken the shock wave above the wing upper surface, either in the field or close to the wing surface. Such devices includes ramps, bumps and passive surface ventilation (using laser-drilled surfaces). The ramps or bumps studied are divided into two types: those intended to reduce drag and others designed to delay buffet, referred to here as 'buffet bumps' (patent pending). Consideration is also given to a novel form of passive control using streamwise slots.

2. Sub Boundary-layer Vortex Generators

Wind-tunnel studies have been undertaken at DERA, Bedford on the use of SBVGs to control the flow over highly-swept wings suitable for stealthy or supersonic combat aircraft or supersonic transport aircraft. Most of the work has been concerned with controlling leading-edge separation [1], [2], [3], although some has been concerned with the control of trailing-edge separation [4]. Here the discussion is restricted to leading-edge separation control, although it should be noted that benefits of a similar magnitude have been obtained with the control of trailing-edge flows [4].

The experiments were performed at low speed* in the 13ft x 9ft Wind Tunnel at DERA Bedford on two models as illustrated in Fig 1. The wings were of delta planform, of 60° leading-edge sweep with a Küchemann tip and streamwise sections of 4% maximum thickness-chord ratio. As shown in Fig 1, these wings have pronounced leading-edge camber. One is of fixed camber, having been designed for efficient sustained manoeuvre at supersonic speeds and the other has a basically symmetrical wing with a variable leading-edge droop.

2.1. PHILOSOPHY OF LEADING-EDGE SEPARATION CONTROL

It is important to understand the flow that one is attempting to control [5]. The two types of flows controlled are illustrated by two sketches of surface flows and spanwise sections through stream surfaces in Fig 2, which are based on oil-flow visualisations of the flow on the wing upper surface.

At angles of incidence or lift coefficients below the threshold level ($C_L \approx 0.4$), the flow is called Type A. In this flow a three-dimensional or 'ordinary' separation [6] occurs on the highly curved part of the upper surface at S_1, reattaching at the line R_1. This separation is distinct from another separation, S_2, which occurs at the wing leading edge

---

* The tests referred to in this paper were performed at a free-stream speed of 61 m/s, corresponding to a Mach number of 0.18 and a Reynolds number based on geometric mean chord of about 4×10^6.

downstream of the point P and reattaches at R_2. Beneath the rolled-up vortex sheet formed by the separation S_2 there is a further separation, S_3. Since the leading-edge separation S_2 is unaffected by boundary-layer conditions upstream of P, this leading-edge separation is said to be 'locally induced'.

At higher angles of incidence the separation S_1 intersects or osculates the leading edge at the point P, where it provokes the leading-edge separation S_2, forming a common vortex sheet with a reattachment at R (Type B). Thus in this type of flow, the leading-edge separation depends on the boundary-layer conditions on the wing upper surface upstream of the separation line S_1. For this reason this leading-edge separation is called 'upstream dependent'.

It follows that, if a boundary-layer control device were placed on the wing upper surface upstream of the section AA and close to the leading edge, no effect on the leading edge flow would be expected in Type A flows. On the other hand, a significant effect would be expected for Type B flows. If, however, the device were downstream of S_1 in Type B flows, no effect on the leading-edge flow would be expected either. Thus, since the point P moves upstream with increasing angle of incidence, there is a limited range of angles of incidence over which a single device at a fixed position can be expected to be effective, and this is borne out by experience as noted below.

2.2. SBVG GEOMETRIES

The possible benefits of controlling Type B flows using SBVGs comprising small wires stuck to the wing upper surface have been studied. These wires were placed in the turbulent boundary layer near the wing leading edge. Each wire was arranged so as to induce a streamwise vortex, the rotational sense of which was such as to a) resist the 'drift' of the boundary-layer flow towards the point P and b) promote the flow towards the separation line deep within the boundary layer. The indications of oil-flow studies are that the wires act to convert boundary-layer vorticity into streamwise vorticity. The wire does this by inducing a separation upstream of the flat front face, the leeward part of this rolled-up sheet providing the necessary component of streamwise vorticity. The side of the wire ensures that the windward part of the rolled-up sheet does not interfere with the leeward sheet. Recent work on a super-scale version of a wire in the Boundary-Layer Tunnel (BLT) at DERA Bedford supports this interpretation [7]. As is evident in the tests in the 13ft x 9ft Tunnel, the rolled-up vortex is found to be remarkably persistent, while being close to the flat surface of the BLT.

The wires were tested singly and multiply, the geometry, position and orientation of the wires being specified in [1]. The data given in the present paper is for wires of diameter approximately equal to the local boundary-layer displacement thickness and set at an angle of about 16° relative to the wing leading edge.

2.3. RESULTS

2.3.1. *Lift-dependent drag*

Fig 3 shows results for the percentage reduction in the lift-dependent drag factor

$$K = (C_D - C_{D0})\pi A / C_L^2$$

due to the wire SBVGs for the fixed-camber wing plotted against lift coefficient for various numbers and arrangements of wires. Here C_D is drag coefficient, A is wing aspect ratio, C_L is lift coefficient and suffix 0 refers to zero lift. Two main points may be made about this figure. The first is that there is a limited range of lift coefficients over which the SBVGs are effective, confirming the remark made in Section 2.1.. The second point is that the maximum reduction of 16% occurs with 28 wires at a lift coefficient of about 0.6. Any further increase in wire number causes a slight decrease in the benefit, due it is believed to an increase in the mutual interference between the SBVGs as they become closer together. Discussion of the mutual interactions between the wires can be found in [1], [2] and [3].

Studies at DERA indicate that a 16% reduction in lift-dependent drag can result in significant savings in fleet costs for combat aircraft and reductions in take-off and landing noise for supersonic transport aircraft.

It is shown in [1] that a major part of the increased leading-edge thrust which provides the reduction in lift-dependent drag comes from the reorganisation of the leading-edge separated flow downstream of the point P. As well as moving the point P downstream, the SBVGs tighten the leading-edge vortex and move it closer to the forward-facing surfaces near the wing leading edge, resulting, in turn, in increased leading-edge thrust.

2.3.2. *Buffet*

Fig 4 illustrates the effect of the wires on the wing-tip acceleration for a particular mode of vibration and for the variable-droop wing, with the leading-edge droop angle δ_f = 25°. It shows that, as well as reducing lift-dependent drag, the wires delay the onset of buffet by about 0.05 or nearly 10% in lift coefficient, a significant benefit.

3. Control of shock waves

In this section consideration is given to methods of controlling the flow in the region of shock waves by locally altering the aerofoil or wing boundary conditions. The change in the surface boundary condition can either be effected by a geometry perturbation (bumps), by surface ventilation using the natural pressure gradients to induce flow through the wing surface or by forced surface suction and blowing. Some studies have been made at DERA of the use of forced suction but the discussion of ventilated surfaces will be mainly concerned with passive systems in this paper.

3.1. SURFACE SHAPE CHANGES

Various studies have been performed at DERA on aerofoils with modifications to the upper surface shape to control the shock wave [8], [9], [10], [11], [12]. This has involved both CFD and experimental investigations.

3.1.1. *Basic considerations*

Our experience has led us to categorise surface shape-change devices into two main groups:

1) <u>Devices aimed primarily at reducing wave drag</u>

These devices are in the form of bumps or ramps usually placed with their crest just downstream of the main limb of the shock. Typically, they are of 10% to 20% chord in length and of amplitude 0.2% to 0.4%. Fig 5 shows a sketch of such an installation indicating that the device functions by producing compression waves that weaken the main shock, reducing the wave drag. Properly positioned, these devices result in no increase in boundary-layer drag and may even reduce this drag component by decreasing the pressure gradients downstream of the shock.

2) <u>Devices aimed mainly at delaying the onset of buffet</u>

Also in the form of a bump or ramp of similar dimensions to those above, these devices are placed with their leading edge just downstream of the main limb of the shock (Fig 6). The effect of the forward face of the bump is to incline the main limb of the shock wave, reducing the pressure rise through the shock and delaying the onset of shock-induced separation as indicated in the figure by plots of local skin friction. The device is shaped so that it does not provoke premature separation at or just downstream of its aft face. This device has been referred to as a 'buffet bump' (patent pending).

In the original concept, bumps or ramps were thought of as variable-geometry devices to be deployed as and when needed [8]. More recently, consideration has been given to these devices as fixtures. Of course shape changes would not be needed if the wing were designed properly in the first place. On the other hand, aircraft weight often increases in mid-life extensions and updates of aircraft projects, and when this happens bumps may have a role to play as retrofits. Additionally, 'buffet bumps' may be useful as fixtures from the outset of the design if it can be shown that, while alleviating buffet, they do not increase drag at the cruise condition. As will be shown later, the indications are that this is the case for a certain class of aerofoil.

3.1.2. *CFD studies*

The emphasis in the theoretical work has been to investigate the prospects for using shape changes to provide drag reductions. These CFD studies have included the investigation of bumps on the upper surface of both natural laminar flow and turbulent (supercritical) aerofoils [8], [9] using the transonic aerofoil code BVGK. Some work has also been performed on the application of bumps to wings for both military and civil transport aircraft [13] using the Viscous Full Potential (VFP) code and the Euler Multiblock solver SAUNA developed at the Aircraft Research Association, Ltd. Bedford.

The aerofoil studies indicated that the concept has considerable promise for reducing drag at drag rise conditions. However, the indications from the studies of wings are that consideration needs to be given to the concept from the inception of the design. Otherwise, practical considerations, such as the position of the spoilers and flaps, could well limit the scope for positioning the device, thereby compromising its performance.

3.2. SURFACE VENTILATION

3.2.1. *Basic considerations*

Fig 7 illustrates the principle of passive ventilation. Air from the region downstream of the shock moves to the lower pressure region further upstream via the plenum beneath the ventilated surface. The resulting recirculation causes compression waves upstream of the shock that are similar in character to those for the drag reducing bump referred to above. It is expected that passive ventilation will reduce wave drag. However, for ventilation with blowing or suction normal to the aerofoil surface, this reduction in drag is offset by the increase in boundary-layer drag due to the combined effect of :

i) the increase in boundary-layer momentum thickness due to transpiration, overcoming the corresponding reduction in skin friction, upstream of the shock;

ii) the increase in skin friction downstream of the shock, offsetting the reduction in boundary-layer momentum thickness, due to the suction in this region.

Of these effects, the latter is generally dominant.

As is shown later, experience with a natural laminar-flow aerofoil at high Reynolds number is that the increase in boundary-layer drag tends to outweigh the reduction in wave drag for normal blowing/suction through holes for lift coefficients of interest [10]*. This has led some consideration to be given to alternative forms of ventilated

* This is in contrast to experience gained elsewhere with turbulent aerofoils at lower Reynolds number [14].

surface. One possibility is to use inclined holes; however, the geometric complexity of this system, arising from the need to ensure that the air through the surface has a positive streamwise component of momentum, probably rules it out for further consideration.

An alternative system recently studied at DERA is to replace normal holes with streamwise slots, as shown in Fig 7 (patent pending). In this concept the recirculation flow is not constrained to pass at right angles to the surface. In addition, streamwise slots allow the surface streamlines to diverge on the solid surfaces between the slots, providing some relief to the boundary layer on these surfaces.

3.2.2. *CFD studies*

A version of BVGK has been developed that models a ventilated surface, connected by a single plenum, with normal blowing/suction through the holes in the surface. Details of the modifications to the viscous-inviscid interaction procedure are given in [12]. Allowance can be included for air to be removed from the plenum so that arrangements with suction can be studied as well as passive systems. BVGK(P) correctly provides the same results as BVGK for systems of zero porosity.

A method to represent a spanwise slot or localised hole suction, with the suction normal to the surface, has been conceived using a code originally developed at the Engineering Sciences Data Unit [15] to represent the effect of an excrescence by an increment in boundary-layer momentum thickness. A suction slot is simulated by a reduction in boundary-layer momentum thickness. For hole suction allowance has to be made for the increment in skin friction due to normal suction which ensures a lower reduction in momentum thickness than that obtained with slot suction [16]. Using the model for slot suction, calculations have been made for aerofoils, showing significant benefits before and after allowance is made for pump drag.

3.3 MEASUREMENTS

A wind-tunnel model that has been used most recently to investigate the effects of shock-control devices at DERA is illustrated in Fig 8. This model has an aerofoil section of 14% thickness-chord ratio that is suitable for a natural laminar-flow design. Tests on this model, as with previous, preliminary tests on a turbulent aerofoil [9], were made in the 8ft x 8ft High Speed Wind Tunnel at DERA Bedford with the 0.635m chord model spanning the working section.

The various modifications to the aerofoil surface are illustrated in Fig 8. The shape-change devices include three geometries: the ramp and asymmetric bumps and a 'buffet bump'. The first two were formed from shape perturbations comprising, respectively, a) two linear segments and b) two circular arcs. The 'buffet bump' was formed so that the modified surface in the region of the bump comprised three flat sections, so made for ease of manufacture. The first two bumps were of maximum height 0.25% chord, with their leading edges at 48.5% chord and of length 20% chord, whereas the 'buffet bump' was of somewhat larger amplitude at 0.31% chord.

Passive ventilation, extending from 50% to 60% chord on the upper surface, was also tested, as shown in Fig 8, the first comprising holes drilled normal to the surface with an open area ratio or porosity 8%. The nominal diameter of the holes was 0.0762 mm and the depth of the holes was 1 mm. The second consisted of streamwise slots of width 0.24% chord and spaced apart by 6% chord, giving a porosity of 4%.

The results described here include measurements at mid-span of surface static pressures and sectional drag inferred from a wake rake of pitot and static tubes situated two aerofoil chords downstream of the model trailing edge. Details of the tests are given in [12]. Here only data for a Mach number of 0.68 and a chord Reynolds number of 19×10^6 are presented.

3.4. EXPERIMENTAL RESULTS AND ASSESSMENT OF PREDICTION

Fig 9 shows comparisons, that are typical of those found for the effect of drag-reduction bumps, between measured and predicted drag polars for the ramp bump and for the datum aerofoil at M = 0.68. BVGK tends to underestimate the drag slightly, probably because of the effect of the (ballotini) transition trip at 5% chord on both surfaces; however, the code accurately predicts the reduction in drag due to the bump for lift coefficients above 0.6, of approximately 12%.

A comparison between measured and predicted pressure distributions for the datum, ramp bump and passive (hole) control aerofoils are shown in Figs 10a, b and c for M = 0.68 and $C_L \approx 0.62$. Except just upstream of the shock for the ramp bump, BVGK or BVGK(P) gives satisfactory predictions for all three cases. Also shown on the figures are the angles of incidence to achieve the required lift coefficient, the drag coefficient and the pitching moment coefficient. Predictions of these quantities are also in satisfactory agreement with measurement.

Figs 11 and 12 show comparisons between measurement and prediction of drag polars and trailing-edge pressure coefficient, C_{pTE}, for the 'buffet bump' and datum aerofoils. Prediction is in good agreement with measurement as regards the drag rise but less satisfactory in respect of trailing-edge pressure. BVGK is in agreement with measurement in showing that the 'buffet bump' does not cause a drag penalty at the 'cruise' lift coefficient, 0.6. BVGK is less robust where shock-induced separation is involved than for attached flows. For flows with shock-induced separation improved methods would seem to be needed to allow confident design of systems for controlling separation as evidently the 'buffet bump' does (Fig 12).

Fig 13 shows a comparison between measured and predicted drag polars for the datum aerofoil and the aerofoil with passive control with holes normal to the surface. BVGK(P) is in agreement with measurement in showing that this form of control results in a significant increase in drag over almost the whole of the range of lift coefficient studied, including values for high drag. This disappointing result stems from the effect of the normal transpiration on boundary-layer drag referred to above.

Fig 14 shows drag polars, indicating that streamwise slots yield lower drag than that of the hole system at high lift. The figure also implies that streamwise slots give a maximum lift that is significantly higher than that of both the other aerofoils, suggesting a possible role for this device in improving high-lift performance. The modelling of this type of flow, yet to be done at DERA, poses significant challenges in view of the three-dimensional nature of the transpiration flow. However, efforts in this direction are justified by the need to understand these flows and how best to exploit them. Work in this area is continuing at DERA.

4. Concluding remarks

The DERA studies have revealed the potential benefits of a number of flow-control devices, and have led to the development of CFD methods for designing them. The following conclusions may be drawn:

1) SBVGs offer significant benefits in controlling flows over highly-swept wings suitable for combat aircraft, providing a reduction in lift-dependent drag of up to 16%.

2) The use of bumps or ramps in the region of the shock, either as fixtures or as active devices, can provide significant reductions in drag, typically about 12%. These reductions are accurately predicted by the CFD code BVGK.

3) A ‘buffet bump’, placed downstream of the shock, can provide significant increases in lift coefficient for buffet onset, while not increasing drag at cruise conditions.

4) Streamwise slots increase the maximum lift coefficient by a significant amount at high subsonic speeds.

5. Acknowledgements

The support of the UK Ministry of Defence and Department of Trade and Industry is acknowledged, as are the efforts of Mr G L Riddle, Mr M J Simmons, Mr A. Shires, Ms C. J. Betts and Mrs P. Baldwin in support of this work.

6. References

1. Ashill, P. R. and Riddle, G. L.: Control of leading-edge separation on a cambered delta wing, AGARD-CP-548, pp 11-1 - 11-13, March 1994.

2 Ashill, P. R., Riddle. G. L. and Stanley. M. J.: Control of three-dimensional separation on highly-swept wings, ICAS-94-4.6.2, September 1994.

3 Ashill, P, R,, Riddle, G. L. and Stanley, M. J.: Separation control on highly-swept wings with fixed or variable camber, The Aeronautical Journal, October 1995, pp 317 - 327.

4 Broadley, Ian, Garry, Kevin: Effectiveness of vortex generator position and orientation on highly swept wings AIAA, Atlanta, 1997.

5 Haines, A. B.: Know your flow: the key to better prediction and successful innovation, AIAA 98-0221, Jan. 1998.

6 Küchemann, D.: *The aerodynamic design of aircraft*, Pergamon Press, 1978.

7 Betts, C. J., Ashill, P. R. and Hackett, K. C.: Fundamental investigations of the flow about devices suitable for use in flow control, Paper to be presented at FLOWCON, Göttingen, Sept. 1998.

8 Ashill, P. R., Fulker, J. L. and Shires, A.: A novel technique for controlling shock strength of laminar-flow aerofoil sections, Paper presented to First European Forum on Laminar Flow Technology, Hamburg, Germany, March 1992.

9 Fulker, J. L., Ashill, P. R. and Simmons, M. J.: Study of simulated active control of shock waves on aerofoil sections, DRA Technical Report TR93025, 1993.

10 Fulker, J. L. and Simmons, M. J.: An experimental study of shock control methods, DRA/AS/HWA/TR94007/1, 1994.

11 Ashill, P. R., Fulker, J. L. and Simmons, M. J.: Simulated active control of shock waves in experiments on aerofoil models, Proceedings ICEFM, Turin, July 1994.

12 Ashill, P. R., Fulker, J. L. , Simmons, M. J. and Gaudet, I. M.: A review of research at DRA on active and passive control of shock waves, ICAS-96-2.1.4, ICAS Sorrento, September 1996.

13 Shires, A.: A theoretical study of active shock control on a transonic combat wing, DRA/AS/HWA/WP94118/1, 1994.

14 Stanewsky, Egon, Délery, Jean, Fulker, John and Geissler, Wolfgang, *EUROSHOCK - drag reduction by passive shock control. Results of the project EUROSHOCK AER2-CT92-0049. Supported by the European Union, 1993-1995*, Notes on Numerical Fluid Mechanics, Volume 56, Vieweg, Braunshwieg/Wiesbaden 1997.

15 Calculation of excrescence drag magnification due to pressure gradients at high subsonic speeds, ESDU TDM 87004, 1987.

16 Fulker, J. L. and Simmons, M. J., DERA unpublished work, 1997.

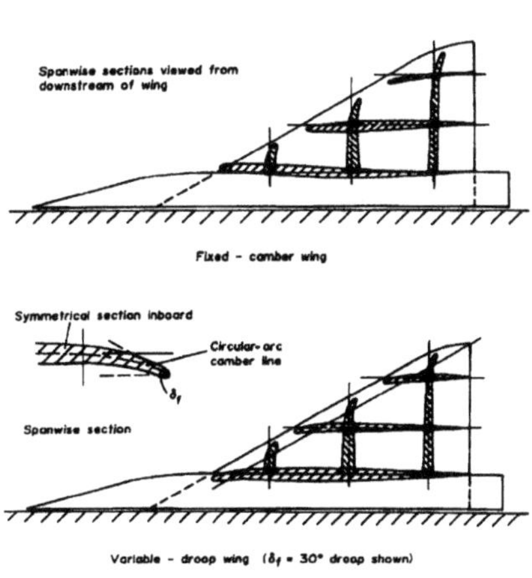

Fig 1 Models for leading-edge separation control studies

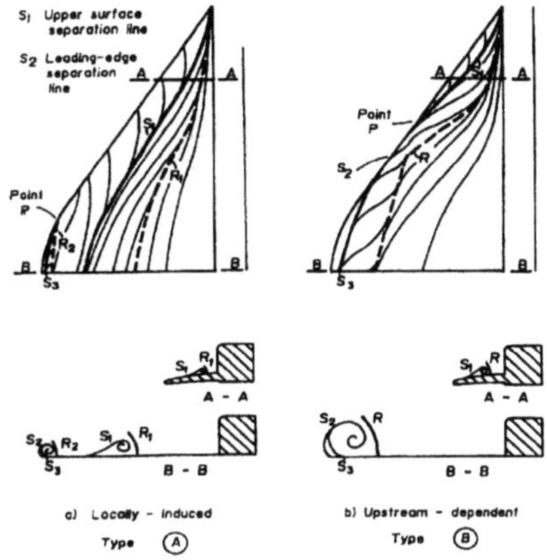

Fig 2 Types of flow studied

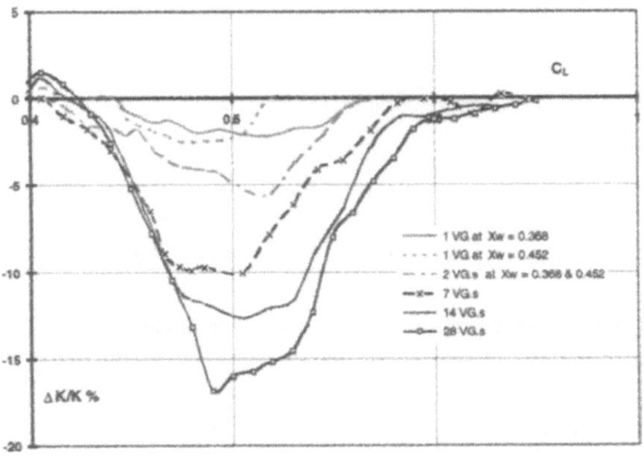

Fig 3 Effect of wire SBVG's on lift-dependent drag factor

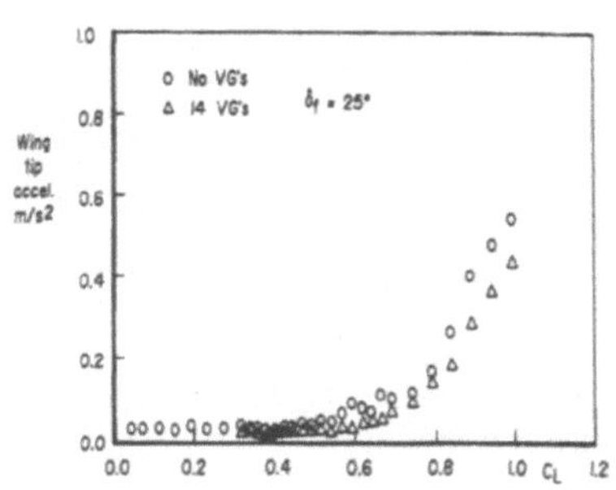

Fig 4 Effect of wire SBVG's on wing-tip acceleration

Fig 5 Sketch illustrating principle of drag-reducing bump

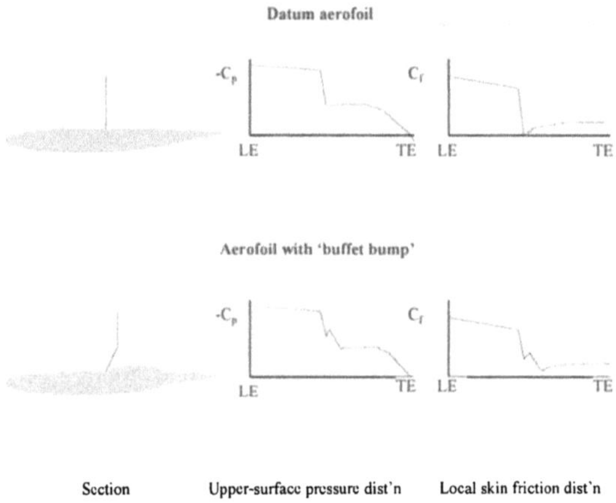

Fig 6 Sketches illustrating principle of 'buffet bump'.

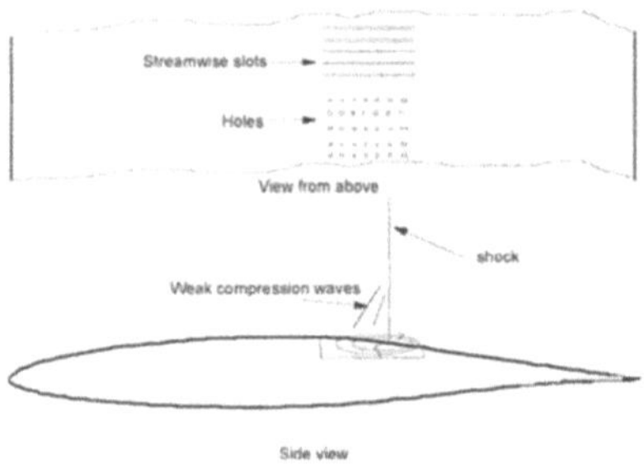

Fig 7 Sketch illustrating principles and geometries of passive ventilation control

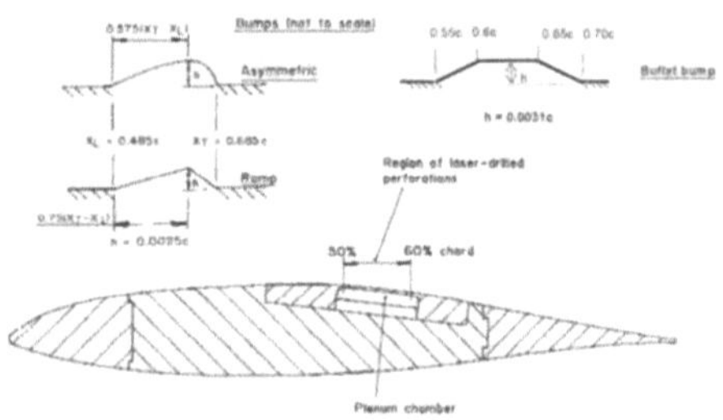

Fig 8 Models used for study of shock control

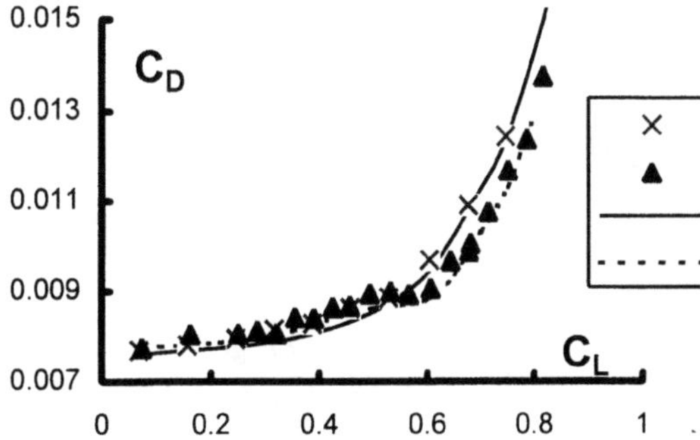

Fig 9 Comparison of predicted and measured drag polars for datum and ramp aerofoils

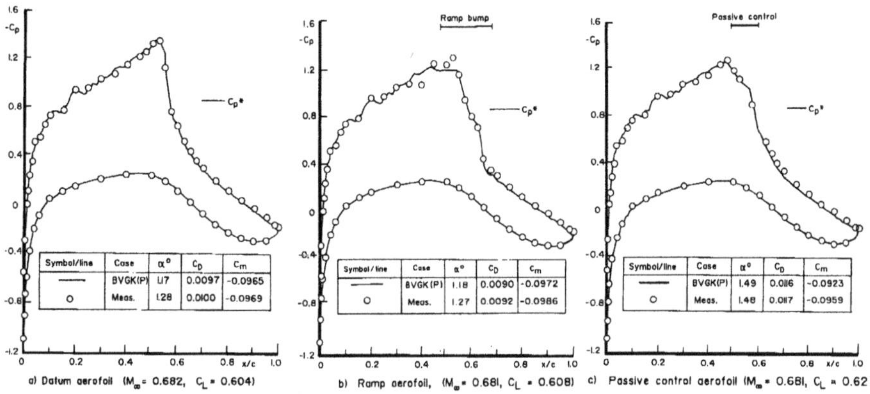

Fig 10 Comparison of predicted and measured pressure distributions, $C_L \approx 0.62$

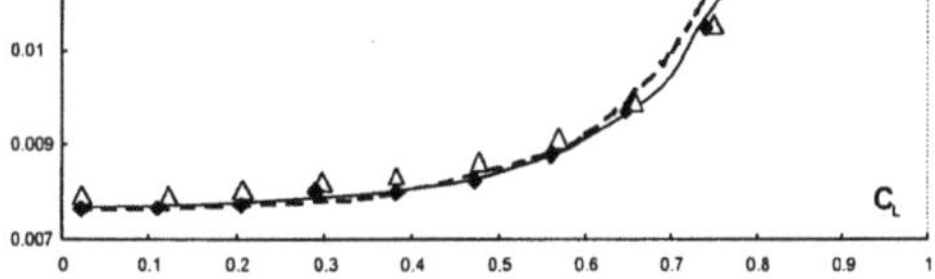

Fig 11 Comparison between measured and predicted drag polars, datum and 'buffet bump' aerofoils

Fig 12 Comparison between measured and predicted trailing-edge pressures, datum and 'buffet bump' aerofoils

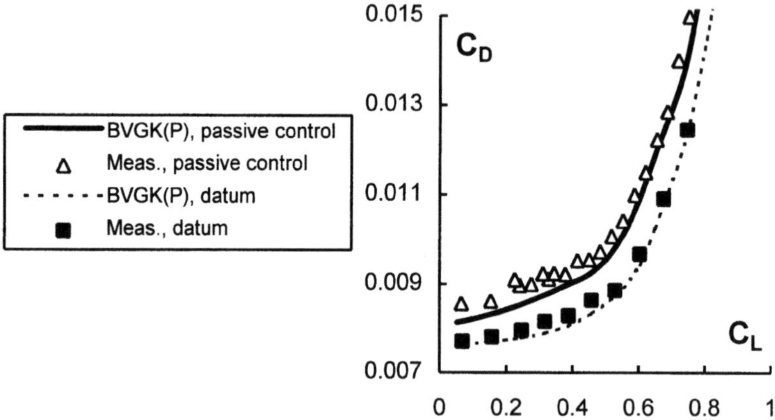

Fig 13 Comparison between predictions and measurement of drag polar illustrating effect of passive control using a surface with holes

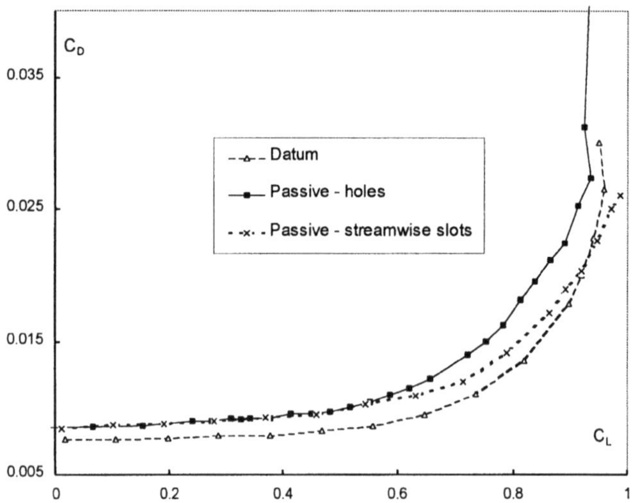

Fig 14 Comparison between drag polars for the datum, passive - holes and passive -slots aerofoils

Assessment of the Capability of Drag Reduction of the Shock Control Device 'SC Bump' on Airfoil Flows and Application Aspects on Wings

G. Dargel
Daimler–Benz Aerospace Airbus GmbH
28183 Bremen, Germany
A.M. Rodde
ONERA
92322 Chatillon, France
J.P. Archambaud
ONERA–CERT
31055 Toulouse, France

Abstract

Based on studies about various concepts to control the shock boundary layer interference the adaptive SC bump seems to be the most efficient shock control device to reduce drag in the off–design condition. The influence of the bump geometry on the drag behaviour is numerically investigated using a DA laminar airfoil and verified by airfoil experiments in the ONERA wind tunnel T2. In spite of encouraging experimental results in drag reduction with a wing–body configuration equipped with a fixed bump in 3D flow condition, the practical application on wings is not solved due to design and integration problems.

1. Introduction

Drag reduction technology is one of the main topics in aircraft industry and research for future commercial aircraft. Besides the laminar flow technology the control of the shock/ boundary layer interference plays a further key problem as the complex interaction of the shock wave with the boundary layer affects the off–design behaviour as well as the drag rise and buffet boundaries.

Various shock control (**SC**) concepts have been investigated – or still going on like the EC project EUROSHOCK II – to control the shock/boundary layer interaction by devices with passive and active flow ventilation, suction strip, local shape modification or combined concepts [1,2].

The capability of drag reduction by a local distensible surface, a '**bump**', on an airfoil shape under the shock was initially suggested by Ashill *et al* [3]. In a cooperative effort of Dasa Airbus (DA) and ONERA the 'SC bump' concept has been evaluated in a numerical investigation using the DA laminar airfoil LVA–1A followed by an experimental verification in the cryogenic ONERA wind tunnel T2, [4].

By transferring the SC experience from the airfoil flow on wings, an experiment with a wing–body configuration fitted with a SC bump device is finally described, giving some application aspects on wings.

G.E.A. Meier and P.R. Viswanath (eds.),
IUTAM Symposium on Mechanics of Passive and Active Flow Control, 57–62.

2. Influence of SC Bump Geometry

The SC bump is defined by its shape f_B with a length l_B, a height h_B and a crest c_B and located on the airfoil surface with a chord length c, Fig. 1.

The bump shapes f_B that are numerically investigated using a viscous–inviscid method [5] show a quite similar level of drag reduction, Fig. 2. For different shapes with fixed bump parameters the wave drag is reduced by about 50% referred to the basic airfoil, yielding a total drag reduction Δc_d by about 17% without viscous drag penalty.

The basic airfoil used is the laminar–type airfoil LVA–1A designed by DA to achieve extended natural laminar flow up to 50% chord at M_∞=0.73 and Re_∞=20*10^6. The use of the bump device requires an airfoil flow with a shock wave and a certain amount of wave drag occurring at the off–design condition with M_∞=0.76 and lift coefficient c_l=0.47. The Reynolds number of Re_∞=6*10^6 is chosen in the view of experimental verification.

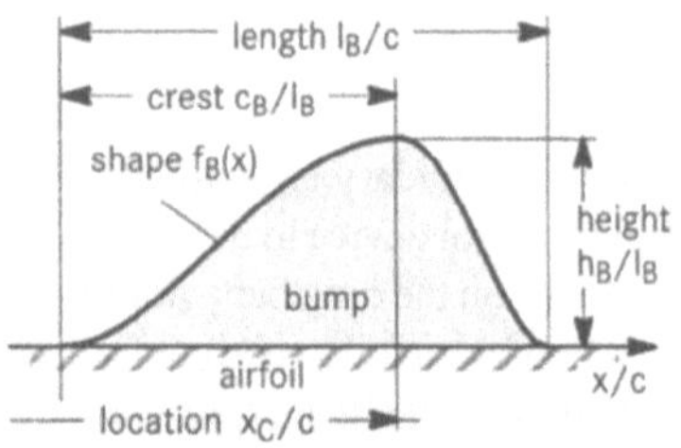

Figure 1: Definition of the SC bump geometry

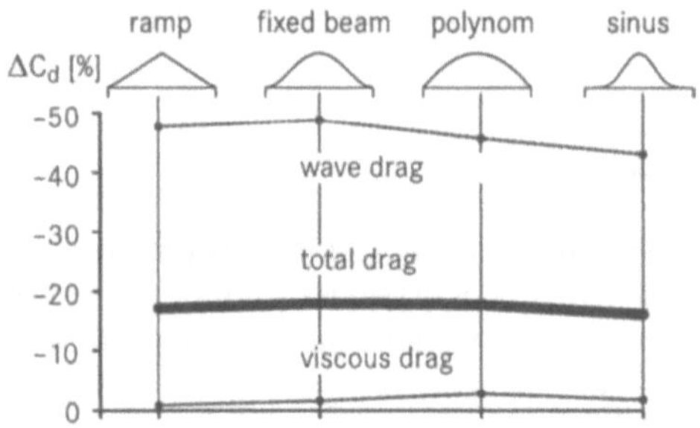

Figure 2: Drag reduction by SC bump shape f_B
l_B/c=0.2, h_B/c=0.003, c_B/l_B=0.5

The bump crest c_B strongly influences the potential of drag reduction and the optimal relation between the crest (x_C) and shock (x_{shock}) position, that is referred to critical flow condition of the basic airfoil, Fig 3. The symmetric bump (c_B/l_B=0.5) shows the largest amount in decreasing drag, while rearward asymmetric bumps (c_B/l_B>0.5) achieve nearly the same level, but generally reduce drag in wider domain, and therefore being less sensitive to off–design condition.

The increase of the bump height h_B up to certain limit is combined with the decrease of total drag, Fig 4, especially for large and rearward asymmetric bumps. If the bump location is fixed on the airfoil surface, a total drag reduction up to 25% is achievable by the adaptation of the bump height at the bump design point.

3. Experimental Verification of Bump Height Adaptation

As a result of the parametric study two bumps of heights h_B/c = 0.002 and 0.004 with a length of l_B/c=0.2 and a crest of c_B/l_B=0.7 have been considered for further shape optimisation. The DA model LVA–1A with a chord length of 150 mm has been equipped with exchangeable inserts carrying two bump shapes and tested in the cryogenic ONERA wind tunnel T2, Fig 5. [4,7].

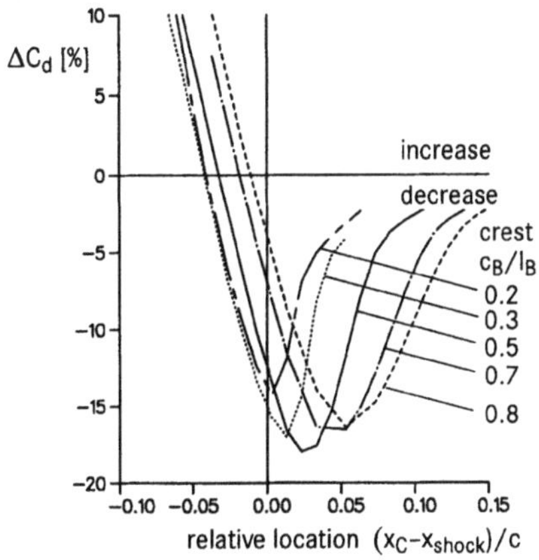

Figure 3: Drag reduction by SC bump crest c_B
f_B=beam, l_B/c=0.2, h_B/c=0.003

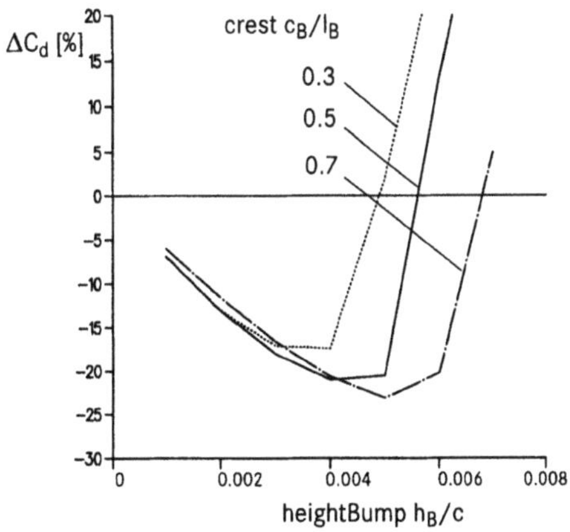

Figure 4: Drag reduction by SC bump height h_B
f_B=beam, l_B/c=0.2

This facility operates by run of about two minutes. Test section top and bottom flexible walls are streamlined for each test [6], removing the wall interference. Tests are performed at $6*10^6$ of chord Reynolds number (total pressure=2.15bar, total temperature=230°K). Boundary layer transition is fixed at 48% of chord on upper and lower surfaces of the airfoil with carborundum grains.

The lift coefficient is computed by integration of the pressure distribution measured on the airfoil. The total drag coefficient is computed from the data measured by a rake moved through the wake, Fig 5.

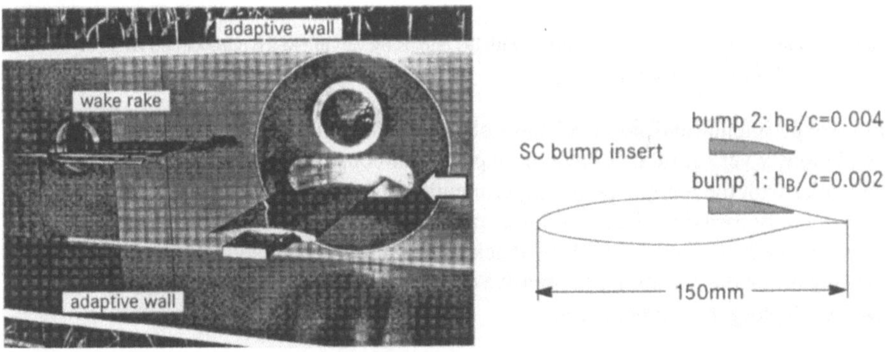

Figure 5: DA model LVA–1A equipped with a bump insert in the ONERA T2 wind tunnel

The effect of the bumps are illustrated at the design point in Fig. 6. On the model the Mach number, just upstream of the shock wave, is significantly reduced by the presence of the bump ascending ramp (left). Consequently the part of the wake profile (right) corresponding to the shock wave is also reduced while the viscous part is unchanged.

Fig. 7 shows drag gains and losses due to the bumps according to the test conditions for various Mach numbers and angles of attack. A fairly large domain of drag reduction can be observed for the two bump shapes. The bump 1 (h_B/c =0.002) has a larger efficiency envelope, indicated by the shift of the 'gains/losses' line to lower Mach number and angle of attack, while the bump 2 (h_B/c =0.004) produces a greater drag decrease. The maximum drag reduction is about 13% for the bump 1 and 18% for the bump 2.

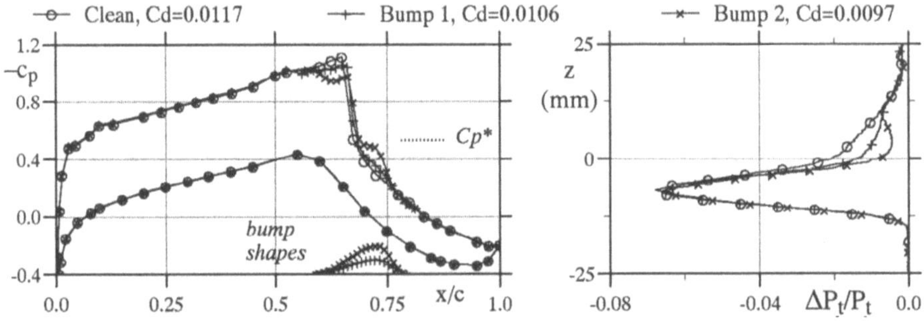

Figure 6: Effect of the SC bumps at the bump design point: Cp on the airfoil (left), wake profile(right) M_{cor}=0.76, $Re_\infty=6*10^6$, $\alpha=1^o$

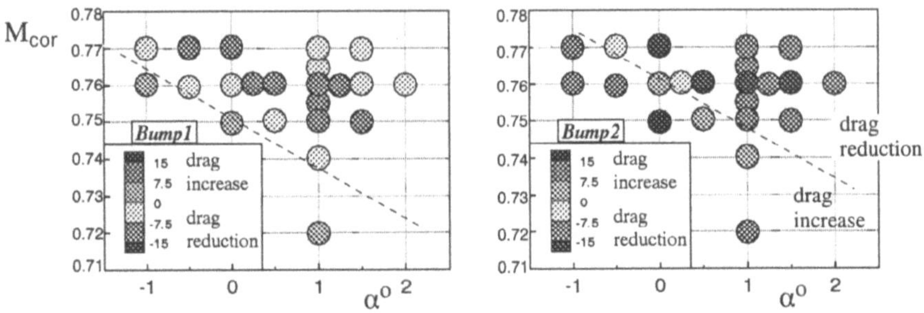

Figure 7: Drag balance of the LVA–1A airfoil with SC bumps by varying Mach number and angle of attack: $(c_{d\,bump} - c_{d\,clean})/c_{d\,clean}$

In Fig. 8 experimental and predicted drag polars relative to M_∞=0.76 are compared. Curves have obviously very similar behaviour, in particular considering the crossing of the clean airfoil polar with the bump 2 polar. Experimental gains are slightly greater than predicted ones. Generally speaking, the evolution of predicted and measured gains and losses versus freestream Mach number and angle of attack are in fairly good agreement. In addition these results underline the possibility to improve drag reduction in a large Mach number and lift range by adapting the bump height.

4. Wing Flow with Fixed SC Bump

Based on the SC experience from airfoil flow with drag reduction of about 20%, a bump device was designed for an existing DA wing–fuselage model in order to investigate the drag behaviour in 3D flow condition.

As bump design point, the flow condition is chosen at M_∞=0.85 and lift coefficient c_L=0.50, showing a shock located on the outer part of the wing for the wind tunnel Reynolds number of $Re_\infty=2.7*10^6$ observed from experiments in the ARA wind tunnel.

Using a viscous–inviscid method for infinite swept wings the bump geometry has been optimised in the outer turbulent–type wing sections [8]. Finally a bump with a length of

$l_B/c=0.2$, a height of $h_B/c = 0.0025$ and a crest of $c_B/l_B=0.7$ was defined between the wing section y/s =.45 and 0.97, following the spanwise location of the shock wave.

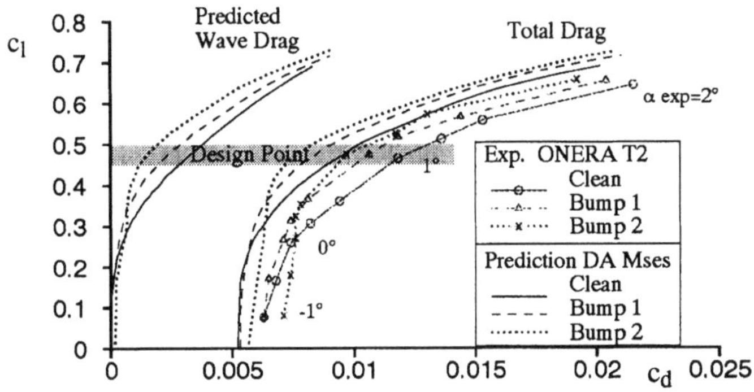

Figure 8: Comparison of experimental and predicted drag polars at the bump design Mach number M_∞=0.76

As the wind tunnel model existed, the SC bump was made out of plastic and glued on the wing surface, Fig 9. The drag balance obtained from force measurements confirms the bump design and shows encouraging gains in drag also in 3D flow condition by reducing the total drag at M_∞=0.84 and at design Mach number of about Δc_D=1.5% in a small lift range, Fig 10. Outside of this lift and Mach number range the fixed bump device leads to a drag increase, caused by the movement of the shock wave hurting the optimal relation between the shock and bump location, see Fig. 3.

Figure 9: Wing–body configuration with SC bump device in ARA wind tunnel

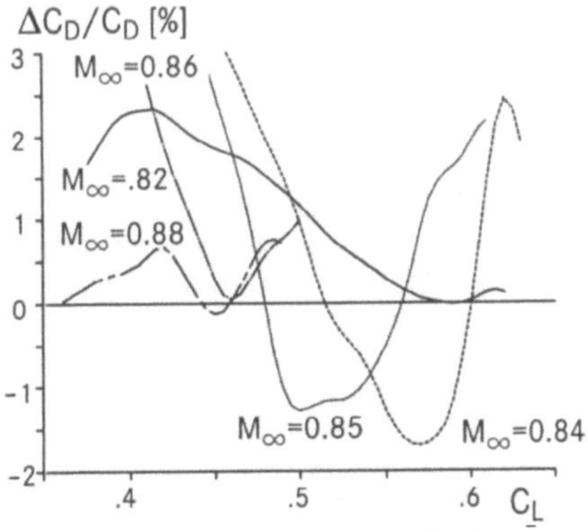

Figure 10: Measured drag balance of the wing–body configuration

5. Applications Aspects

The potential in drag reduction is tremendously influenced by the shock movement that is of course stronger for a transonic wing with turbulent–type airfoils as used in the 3D experiment than for a laminar wing as used in the airfoil experiment. This requires a wing design with less shock movement in chordwise and spanwise direction or/and the introduction of variable camber by trailing edge flaps in order to stabilize the shock location.

An adaptation of the bump height is nevertheless required in order to avoid drag rise when no or weak shocks are present. The integration of an adaptive SC bump in the wing structure is complicated as the rear part of the wing box and the leading edge of the spoiler are located in area of the bump which position is determined by the shock of the clean wing. As a bump length of 20% chord is desirable, but spoilers normally have a length of about 15% chord, only a partial integration of the bump into the spoiler is possible and also a rearward shock location is needed.

6. Conclusion

The drag reduction by the SC bump device depends on the bump geometry and the relative location between the shock and the bump. The numerical study indicates that large rearward asymmetric bumps with height adaptation give the largest profit in drag decrease not only at the bump design point but also in off–design condition.

Airfoil experiments in the ONERA wind tunnel T2 with the laminar DA airfoil model LVA–1A equipped with two bumps of different heights confirmed the predicted potential of drag reduction, showing a drag decrease of about 13% for the lower and 18% for the higher bump. Wind tunnel experiments with a wing–body configuration equipped with a fixed bump on the outer wing showed a drag reduction of about 1.5% in 3D flow condition.

The practical application of the SC bump device on wings is not solved for reducing drag in the off–design region of the flight envelope of commercial aircraft. Further investigations are necessary in order to assess the SC bump device integration into the wing structure compared to the profit in fuel consumption caused by the drag reduction due to the SC bump.

References

[1] E. Stanewsky, J. Fulker, J. Delery, J. Geißler: 'Drag Reduction by passive Shock Control – Results of the Project EUROSHOCK, AER2–CT92–0049',Notes on Num. Fluid Mech., Vol 56, Vieweg Verlag, 1997

[2] E. Stanewsky: 'EUROSHOCK II –Drag Reduction by Shock and Boundary Layer Control', IMT, Area–3 AERONAUTICS, Project BE–1316, Technical Annex, 1996

[3] P.R. Ashill, J. Fulker, A. Shires: 'A novel technique for controlling shock strength of laminar–flow aerofoil sections', Proceed. of the First European Forum on Laminar Flow Technology, DGLR–Bericht 92–06, 1992

[4] G. Dargel, A.M. Rodde, J.P. Archambaud: 'Evaluation of the Shock Control Device 'SC Bump' using the Laminar Airfoil DA LVA–1Ae –Design, Experiments and Analysis–', DA–Rep. No. EF03/98, 1998.

[5] M. Drela, M.B. Giles: 'ISES: A two–dimensional viscous aerodynamic design and analysis code', AIAA–Paper 87–0424, 1987

[6] J.P.Archambaud,A. Mignosi: 'Two–dimensional and three–dimensional adaptation at the T2 transonic wind tunnel', AIAA 4th Flight test Conference & 15th Aerodynamic Testing Conference, San Diego Cal., 1988.

[7] J.P. Archambaud, J.B. Dor, J.F. Breil: 'Qualification of the passive control with bumps on LVA–airfoil in the ONERA CERT T2 wind tunnel', ONERA–Rep. No. 21/5017.26 (OA No. 110/1685 AN 241 D), 1997.

[8] L. Gildemeister, G. Dargel,, H. Hansen::'Theoretische Grundsatzuntersuchungen zum Thema 'Adaptive Hautstrukturen im Verdichtungsstoß' in 'Beeinflussung von Tragflügelströmungen', W. Nitsche, R. Henke (Editor), ILR Mitt. 311, TU Berlin, 1996.

PASSIVE CONTROL OF BUFFET EXCITATION

G. A. FLYNN AND J. F. MORRISON

Department of Aeronautics, Imperial College
London SW7 2BY

1. Introduction

It has recently been shown by Heenan and Morrison (1998a,b) that buffet due to separated flow downstream of a backward-facing step can be attenuated more or less completely by use of a permeable reattachment surface vented to an airtight plenum. With an impermeable surface, the global unsteadiness is driven by the feedback from the reattachment point of pressure and vorticity fluctuations towards the step. Their work also suggests that a permeable surface immediately downstream of separation also affects the pressure field. In this work therefore, the intention is to use a permeable surface to reduce buffet excitation of a *fully* separated flow, thereby reducing buffeting, the structural response. Buffeting is unlikely to be a serious problem with primary structures on aircraft. However, it produces deleterious effects on flaps with potentially serious problems caused by fatigue. The present work is not aimed at the extension of the buffet boundary, but rather seeks to make buffet onset less damaging to secondary structures. The initial objective of this project is to establish proof of concept, showing that a permeable surface can reduce the buffeting of a wing excited in the first bending mode. Initial positive results have been obtained and further, initial, results from a "segmented" model described below are also presented.

For a semi-rigid, cantilevered wing, Huston (1957) and Davis and Wornom (1958) define the buffet excitation parameter:

$$\sqrt{nG(n)} = \frac{2}{\sqrt{\pi}}\frac{m\ddot{z}}{qS}\varsigma^{\frac{1}{2}}, \tag{1}$$

where n is the resonant frequency reduced by the chord, c, and the free-stream velocity, U_∞. m is the modal mass, ζ the damping ratio, $\ddot{z}$ the rms tip acceleration, q the free-stream dynamic head and S the wing plan area. Thus Equation (1) allows calculation of the (non-dimensional) resultant generalised force where all the terms on the right-hand side are known. This formulation is particularly useful in that it permits direct comparison with a dimensionless surface-pressure-fluctuation spectrum in the form proposed by Owen (1958):

G.E.A. Meier and P.R. Viswanath (eds.),
IUTAM Symposium on Mechanics of Passive and Active Flow Control, 63–68.

$$\frac{\overline{p^2}}{q} = \int_0^\infty \phi(n)dn = \int_0^\infty n\phi(n)d(\log n), \qquad (2)$$

where n here is the reduced frequency and $\sqrt{n\phi(n)}$ is the (dimensionless) rms intensity of the aerodynamic excitation – a measure of the energy of the excitation at the measurement point. Therefore $\sqrt{nG(n)}$ may be interpreted as the 'integrated' response of the wing.

2. Apparatus

Two models have been used for buffeting measurements: both are wings of rectangular planform, of aspect ratio 3 and GA(W)-1 section. This profile was selected for its benign stall over a wide range of Reynolds numbers. For the smaller, preliminary model, $Re_c = 5.5 \times 10^5$, and for the larger one, $Re_c = 9.5 \times 10^5$. In both cases, transition is fixed near the leading edge using sand-grain roughness, optimised by trial and error. The wings are cantilevered at the root using a seismic mass in order to ensure that the motion of the wing is uncoupled from that of the support. Ideally, the natural frequency of the seismic mass is $\mathcal{O}(1)$ while that of the wing is $\mathcal{O}(10)$. Having been calibrated in a 'wind-off' test using an accelerometer, root strain gauges were used to measure the rms tip acceleration. The damping ratio is estimated from the application of the random decrement method (Cole 1973) to the raw data of wing-tip displacement: these measurements and those of $\ddot{z}$ are made simultaneously. It is therefore assumed, sensibly, that the aerodynamic damping is non-negligible and varies with the severity of the buffeting. Zan and Maull (1992) discuss the significance of aerodynamic damping. On scaling up to full-size conditions, it is likely that the aerodynamic damping will increase.

The modal mass, m, is calculated assuming a first-bending mode shape given by:

$$\frac{z}{s} = \eta^2 - \frac{1}{3}\eta^3, \qquad (3)$$

where η is the fraction of the span, s. For the smaller model, the actual mode shape is ill-defined because the wing is made from mahogany making it much stiffer than the steel root spar. Thus most of the bending actually occurs in the spar and this is the main reason for the development of the larger, segmented model. The actual mode shape may be closer to that for a rigid body mode and estimates of m may therefore be inaccurate. The experimental techniques are fully reported by Flynn *et al.* (1999).

For the smaller model, the permeable surfaces consist of two strips of perforated plate with an open-area ratio (OAR) of 33% and circular holes of 3mm diameter. The strips extend nearly the whole span of the wing, are $0.06c$ in the chordwise direction and with leading edges at $0.37c$ and $0.68c$. These perforated strips are unlikely to be the optimum, in terms of position, open area ratio or plenum depth. Two pressure-fluctuation transducers were also mounted on the surface at $0.31c$ and $0.55c$, so that the former is just upstream of the first strip and the latter is about midway between the two strips. These are driven by purpose-built chips which also amplify the output signal: for a fuller description, see Heenan and Morrison (1998b). The natural frequency of the model (first bending mode) is about 20Hz while that of the seismic mass is about 1Hz.

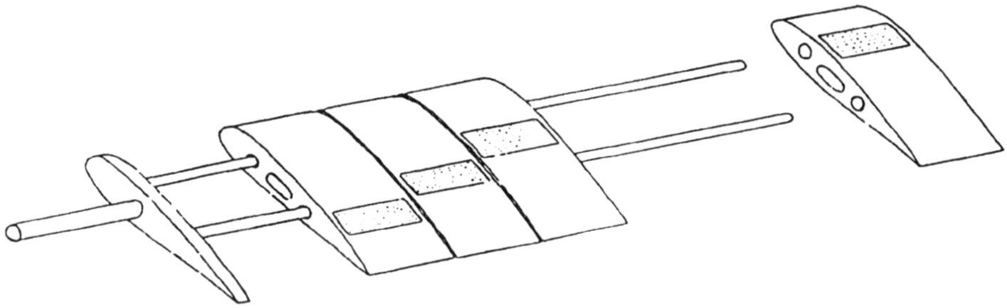

Figure 1: Schematic of segmented model

In contrast with the smaller, integral model, the larger one is segmented, comprising seven moulded segments with a contoured tip. Figure 1 outlines the basic construction. In addition to providing a mode shape that is more accurately represented by Equation (3), the segmented model also allows the chordwise position of the perforated strips to be varied along the span. As in the case of the integral model, each perforated strip vents to an air-tight plenum. For this model, the OAR of the perforated sheet is 25%, and is the same as that identified by Heenan and Morrison (1998b) as the value above which the effects of permeability (in a reattaching flow) tend to 'saturate'. Unfortunately, initial tests indicate that the initial design is not sufficiently stiff, the wing having a natural frequency of about 8Hz only, which is too close to that of the seismic mass. However initial results are illuminating and are presented here: a fuller description of the model, together with further results, obtained using a stiffer model, are presented by Flynn *et al.* (1999).

3. Results

Figure 2 shows the buffet excitation parameter, $\sqrt{nG(n)}$, as a function of incidence, obtained with the smaller model. Results are shown for both configurations, with and without permeable strips. They show an attenuation of $\sqrt{nG(n)}$ in the heavy buffet range, $18.5° < \alpha < 22°$. At even higher incidences, the use of the permeable strips increases it slightly. An earlier test using the same model (but performed by a different worker) gave very similar results and both tests established that attenuation of $\sqrt{nG(n)}$ is possible.

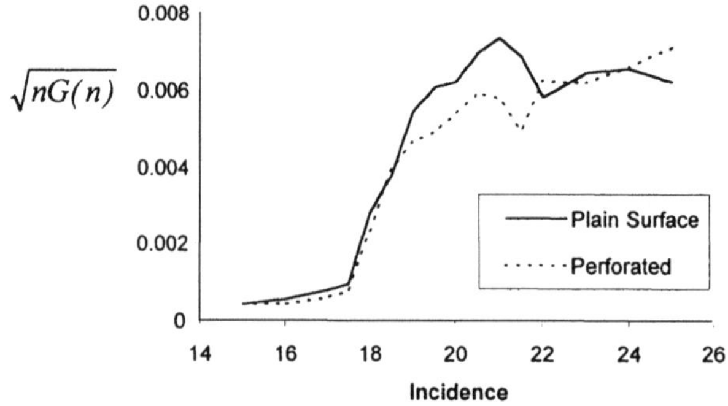

Figure 2: Small model, $\sqrt{nG(n)}$ vs. Incidence, α.

Figure 3 shows the corresponding frequency spectrum obtained from the downstream transducer at $0.55c$ with the wing at an incidence of 20°, in the heavy buffet region. It shows that the pressure fluctuation intensity is reduced across a wide range of frequencies. At this incidence, the upstream transducer provides similar results and both are in the fully stalled region of the wing. However, for either $\alpha < 18°$, or for $\alpha > 22°$, the buffet excitation appears to increase in the region of the transducers. It is not yet clear how the perforated strips reduce the pressure fluctuation intensity - in attached flow they *increase* $\sqrt{n\phi(n)}$ in their immediate vicinity (Heenan and Morrison 1998b).

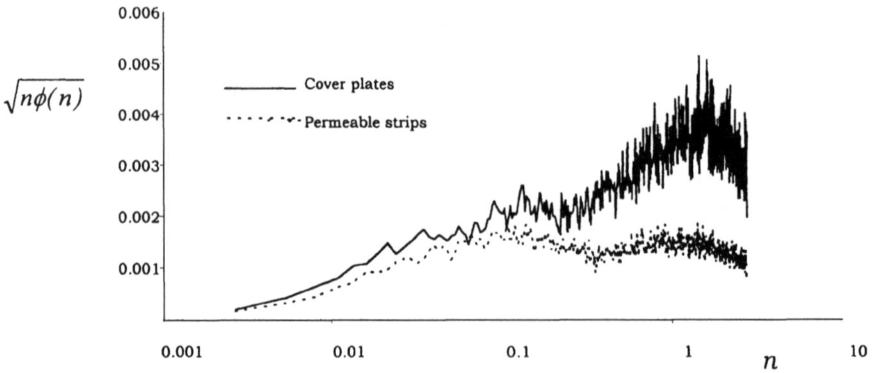

Figure 3: Small model, $\sqrt{n\phi(n)}$ at $x/c = 0.55$.

Flynn *et al.* (1999) show surface oil-flow visualizations obtained at the higher Reynolds number using an integral model with dimensions the same as those of the segmented model shown in Figure 1. Large variations in the chordwise position of the mean separation line are apparent leading to the formation of mushroom-shaped "stall cells". These have received little attention in the published literature, Gregory *et al.* (1971) and Winkelmann and Barlow (1980) being the exceptions. These visualizations have enabled some qualitative interpretation of the mechanisms that give rise to the buffet excitation. Figure 4 compares measurements of $\sqrt{nG(n)}$ for the unmodified wing with those obtained using permeable strips, their distribution being guided by the surface flow visualizations. For $\alpha > 21°$, configuration 'C' of the permeable strips is clearly effective in reducing the buffet excitation. However, if the strips towards the root are placed nearer the leading edge, as in configuration 'E', $\sqrt{nG(n)}$ increases above the datum value for $\alpha < 22°$. Therefore, it would appear that, below this incidence, the permeable strips cause separation to occur earlier: it appears that an effect of the strip is to remove the large favourable pressure gradient present in the datum case.

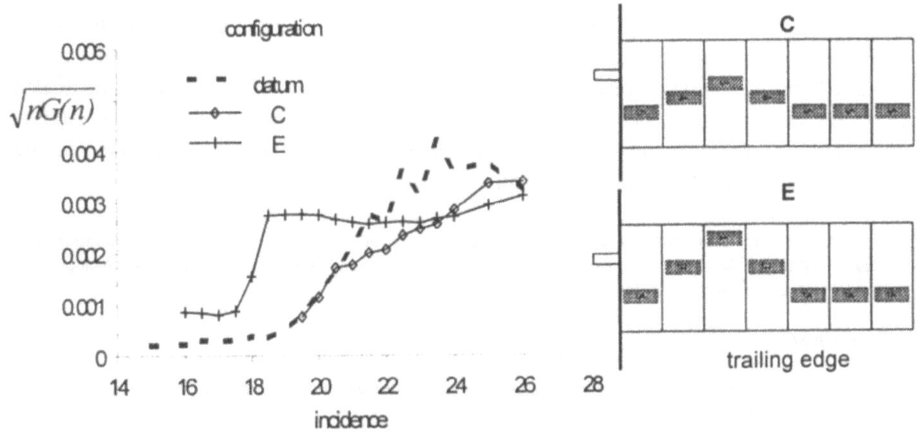

Figure 4: Segmented model, $\sqrt{nG(n)}$ vs. Incidence, α.

4. Discussion

It is clear that the key mechanisms which lead to large buffet excitation are related to the nature of the separation, and in particular, to the behaviour of the stall cells since they determine the plan area of separated flow immediately adjacent to the wing. The change from a number of stall cells (three for this aspect ratio when a tip endplate is used) to a single large one at $\alpha = 18°$ coincides with the large increase in $\sqrt{nG(n)}$. This effect has also been noted by Winkelmann and Barlow (1980). Sharples (1998) has confirmed that the wavelength of the cells does not change when the wing aspect ratio is changed, suggesting that the wavelength of the stall cell is determined by the wing chord. There is some visual evidence that the attenuation of $\sqrt{nG(n)}$ is largest when the foci (of surface-normal vorticity) lie over the permeable strip. This requires further investigation – Zan and Maull (1992) suggest that the foci are responsible for frequencies that are typically an order of magnitude smaller than those associated with buffeting. It should be noted that the change from one stable pattern of stall cells to another is usually accompanied by a 'flicking' between the two. Moreover, just as the rapid rise in buffeting occurs, around $\alpha = 20°$, the static pressure distributions, and therefore the behaviour of the stall cells, exhibit hysteresis with changes in either α or free-stream velocity. Similar behaviour is evident in the work of Meyer *et al.* (1997).

5. Conclusions

Initial results confirm that buffet excitation of wings (and their components) may be attenuated by the correct positioning of permeable strips above an airtight plenum on the suction surface. The principle mechanisms generating pressure fluctuations at the wing surface and therefore the way in which a permeable surface produces buffet alleviation are not yet fully understood. Repeatability of buffeting measurements is not easy to achieve when the wing amplitude is large and motion in the plane of the wing is also large. A stiffer wing is therefore likely to improve the accuracy of the data as well as increasing its natural frequency. Simulations (albeit at low Reynolds number) to provide more details of the flow field are currently under way.

Acknowledgements: This work is supported under contract ASF/2601U with Defence Evaluation and Research Agency, Bedford. We are indebted to both its monitor, J. L. Fulker, and D. G. Mabey who originally suggested the work to us.

6. References

Cole, H. A. 1973 On-line failure detection and damping measurement of aerospace structures by random decrement signatures. NASA CR-2205.

Davis, D. D., Jr. and Wornom, D. E. 1958 Buffet tests of an attack-airplane model with emphasis on analysis of data from wind-tunnel tests. NACA Research Memo. L57H13.

Flynn, G. A., Morrison, J. F. and Mabey, D. G. 1999 Buffet alleviation on an unswept wing at high incidence. AIAA paper 99-0791.

Gregory, N., Quincey, V. G., O'Reilly, C. L. and Hall, D. J. 1971 Progress report on observations of three-dimensional flow patterns obtained during stall development on aerofoils, and on the problem of measuring two-dimensional characteristics. ARC CP No. 1146.

Heenan, A. F. and Morrison, J. F. 1998a Passive control of backstep flow. *Experimental Thermal and Fluid Science,* **16**(1/2). 122-132.

Heenan, A. F. and Morrison, J. F. 1998b Passive control of pressure fluctuations generated by separated flow. *AIAA J.* **36**(6), 1014-1022.

Huston, W. B. 1957 A Study of the correlation between flight and wind-tunnel buffet loads. AGARD Report 111.

Meyer, R., Bechert, D. W., Hage, M. and Montag, P. 1997 BMBF-Vorhaben 13N6537-8 Aeroflexible Oberflächenklappen als "Rückstrombremsen" nach dem Vorbild der Deckfedern des Vogelflügels. Abschlußbericht DLR IB 92517- 97/B5.

Owen, T. B. 1958 Techniques of pressure-fluctuation measurements employed in the RAE low-speed wind-tunnels. AGARD Report 172.

Sharples, N. J. 1998 The flowfield of a rectangular planform wing beyond stall. Final-year undergraduate project. Dept. Aeronautics, Imperial College.

Winkelmann, A. E. and Barlow, J. B. 1980 Flowfield model for a rectangular planform wing beyond stall. *AIAA J.* **18**, 1006 -1008.

Zan, S. J. and Maull, D. J. 1992 Buffet excitation of wings at low speeds. *J. Aircraft*, **29**(6), 1137-1143.

ENHANCED THRUST-EFFICIENCY OF OVEREXPANDED NOZZLES BY PASSIVE VENTING

JÜRGEN GRÄSEL AND ALFRED E. BEYLICH
Stoßwellenlabor, RWTH Aachen, D-52056 Aachen, Germany

1. Introduction

The design of thrust nozzles for airbreathing, hypersonic vehicles requires an efficient propulsion airframe integration. Single expansion ramp nozzles (SERN) have a unique installation advantage because the vehicles aft end can be used as an external expansion ramp to achieve high area ratios [1]. Low thrust efficiencies have been measured at transonic flight regime due to the overexpansion of the nozzle. One approach to improve the off-design performance penalities is the passive cavity concept. The basic idea behind this concept originates from the design of supercritical airfoils [2]. The drag reduction at transonic speeds is achieved by replacing a part of the airfoils upper surface by a porous plate with a vent chamber beneath it. Applied to the overexpanded nozzle the passive ventilation along the ramp enables the self-adjusting capabilty of the jet to low pressure ratios by encouraging flow separation. The objective of the present study is to investigate the potential benefit in thrust efficiency by this method. A parametric study varying the percentage of porosity and cavity geometry is conducted to find an optimal configuration.

2. Computational Model

The computations in this study are based on the two-dimensional, Favre-averaged Navier-Stokes equations. An implicit time marching scheme with a lower-upper factorization of the implicit operator and a second order total variation diminishing (TVD) discretization of the convective fluxes is implemented. A two-equation model has been employed for turbulence closure. The complete description of the numerical method and the boundary conditions is given in Ref. [3].

G.E.A. Meier and P.R. Viswanath (eds.),
IUTAM Symposium on Mechanics of Passive and Active Flow Control, 69–74.

2.1. SIMULATION OF THE POROUS SURFACE

The transpiration velocity v_w normal to the porous surface is evaluated from the linear form of the Darcy pressure law:

$$v_w = C\,(p_c - p_w)\,. \tag{1}$$

According to Idelchik [4] the constant C depends on the specific geometry of the porous plate and must be calibrated for each individual plate. The plenum chamber pressure p_c is determined from the mass averaged wall pressure p_w along the porous surface. The production of turbulent kinetic energy in the viscous sublayer of the boundary layer is strongly affected by the mass transfer over the porous surface. A suitable turbulence model to capture the distortion of the velocity profiles is the k-ω model of Wilcox [5]. The model requires no additional damping terms in the sublayer. The scaling of the law of the wall constants with v_w is incoporated through the definition of the specific dissipation rate ω at the wall :

$$\omega = \frac{u_\tau^2}{\nu}\,\frac{10}{v_w^+(1+5\,v_w^+)} \qquad \text{and} \qquad v_w^+ = \frac{v_w}{u_\tau}\,. \tag{2}$$

The level of agreement with the analytical solution of Rotta [6] displayed in Fig.1 confirms the model ability to desribe transpired boundary layers. A more severe test case is the supersonic flow past a closed cavity [7]. The flowfield characteristics like the expansion fan at the cavity's leading edge and the compression waves at the shear layer impingment and separation are depicted in Fig. 3. It is believed that the discrepancies in the computed c_p-distribution in Fig. 2 are due to three-dimensional effects such as the lateral pressure gradient inside the cavity. A switching of the flow structure from closed to open cavity type can be seen when the solid floor is replaced by a porous plate. The surface pressure for this case is correctly predicted.

2.2. BASELINE CONFIGURATION

A 1/100-scale nozzle (area ratio: 5.2, nozzle length: 10.5 m, throat height: 0.83 m) based on the ELAC-propulsion system concept [8] is used as the reference model for the present study. The stagnation temperature and pressure of the gas are chosen as T_0=300 K and P_0=0.3 MPa for the nozzle and P_0=0.1 MPa for the freestream respectively. The nozzle is considered for a freestream Mach number of M_∞=1.2 and a nozzle pressure ratio (NPR) of 7. All forces acting on the upper external surface and the lower flap cowling of the nozzle as well as cavity internal drag and skin friction force are not contained in the predictions.

3. Results and Discussion

The baseline configuration is investigated at first to characterize the dominant flow features of the asymmetric nozzle. For comparison changes in the nozzle flowfield due to the passive venting are shown. In the next step the sensitivity of the nozzle performance with respect to the cavity parameters is discussed.

Due to the low NPR, the nozzle operates at off-design conditions where the exhaust flow is insufficient to fill the large nozzle exit area. As can be seen from the calculated Mach number distribution in Fig.4 the external flow expands around the aerodynamic shape which is formed by the curved portion of the cowl, the dead water region, and the shear layer and is finally deflected by an external plume shock. The resulting lower pressure level of the external flow encourages the overexpansion of the nozzle by delaying the boundary layer separation on the expansion ramp. A rotation of the thrust vector to negative angles and an increased base drag is the consequence.

The Mach number contours in Fig.4 illustrate the mechanism, whereby the nozzle flowfield is affected by the passive ventilation. In the case shown, the leading edge of the porous plate is fixed at x=3.5 cm and the plate extends up to x=11 cm. The permebiality constant $C=5.5\times10^{-3}\mathrm{m}^2\mathrm{s/kg}$ is calculated with a porosity of 3.8%, a plate thickness of 5 mm and a hole diameter of 0.3 mm. The relative length of the suction and bleeding portion of the porous plate depends on the porosity and the cavity position. In all cases shown the suction is concentrated over a much smaller portion of the plate than is the bleed.

The most prominent deviations in the iso-Mach field are the jet contraction and its deflection in downward direction. The deflection is caused by the blowing along the rear part of the expansion ramp. The boundary layer considerably thickens and separates at x=7 cm without reattachment. Up to the separation point only minor differences exist between the two flow fields, which can be attributed to their supersonic nature. The oblique shock wave at the ramp trailing edge is replaced by a series of weak compression waves to match the jet static pressure to the external flow. The boundary layer of the ramp upper surface merges with the shear layer and the separated region. The subsonic portion of the boundary layer is significantly increased by the bleeding, through which the external static pressure can creep upstream and influences the onset for flow detachment. The computed transpiration velocities are of the order of 10% of the outer flow velocity. The variation of the ramp surface pressure with plate porosity σ is plotted in Fig.5. The maximum porosity in the computations is limited to 10% in order to exclude possible cavity pressure fluctuations. The predicted surface pressure distributions of the porous wall compared with the solid wall show a weakening of the recompression shock and a smoothening of the pressure gradients. The leading edge of the porous plate is indicated by the strong pressure drop of the expansion fan at the cavity inflow. The separation point is shifted upstream with higher porosity. A comparison of the normal thrust coefficient Δc_{f_y} for the solid and the porous wall in Fig.5 with varying porosity clearly demonstrates the positive impact of the passive ventilation on the thrust vector. The negative normal force was reduced by one third due to the passive venting. Apparently the normal thrust coefficient is most sensitive to porosity variations between 3.5% and 4.5% porosity, whereas this effect is less pronounced for the axial coefficient. For lower porosities the passive venting is too weak to separate the boundary layer considerably. For higher porosities the increase in wall pressure at the rear part of the ramp is neutralized by the stronger pressure drop at the cavity inflow. For a given porosity of 4% the position of the leading- and trailing edge of the porous plate have been changed independently to find the optimum location. It is found that an upstream shifting of the cavity leading edge strenghtens the passive effect. This trend is due to the resulting higher pressure gradient over the porous surface. In general the leading edge should always be located in the high pressure, i.e. the forward part of the ramp. Higher back pressure levels can be imposed along the ramp surface by extending the cavity to the ramp trailing edge.

4. Concluding Remarks

A computational investigation has been performed to study the effects of passive control on the flowfield of an overexpanded SERN nozzle. It is shown that besides the typical features of passive control like relaxation of the flow gradients and suppression of the shock strength, this method additionally enables a self-correcting adjustemt of the jet to low pressure ratios. The effectivness of the passive venting depends on the porosity and the location of the cavity edges relative to the wall pressure gradient. The negative normal component of the thrust vector can be reduced by more than 30 % in comparison to the baseline configuration.

5. Acknowledgement

This work is related to Project C6 of Sonderforschungsbereich 253, which was financially supported by the Deutsche Forschungsgemeinschaft.

References

1. A.E. Beylich. Performance of a Hypersonic Twin-Nozzle System. *AIAA Journal, Vol. 34, No. 5*, pages 953–960, 1996.

2. P. Thiede, P. Krogmann, and E. Stanewsky. Active and Passive Shock/Boundary Layer Interaction Control on Supercritical Airfoils. *AGARD CP-365*, pages 24,1–13, 1984.
3. J. Gräsel and A.E. Beylich. Passive Control for The Off-Design Performance of a Single Expansion-Ramp Nozzle. *AIAA-Paper 98-1602*, pages 1–10, 1998.
4. I.E. Idelchik. *Handbook of Hydraulic resistance.* Springer Verlag, 1986.
5. D.C. Wilcox. *Turbulence Modeling for CFD.* DCW Industries, 1994.
6. J.C. Rotta. Control of turbulent boundary layers by uniform injection or suction of fluid. *DGLR Jahrbuch 1970*, pages 91–104, 1970.
7. L.R. Stallings and F.J Wilcox. Experimental Cavity Pressure Distributions at Supersonic Speeds. Technical paper 2683, NASA Langley Research Center, 1987.
8. D.A. Steinebach, W. Kühl, and H.E. Gallus. Design aspects of the propulsion system for aerospace planes. *AIAA Paper 93-5127*, 1993.

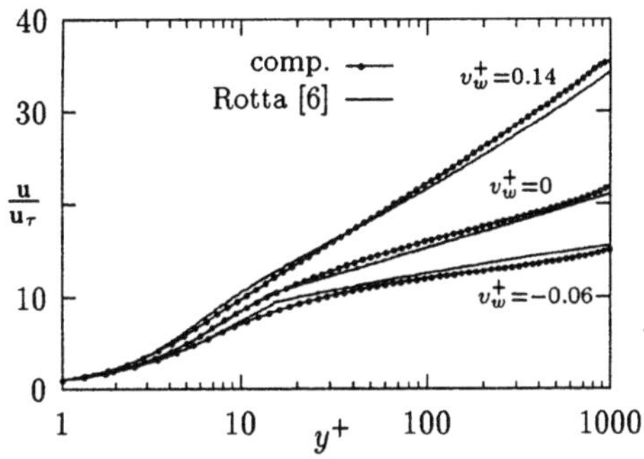

Figure 1. Velocity profiles for variable v_w^+

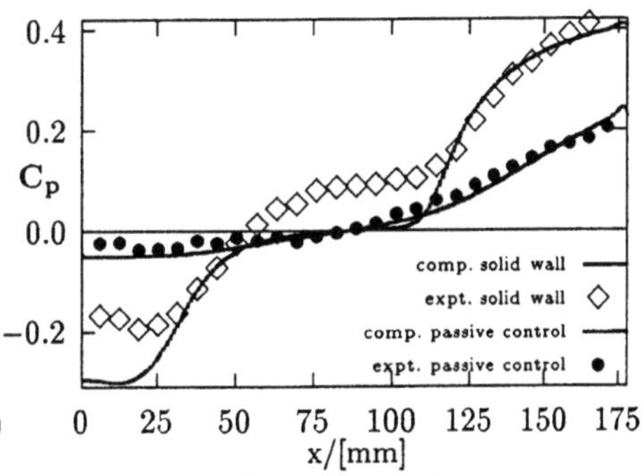

Figure 2. C_p-distribution of closed cavity

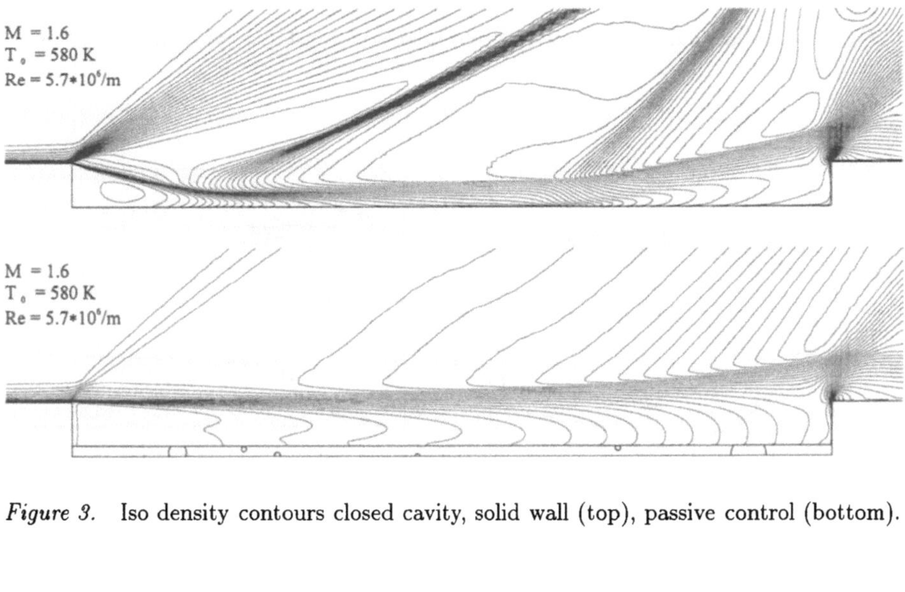

Figure 3. Iso density contours closed cavity, solid wall (top), passive control (bottom).

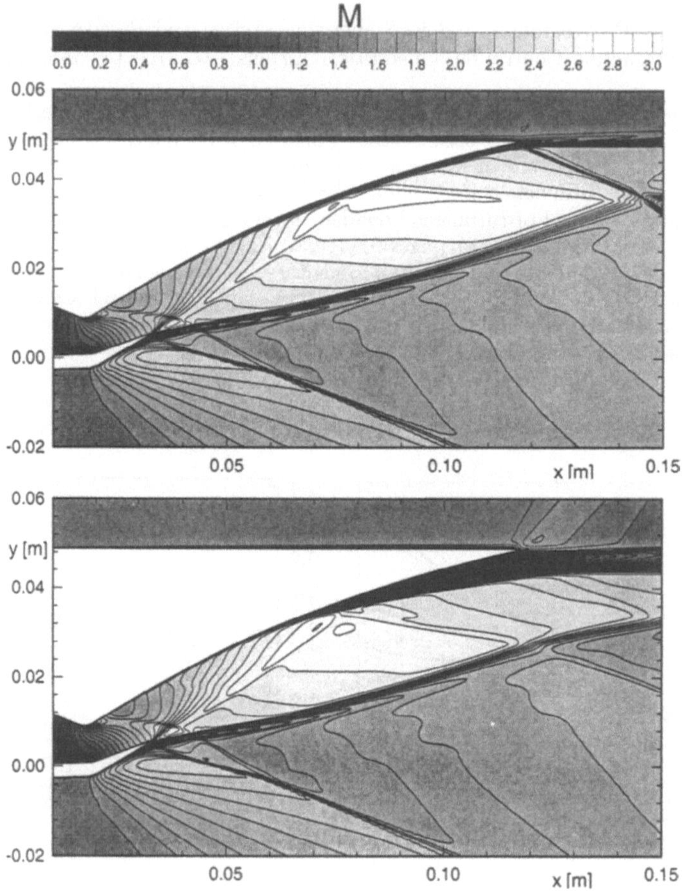

Figure 4. Mach number contours SERN, solid wall (top), passive control (bottom).

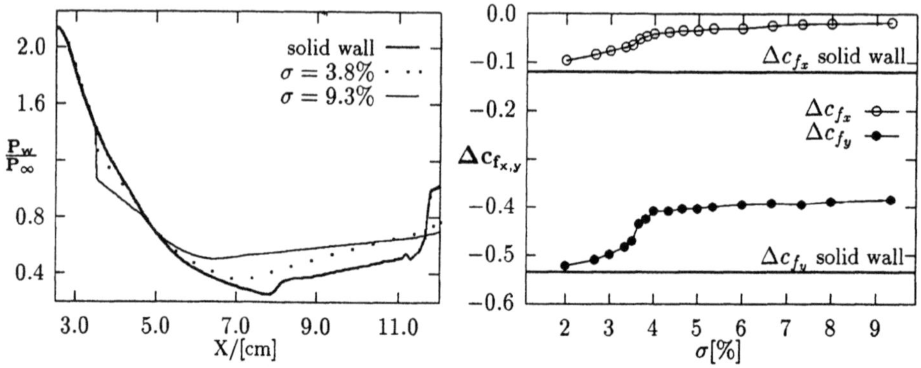

Figure 5. Ramp surface pressure (left) and thrust coefficient (right) with variable σ.

THE CONTROL OF COMPRESSIBLE DYNAMIC STALL USING ADAPTIVE AIRFOILS

M.S.CHANDRASEKHARA*
Navy-NASA Joint Institute of Aeronautics, Department of Aeronautics and Astronautics, Naval Postgraduate School, Monterey, CA 93943, USA

M.C.WILDER
MCAT Inc., Mountain View, CA, 94043, USA

L.W.CARR
US Army ATCOM, AvRDEC & Fluid Mechanics Laboratory Branch NASA Ames Research Center, Moffett Field, CA 94035, USA

1. Introduction

Research [1] has shown that the mechanisms of compressible dynamic stall onset can be vastly different even for *small* changes in flow conditions. At low Mach numbers ($M < 0.3$) and Reynolds numbers ($Re < 1x10^6$), laminar separation bubble bursting induces dynamic stall. At $M = 0.35$, and $Re = 1.3x10^6$, the local flow becomes mildly supersonic and encompasses the laminar bubble over the airfoil and the complex interactions that ensue eventually leads to dynamic stall. Tripping the airfoil again changes dynamic stall onset from such interactions to pressure gradient induced. At a slightly higher Mach number ($M = 0.45$) and approximately the same Reynolds number, shock induced separation causes dynamic stall once the local Mach number at the foot of the leading edge shock exceeds 1.2. Further, in high (flight) Reynolds number subsonic flow, the mechanism changes from laminar separation bubble bursting to that due to excessive adverse pressure gradient. Also, at the Mach numbers of interest for a helicopter or a fighter aircraft, a higher reduced frequency can produce suction peak values sufficient to cause supersonic flow which can alter the mechanism of dynamic stall onset. Thus, since all these mechanisms may be encountered by a rotor-blade during a single cycle, there is a strong need to develop a reliable flow control method for use under such dramatically varying conditions. Although this is a significant challenge, the identification of the fact that these mechanisms are all in some way related to the airfoil leading edge adverse pressure gradient has lead to an effective method of flow control to be described below.

* **Mailing Address: NASA Ames Research Center, M.S. 260-1, Moffett Field, CA 94035-1000**

G.E.A. Meier and P.R. Viswanath (eds.),
IUTAM Symposium on Mechanics of Passive and Active Flow Control, 75–80.

When the flow separates, the vorticity is no longer diffused through the boundary layer, but instead is shed abruptly by convection into the wake as the flow detaches. Thus, active flow separation control requires management of vorticity. It has been shown [2] that in unsteady flows, the vorticity flux is related to the adverse pressure gradient through the surface acceleration for the case of no surface transpiration as follows:

$$\nu \frac{\partial \Omega}{\partial n} = \frac{\partial U_s}{\partial t} + \frac{1}{\rho}\frac{\partial p}{\partial s} \tag{1}$$

This equation suggests that dramatic changes in the flow vorticity field can be introduced by modifying the pressure gradient field, which requires altering the airfoil geometry in real-time. The concept of the *Dynamically Deforming Leading Edge (DDLE)* airfoil enables achieving this very difficult goal and provides a potentially powerful flow control method.

2. The Dynamically Deforming Leading Edge (DDLE) Airfoil

The DDLE airfoil is a derivative of a 15.2cm-chord NACA0012 airfoil. Its leading edge is fabricated (Fig. 1 [3]) from a 50μm thick composite fiber skin. Its nose curvature can

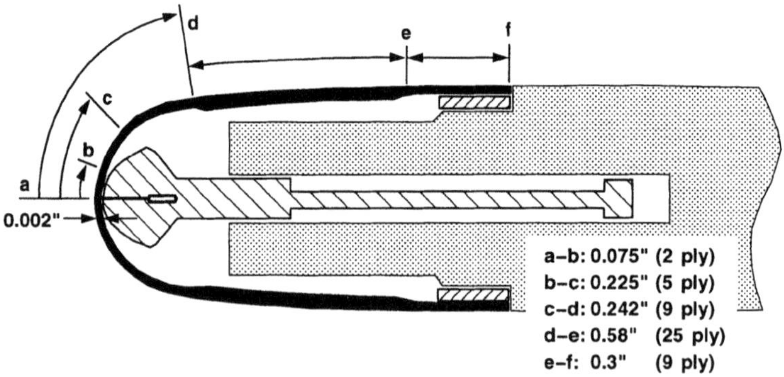

Fig. 1. Construction Details of the DDLE Airfoil Model

be dynamically changed (by as much as 320%) by retracting the leading edge a **small** amount (2mm). The result is an *adaptive airfoil* that can be made to adapt to the instantaneous flow over the entire oscillation cycle. The airfoil deformation schedule is controlled by a computer to generate a large number of shapes with fine control, at variable rates starting from arbitrary angles of attack. An integer shape change occurs for a leading edge retraction of 75μm ($x/c = 0.0005$), Fig. 2. In the present experiments, the maximum movement was less than 1.5mm.

The experimental technique of Point Diffraction Interferometry (PDI) [4] was used to acquire interferograms for $\alpha = 8$-18 deg, $M = 0.3$-0.45, for DDLE airfoil shapes 0-22 in steady flow. When the airfoil was oscillated as $\alpha = 10° + 10° \sin\omega t$ at a reduced frequency, $k = \pi fc/U_\infty = 0.05$, unsteady flow data was acquired with the leading edge fixed at selected shapes 7.5, 8 and 8.5 for M = 0.3 and shapes 6 and 8.5 for M = 0.4. This enables exploration of the fluid physics of unsteady flow control over a large portion of the operational envelope of a rotor-blade. Some results from an analysis of these images are discussed below.

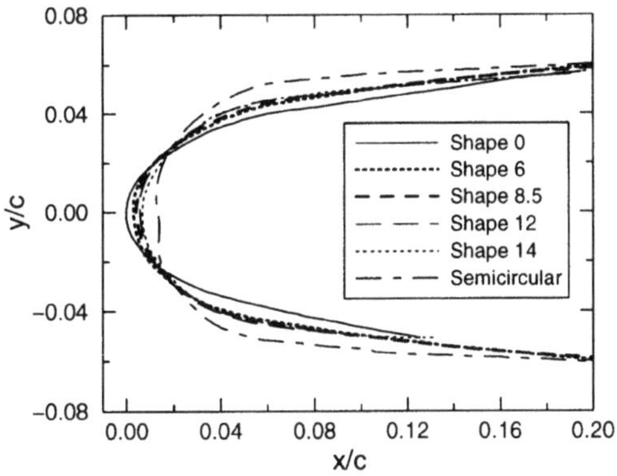

Fig. 2. DDLE Airfoil Shape Profiles

3. Results and Discussion

3.1. STEADY FLOW REGIMES AT M = 0.3

Figure 3 shows the DDLE airfoil flow regimes [5] in steady flow at M = 0.3. Here, shape 0 is nearly identical to the NACA 0012 profile (α_{ss}=14 deg). Increasing the nose radius (shape number) increases the static stall angle, with shapes around 8 showing fully attached flow (region "A") even at α = 17.5 deg. Some trailing edge separation is present for shapes 2-14, region "T1", but the leading edge flow is attached and the airfoil continues to generate large peak suction pressures and thus, a sizeable suction lift. For sharp nose radii (shapes 0-6), as trailing edge separation propagates towards the leading edge, the suction peak begins to drop in the region "T2". Eventually, total loss of lift occurs through leading-edge stall, region "L", similar to the basic NACA 0012 airfoil flow for M > 0.2. It is important to note here that small geometry changes have produced new overall pressure distributions by inducing gross changes in the potential flow and wake characteristics of the resulting airfoil because of large nose radius changes. For shapes 2-10, this has caused the flow to remain attached up to higher angles of attack. The vorticity distributions have also been favorably modified by these **small** leading edge movements (see eq.(1)). Fig. 3 also shows that if an airfoil flow was initially separated, say at α = 16 deg, simply pulling back the nose about 300-400μm (a range of motion within the reach of present day smart material actuators) will *fully reattach* the flow. The considerably expanded attached flow envelope points to the effectiveness of this flow control method. The figure also shows that attached flow is maintained up to $\alpha \approx 18$ deg for shapes 6-9 opening a "window of opportunity" for implementing flow control. It is noted here that the rate at which the surface is adapted plays a major role in establishing the size of this window because of the influence of the unsteady term in eq. (1). The window size is inversely related to the rate of adaptation, [5]. Also, it decreases with increasing freestream Mach number. This is partly attributable to the strong shocks forming in the flow at high angles that can cause separation.

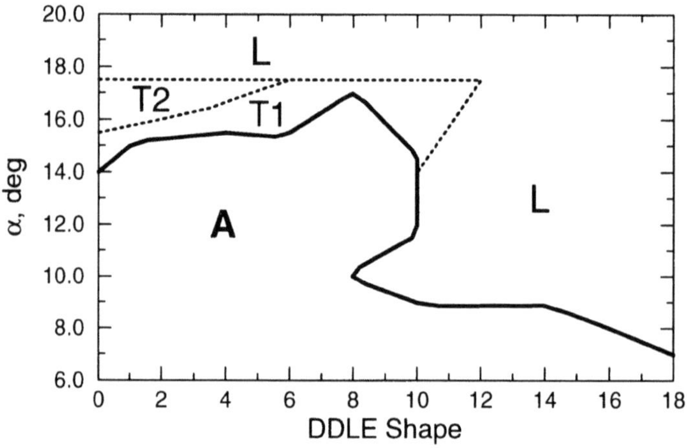

Fig. 3. Flow Regimes over the DDLE Airfoil, M = 0.3, k = 0

3.2. UNSTEADY FLOW RESULTS, M = 0.03, k = 0.05, $\alpha=10^0+10^0 \sin \omega t$

The better performance of airfoil shapes 6-8 discussed above implies that even in unsteady flows they may exhibit superior flow characteristics. The presence of a phase lag (of about a degree in α, [6]) in the pressure field development due to the imposed unsteadiness lends further support to this statement. Unsteady PDI data acquired for shapes 7.5, 8 and 8.5 showed the most remarkable result that the dynamic stall vortex was **absent** even at $\alpha = 20$ deg and the leading edge flow was attached throughout the cycle. Fig. 4 plots the peak suction values over the DDLE airfoils for the entire cycle and for the upstroke over the NACA 0012 airfoil. It is clear that, for $\alpha > 8$ deg, the peak suction for the DDLE airfoils lags that of the NACA 0012 airfoil by about 2 degrees. Also, it drops gradually over the DDLE shape 8.5 airfoil after 17 deg (possibly due to some trailing edge separation) unlike the abrupt decrease seen for the NACA 0012 airfoil. A high C_p value (≈ -4.0) is measured at $\alpha = 20$ deg and by about $\alpha = 8$ deg on the downstroke the hysteresis effects are lost. DDLE shape 7.5 and 8 airfoils show more sudden drops in comparison, but C_p is still high at (≈ -4.0) at $\alpha = 20$ deg. These data confirm that the shape 8.5 airfoil is best for M = 0.3, k = 0.05. As can be expected, these resulting **stall-free** shapes depend upon the flow conditions.

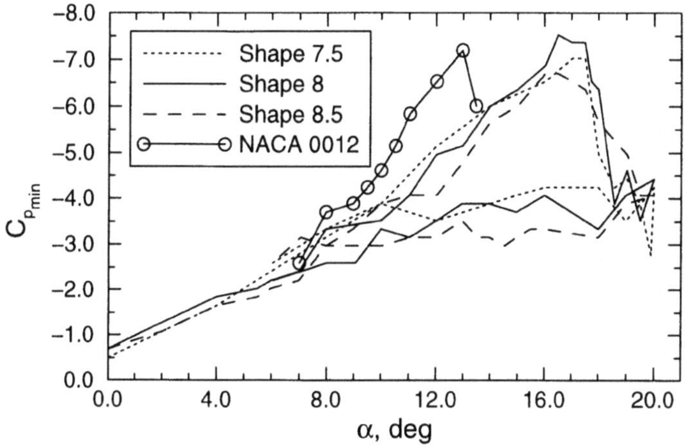

Fig. 4. Development of Peak Suction Pressure over Different DDLE Airfoil Shapes; M = 0.3, k = 0.05

The peak vorticity flux values (obtained from the PDI derived pressure distributions by neglecting the surface acceleration term) shown in Fig.5a for the shape 8.5 airfoil are considerably less and the distribution in x of these is broader than those for the NACA 0012 airfoil (Fig. 5b). This explains the **absence** of the dynamic stall vortex for the DDLE airfoil case. A large, sharp, and narrow peak and another smaller, but wider peak (about 250 & 100-125 respectively, for α = 17 deg) are seen for this case. Since the

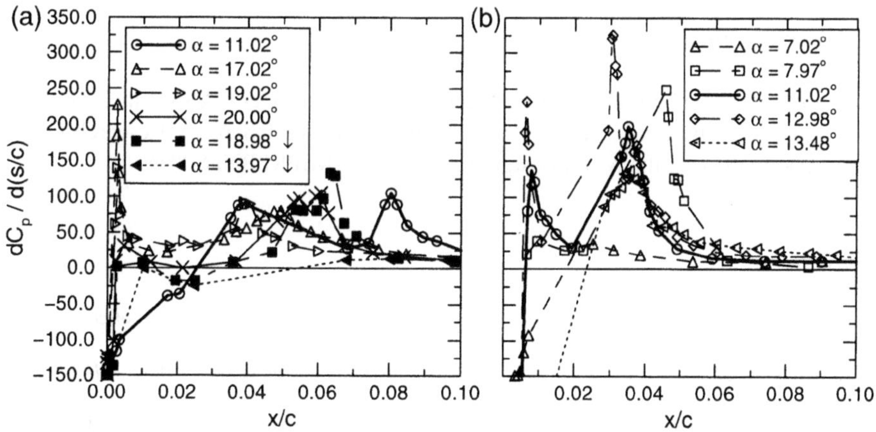

Fig. 5. Vorticity Flux Distributions; M = 0.3, k = 0.05, (a) Shape 8.5 Airfoil, (b) NACA 0012 Airfoil

area under the curve gives the total vorticity production, the broader distributions provide a greater contribution to this term. The second peak moves progressively closer to the leading edge with increasing angle of attack (from x/c = 0.08 at α = 11.02 deg to x/c = 0.04 at α = 19 deg). The first peak decreases for $\alpha \geq 17$ deg due to the trailing edge separation mentioned earlier. Also, there is a reduction in the peak vorticity flux at α = 20 deg, where the degree of unsteadiness decreases to zero. The peaks move toward the trailing edge as $\alpha \rightarrow 20$ deg and during the downstroke when the values drop as the excess vorticity is shed (and light dynamic stall occurs) while the flow adjusts to the lower angles of attack. In Fig. 5b the distributions for the NACA 0012 airfoil show a peak value of around 250 at $\alpha = 8$ deg near the suction peak location ($x/c \approx 0.04$), which increases to 350 around $x/c = 0.025$ for $\alpha = 13$ deg. The PDI images of Chandrasekhara et al [1] show that dynamic stall ensued at $\alpha = 14$ deg for this case, with *no* trailing edge separation. It appears that trailing-edge separation provides a mechanism for shedding the vorticity at the high angles at which the DDLE airfoil was tested. For the vorticity to coalesce, these sharp high peak values are necessary. By lowering their levels, and distributing the total vorticity over a greater area on the airfoil upper surface, it becomes possible to keep the flow attached and produce lift at low drag values until higher angles of attack.

4. Conclusions

A novel compressible, unsteady flow separation control method has been developed using the dynamically deforming leading edge airfoil concept. Results have demonstrated that considerable delay of both static and dynamic stall can be achieved due to the effects introduced on the flow vorticity field. This ability to manipulate the flow vorticity field appears to be the key for successful control of dynamic stall.

5. Acknowledgements

This work was carried out in the Fluid Mechanics Laboratory at NASA Ames Research Center (supported by Dr. S.S.Davis) with funding from the US Army Research Office (Dr. T.L.Doligalski) to the Naval Postgraduate School (MIPR8BNPS007).

6. References

1. Chandrasekhara, M.S., Wilder, M.C., and Carr, L.W., (1998) Competing mechanisms of compressible dynamic stall, *AIAA J.*, **36**, 387-393.
2. Reynolds, W.C., and Carr, L.W., (1985) Review of unsteady, driven, separated flows, *AIAA PAPER 85-0527.*
3. Chandrasekhara, M.S., Carr, L.W., Wilder, M.C., Paulson, G.N., and Sticht, C.D., (1997) Design and development of a dynamically deforming leading edge airfoil for unsteady flow control, *IEEE Publication 97CH36121*, 132-140.
4. Brock, N., Chandrasekhara, M.S., and Carr, L.W. , (1991) A real time interferometry system for unsteady flow measurements, *IEEE Publication 91CH3028-8*, 423-430.
5. Chandrasekhara, M.S., Wilder, M.C., and Carr, L.W., (1998, In press) Unsteady stall control using dynamically deforming airfoils, *AIAA J.*
6. Chandrasekhara, M.S., and Carr, L.W. , (1994) Compressibility effects on dynamic stall of oscillating airfoils, *AGARD-CP-552*, 3.1-3.15

APPLICATION OF MICRO-VORTEX GENERATORS FOR TURBULENT FLOW SEPARATION CONTROL

John C. Lin
NASA Langley Research Center
Hampton, Virginia USA 23681-2199

1. Introduction

Flow separation on multi-element high-lift airfoils can be a complicated function of geometry and flight conditions, and can not be predicted reliably at this time. Previous reports have shown that certain high-lift configurations exhibit boundary-layer separation on the flap at low angles of attack, while fully attached flap flow occurs near maximum-lift conditions [1],[2]. In these cases, altering the geometrical configuration to avoid low angle-of-attack flap separation had the detrimental effect of reducing maximum lift. In situations like this, one possible means of maintaining high maximum-lift values while attenuating boundary-layer separation at low angles of attack is to employ flap-mounted, low-profile, vortex generators.

Conventional, vane-type, passive vortex generators (VGs) with device height, h, on the order of the boundary-layer thickness, δ, have long been used to increase near-wall momentum through the momentum transfer from the outer flow to the wall region. First introduced by Taylor [3], these vortex generators consist of a row of small plates or airfoils that project normal to the surface and are set at an angle of incidence to the local flow to produce an array of streamwise trailing vortices. These vortex generators are commonly used to delay boundary-layer separation [4], to enhance aircraft wing lift [5],[6], and to tailor wing-buffet characteristics at transonic speeds [6]. A wide variety of VGs are in use, and most aircraft employ them for separation control. Many applications, however, use relatively large VGs which cost residual drag through conversion of aircraft forward momentum into unrecoverable turbulence in the aircraft wake. By reducing the height of conventional vane-type vortex generators to only a fraction of δ, vortex generators can still provide wallward momentum transfer over a region several times their own height, especially in turbulent boundary layers where the velocity profile is relatively full [7].

The objective of this study was to determine the effectiveness of the sub-δ-scale VGs in reducing or eliminating separation on a typical single-flap, three-element high-lift system at near-flight Reynolds numbers. These sub-δ-scale VGs, because of their small size, are often referred as "micro-vortex generators" or "micro VGs". An additional practical goal was to determine a size and placement for the micro VGs to allow stowage within the flap well during cruise, to avoid any cruise drag penalty.

G.E.A. Meier and P.R. Viswanath (eds.),
IUTAM Symposium on Mechanics of Passive and Active Flow Control, 81–88.

2. Experimental Apparatus

2.1. TEST FACILITY

The experimental results were obtained from tests conducted in the Low-Turbulence Pressure Tunnel (LTPT) at NASA Langley Research Center. The LTPT is a single-return, closed-circuit wind tunnel. The testing was conducted at a freestream Mach number, M_∞, of 0.20 and Reynolds numbers based on reference airfoil chord, Re_c of 5 and 9 $\times 10^6$. These two Reynolds numbers represent typical three-dimensional wind-tunnel and flight conditions, respectively, at the critical wing station for low-speed stall of a narrow-body transport.

To ensure spanwise uniformity of the flowfield near maximum-lift conditions, a suction system employing porous endplates was used to remove the sidewall boundary layer in the vicinity of the airfoil roots. The suction flow rate was selected to ensure spanwise pressure uniformity at several chordwise locations [8]. The trailing-edge flow was found to be substantially two-dimensional, with a maximum spanwise variation in pressure coefficient, C_P, of 0.05 in the region between the flap brackets [1],[8],[9].

2.2. MODEL AND MEASUREMENTS

The high-lift model investigated was a McDonnell Douglas Aerospace (now Boeing-Long Beach) two-dimensional, single-flap, three-element airfoil [1]. The 11.55% thick model spanned the width of the test section (36 inches) and had a reference (stowed) airfoil chord, c, of 22 inches. The model was configured for landing with a slat deflection of -30° and a flap deflection of 35°. Typically, the maximum-lift coefficient, $C_{l,max}$, for this configuration was approximately 4.5 and occurred at angle of attack, α, of approximately 21°. Flow separation occurred on the flap over a broad range of angle of attack ($-4° \leq \alpha \leq 18°$) below $C_{l,max}$ [1]. Measurements in the LTPT included lift, drag, surface pressure, mean-velocity profiles in the airfoil wake, and fluctuating surface heat fluxes.

Surface pressure measurements were made with over 140 pressure taps that were located along the centerline of the model, as well as in a spanwise row along the trailing edge of each airfoil element to monitor two-dimensionality of the flow. Integration of the pressure measurements yielded the lift data. The uncertainty in $C_{l,max}$ was calculated to be approximately ±0.02 (or less than 1% for a $C_{l,max}$ value of 4.5). Drag data were computed by integration of the local wake profile taken by a 5-hole-probe wake rake located 1.35c downstream of the model, at the centerline station. The uncertainty in drag coefficient, C_d, was calculated to be approximately ±0.0010 for the large wakes of high-lift models (e.g., 2.5% for a typical C_d value of 0.040). Hot-film sensor arrays in the form of streamwise strips, like those reported in References [10] and [11], were used to determine transition and to confirm separation locations on the flap. Each of the 128 sensors was driven by a constant-temperature anemometer. The anemometers were arranged to monitor only the time-varying (fluctuating) component of the hot-film output voltage. Since the films were uncalibrated, the data are presented in the form of root-mean square (RMS) values of digitized counts (100 counts = 5.5 mV).

2.3. MICRO VORTEX GENERATORS

Two variants of the vane-type micro VGs were chosen for this study: (1) a delta-wing with a height of 0.1 inches; and (2) a trapezoid-wing with a height of 0.04 inches. Each of these were arranged to produce either counter-rotating (CTR) or co-rotating (COR) streamwise vortices. The geometry of the micro VGs are presented in Figure 1.

The confluent layer over the flap consists of the flap boundary layer, the jet-like flow through the flap gap, the wake of the main element, and sometimes the wake of the slat. Thus, it is difficult to define a "boundary-layer thickness" for the upper surface of the flap. For example, the velocity profile of a similar 3-element airfoil tested in the LTPT [9] is shown in Figure 2. This profile was measured at a position located 32% of the flap chord, and clearly shows the flap near-wall viscous region, the gap jet, and the main-element wake. The heights of the two micro VG types are shown for reference. The source of high-momentum flow which the micro VGs transfer toward the flap surface is the gap jet, which peaks at about 1% of the reference airfoil chord, c. The micro VGs are sized to occupy only approximately 20% (trapezoid-wing, h/c = 0.18%) or 50% (delta-wing, h/c = 0.45%) of the near-wall flow beneath the velocity-peak from the flap gap.

Typical baseline (VG off) separation on the flap occurred at approximately 45% of the flap chord and the micro VGs were placed at 19%, 25%, or 33% of the flap chord (FC). Micro VGs were also investigated on the lower surface of the main element just upstream of the backward-facing step of the flap well, and on the upper surface at 10%, 25%, 75%, and 100% of the main-element chord (MEC).

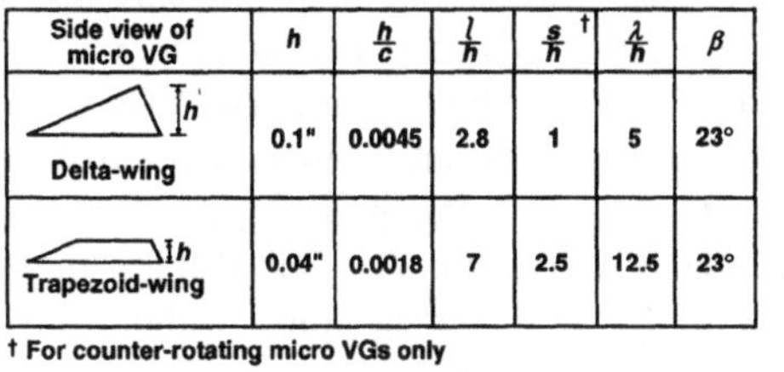

Side view of micro VG	h	$\frac{h}{c}$	$\frac{l}{h}$	$\frac{s}{h}$ †	$\frac{\lambda}{h}$	β
Delta-wing	0.1"	0.0045	2.8	1	5	23°
Trapezoid-wing	0.04"	0.0018	7	2.5	12.5	23°

† For counter-rotating micro VGs only

Figure 1. Micro VG geometry.

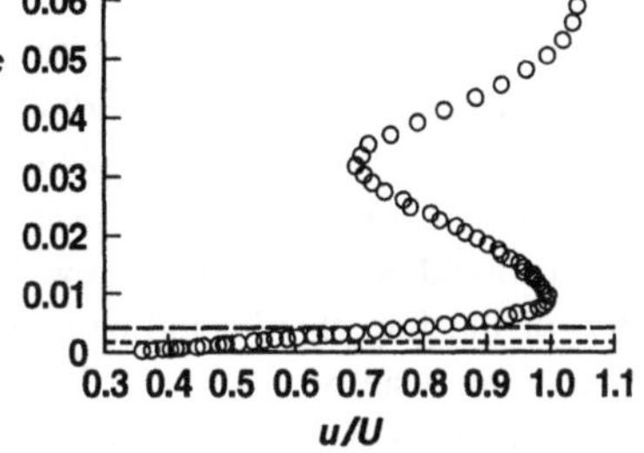

Figure 2. Measured velocity profile from Chin et al. [9] (Taken at 32% flap chord, $\alpha = 8°$, $M_\infty = 0.2$, $Re_c = 9 \times 10^6$).

3. Results and Discussion

For clarity in summarizing the results, the effectiveness of the various micro VG configurations tested are first compared, followed by a more detailed discussion of the observations. The relative separation-control effectiveness for all micro VG configurations and locations examined is summarized in Figure 3.

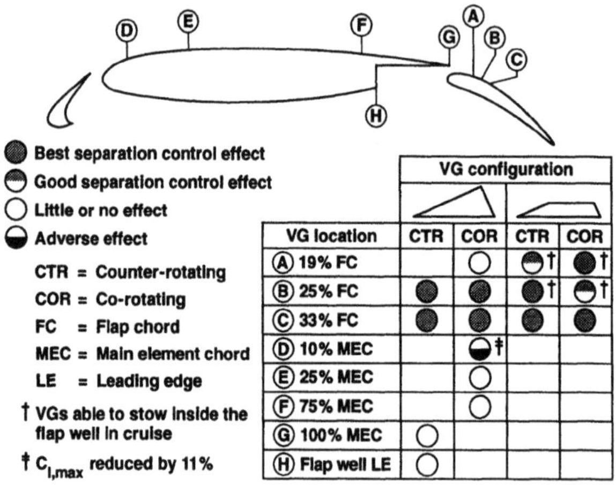

Figure 3. Summary: relative effects of micro VGs for approach conditions.

At a typical landing approach condition ($\alpha = 8°$), both the delta-wing and trapezoid-wing micro VGs were effective in alleviating boundary-layer separation on the flap when placed at 33% and 25% of the flap chord. However, when they were moved upstream to 19% of the flap chord, only the shorter trapezoid-wing micro VGs still maintained some effectiveness. Perhaps the higher (and larger) streamwise vortices produced by the taller delta-wing micro VGs migrated away from the surface, thus becoming less capable of transferring streamwise momentum to the wall region. The most effective streamwise location for the counter-rotating trapezoid-wing micro VGs was at 25% of the flap chord. The co-rotating trapezoid-wing micro VGs were slightly less effective than the counter-rotating configuration when placed at 25% of the flap chord. However, this trend was reversed when the co-rotating micro VGs were moved upstream to 19% of the flap chord. This may be because streamwise co-rotating vortices generally persist longer than counter-rotating vortices in a turbulent boundary layer [6]. Delta-wing micro VGs were also investigated on the main element but yielded no significant positive effect in flap separation control and performance enhancement, such as increasing the $C_{l,max}$.

It should be noted that the low-profile, trapezoid-wing micro VGs placed at less than 26% of the flap chord could be stowed inside the flap well during cruise without any device drag penalty. Since the flow physics associated with high-lift separation control are generally similar, the following discussion primarily concentrates on the results associated with the best-case micro VG configuration, which is the counter-rotating, trapezoid-wing, micro VGs located at 25% of the flap chord.

Separation-control effectiveness was evaluated in terms of lift enhancement and drag reduction. Typically, attenuating flap separation resulted in a significant narrowing of the wake of the three-element airfoil, as shown in Figure 4. The figure shows wake profiles with and without micro VGs for both $Re_c = 5$ and 9×10^6 at $\alpha = 8°$ and $M_\infty = 0.2$. Flow separation on the flap resulted in a wake that covered approximately 0.5c in the vertical direction, while generator-induced attached flow resulted in a wake that covered approximately 0.2c. The narrowing of the wake is evidence of significant drag reduction. There is little difference in wake profile between the 5 and 9 million Reynolds number cases, except that the former had a slightly larger wake because of a slightly larger separated-flow region. Also, the VG-on wake had a slightly higher core-velocity and significantly lower vertical position than the VG-off wake, suggesting a higher downwash velocity associated with increased circulation and, therefore, enhanced lift for the VG-on case.

Separation alleviation on the flap can have a global effect on the entire high-lift system, as shown in the surface pressure distribution of Figure 5 for chord Reynolds number of 9 million. The micro VGs increased the suction pressure on the upper surfaces of the slat and the main element in addition to that of the flap. The increase in the suction-pressure levels resulted in significant lift enhancement at a typical approach angle of attack (i.e., $\alpha = 8°$). For both Reynolds numbers tested, the main element contributed approximately 60% of the increase in lift, while the flap and the slat contributed approximately 25% and 15%, respectively. Just before airfoil stall at $\alpha \approx 21°$, the flow over the flap reattached, and the micro VGs no longer affected the surface pressure distribution.

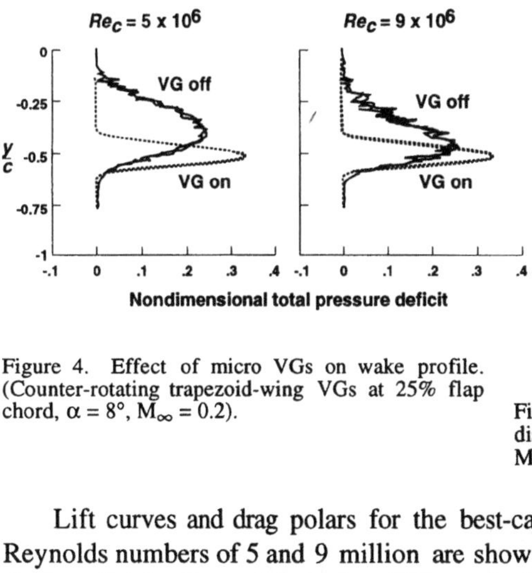

Figure 4. Effect of micro VGs on wake profile. (Counter-rotating trapezoid-wing VGs at 25% flap chord, $\alpha = 8°$, $M_\infty = 0.2$).

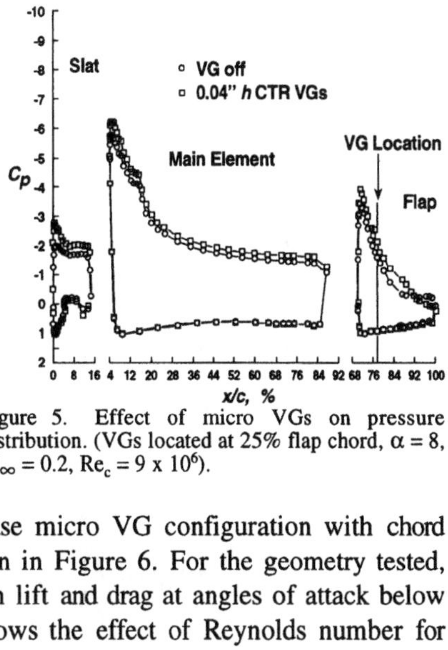

Figure 5. Effect of micro VGs on pressure distribution. (VGs located at 25% flap chord, $\alpha = 8$, $M_\infty = 0.2$, $Re_c = 9 \times 10^6$).

Lift curves and drag polars for the best-case micro VG configuration with chord Reynolds numbers of 5 and 9 million are shown in Figure 6. For the geometry tested, the beneficial effects of the micro VGs on both lift and drag at angles of attack below maximum lift are dramatic. The figure also shows the effect of Reynolds number for baseline (VG off) drag at $12° \leq \alpha \leq 16°$, where the flow moves towards reattachment only for the $Re_c = 9 \times 10^6$ case. The lift coefficients are unaffected by Reynolds number.

At typical approach angles of attack of 4° to 8°, the micro-VG-induced attached flow on the flap could increase the lift on the order of 10%, reduce the drag on the order of 50%, and increase the lift-to-drag ratio, L/D, on the order of 100% (Figure 7). A summary plot of L/D curves for all flap-mounted micro VGs examined is shown in Figure 8. Figure 3 uses information from this figure to define separation control effectiveness with $L/D \geq 90$ as having the best controlling effect, $60 \leq L/D < 90$ as having good controlling effect, and L/D < 60 as having little or no controlling effect at α = 8°. These flap-mounted micro VGs (L/D $\geq$ 60) also did not adversely affect the $C_{l,max}$, where the baseline flow reattached over the flap. The $C_{l,max}$ values for these cases all came within the experimental uncertainty of 1%.

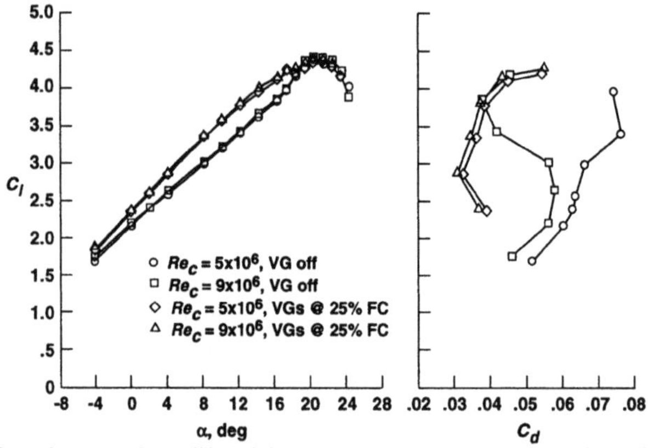

Figure 6. Effect of micro VGs on lift and drag. (Counter-rotating trapezoid-wing VGs, M_∞ = 0.2).

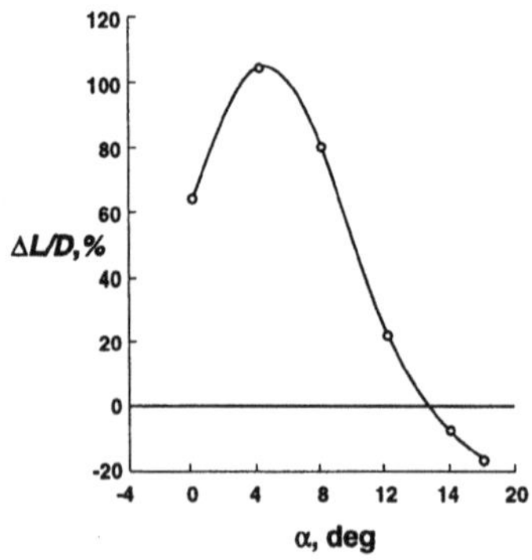

Figure 7. Effect of micro VGs on L/D. (Counter-rotating trapezoid-wing VGs at 25% flap chord, M_∞ = 0.2, Re_c = 9 x 10^6). $\Delta L/D = [(L/D_{VG,on} - L/D_{VG,off})/(L/D_{VG,off})] \times 100$.

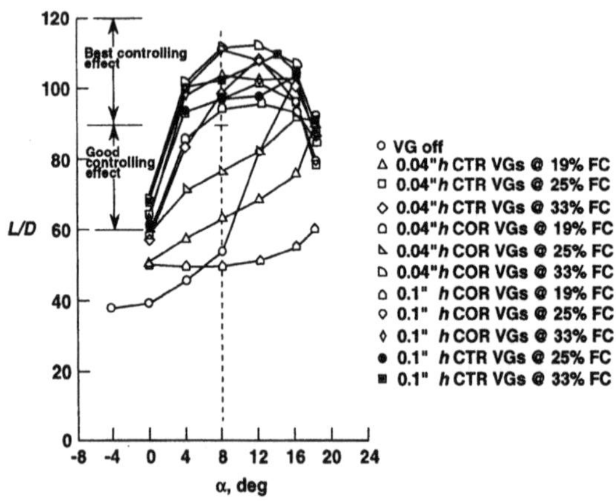

Figure 8. Effect of micro VG configuration and location on L/D. (M_∞ = 0.2, Re_c = 9 x 10^6).

In addition to improving lift and drag, the hot-film results indicated that the best-case micro VGs significantly reduced the separation-induced fluctuating shear-stress (heat flux) levels on the upper surface of the flap by as much as 70%, as indicated by the RMS distributions of Figure 9 for both 5 and 9 million chord Reynolds numbers. As expected, the larger Reynolds number produced a higher level of surface heat-flux fluctuations as well as moving the transition location forward. The first peak in the RMS distribution indicates the transition location, while the region just downstream of the second peak indicates the approximate location of turbulent separation. The turbulent separation location was also confirmed by examining the pressure distribution on the upper surface of the flap (i.e., lack of pressure gradient). Both the pressure and RMS distributions confirmed that micro VGs delayed the separation location from approximately 45% flap chord to at least 85% flap chord. Using the wake profiles as a guide, a conceptual sketch of the flow physics associated with and without the flap-mounted micro VGs is illustrated in Figure 10. The reduction in large-scale flow separation, wake-profile size, and fluctuating shear-stress levels are directly linked.

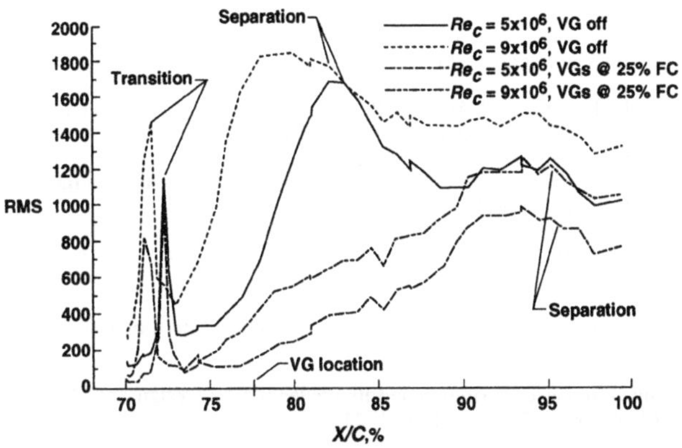

Figure 9. Effect of micro VGs on RMS anemometer output from hot-film sensors on flap upper surface. (Counter-rotating trapezoid-wing VGs, $\alpha = 8°$, $M_\infty = 0.2$).

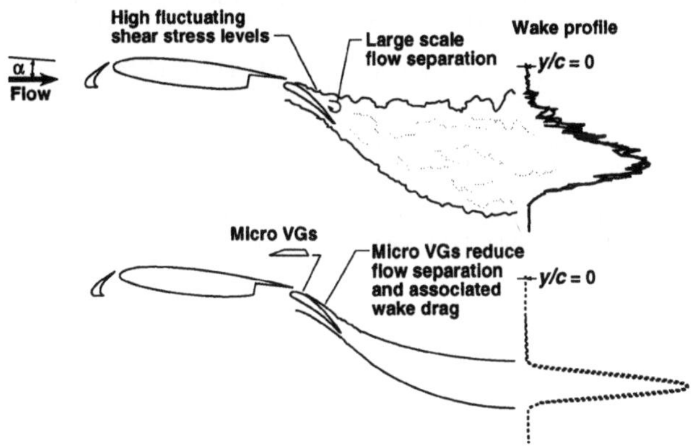

Figure 10. Flow physics associated with flap-mounted micro VGs at approach conditions.

4. Conclusions

Micro VGs as small as 0.18% of reference airfoil chord can effectively control flap separation at landing-approach conditions for a three-element high-lift airfoil which have been optimized for maximum lift. Both counter-rotating and co-rotating streamwise vortices were effective in reducing flow separation on the flap at near-flight Reynolds numbers. Separation alleviation on the flap significantly increased the lift on the order of 10%, reduced the drag on the order of 50%, and increased L/D on the order of 100%. Relative fluctuating shear-stress levels on the upper surface of the flap were reduced by as much as 70% through separation control. The optimum chordwise location of micro VGs (≈25% flap chord) allows them to be hidden inside the flap well during aircraft cruise, thus incurring no cruise drag penalty. Micro VG applications did not adversely affect the maximum lift nor lead to performance improvements for cases where the flow was already attached.

References

1. Valarezo, W.O.: Topics in High-Lift Aerodynamics, AIAA 93-3136 (1993).

2. Woodward, D.S., Hardy, B.C., and Ashill, P.R.: Some Types of Scale Effect in Low-Speed High-lift Flows, International Council of the Aeronautical Sciences Paper 88-4.9.3 (1988).

3. Taylor, H.D.: The Elimination of Diffuser Separation by Vortex Generators, United Aircraft Corporation Report No. R-4012-3 (1947).

4 Schubauer, G.B. and Spangenber, W.G.: Forced Mixing in Boundary Layers, *J. Fluid Mechanics* **8** (1960), 10-32.

5. Bragg, M.B. and Gregorek, G.M.: Experimental Study of Airfoil Performance with Vortex Generators, *Journal of Aircraft* **24** (1987), 305-309.

6. Pearcey, H.H.: Shock Induced Separation and Its Prevention by Design and Boundary-Layer Control, *Boundary Layer and Flow Control* **2** (1961), 1166-1344.

7. Lin, J.C., Howard, F.G., and Selby, G.V.: Small Submerged Vortex Generators for Turbulent Flow Separation Control, *J. Spacecraft and Rockets* **27** (1990), 503-507.

8. Paschal, K., Goodman, W., McGhee, R., Walker, B., and Wilcox, P.: Evaluation of Tunnel Sidewall Boundary-Layer-Control Systems for High-Lift Airfoil Testing, AIAA 91-3243 (1991).

9. Chin, V.D., Peters, D.W., Spaid, F.W., and McGhee, R.J.: Flowfield Measurements about a Multi-Element Airfoil at High Reynols Numbers, AIAA 93-3137 (1993).

10. Sewall, W.G., Stack, J.P., McGhee, R.J., and Mangalam, S.M.: A New Multipoint Thin-Film Diagnostic Technique for Fluid Dynamic Studies, SAE 881453 (1988).

11. Nakayama, A., Stack, J.P., Lin, J.C., and Valarezo, W.O.: Surface Hot Film Technique for Measurements of Transition, Separation, and Reattachment Points, AIAA 93-2918 (1993).

FUNDAMENTAL INVESTIGATIONS OF THE FLOW ABOUT DEVICES SUITABLE FOR USE IN FLOW CONTROL

C J Betts, P R Ashill and K C Hackett
Defence Evaluation and Research Agency,
High Speed and Weapon Aerodynamics Department,
Bedford, MK41 6AE, United Kingdom

1. Introduction

The aim of obtaining improved cruise and manoeuvre performance of both military combat and civil transport aircraft can only be achieved by a thorough understanding of the flows over their lifting surfaces. At manoeuvre conditions, separation of the flow from these surfaces can occur that can result in an increase in unsteady forces (buffet), higher drag and a loss of control. There is, therefore, a need to be able to control these separations in a way that does not adversely effect the aircraft performance in other parts of the flight envelope.

The work described herein is concerned with studies of miniature, sub-boundary-layer devices which might be used to control separations on lifting surfaces, such as wings. [1,2] Such devices could be used to improve the performance of civil aircraft and advanced combat aircraft with minimum parasitic drag penalty. These devices could be passive or 'active', being deployed only when required and the latter possibility may be exploited using the technology developed for Micro-Electro Mechanical Systems (MEMS). Studies have been concerned, initially, with flows around a single device in a zero streamwise pressure gradient. Tests have been performed in the DERA Bedford Boundary Layer Tunnel at low speed (up to 40m/s) using three-component Laser Doppler Anemometry (LDA). The LDA measurements have allowed the flowfield to be mapped downstream of the device as well as determination of the circulation around the vortices in different cross-flow planes downstream. From these measurements, the vortex decay and flow characteristics of each device can be determined.

Scaled-down devices have been mounted in three different configurations on an aerofoil model that has been tested at high subsonic speed, in the DERA Bedford 8ft Tunnel, to assess whether the devices increase the buffet-onset lift coefficient with respect to the datum configuration.

The objectives of the work are to provide flowfield data for the assessment of Computational Fluid Dynamics (CFD) methods, to provide correlations allowing extrapolation of data from wind-tunnel to flight-scale and to determine suitable devices for use in controlling shock-induced separation at high subsonic speeds.

G.E.A. Meier and P.R. Viswanath (eds.),
IUTAM Symposium on Mechanics of Passive and Active Flow Control, 89–94.

The experimental details are given in Section 2 followed by a presentation of results in Section 3. Some concluding remarks are given in Section 4.

2. Experimental Details

2.1 LOW-SPEED INVESTIGATIONS

2.1.1. *Boundary Layer Tunnel*

This is an open-return tunnel (Fig. 1) and, for this work, has three 1.83m long sections, 1.2m wide with a flexible floor allowing a maximum height of 0.7m (the standard height being 0.3m). This allowed the streamwise pressure gradient to be set at zero. A jet catcher was secured at the open jet to diffuse the high velocity air to an acceptable level. The devices were mounted on the centre-line of the tunnel roof close to the join between the second and third sections, see Fig 1. This meant that optical access was possible from the trailing edge of the device to about 0.5m downstream of the trailing edge enabling LDA to be made from 0 device heights to 15 device heights for a device height, h = 30mm.

2.1.2. *Laser-Doppler Anemometry Measurements*

Three component Laser Doppler Anemometry is a well established technique for use in wind tunnels [3]. In this experiment, three beam pairs (green, blue and violet) were produced by a 4W laser and two optical probes with fibre-optic delivery were used with beam expansion. The measurement volume was 170 μm x 170 μm x 4mm. Each set of signals was analysed using a signal processor (Burst Spectrum Analyser) connected to a personal computer. The optical probes were mounted on a three-component traverse controlled automatically. Seeding was introduced via a pipe through a hole in the settling chamber roof ahead of the contraction to produce a seeding plume. Smoke oil particles, 1 micron in diameter, were emitted from a six-jet atomiser under compressed air. It was found that a record length of 10 seconds was sufficient to determine the velocity components. A commercially available LDA data-processing package was used to calculate the three components of velocity and the turbulence intensity and the data was exported as files for further visualisation using other software tools.

2.1.3. *Devices*

Following a review of means of active and passive flow control, a selection of devices were chosen for a study of the fundamental flow characteristics, (Fig 2). The wedge-type devices, for example, produce a counter-rotating vortex pair and have been tested both forwards (as in the figure) and reversed. A vane, in the form of a thin blade orthogonal to the surface producing a single vortex, has been tested at different orientations to the freestream. The vane is a conventional vortex generator and will be used as a datum for comparisons. A dowel has also been tested at different orientations to the freestream.

2.2 HIGH-SPEED INVESTIGATIONS

2.2.1. *8ft Tunnel Model*

The model used to test different methods of passive control [4] was mounted between the sidewalls of the 8ft x 8ft high speed, pressurised wind tunnel. The aerofoil section was of 14% maximum thickness-to-chord ratio and was designed to be suitable for natural laminar flow with favourable pressure gradients on the upper and lower surfaces between the leading edge and 50% chord. The chord length of the aerofoil section was 0.635m. Boundary layer transition was fixed at 5% chord on both surfaces.

2.2.2. *Measurements*

The measurements included surface static pressures at three stations across the span, 25%, 50% and 75% span. In addition, at one spanwise station, unsteady pressures were measured using fast-response pressure transducers. Sectional drag at the mid-span was determined from measurements made with a wake rake of pitot and static tubes placed about two aerofoil chords downstream of the trailing edge.

2.2.3. *Device configurations*

Over the Mach number range tested (0.67-0.71), the foot of the shock on the aerofoil upper surface is largely fixed at about 55% chord. The wedge devices, tested at low-speed, were scaled down to 0.5mm height (about one boundary-layer displacement thickness in this region) and were positioned upstream of the shock. Tests were performed at chord Reynolds numbers of 13.1 and 29.7 million. All wedge configurations consisted of a row of wedges between the port and starboard pressure-plotting stations (Fig 3) and were as follows:

Configuration	Leading-edge position	Spanwise spacing
1	52% chord	1 wedge height
2	52% chord	7 wedge heights
3	53.4% chord	7 wedge heights

3. Results

3.1 LOW-SPEED INVESTIGATIONS

3.1.1. *Velocity Vectors in the cross-flow plane*

Fig. 4 shows a comparison between measured and computed velocity vectors (velocity profiles have also been analysed but are not presented here). Fig 4a shows the velocity vectors and streamtraces in a cross-flow plane at 5 device heights downstream of the 30mm height wedge at a freestream speed of 30m/s predicted by Computational Fluid Dynamics (performed at British Aerospace Sowerby Research Centre) using a Reynolds-Averaged Navier-Stokes code. Fig 4b shows the corresponding results from LDA measurements. The experiment and theory agree well. The equivalent experimental data for the vane (at 20°) and the reversed wedge are given in Figs 5 and 6, respectively. The vortex pair arising from the reversed wedge is much more widely spaced than that of the wedge. These results allow comparison of the characteristics of the different devices tested and provide detail to allow an improved understanding of the physics of the flow.

3.1.2. *Circulation*

The circulation around the vortex in a specified cross-flow plane was obtained by surrounding the vortex core within a rectangular contour, typically, 2.5h x 2h. The circulation was then determined by performing the appropriate line integration around the contour. The non-dimensional circulation is shown in Fig 7 against the non-dimensional height h+ for some of the different devices.

3.1.3. *Vortex Decay*

By making measurements in the cross-flow plane at different positions downstream of the devices, it is possible to calculate the vortex decay by comparing the circulation at a given station with that at x/h=0.5. The 30mm height vane set at 45° to the flow, initially has a stronger vortex than the other devices. All the vortices have been found to be of a similar strength at x/h=15.

3.2 HIGH-SPEED INVESTIGATIONS

The pressure distribution for M=0.71, R_c=19 million, C_L=0,68 is given in Fig. 8a. Chordwise pressure distributions of the rms value of the pressure fluctuations, made non-dimensional by the freestream dynamic pressure, p_{rms}/q for the datum configuration and wedge configuration 3 are given in Fig 8b. The effect of the wedges would appear to be to reduce the level of p_{rms}/q in the region of the shock and further downstream. The results for wedge configurations 1 and 2 have the same effect with respect to the datum but have been omitted for clarity. From this point of view, the wedges would appear to have a favourable effect. However, analysis of the variation of unsteady pressure with lift coefficient at each transducer location does not appear to indicate that any of the wedge configurations produces an increase in the buffet-onset lift coefficient at any of the flow conditions tested.

4. Concluding Remarks

This investigation is providing detailed information concerning the flowfield downstream of sub boundary-layer devices for the assessment of CFD, for the prediction of flight characteristics of this device and for design purposes. The main conclusions are:

1) Detailed information of the flowfield downstream of different devices has been obtained
2) CFD gives fair prediction of the velocity field in a cross-flow plane downstream of a wedge.
3) Although wedge devices reduce the unsteady pressure fluctuations locally in the region of a shock on an aerofoil they do not delay the onset of buffet.

Future work will involve fundamental investigations of multiple devices in zero streamwise pressure gradient and single devices in varying pressure gradients. High-speed tests will be conducted on alternative devices for shock-induced separation control.

5. References

1 H H Pearcey, 'Shock-induced separation and its prevention by design and boundary layer control'; Part IV, Vol 2 Boundary Layer and Flow Control (ed. G V Lachmann) Pergamon press (1961).

2 J M McMichael, 'Progress and prospects for active control using Micro-fabricated Mechanical Systems', AIAA96-0306 (1996).

3 W J Yanta, 'A three-dimensional laser Doppler velocimeter (LDV) for use in wind tunnels', Proc of 8th Int. Congress of Instrumentation in Aerospace Simulation Facilities, Monterey, California, USA, pp294-301 (1979)

4 P R Ashill and J L Fulker, 'A review of flow control research at DERA', Proc. of IUTAM Symposium, FLOWCON (1998)

6. Figures

Figure 1. Photograph of Boundary Layer Tunnel

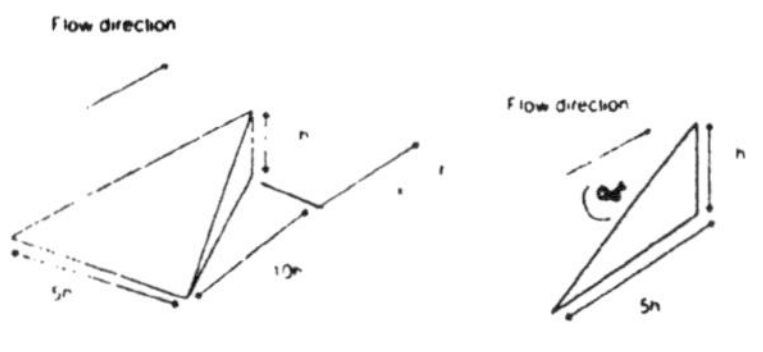

a) Wedge b) Vane

Figure 2. Sketch of types of devices studied

Figure 3. Wedges mounted on aerofoil model

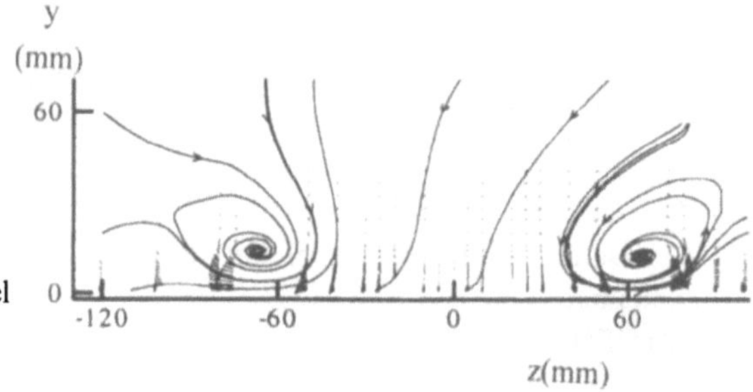

Figure 6. Measured velocity vectors in cross-flow plane, x/h=5, U=30m/s, reversed wedge h=30mm

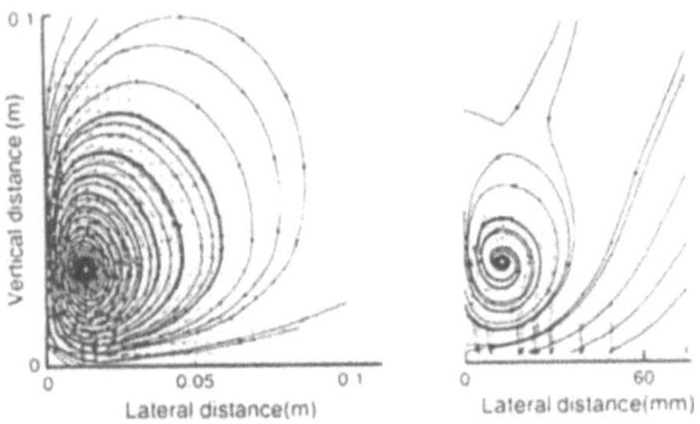

a) Predicted b) Measured

Figure 4. Velocity vectors in cross-flow plane, x/h=5, U=30m/s, wedge h=30mm

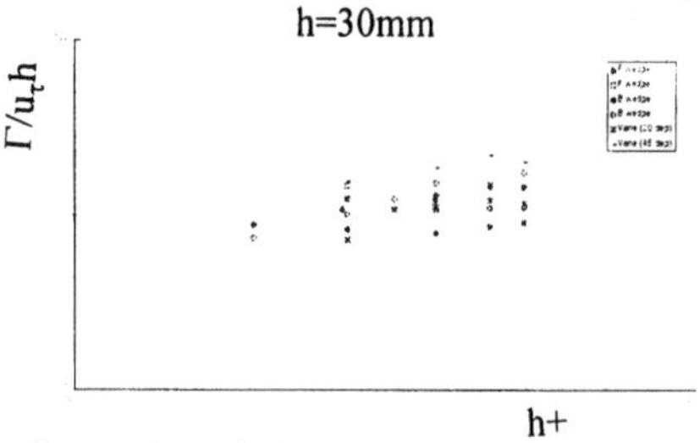

Figure 7. Variation of circulation with non-dimensional height h+, x/h=5

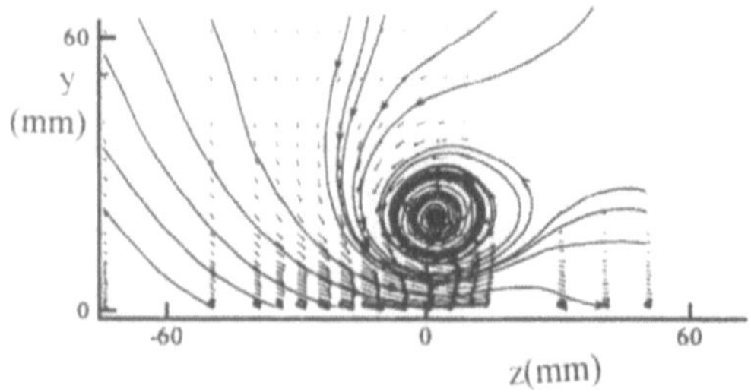

Figure 5. Measured velocity vectors in cross-flow plane, x/h=5, U=30m/s, vane h=30mm

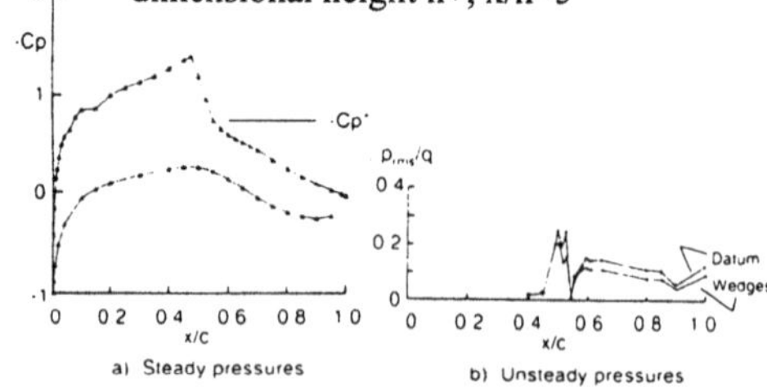

Figure 8. Pressure distribution M=0.71, R_c=19 x 10^6, C_L=0.68.

7. Acknowledgements

The authors wish to thank staff at British Aerospace (Operations) Ltd Sowerby Research Centre for performing the Computational Fluid Dynamics calculations. This work has been supported by the UK Ministry of Defence and Department of Trade and Industry.

A novel use of vortex generators for separation control in the corner between a wall and a wing

W. Hage[1], D.W. Bechert[1] & R. Meyer[2]

[1] Deutsches Zentrum für Luft- und Raumfahrt (DLR), Abt. Turbulenzforschung, Müller-Breslau-Str. 8, D-10623 Berlin, Germany email:wolfram.hage@dlr.de
[2] Hermann-Föttinger-Institut für Strömungsmechanik (HFI), Technische Universität Berlin, Straße des 17.Juni 135, D-10623 Berlin, Germany

ABSTRACT

It is a well established fact that vortex generators are a very effective means to control flow separation. Vortex generators enhance the mixing between the mean flow and the boundary layer. They provide a transport of fluid with high momentum to regions near the wall, where the retarded flow tends to separate.
In our windtunnel experiments with the (laminar) airfoil HQ 17 we obtained a higher drag from force balance measurements as compared to wake rake data of previous investigations. This difference of the measured drag coefficient is caused by separation in the corner between the side wall and the wing (see Figure 1).

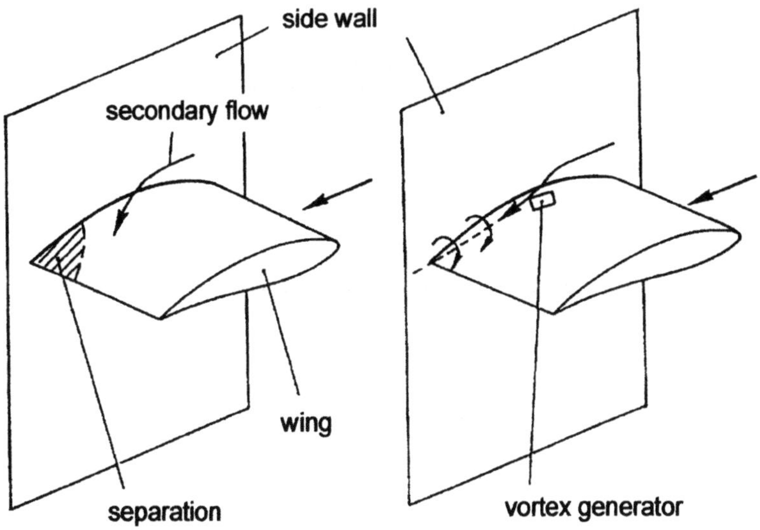

Figure 1: separation in the corner between airfoil and wall

G.E.A. Meier and P.R. Viswanath (eds.),
IUTAM Symposium on Mechanics of Passive and Active Flow Control, 95–97.

The corner flow separation is caused by the interference between the flow on the wall and of that on the airfoil. On the side wall one observes a developed boundary layer with retarded flow with low kinetic energy in the vicinity of the wall.
The low pressure on the convex side of the airfoil causes a suction of the retarded fluid from the wall towards the airfoil. This slower fluid is unable to overcome the increasing pressure on the airfoil and the flow separates in the corner between wall and airfoil. The separation causes a decrease in lift and an increase in drag.
There are several known techniques to control flow separation. Means like wall suction or tangential blowing were not considered, because they require a comparatively larger technical apparatus.

It turned out, that a single vane type vortex generator placed in each corner on the wing reduces the drag coefficient by up to 10% and it also increases the lift coefficient. This vortex generator is inclined to the flow (see Figure 1) and thus it hampers the invading of retarded fluid from the wall into the area of lower pressure on the airfoil. The vortex generator protrudes out of the boundary layer into the region of higher velocity. This causes vortex shedding, which feeds fluid with higher kinetic energy into the corner between airfoil and wall.

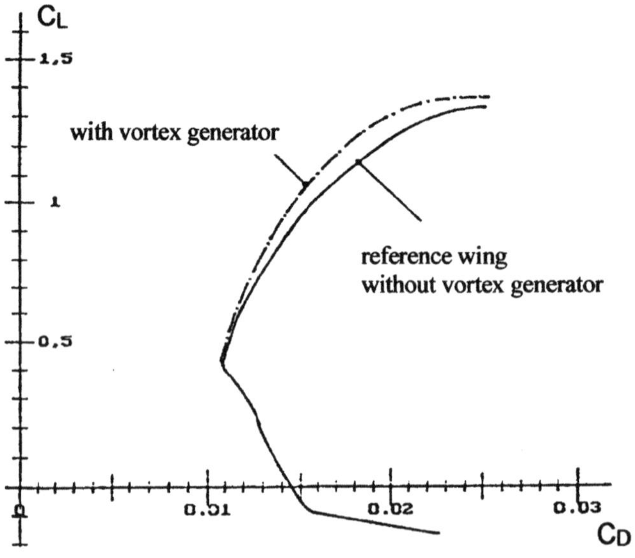

Figure 2: C_L/C_D polar with and without vortex generator on airfoil HQ 17

It is remarkable that, although a parasitic drag of the vortex generator is unavoidable, a net drag reduction of the airfoil over the whole C_L/C_D-polar is obtained (see Figure 2). This finding with our simplified configuration suggests a possible further application of vortex generators in the field of turbomachines.

It seems very likely that vortex generators placed on the blades as well as on the guide vanes of turbomachines should be able to control the separation which occurs in the corner between the casing or hub and the blades. This type of corner separation is responsible for the major part of the losses in axial turbomachines.

Windtunnel Experiments with artificial bird feathers for passive separation control on airfoils

R. Meyer[1], D.W. Bechert[2] and W. Hage[2]

[1] Hermann-Föttinger-Institut, TU Berlin, Strasse des 17. Juni 135, 10623 Berlin, Germany
[2] DLR, Abt. Turbulenzforschung, Müller-Breslau-Str. 8, 10623 Berlin, Germany

During the landing approach of birds and in flight through very turbulent air one observes that the covering feathers on the upper side of bird's wings tend to pop up. The aerodynamics of this particular effect is investigated. Our experiments have been carried out with a real glider wing section in a wind tunnel. Lift and drag have been measured with a balance. We also measured the pressure distribution on the airfoil. In addition, flow visualisation photographs with TiO_2-oil-benzine paint on the wing surface were taken. The wing section is a so-called „laminar" airfoil, type HQ41 developed by Horstmann & Quast, DLR Braunschweig. The Reynolds number range of our experiments is $1.0 - 2.0 \times 10^6$, so that the lower flight velocities of the real aircraft are covered. This particular regime is relevant for the high-lift condition considered here.

The artificial bird feathers in our experiments look different from the original biological ones. We have tested a large variety of movable plastic and metal sheets attached to the upper surface of the airfoil. These devices are actuated by the flow itself and are in operation (i.e., bristled) only if flow separation on the wing starts or takes place and they stabilise and limit the separation. A design which permits pivoting of the movable sheets on their leading edges (see Fig. 1) and with tether threads (to prevent tipping over) seems to be easier to operate than any devices which emulate real bird's feathers. A perforation of the sheets is necessary in order to prevent lifting of the devices by the pressure distribution of attached flow.

These devices are compatible with laminar and turbulent airfoils as well as with various conventional flaps on aircraft wings.

Under attached flow conditions, the devices are attached to the airfoil and thus the production of parasitic drag is avoided. When flow separation starts to occur, the local flow reversal lifts the devices more and more for increasing angles of attack. In Fig. 2 we provide data at $Re = 10^6$ for <u>one</u> device on the airfoil. The increase in maximum lift is 11%. The rear device in Fig. 1 produces a direct extension of the lift curve as can be seen in Fig. 2.

G.E.A. Meier and P.R. Viswanath (eds.),
IUTAM Symposium on Mechanics of Passive and Active Flow Control, 99–100.

The gain in lift can be further increased beyond that value either by extending the rear device size in flow direction or by additional devices attached further upstream on the wing.
With one rear device we reached up to 19% increase of maximum lift. The rear device seems to operate mainly by impeding the reversed flow during separation. It also reduces gradually and appropriately the effective camber of the wing so that separation is limited.
Additional devices upstream do produce additional beneficial effects for higher angles of attack. We have tested up to three devices which produce as much as 23% increase of maximum lift. They seem to operate with a slightly different mechanism: Fluttering of these additional devices is important and energises the near-wall part of the boundary layer. Thus, separation is delayed and lift is enhanced.

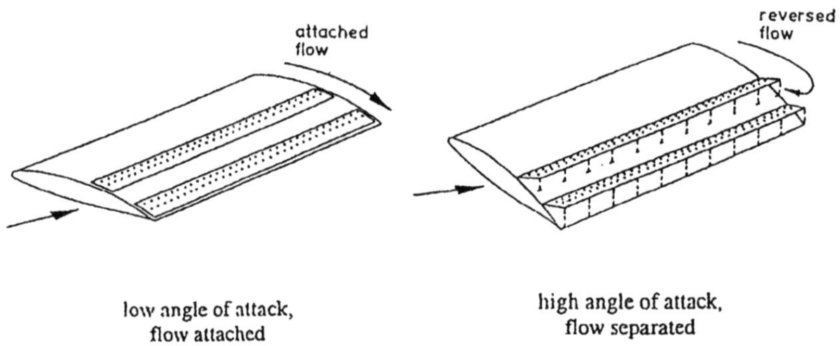

Fig. 1. Operation of the artificial bird's feathers.

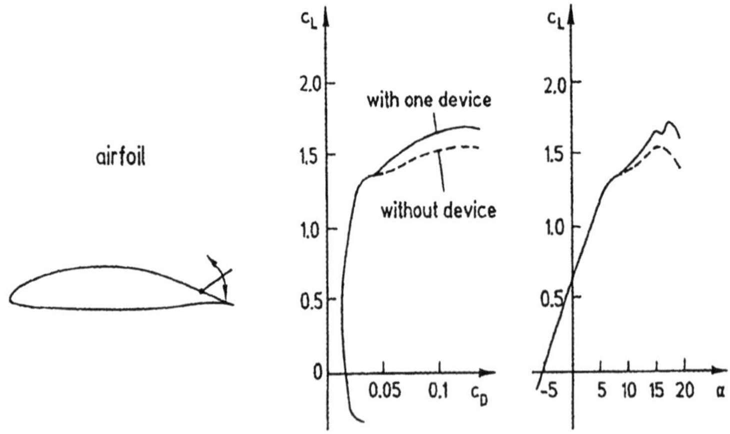

Fig. 2 Drag polar and lift coefficient C_L versus angle of attack α for our airfoil with one device.

A METHOD OF MIXING PROCESS CONTROL IN A SHEAR LAYER OF A SUPERSONIC NONISOBARIC JET

V.I.ZAPRYAGAEV
Institute of Theoretical and Applied Mechanics, Russian Academy of Sciences, Siberian Division, Novosibirsk, RUSSIA, 630090

A possibility of controlling the process of mixing in the shear layer of the initial section of a supersonic nonisobaric jet is based on the relatively recently discovered new physical phenomenon connected with the formation of stationary streamwise vortex structures at the boundary of a supersonic underexpanded jet.

1. Introduction

It has been experimentally stated that the initial section of the shear layer of a supersonic underexpanded jet issuing from an axisymmetrical nozzle has a 3-D structure. At the present time, the 3-D character of flow in the jet shear layer is associated with the appearance and development of stationary streamwise vortex structures. The presence of the structures at the jet boundary affects the processes of turbulent mixing, and that, mainly, determines the importance of investigating their physical nature and revealing the methods of controlling the mixing process.

A systematic study of streamwise vortex structures in the initial section of the free supersonic nonisobaric jet issuing from the axisymmetrical nozzle into the ambient space at large Reynolds numbers was started in [1-2]. Soon, the main scientific result was confirmed by papers [3-4]. The 3-D structure of the initial section of the jet was also discovered for the jets of rarefied gas [5-6]. This points out to a rather general character of the phenomenon of appearing the 3-D structure of the flow in the shear layer of the supersonic jet.

2. The spatial structure of flow in the initial section of a supersonic jet

A hypothesis about the presence of the Gortler vortices in a free supersonic underexpanded jet was put forward by G.F.Glotov in 1984. It was based on the analysis of Schlieren pictures of the jets issuing into the ambient space. A sketch of supersonic underexpanded jet in the presence of 3-D flow in the jet shear layer is represented in Fig.1. [1-2].

G.E.A. Meier and P.R. Viswanath (eds.),
IUTAM Symposium on Mechanics of Passive and Active Flow Control, 101–106.

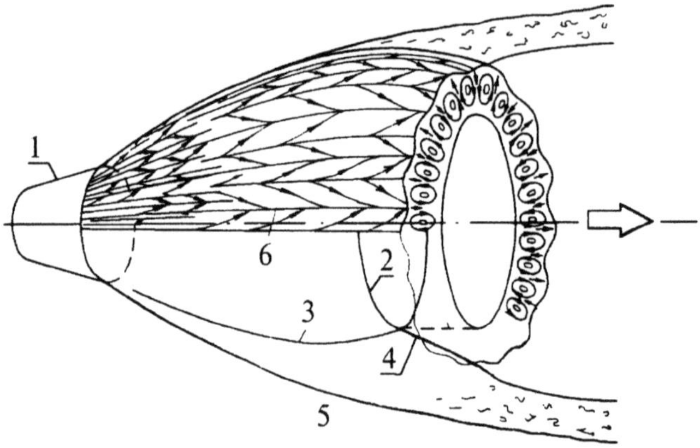

Figure 1. A sketch of supersonic underexpanded jet issuing from an axisymmetrical nozzle with the picture of streamwise vortex structures in the initial section of the jet; 1 - nozzle, 2 - the Mach disc, 3 -barrel shock; 4 - reflected shock; 5 - jet boundary, 6 - streamwise vortices.

This flow pattern is based on the results of flow visualization with the use of the Schlieren method, the laser sheet method, and the results of probing the supersonic jet at large Reynolds numbers with the help of a small-sized Pitot pressure probe [1-4,7]. The data presented in these papers on the character of dependence of the Pitot pressure on the azimuthal angle differ significantly by the amplitude of perturbations registered in the shear layer and their spectral composition. One of the uncontrolled parameters is the roughness of the nozzle surface. Thus, it seems reasonable to consider the influence of a local microrouphness of the nozzle inner surface on the development of streamwise vortex structures in the mixing layer of a supersonic underexpanded jet.

2.1. EFFECT OF THE NOZZLE ROUGHNESS ON THE GENERATION AND DEVELOPMENT OF STREAMWISE VORTICES

The results of an experimental investigation of the influence of artificial microroughness of the nozzle surface on the generation and development of streamwise vortex structures in the shear layer of the initial section of a high-speed jet are described in [8]. The characteristics of three nozzles used in experiment are listed in Table 1.

TABLE 1.

Nozzle number	D_a (mm)	Re_d*10^{-6}	np=P_a/P_h	M_a	Paper
1	20.0	5.05	4.12	1.5	[8]
2	40.0	0.83-2.02	1.0-2.0	1.0	[8]
3	20.0	1.3-1.95	2.12-3.18	1.0	[9]

The Reynolds number Re_d was based on the flow parameters at the nozzle exit and the nozzle diameter D_a, and the Mach number at the nozzle exit M_a was determined from the nozzle geometry, np= P_a /P_h is the jet pressure ratio of the pressure at the nozzle exit P_a to that in the ambient space P_h.

The experiments were carried out in a jet facility of the Institute of Theoretical and Applied Mechanics SB RAS. The object of the study was a high-speed jet of cold air with stagnation temperature T_o=273K, which was issuing from an axisymmetric nozzle into the ambient space with atmospheric pressure P_h and temperature P_h=293K.

The measured velocity profile for $x/R_a = 0.015$ is in good agreement with the Blasius profile for nozzle No.2. We can conclude from the analysis of this result that the flow in a convergent nozzle No.2 for $Re_d \sim 10^6$ is laminar [8]. The measurements and estimates for the displacement thickness at the nozzle exit yield $\delta^*=0.05$ mm for nozzle No.1 and $\delta^*=0.13$ mm for nozzle No.2 for $Re_d \sim 10^6$.

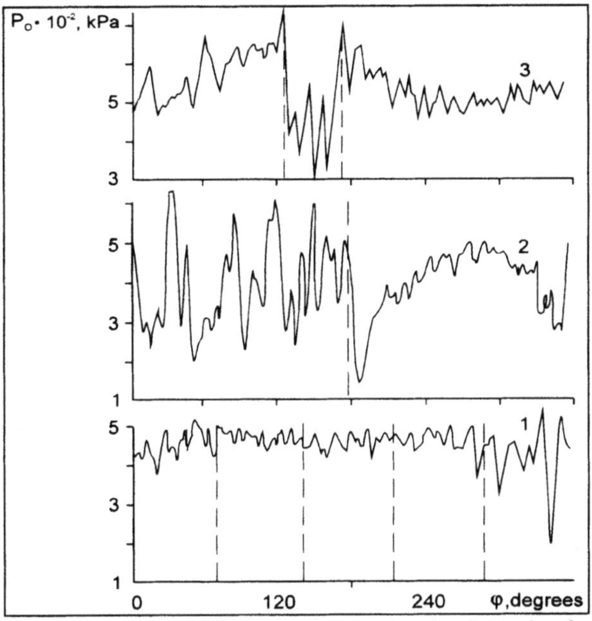

Figure 2. The Pitot pressure distribution $P_o(\varphi)$ versus the azimuthal angle φ for different roughness of the internal surface of nozzle No.1.

The results of the measured nonuniformity of the Pitot pressure distribution $P_o(\varphi)$ are presented in Fig.2 versus the azimuthal angle φ for different roughness of the internal surface of nozzle No.1. Experimental data correspond to the distance $x/R_a=2.0$ between the probe and the nozzle exit. Curve 1 is characteristic of the nozzle with scratches 0.05 mm deep on its internal surface with a step of 1 and 2° . The scratches were applied to different sections of the internal surface of the nozzle.

Curve 2 corresponds to sand roughness applied onto the throat region within the range $\varphi=0\div180^o$ ($\varphi=180\div360^o$ is the region with natural roughness). In this case, the size of particles glued onto the surface was (0.08±0.02) mm. The particles were glued as a band 1 mm wide. Curve 3 refers to smaller particles with the transverse size of 0.05 mm, which were glued onto the internal surface of the nozzle near to its exit within the range of azimuthal angles $\varphi=130\div180^o$.

It follows from comparison of the Pitot pressure distributions $P_o(\varphi)$ (with artificial roughness) and similar plots for natural roughness that the amplitude of variation of the Pitot pressure in the shear layer of the supersonic underexpended jet is increased by additional roughness.

2.2. EFFECT OF THE LOCAL NOZZLE MICROROUGHNESS ON THE DEVELOPMENT OF STREAMWISE VORTICES

It seems important to study how a local single microroughness on the nozzle surface affects the generation of streamwise vortices in the mixing layer, which is closely connected with the flow receptivity to the value and spectral composition of surface microroughnesses. Some data on the structure of stationary disturbances in the mixing layer, which are generated by a local microroughness located on the inner surface of the nozzle, are presented below, see also [9]. The geometric parameters of nozzle No.3 are presented in Table 1. The gasdynamic parameters of three examined regimes for a convergent nozzle are presented in Table.2.

TABLE 2.

N	P_s(kPa)	x_l/Ra	M_j	np=Pa/Ph	Re/10^6
1	392.0	3.25	1.56	2.12	1.30
2	490.0	3.75	1.71	2.65	1.71
3	588.0	4.0	1.83	3.18	1.95

Here N is the regime number, P_s corresponds to the stagnation pressure in the plenum chamber of the jet facility, x_l/R_a is the length of the first cell of the jet normalized to the nozzle exit radius R_a; M_j is the Mach number corresponding to isentropic expansion of the flow in an underexpanded jet from the nozzle exit pressure P_a to the ambient pressure P_h; (D_a=2R_a).

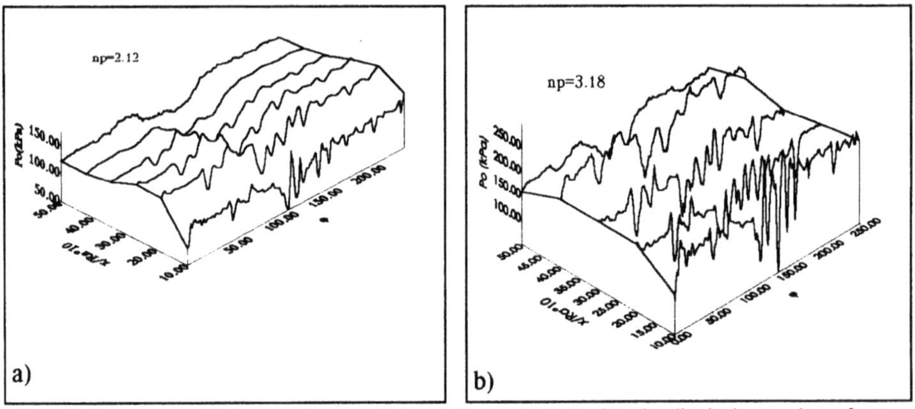

Figure 3. Pitot pressure $P_o(\varphi)$ versus the azimuthal angle for nozzle No.3 with a localized microroughness for two regimes (N = 1, 3 - a, b respectively).

The dependence of the measured pressure Po(φ) on the azimuthal angle for two regimes (see Table 2) in five various cross-sections of the jet is shown in Fig. 3. The presented plots Po(φ) are obtained for a radius corresponding to the middle of the mixing layer, $r = r_{0.5}$, where $Po(r_{0.5}) \sim 0.5{*}Po(r_m)$, r_m is the radius corresponding to the internal boundary of the mixing layer of the jet. The position of a spherical particle of diameter k~0.150 mm corresponds to φ_o=122°. The microroughness diameter k is comparable with the measured displacement thickness δ* of the boundary layer at the nozzle exit, i.e., $k/\delta^* \approx 1$.

The increased amplitude of variation of the Pitot pressure Po corresponds to the range of azimuthal angles $\Delta\varphi$=90-190°, which is in agreement with disturbances in the mixing layer that resulted from the action of a localized microroughness.

The curvature of the jet boundary is not constant. The curvature changes its sign at the end of the first cell. According to available data, the sign and the value of the curvature of streamlines are determining parameters for the development of the Goertler-type disturbances [10]. An increase in curvature favors the amplification of this type of disturbances. This is confirmed by the comparison of the shape of the functions Po(φ) for the cross-section $x/R_a = 4.0$ for two jet pressure ratios np = 2.12 and 3.18 (Fig. 3,a and 3,b). The change of the shape of the curves Po(φ) with streamwise distance correlates with the gasdynamic structure of the initial section of a nonisobaric jet. The spectra of the azimuthal wave number is represented in [9].

3. Laminar-turbulent transition in the shear layer of supersonic jet

The conclusion that the generation and development of streamwise structures are conditioned by hydrodynamic instability in the regime of transition from the laminar form of the flow to the turbulent one by means of the formation of vortices of the Taylor-Gortler type is made on the basis of the analysis of the results of an experimental investigation of three-dimensional structure of the flow in a supersonic nonisobaric jet, as well as the available calculated data on the characteristics of stability of curved [11-12] and compressible [13] shear layers. The transition from the laminar form of the flow to the turbulent one in the jet shear layer depends not only on the characteristics of the flow stability, but on the level of the initial perturbations. It provides a possibility of passive control of the mixing process at the initial section of a supersonic nonisobaric jet.

A transitional flow regime from the laminar to turbulent state is observed in the shear layer of the initial section of a supersonic cold air underexpanded jet at Mach numbers of the jet flow $M_j = 1.0\text{-}3.0$ and Reynolds numbers $(1\text{-}10)*10^6$. The basis for this statement is the following observations:

- Laminar flow regime at the nozzle exit. It is confirmed by the measurement of the velocity profile in the immediate vicinity of the nozzle exit [8].
- Almost periodical structure of radiating acoustic waves conditioned by the Kelvin-Helmholtz instability is observed on the photos of supersonic jet with small exposure time.
- Regular structure of stationary Taylor-Gortler waves in the jet shear layer.
- High receptivity of the flow near the nozzle exit to the phenomena of various physical kind, namely:
    - The surface roughness at the nozzle exit [8, 9].
    - Substantial influence of acoustic situation near the nozzle exit on the flow evolution in the jet, the presence of discrete tone in the noise spectrum of the supersonic underexpanded jet [14].
- The results of an experimental investigation of additional mass of a supersonic jet issuing into the ambient space, which give grounds to state the presence of the laminar flow regime in the initial section of the jet at Reynolds numbers $5*10^6$ [15].

- Decrease in maximum values of the calculated growth rates of the Kelvin-Helmholtz disturbances with increasing both the flow Mach number and the relative density of the flow [13].

4. Conclusion

It is shown in the paper that the flow structure in the shear layer of the initial part of supersonic jet at large Reynolds numbers has a 3-D character. The appearance of 3-D character of the flow in the shear layer of a supersonic jet is associated with the development of hydrodynamic instability of the flow of the Taylor-Gortler type. An essential effect of roughness of the nozzle inner surface on the development of stationary streamwise vortex structures is shown. The thesis about the connection of the process of appearance of streamwise vortex structures in the shear layer of supersonic jet with the process of laminar-turbulent transition is stated. In turn, the understanding of physical mechanisms responsible for the transition from the laminar flow in the nozzle to the turbulent one in the compressible shear layer provides a means for controlling these processes and opens a possibility of intensifying the mixing in the initial section of a supersonic jet.

5. References

1. Zapryagaev, V.I., Solotchin, A.V. Spatial structure of flow in the initial section of a supersonic underexpanded jet, *Preprint No.23-88, Academy of Sciences USSR, Siberian Division, ITAM*, Novosibirsk, 1988, 28p., (in Russian).
2. Zapryagaev, V.I., Solotchin, A.V. Three-dimensional feature of the flow structure in a supersonic underexpanded jet, *J. Appl. Mech. Tech. Phys.*, **32**, No. 4, (1991), 503-507.
3. Krothopalli, A., Buzuna, G., Lourenco, L. Streamwise Vortices in an Underexpanded Axisymmetric Jet, *Phys. Fluids A*, **3**, No. 8, (1991), 1848-1851.
4. Arnette, S.A., Sammimy, M., Elliott, G.S. On streamwise vortices in high Reynolds number supersonic axisymmetric jets, *Phys. Fluids A*, **5**, No.1, (1993), 187-202.
5. Novopashin, S.A., Perepelkin, A.L. Axial Symmetry Loss of a Supersonic Preturbulent Jet, *Phys. Lett.A*, **135**, (1989), 290-293.
6. Teshima, K. Three-Dimensional Characteristics of Supersonic Jets, *Proceedings of 17-th Conf. on Rarefied Gas Dynamics*, Aahen, BRD, (1990), 1042-1048.
7. Zapryagaev, V.I., Mironov, S.G., Solotchin, A.V. Spectral compositiòn of wavenumbers of streamwise vortices and specific features of the flow structure in a supersonic jet, *J. Appl. Mech. Tech. Phys*, **34**, No.5 (1993), 634-640.
8. Zapryagaev, V.I., Solotchin, A.V. An experimental investigation of the nozzle roughness effect on streamwise vortices in a supersonic jet, *J. Appl. Mech. Tech. Phys*, **38**, No. 1 (1997), 78-86.
9. Zapryagaev, V.I., Solotchin, A.V. Evolution of streamwise vortices at the initial part of a supersonic nonisobaric jet in the presence of microroughnesses of the internal surface of the nozzle, *Fluid Dynamic* **32**, No.3 (1997), 465-469.
10. Saric, W.S. *Gortler vortices*, Annual Rev. Fluid Mech. **26** (1994), 379-409.
11. Liou, W.W. Linear instability of curved free shear layers, *Phys. Fluids* **6**, No. 2 (1994), 541-549.
12. Terekhova, N.M. Streamwise vortices in axisymmetric jets, *J.Appl. Mech. Tech. Phys*, **37**, No. 3 (1996), 339-349.
13. Jackson, T.L., Grocsh, G.E. Inviscid spatial stability of a compressible mixing layer, *J. Fluid Mech.*, **208** (1989), 609-637.
14. Antonov, A.N., Komarov, V.M., Kuptsov, V.V. *Pressure pulsations in a jet and separated flows*, Machinostroenie, Moscow, 1990, (in Russian).
15. Glotov, G.F., Gozhenko, G.F., Feiman, M.I. A study of the additional mass of underexpanded gas jets, *Trudy TsAGI, Iss.1377, TsAGI Publications*, Moscow (1971), 1-13, (in Russian).

PASSIVE CONTROL METHODS: RIBLETS, VORTEX GENERATORS AND SELF-ACTIVATING FLAPS

DIETRICH W. BECHERT
DLR
Turbulenzforschung, Institut für Antriebstechnik
Müller-Breslau-Str. 8
10623 Berlin

At the outset, the mechanism of *riblet* drag reduction is outlined [1]. This particular approach to turbulent wall shear stress reduction is about to be implemented on long-range commercial aircraft. A few issues in that context are discussed.

Besides riblets, other concepts for passive wall shear stress reduction have been suggested. The following devices have been tested in our oil channel:

(i) Walls with slits in the surface. Through the slits, fluid is ingested and expelled passively, driven by the fluctuating pressures of the turbulent boundary layer [1]. Experimentally, a shear stress reduction by this type of surface is not observed. Any beneficial effects are more than cancelled by the fact that the turbulent momentum transfer normal to the surface is enhanced and thus the wall shear stress is increased.

(ii) Walls with omnidirectional roughness (sand roughness, spheres) [2] and with unidirectional roughness (scratches). In our experiments with omnidirectionally rough surfaces, no shear stress reduction has been found. Actually, Tani's hypothesis [2] of drag reduction by rough surfaces boils down to a rearrangement of Nikuradse's experimental data which per se did not exhibit drag reduction either. On the other hand, scratches on the surface which were arranged in the streamwise direction did produce a marginal drag reduction in the order of 1%. These scratches, however, are assumed to operate like (poorly shaped) riblets.

(iii) Walls with "chevrons" arranged in an orderly or in a stochastic array [3]. Our oil channel tests with "chevrons" manufactured according to the dimensions given by Sirovich and Karlsson did not exhibit any drag reduction.

(iv) Walls with a belt surface which is moved passively by the wall shear stress itself in the streamwise direction (slip wall) [4]. Our estimations on the drag reduction potential of this novel device were not yet verified in the experiments. However, a drag reduction of about 10% did indeed occur.

Whereas the above devices are aimed at turbulent wall shear stress reduction, *vortex generators* are devices with the ability to suppress or delay separation. Vortex generators are widely and successfully used on the majority of existing aircraft. However, general appreciation among scientists is hampered by prejudices and by proprietary restrictions on general information. At present the situation is further aggravated by the increasing reliance on numerical prediction schemes which still lack the ability to predict the effects of vortex generators.

G.E.A. Meier and P.R. Viswanath (eds.),
IUTAM Symposium on Mechanics of Passive and Active Flow Control, 107–108.

The operation and few design guidelines of the most often used vortex generator devices are outlined. We rely here on the various pieces of information available in the open literature [5,6] and on own experiments. Whereas the use on aircraft and on wind turbines is fairly common, other applications as e.g., in turbomachinery are possible and indeed most promising.

A more recent development is the use of *self-activating flaps* on wings in order to delay separation. This device is related to the operation of the covering feathers on the upper surface of the wings of birds [7,8]. Once reversed flow occurs on the suction side of the airfoil, the flaps begin to rise. They limit the separation regime and reduce the camber of the wing at high angles of attack. With suitable flap designs, the maximum lift of an airfoil can be increased by as much as 10-20%. On the other hand, under cruise conditions with attached flow, the self-activating flaps also are attached and do not produce additional drag.

References

[1] D. W. Bechert, M. Bruse, W. Hage, J. G. T. van der Hoeven & G. Hoppe: *Experiments on drag-reducing surfaces and their optimisation with an adjustable geometry*. J. Fluid Mech. **338,** pp. 59 - 87, (1997)

[2] I. Tani: *Drag reduction by riblet viewed as roughness problem*. Proc. Japan Acad., **64**, Ser. B (1988)

[3] L. Sirovich & S. Karlsson: *Turbulent drag reduction by passive mechanisms*, NATURE, Vol. 388, pp. 753 - 755 (1997).

[4] D. W. Bechert, W. Hage, & M. Bruse: *Drag reduction with the slip wall*, AIAA Journal **34**, No. 5, pp. 1072 - 74 (1996).

[5] G. V. Lachmann (editor): *Boundary layer and flow control*, Pergamon Press, Oxford (1961)

[6] P. K. Chang: *Control of flow separation*, McGraw-Hill Book Company (1976).

[7] W. Liebe: *Der Auftrieb am Tragflügel: Entstehung und Zusammenbruch*, Aerokurier, Heft 12, pp. 1520-23, (1979)

[8] D. W. Bechert, M. Bruse, W. Hage & R. Meyer: *Biological surfaces and their technological application - Laboratory and flight experiments on drag reduction and separation control.* AIAA-Paper 97-1960 (1997).

NATURAL RIBLETS

A. Gyr

Institute of Hydromechanics and Water Resources Management,
Swiss Federal Institute of Technology Zurich, 8093 Zurich, Switzerland

1. Introduction

Near-wall turbulence is dominated energetically by a few large-scale, coherent flow structures which are responsible for up to 80% of the production of turbulent fluctuations in the boundary layer. According to observations, these structures are vortices which lie parallel to and close to the wall, and are oriented in the streamwise direction.

The coherent structures are observed to burst, transferring energy from the large to the small scales and producing turbulent fluctuations. Interfering with the dynamics of the coherent structures therefore provides a possible mechanism for the control of turbulence. A drag reducing process has to reduce the momentum exchange in time. This implies that drag reduction can occur even if the instability process ending up in a burst event is of higher strength as long as the occurrence of such instabilities decreases more drastically, or if the bursting occurs more often but the released momentum in the instability break-up is much lower. As we will see this is the case for natural riblets.

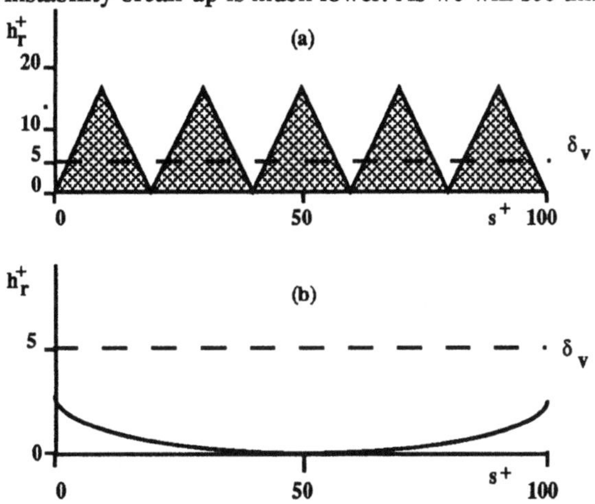

Figure 1. Sketch of a typical drag reducing V-groove riblet (a) and a natural sand pattern "riblet" (b) on the same scale. δv the boundary of the viscous sublayer.

The regions of low- and high speed fluid are not fixed in space and in time they cover the whole floor with the same probability. In the mean they have a spanwise length scale of about $\lambda^+ \approx 100$ in viscous units. ($\lambda^+ = l u_\tau / \nu$ with u_t the wall shear velocity, $u_\tau = (\tau_w / \rho)^{1/2}$ given by the wall shear stress, τ_w, and the fluid density, ρ.) Based on this observation it has been suggested that so-called *riblets* , longitudinal surface roughness elements, could be imprinted on a wall to stabilise these regions and produce drag reduction, e.g. Walsh (1990). This idea had its basis in the fact that a stabilisation of the low-speed

G.E.A. Meier and P.R. Viswanath (eds.),
IUTAM Symposium on Mechanics of Passive and Active Flow Control, 109–114.

regions of the near-wall turbulent boundary layer could suppress or postpone the growth of the secondary instability and therefore a drag reduction would result. Initially this was thought to be achievable with riblets with a width of the same order of magnitude as that of the streaky flow pattern, (Liu et al., 1966). Such widely spaced longitudinal grooves, hereafter termed 'low-frequency riblets' (LFR), are indeed drag reducing. However, it was found that the most effective groove spacing for drag reduction was a quarter of this low frequency spacing, see Bechert et al. (1997). These riblets with much smaller spacing are termed 'high frequency-riblets' (HFR). The result that the high-frequency riblets produce a higher drag reduction than low frequency ones apparently seems to contradict the stabilisation hypothesis outlined above. Low-frequency type riblets exist as sand pattern.

2. Formation of riblets by sediment transport

Erodible material on river beds is largely exposed to a turbulent flow of boundary layer type. The riblet like streaky forms which are drag reducing are a dynamic response to this interaction.

To apply ideas gained by smooth wall investigations require that we restrict ourselves to the investigation of flows over hydraulically smooth walls. That means the grains have to be so small that they are embedded in the viscous sublayer.

In cases where the bed consists of very fine sand, riblets can indeed be formed, and they belong to the class of the low-frequency riblets as expected, although this configuration can be understood only if the sweeps are spatially stabilised. It is this stabilisation which is of utmost interest for turbulence research, because the formation of stripes must be the result of a self-organisation based on a feedback mechanism between flow, sediment transport and the bed configuration. For sand of size $k^+<5$ riblet-like "roughness" can be observed to form, Fig. 2 on a flat bed at $Re_d = 2.56$.

To understand the formation mechanism it is first necessary to know how grains are transported by the flow.

Figure 2. A sand-bed with its typical riblet like roughness. $u_\tau = 13$ mm/s.

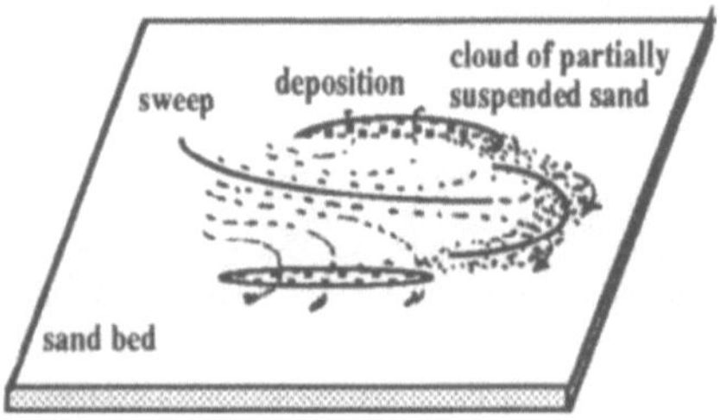

Figure 3. Sketch of the interaction of a sweep with a sand bed.

When a sweep, consisting of a patch of spanwise limited fast outer fluid, hits the bed, it is decelerated at the bed and produces very concentrated vorticity at its front edge. This vortex core has a high underpressure necessary for the sediment transport; it can be observed that sand is transported in a patchy fashion as shown in Fig. 3. The streaky sand pattern on the bed originates from settled grains which are transported in the side-edge vortices of the sweep in which they are embedded. The vorticity pushes the grains down towards the bed at a distance of a core diameter of the vortex. This process ends

after the sweep has decelerated. The stripes have a height of about one sand grain and a transverse extent of a few sand grains.
It is therefore not surprising that the streaks scale with the mean scale of the lateral distance of the burst events but it is surprising that they become fixed in space because the next sweep will hit the bed in an area where sand grains were already deposited by the preceding sweep event. Indeed, the appearance of a streaky sand pattern starts as a meandering pattern. The stabilisation in aligned sand streaks is explained in terms of a stabilisation of the sweeps in space.

3. Experiments and methods of analysis

Experiments were conducted in a rectangular open channel, see Schmid (1985). The bottom was covered with uniform quartz sand of mean diameter 0.2 mm and a density, ρ_s, of 2.65×10^3 kg/m^3. The initial configuration was a flat bed. LDA- velocity measurements were carried out at about 200 hydraulic radii downstream of the inlet contraction. Uniform flow for the individual experiments was maintained over the length of the flume. Measurements were taken under the condition: J= 0.06%, h= 39.5mm, Q= 1.95 l/s, ν= 1.01 mm^2/s, Re= 27086 and u_τ= 13 mm/s at which riblets could be observed.
Sediment transport rate was measured at the end of the channel by weighing the grains caught in a sediment trap at regular time intervals. It was found that the transport reached a stable value for a given slope after a short time. It was therefore assumed that the value found in this way could also be used as a good estimate for the local transport.
The error in measuring the bed shear velocity leads to a considerable uncertainty in the value of u_τ. This is a well known problem for all open channel flows and has been discussed e.g. Gust & Southard (1983). In this investigation, u_τ was evaluated from the flow by four different methods, and the value of u_τ used in the flow characterisation in the riblet regime was 13 mm/s, the average of the values obtained.
The friction factor f and the Reynolds number Re_Q are defined as

$$f = 8\frac{u_\tau^2}{u_Q^2}, \quad and \quad Re_Q = \frac{u_Q h}{\nu}$$

with u_Q the mean volumetric flow velocity (u_Q= Q/A).

4. Results

The main result is that during the time that the sand is transported by sweeps and producing nearly stable longitudinal stripes, the friction factor f has a minimum value This regime is thus drag reducing as shown in Fig. 4. The drag reduction can occur due to the solid transport, or the sand stripes alter the turbulent production. However, since the pattern and the transport are present simultaneously, and, the transport is necessary to maintain the pattern, these two processes cannot be investigated separately. One possible way to address this is to study the transition from one regime to the other. From regime (a) to (b) in Fig. 4, the transport starts from zero and the bedform changes simultaneously, and this transition thus provides no specific information. The transition from (b) to (c) in Fig. 4, however, associated with a dramatic increase of f, is helpful in deciding which mechanism is relevant. Measurements of the grain transport showed that the sediment transport gradually increases and shows no step-like increase at the boundary from (b) to (c). What changes suddenly is the observed pattern.

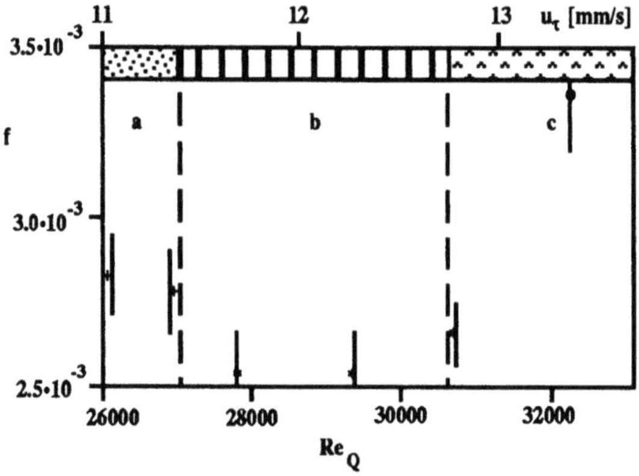

Figure 4. The friction coefficient f as a function of Re_Q in the range of (a) a stable orange-peel pattern, (b) a streaky pattern, and (c) a fully developed arrow-like pattern.

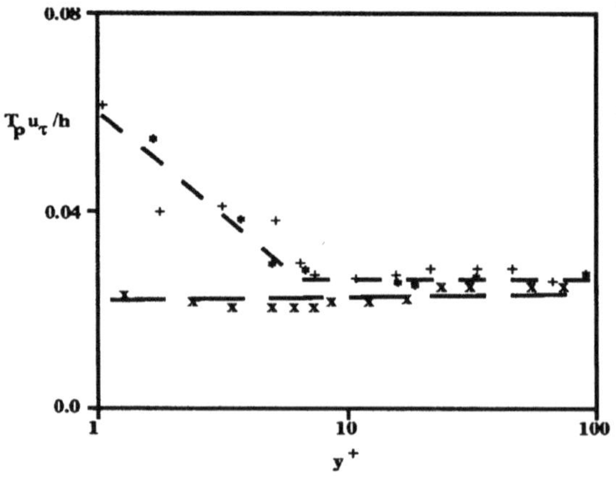

Figure 5. Mean periods Tp of the sweeps as a function of wall distance for the smooth wall at Re= 29200 (*), the „orange peel" at Re= 20400 (+) and the "riblet" bedforms at Re= 27100 (x) in mixed parameters.

The sediment transport as well as the longitudinal vortices are part of the same event, a decelerating sweep. The sand grains are injected into the fluid at the front, and only there, see Fig. 3, and they settle in the area where fluid is transported towards the bed by the developing longitudinal vortices. Therefore the stripes have to have the same spacing as the longitudinal vortices, although, the vorticity of these longitudinal vortices is too small to move the grains from the bed. Vice versa the explanation why the sweeps become spatially fixed stems from the instability process responsible for the sweep production. The streaky form is due to a channelization of the sweeps, which, when lost, leads to a higher drag.

Let us define the time between two consecutive appearances of sweeps as the' period' of a sweep. The period of the sweeps is normalised by mixed co-ordinates (u_τ,h), $T_P^* = T_P u_\tau / h$. In these co-ordinates T_P^* for a sand-riblet covered wall is, close to the wall, only about a third of the value found in the case without sediment transport, see Fig. 5.

The mean duration time of a sweep, T_B, is of the order one half to one third of T_P. The duration time together with the angle of impact of the sweeps with the bed are relevant since the time of interaction and the vorticity production during this time are responsible for the transport. The angle of impact is not changing, therefore the magnitude and the period of occurrence of the burst are responsible for the drag reduction. In other words, the intermittency of the sweeps, shown in Fig. 6 have to be studied. In case of the existence of the natural riblets the intermittency of the sweeps has the same behaviour as for a flow on smooth beds. They have a much shorter intermittency than otherwise. With respect to the much shorter period of occurrence one can conclude that the number of sweeps reaching the bed is much larger than under flat bed conditions.

In a more statistical sense this shows up in the momentum exchange by the sweeps as represented by the Reynolds stresses in the fourth quadrant. These values show that in case that riblet forms the momentum exchange of sweep decreases in relation to the actual wall shear stress. We can speak of a drag reduction, and since the momentum exchange by this kind of events is distributed over a higher number of them we conclude that the strength of the single burst events has decreased.

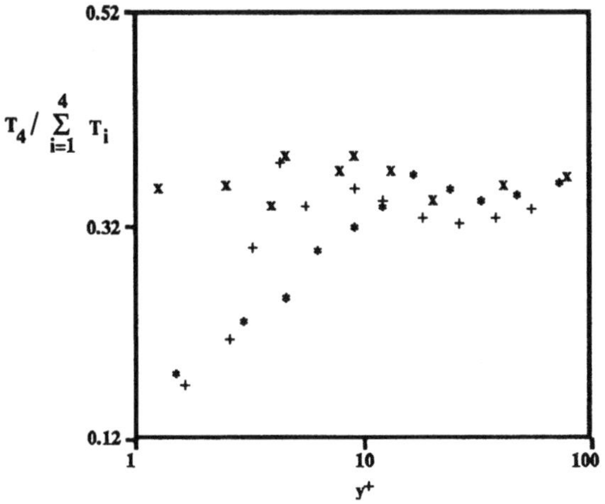

Figure 6. Intermittency factor of the sweeps as a function of wall distance for the smooth wall at Re= 29200 (*), the „orange peel" at Re= 20400 (+) and the "riblet" at Re= 27100 (x) bedforms.

5. Sediment and drag reduction

The spanwise distance of the streaks, s^+, is about 100, Fig. 1, and agrees with λ_z^+ of the near-wall coherent structures. These values can be compared with the values of high frequency symmetric V-groove riblets, e.g. Savill (1990). Drag reduction occurs in a range of $s^+<50$ and an h_r^+ of about 0.4- 0.8 s^+. The h_r^+ of natural riblets is about one half the value, and s^+ about five times larger than that of HFR. The ratio h^+/s^+ for natural riblets is about 0.05, which is an order of magnitude smaller than for the artificial ones.

Drag reduction by HFR has been explained, due to a mechanism inhibiting spanwise motions and influencing the near-wall turbulent structure, Luchini et al. (1991). The spanwise distance of natural riblets suggests that they probably could be due to a mechanism as proposed by Kline and mentioned in the introduction. However, the suggested concept was that the bursting mechanism becomes more stable which contradicts the findings that the number of bursts increases.. Based on the found stabilisation of the flow structures in space it can be concluded that the sweeps become channelled and drag reduction has to be explained by this process. The stabilisation of this events is thought to be the result of a stabilisation of sand streaks due to a feedback system which organises the sand deposition in such a quasi stationary pattern. The self organisation of the coherent flow structures and the bed forms is analogous to the explanation given by Studerus (1982) for stripes forming by secondary motion, however, the structural influence is based on a viscous effect instead of a roughness effect.

The instability cycle starts with the remnant of the previous sweep consisting of two very long counterrotating side vortices. They belong to a horseshoe like structure as shown in Fig. 3 from which the sand is deposited and the vorticity of the front loop is almost annihilated. Beneath the longitudinal vortices vorticity of the opposite sign is produced at the wall. By vorticity annihilation the vorticity distribution in the longitudinal vortices becomes inhomogeneous, their centres move upwards and towards each other, see Fig. 7. By induction the vortices are lifted and approaching each other by squeezing the upwelling slow fluid which produces a very local area in which the mean velocity profile becomes inflected and unstable. This produces a burst which is the source of the next sweep.

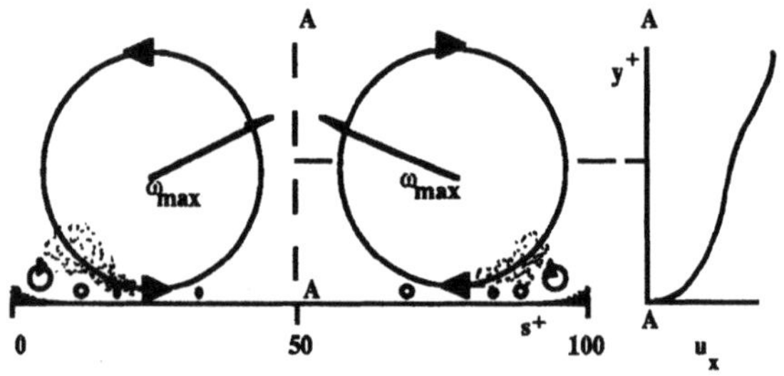

Figure 7: Sketch of the stabilisation of the sweep process by a channelling due to the deposition of sand in streaks. On the streak side, vorticity of the longitudinal vortices is annihilated and therefore the vorticity distribution in the vortices made inhomogeneous. The centre of the vorticity is shifted upward and towards the counterpart producing a well defined location at which an inflection instability of the mean velocity profile occurs. The lateral position of this location becomes stabilised and therefore also the resulting sweeps.

Two mechanisms help to stabilise the sweeps in spanwise direction: 1. The sand streaks help to localise the instability zone more pronounced in the centre between the longitudinal vortices. 2. The selfinduction of the vortex pair dominates the induction by the neighbouring one.

The faster up lifting of vortex pairs results in a faster burst sequence, which was found, see Fig 5. The channelization of the instability process is equivalent to an averaging process of the momentum transport producing a lower u_τ and therefore a lower friction factor.

6. References

Bechert, D.W., Bruse, M., Hage, W., van der Hoeven, J.G.T. & Hoppe, G. (1997) Experiments on drag reducing surfaces and their optimization with adjustable geometry. J. Fluid Mech. **338**, 59-87.

Gust, G. and Southard, J.B. (1983) Effects of weak bed load on the universal law of the wall. *J. of Geophysical Res.* **88**, 5939-5952.

Liu, C. K., Kline, S. J. and Johnston, J. P. (1966) An experimental study of turbulent boundary layers on rough walls. *Thermoscience Div., Mech. Eng. Dept., Stanford Univ.*, Rep. **MD-15**.

Luchini, P., Manzo, F. and Pozzi, A. (1991) Resistance of a grooved surface to parallel flow and cross flow. *J. Fluid Mech.* **228**, 87-109.

Savill, A. M. (1990) Drag reduction by passive devices- a review of some recent developments. In: *Structure of turbulence and drag reduction* (Ed. by A. Gyr), pp. 429-465.Springer Verl. Berlin, Heidelberg, NY.

Schmid, A. (1985) Wandnahe turbulente Bewegungsabläufe und ihre Bedeutung für die Riffelbildung. *Ph. D. Thesis ETH Zürich.* **7697** or Rep. R22-85 Inst. Hydromech. and Water Res. Management ETH Zürich.

Studerus, X. (1982) Sekundärströmungen im offenen Gerinne über rauhen Längsstreifen. *Ph.D. Thesis ETH Zürich* **7035**, or Rep. R19-82 Inst. Hydromech. and Water Res. Management ETH Zürich.

Walsh, M. J. (1990) Riblets.Viscous drag reduction in boundary layers. In: *Viscous drag reduction in boundary layers* (Ed. by D. M. Bushnell and J. N. Hefner), *Progr. in Astronautics and Aeronautics*, **123**, 203-261.

THE FLOW OVER RIBLETS: VELOCITY MEASUREMENTS WITH HOT-FILM PROBES

M. BRUSE, D.W. BECHERT AND W. HAGE

The German Aerospace Center (DLR), Institute of Propulsion Technology, Department of Turbulence Research, Mueller-Breslau-Str. 8, D-10623 Berlin, Germany

The fact that riblets do reduce shear-stress has been established for more than ten years beyond any reasonable doubt. A plausible explanation is, that the longitudinal ribs rectify the turbulent flow in mean flow direction by hampering the fluctuating cross-flow velocity component w'. In a turbulent boundary layer the cross-flow w' and the flow component normal to the wall v' are connected through the burst and sweep phenomena. If the cross-flow fluctuation w' close to the wall can be reduced, the turbulent momentum transfer close to the surface will be reduced as well and consequently, the shear stress $-\rho \cdot (\overline{u'v'})$ will be decreased.

The optimization experiments result in the optimal geometry of very thin blade-ribs (see. Fig. 1 and Refs. [2], [3] & [6]). In this case the maximum shear-stress reduction is almost 10%. However, a technologically more suitable design are riblets with a trapezoidal cross-section of the grooves (8.5%).

Assuming a viscous flow over the riblets, the theory has shown that the origins of the velocity profiles of longitudinal- and cross-flow are located underneath the rib tips (see Fig. 2 and Refs. [1] & [8]). The vertical distance to the rib tips (the protrusion height difference Δh_p) seems to affect the gradient of the shear-stress data $\Delta\tau$ for small s^+-values [$\Delta\tau = (\tau_{riblet} - \tau_{ref.})/\tau_{ref.}$ & $s^+ = s \cdot u_\tau/\nu$ with s ≡ lateral rib spacing] (see Figs. 1 & 2). A higher gradient leads to more shear-stress reduction. With increasing s^+ the ribs start to protrude out of the viscous sublayer and the riblet-surface starts to act like a rough one.

To check the theory we wanted to measure the velocity profiles over the rib tip and on the centerline of a groove. In the Berlin oil channel (Ref. [4]), where all the experiments were carried out, the lateral rib spacing is in the order of some millimeters. That is why we can use commercial

G.E.A. Meier and P.R. Viswanath (eds.),
IUTAM Symposium on Mechanics of Passive and Active Flow Control, 115–120.

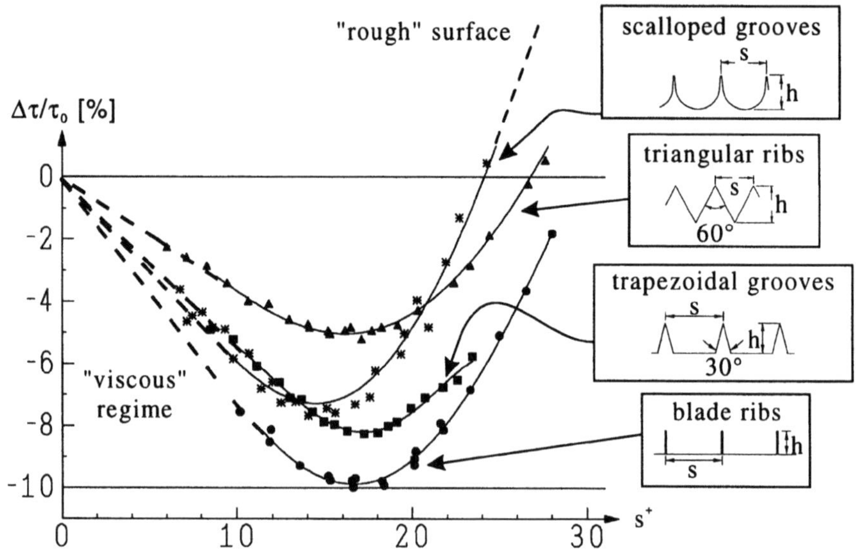

Figure 1. Shear-stress data for different riblet geometries

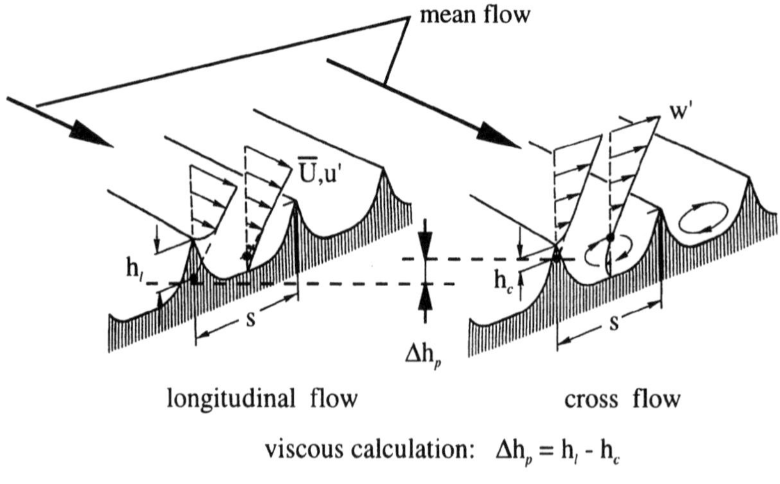

Figure 2. Velocity profiles over riblet structured surface

hot-film probes for the velocity measurements very close to the surface. When we started these measurements we tried to use hot-wires, but due to the high roughness of the wires small dirt particles (diameter < $1\mu m$) stuck to the wire and deteriorated the calibration. By contrast, using hot-film sensors, the time for this problem to occur is long enough to take data. A cleaning mechanism consisting of a small tube was installed several millimeters downstream of the sensors. It generates an oil jet which cleans the sensors during the acquisition breaks.

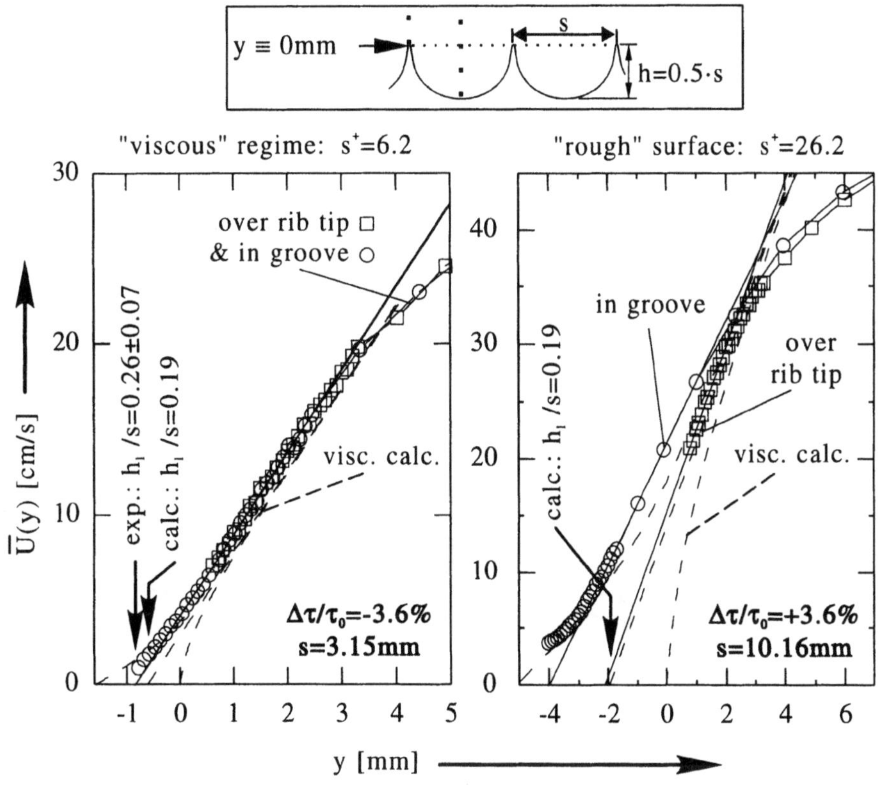

Figure 3. Velocity measurements over riblets for small and high s^+-values

In addition to the velocity calibration of the sensors we had to carry out an angular calibration of the sensors. This is because the probes are installed slightly slanted to the wall to take data for small wall or riblet distances. For the angular calibration a nozzle generating an oil jet (diameter 20mm) of constant speed moves in a circle around the sensors. After having solved the above measuring problems, data could be collected.

For low s^+-values a uniform gradient of the mean-flow velocity ($U \propto y$) can be observed (see Fig. 3, left). Here the ribs are obviously submerged in the viscous sublayer. The protrusion height calculated with this gradient agrees within the error bars of the measurement with the protrusion height of the viscous calculation ($h_{l,visc} = 0.192 \cdot s$; see Figs. 2 & 3 and Ref. [1]). To obtain a more accurate protrusion height measurement a more precise determination of the wall distance would be necessary. However, the finite size of the sensitive volume of the probes produces an uncertainty of the same 0.2mm order.

For higher s^+-values (see Fig. 3, right) the rib tips protrude out of the viscous sublayer and no locally constant gradient comparable to the one of the viscous calculation can be found. Actually, this situation corresponds to the conditions of "rough" surfaces.

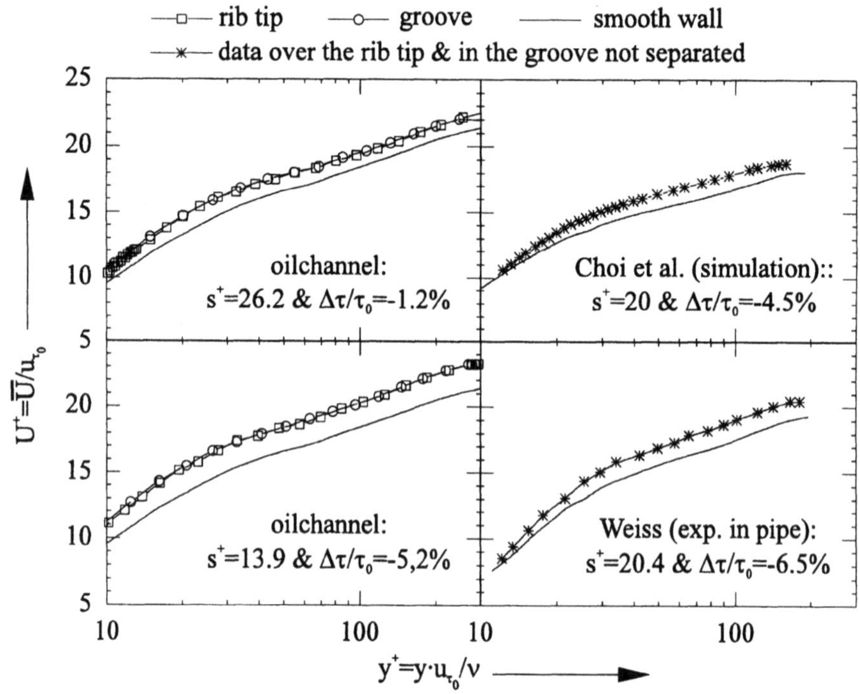

Figure 4. Mean-flow velocity U^+ over riblets and a smooth wall in the regime of the logarithmic law

Plotting the mean-flow velocity and the wall distance in wall units ($U^+ = \overline{U}/u_{\tau_0}$ and $y^+ = y \cdot u_{\tau_0}/\nu$), it is obvious that in the regime of the logarithmic law the velocity profiles over the riblets in comparison to the ones over the smooth wall are slightly shifted away from the wall (see

Fig. 4). This was also found by Choi et al. (Ref. [7]) in a simulation and by Weiss in a pipe-flow experiment (Ref. [9]). In our measurements, data both above the rib tips and on the centerline of the grooves have been collected. Our explanation for the shifted profiles in the regime of logarithmic law is, that the riblets, when they are submerged in the viscous sublayer and do reduce the wall shear-stress, increase the thickness of the viscous sublayer. This idea is supported by the fact that the linear gradient in the logarithmic law is not changed for the flow over a smooth wall or a riblet structured surface, respectively.

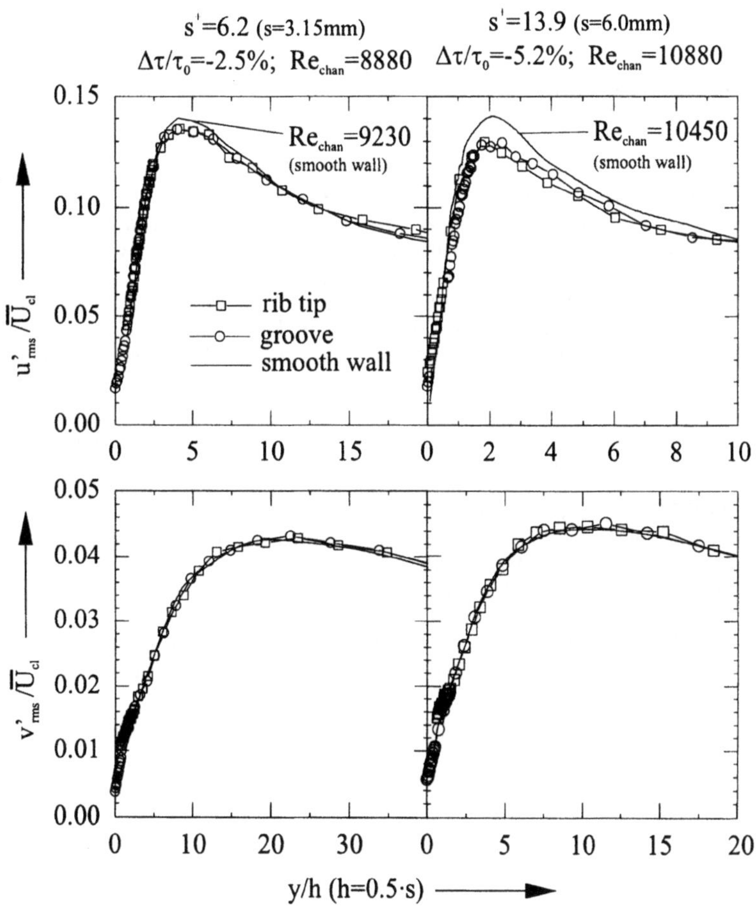

Figure 5. Fluctuating velocity over riblets and a smooth wall in mean-flow direction u' and normal to the wall v' close to the surface

With the hot-wire probes which we used so far we only could carry out measurements of the fluctuating velocities in the mean-flow direction (u')

and normal to the wall (v'). Like for the mean-flow velocity data, in the region far away from the wall the velocity profiles are nearly not affected by the riblets. But close to the wall the fluctuations in the mean-flow direction are reduced when the surface is structured with riblets (see Fig. 5, upper part). However, the fluctuating velocity normal to the wall is not affected by the ribs (see Fig. 5, lower part). From theoretical considerations we also would expect a decrease of the fluctuating velocity normal to the wall. On the other hand we do not know how much the velocity fluctuations are affected by the ribs and if the hot-film technique (or any other velocity measurement) is accurate enough to detect those variations.

References

1. Bechert, D.W. & Bartenwerfer, M. (1989) The viscous flow on surfaces with longitudinal ribs, *J. Fluid Mech.* **206**, pp. 105–129.
2. Bechert, D.W., Bruse, M., Hage, W., & Meyer, R. (1997) Biological surfaces and their technological application – laboratory and flight experiments on drag reduction and separation control, *AIAA paper.* **97–1960**.
3. Bechert, D.W., Bruse, M., Hage, W., van der Hoeven, J.G.Th. & Hoppe, G. (1997) Experiments on drag-reducing surfaces and their optimization with an adjustable geometry, *J. Fluid Mech.* **338**, pp. 59–87.
4. Bechert, D.W., Hoppe, G., van der Hoeven, J.G.Th. & Makris, R. (1992) The Berlin oil channel for drag reduction research, *Exp. in Fluids* **12**, pp. 251–260.
5. Bruse, M. (1998) Zur Strömungsmechanik wandreibungsvermindernder *riblet* Oberflächen (1998), Dissertation at the Technical University Berlin (Dept. 10, Transport and Applied Mechanics).
6. Bruse, M., Bechert, D.W., van der Hoeven, J.G.Th., Hage, W. & Hoppe, G. (1993) Experiments with conventional and novel adjustable drag-reducing surfaces, in: *Near-wall turbulent flows* (eds. R.M.C. So, C.G. Speziale & B.E. Launder), pp. 719–738, ELSEVIER, Amsterdam.
7. Choi, H., Moin, P. & Kim, J. (1993) Direct numerical simulation of a turbulent flow over riblets, in: *J. Fluid Mech.* **255**, pp. 503–539.
8. Luchini, P., Manzo, F., & Pozzi, A. (1991) Resistance of a grooved surface to parallel flow and cross-flow, *J. Fluid Mech.* **228**, pp. 87–109.
9. Weiss, M.H. (1993) Drag reduction with riblets in a pipe flow, Ph.-D. thesis at the University of Calgary (Canada), Department of Mechanical Engineering.

NUMERICAL INVESTIGATION OF FLOW AROUND A VENTED SPHERE

A. KUMAR[1] and G. K. SURYANARAYANA[2]
[1] *CSIR Centre for Mathematical Modelling & Computer Simulation, Bangalore 560 037 India*
[2] *National Aerospace Laboratories, Bangalore 560 017 India*

Abstract

Numerical simulations of flow past a vented sphere have been presented. Interesting interactions between the external shear layer and the vent-jet are reported. Computations using a $k-\epsilon$ turbulence model at supercritical Re show that there is no drag reduction due to venting, consistent with the experimental findings with a tripped boundary layer. Features of the near-wake flow leading to some understanding of the vortex shedding at a subcritical Re have been observed.

1. Introduction

Several attempts comprising passive and energetic methods have been made in the past to control the flow past bluff bodies in order to achieve drag reduction. Most of these pertain to two-dimensional bluff bodies, such as circular cylinders and airfoils with truncated trailing edges. On a sphere, which represents an idealised three-dimensional bluff body, there have been only a few attempts. Further, even the flow past a basic (unvented) sphere has been much less investigated as compared to the flow past a circular cylinder, experimentally or computationally.

Meier et al. [1] have introduced the concept of natural ventilation to reduce the pressure drag of a sphere. When the stagnation and the base regions of a sphere are interconnected through an internal duct, mass, momentum and energy are passively added into the near-wake. Suryanarayana et al. [2-4] have shown that the effects of ventilation over a wide range of Reynolds numbers Re can be broadly classified under two regimes, viz., strong (supercritical) and weak (subcritical) interaction regimes. A substantial drag reduction of as much as 65% was measured in the strong interaction regime, caused by the increased base pressure, stablisation and

G.E.A. Meier and P.R. Viswanath (eds.),
IUTAM Symposium on Mechanics of Passive and Active Flow Control, 121–126.

symmetrisation of the near-wake flow when the external boundary layer on the sphere had a natural transition. The concept has several interesting applications, including in the improvement of stability characteristics of a descending parachute [5]. However, when the external boundary layer was tripped, there was practically no reduction in the drag.

The present paper investigates the flow past a vented sphere using numerical simulations of Navier-Stokes equations. It must be noted that even though the flow past a sphere at high Re is known to be three-dimensional (e.g., see [6]), axisymmetry of the flow has been assumed in order to simplify the solutions. The simulations have been done using a $k - \epsilon$ turbulence model; this enables comparison of the present results with the flow past a vented sphere with a tripped boundary layer and not a natural flow. Notwithstanding these limitations, an attempt has been made to gain insight into the mechanics of the flow past a vented sphere.

2. Results and Discussion

Flow around a sphere of diameter $D = 200\ mm$ having a venting duct of diameter $d = 30\ mm$ has been considered; the projected area of the vent thus being 2.25% of that of the sphere. These dimensions correspond to those of the experimental model [2-4]. We have employed three grids, denoted as, medium, standard and fine. The standard grid consists of a 201×51 grid for the domain outside the sphere, with a 61×21 grid for the duct region. Similarly, the fine grid consists of a 301×76 grid and a 91×31 grid, and the medium grid consists of a 121×41 grid and a 41×21 grid. The outer boundary of the computational domain is located at $15D$. Computations at a supercritical Re of 0.85×10^6, corresponding to free-stream speed of $55\ m/s$, have been carried out using compressible Navier-Stokes equations with $k - \epsilon$ turbulence model. Computations at a subcritical Re of 2000 have employed laminar incompressible Navier-Stokes equations. The computations have been carried out in a time accurate manner, using a code described in [7].

2.1 SUPERCRITICAL Re FLOW-FIELD

Features of the flow at the supercritical Re of 0.85×10^6 are shown in Figs. 1-4. The effect of the vent is felt from about $0.1D$ upstream of the sphere; the near-axis flow begins to accelerate as it enters the duct. Some of the surrounding fluid also enters the duct due to the extremely favourable pressure gradient near the entrance, thus forming a stagnation ring above the vent entrance. Fig. 1 shows the flow-field close to the entrance for two entry shapes, viz., a sharp edge and a bell-mouth. Separation of the flow due to the sharp entry (Fig. 1a) can be avoided when the entry is rounded (Fig. 1b). The vent mass-flow coefficient, C_μ, defined as the mass-flow through the vent divided by $\rho_\infty u_\infty \pi d^2/4$, is 0.8253, 0.9785 and 0.9747 for rounding radii of 0, 1.5 mm and 3 mm, respectively. The reduced value of C_μ for the sharp

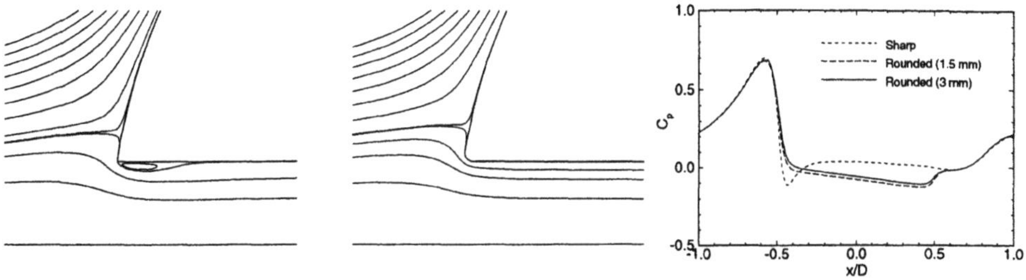

Fig. 1: Flow near (a) a sharp, and (b) a bell-mouth (1.5 mm) vent entrance, and (c) comparison of C_p on the axis; $Re = 0.85 \times 10^6$, Medium grid

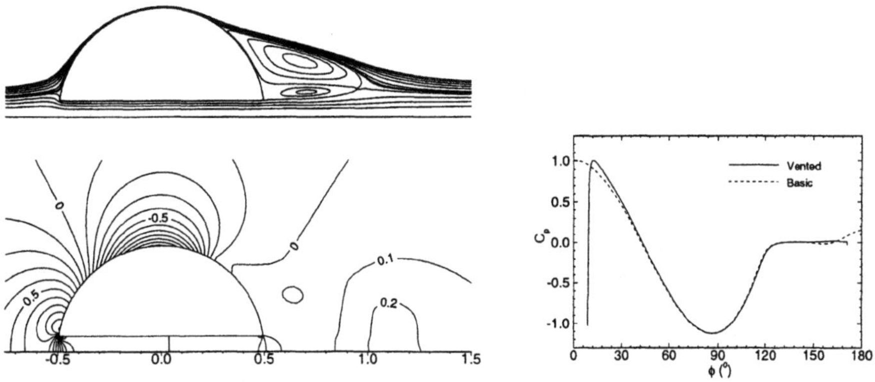

Fig. 2: (a) Flow-field, and (b) C_p contours around the vented sphere, and (c) C_p on the sphere, Fine grid

entry is due to flow separation. Certain variations caused by the sharp entry are seen inside the duct. For further computations, a radius of 1.5 *mm* at the vent entrance and exit has been chosen.

As shown in Fig. 2a, the jet emerging from the base gets partially reversed due to the main clockwise rotating vortex. It reattaches on the sphere, forming an anti-clockwise rotating vortex. The partially reversed jet-flow passes between the sphere and the main vortex, joins the shear layer separating from the sphere and reattaches on the vent-jet. This gives rise to a saddle point in the near-wake. The main vortex is thus *not* wall-bouded. Figs. 2b-c show the pressure contours and the pressure distribution around the vented sphere. An increase of the base pressure due to venting which can reduce the drag is *not* indicated (Fig. 2c). In the experiments [3-4], a significant base pressure rise occured when the external boundary layer had a natural transition (featuring laminar boundary layer separation, transi-

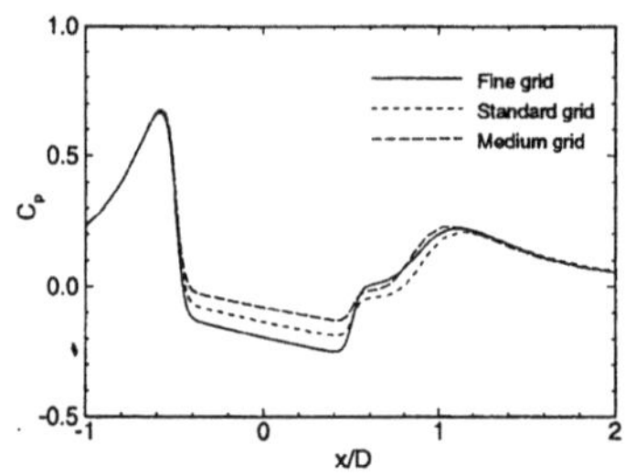

Fig. 3: C_p on the axis

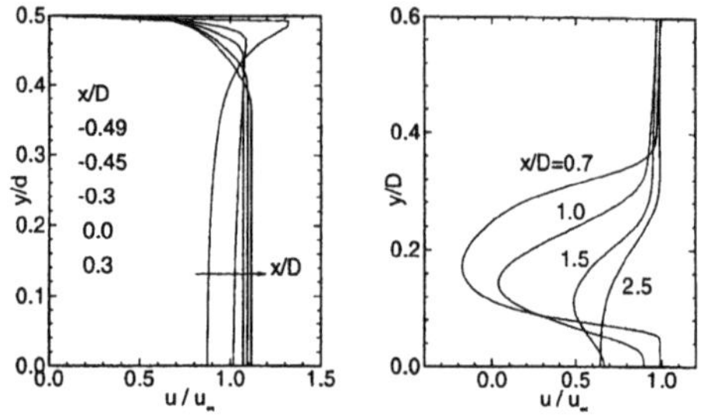

Fig. 4: Velocity profiles in (a) the vent, and in (b) the wake

tional/turbulent shear layer reattachment and an ultimate turbulent boundary layer separation), with the ultimate separation occuring around 155° at $Re = 0.8 \times 10^6$. However when the flow was tripped, there was no significant reduction in drag. In the present simulations also, which involve fully turbulent flow, there is no base pressure rise due to venting. A suitable modelling of the boundary layer transition is thus essential to examine the various aspects of drag reduction.

Fig. 3 shows variation of pressure along the axis on the three grids. The pressure is seen to drop linearly in the vent. This linear slope $dC_p/d(x/D)$ is -0.154 (fine grid) as compared to the measured [4] value of -0.1808 at $Re = 0.653 \times 10^6$ for externally tripped flow with smooth internal duct. Fig. 3 also shows that flow is essentially well captured on the grids used.

The developments of the velocity profiles in the vent and in the wake are shown in Fig. 4. Profiles typical of a developing pipe-flow are evident in Fig. 4a. A boundary layer develops on the wall, which accelerates the near-axis flow in the vent. In the wake velocity profile at $x/D = 0.7$, which is approximately the location of the centre of the two recirculating regions, the plug-like vent-jet and a substantial reversed flow region may be noted. The near-axis flow first deccelerates and then accelerates to reach the free-stream condition.

To study the effect of vent divergence, the flow around a sphere with a 30-50 mm vent and with a 50-30 mm vent have been considered. While the external flow-field (Fig. 5) is found to be generally unaffected by the different vent configurations, substantial differences are seen in the pressure and velocity distribution inside the vent (Figs. 6-7). The diverging (30-50) duct shows a high acceleration near the vent entrance, which interestingly leads to a higher C_μ, compared to the converging (50-30) duct. The velocity profile in the ducts at $x = 0$ are shown in Fig. 7, d_o in the figure is the local vent diameter. The value of C_μ, based on the average diameter of 40 mm, for the two ducts are 1.115 and 0.570, respectively; in addition to higher acceleration, the divergence may provide relief against the axial boundary layer growth.

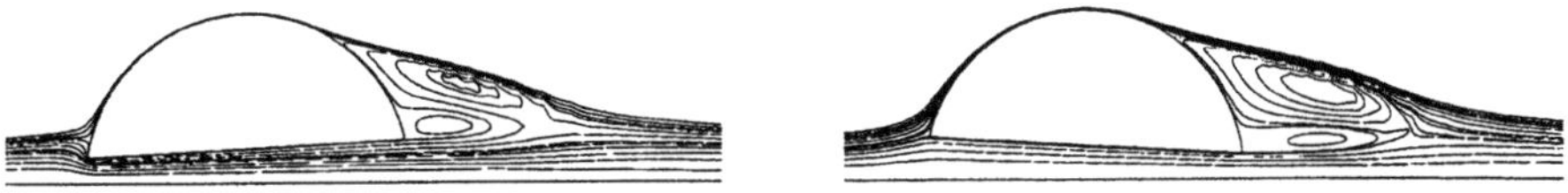

Fig. 5: Flow around (a) a 30-50, and (b) a 50-30 vented sphere; Standard grid

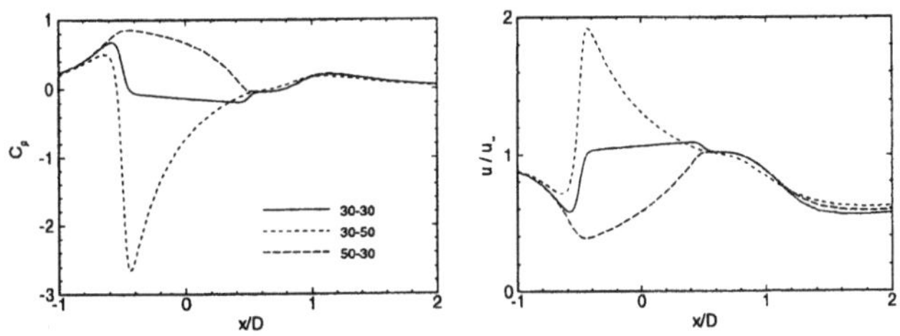

Fig. 6: (a) C_p, and (b) u on the axis for different vent configurations

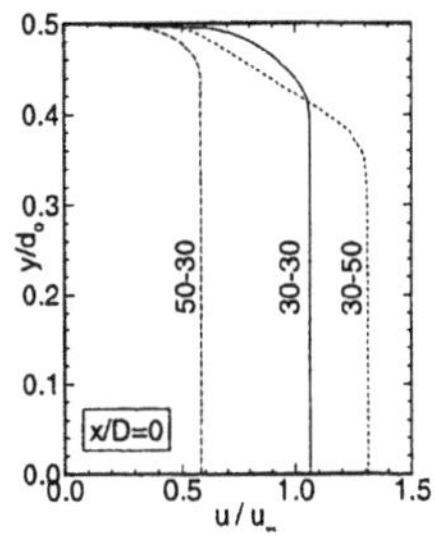

Fig. 7: Velocity profiles in the vent

2.2 SUBCRITICAL *Re* FLOW-FIELD

Computations at $Re = 2000$ show vortex shedding. Figs. 8 shows the unsteady flow at time interval $\Delta\tau = \Delta t u_\infty / D$ of 1. For the sake of presentation, let $\tau = 0$ for the first diagram in Fig. 8. At $\tau = 0$ the main clockwise rotating vortex, to be referred as C1, and a similar vortex, C2, near the separation point can be identified. Under the effect of C1 the vent-jet is partially reversed and passes between the sphere and C1-C2, thus forming a pair of anti-clockwise rotating vortices, A1 and A2. Saddle points between C1 and C2, between A1 and A2, and between C1 and the vent-jet can be seen. As the main vortex C1 grows the vent-jet gets further reversed; stagnation of the vent-jet at the axis can be seen from $\tau = 1$. The vortex A1 and A2 grow at first, see the flow $\tau = 1$, and then while A1 continues to grow, A2 diminishes and merges with A1, $\tau = 4$. This forces C1 to break away from C2, see the flow at $\tau = 5$. The flow at $\tau = 5$ further shows that a part of the shear layer from the sphere now passes around C2 and reattaches on the sphere. A part of the reattaching flow passes underneath A1, and takes part in the shedding process. The vortex C2 now takes the role of C1. As it grows a new A2 and a new C2, see the flow at $\tau = 6$, and a new A1, $\tau = 7$, are formed and the vortex shedding cycle continues. The time period of vortex shedding is approximately 7.5, or a Strouhal frequency of 0.133; the experimentally observed value at $Re = 6500$ being 0.157 [4].

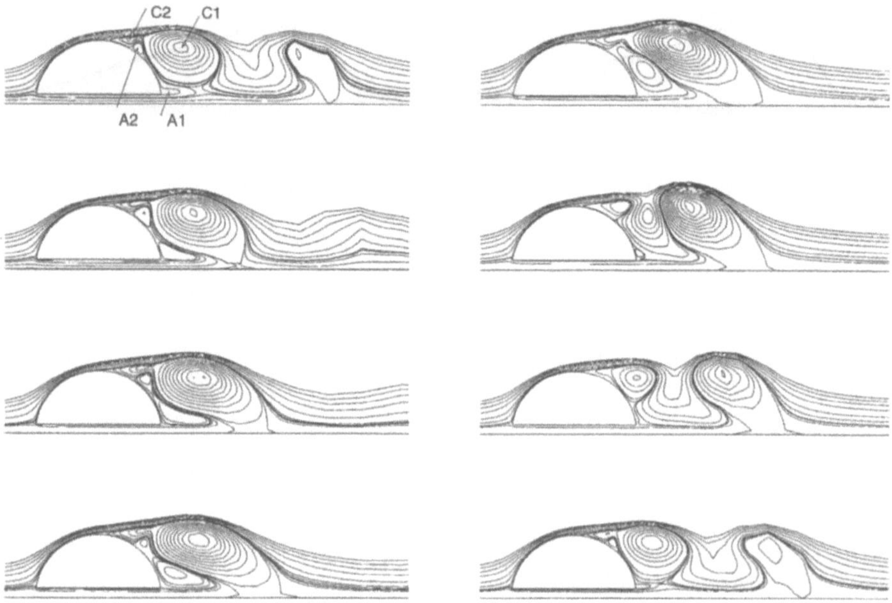

Fig. 8: Unsteady flow at time interval $\Delta\tau = 1$; $Re = 2000$, Standard grid

3. Concluding Remarks

The numerical simulations presented here highlight the interesting flow features associated with a vented sphere. Similar to the experimental observations with a tripped boundary layer at a supercritical Re, the computed flow, employing a $k-\epsilon$ turbulence model, was found to offer no drag reduction due to venting. Transition to turbulence appears to play an important role in drag reduction associated with natural flow and must be included in the numerical simulation to understand this aspect.

References

1. Meier, G.E.A. *et al.*, DGLR Bericht 90-06, pp 311-315, 1990
2. Suryanarayana, G.K. *et al.*, Expts. in Fluids, 16:73-81, 1993
3. Suryanarayana, G.K. *et al.*, Expts. in Fluids, 19:78-88, 1995
4. Suryanarayana, G.K., Ph. D. Thesis, I.I.Sc., Bangalore, India, 1996
5. Peterson, C.W. *et al.*, Ann. Rev. Fluid Mech., 28:361-387, 1996
6. Taneda, S., J. Fluid Mech., 85(1):187-192, 1978
7. CFD-ACE, Version 4.0, CFD Research Corporation, Huntsville, USA

DRAG REDUCTION OF A SPHERE AT SUPERCRITICAL REYNOLDS NUMBERS USING PASSIVE VENTILATION

M. LU, F.-R. GROSCHE, G.E.A. MEIER
German Aerospace Center (DLR), Institute of Fluid Mechanics, Bunsenstr. 10, 37073 Goettingen, Germany

Abstract:

This paper describes an experimental investigation of drag reduction of a sphere at supercritical Reynolds numbers up to Re = 4.5×10^6 using passive ventilation. This ventilation was created by a channel connecting the stagnation region with the wake region. Passive ventilation is a very effective method of decreasing drag of smooth spheres in the named Reynolds numbers range. A drag reduction of about 40%-60% can be achieved with a cross-section area of the venting channel of about 2% of the sphere. Pressure distributions on the sphere show clear changes of base pressure and point of separation. In addition, the flow on the vented sphere showed smaller oscillations at supercritical Reynolds numbers than those on the basic sphere. The reduction in oscillation amplitude came basically from a stabilization of the wake.

Flow visualization in a water tunnel was made for the basic and vented spheres. The visualization of the vented sphere showed more stabilization of the wake than that of the basic sphere.

1. Introduction

The flow around a bluff-body e.g. a sphere or a cylinder causes a stagnation region around the front and a low pressure wake region due to separation. Those bodies have a relatively high drag. The size of the separation region depends on Reynolds number, roughness and turbulence intensity. A drag reduction can be reached by reducing the separation region. More than 90% of the total drag of bluff-bodies is due to the separation [1].

The present concept of passive ventilation was first described in detail by Meier et al. [2]. Here a connection between the stagnation region and the wake region is established by one or many channels in order to change the wake structure, so that the drag can be reduced.

G.E.A. Meier and P.R. Viswanath (eds.),
IUTAM Symposium on Mechanics of Passive and Active Flow Control, 127–132.

G.E.A. Meier et al. [3] and G.K. Suryanarayana et al. [4] have reported on drag measurement in wind tunnel. The results show that the drag of a sphere can be reduced by passive ventilation to 50%-60% at supercritical Reynolds numbers. The previous measurement is up to a Reynolds number of 8.5×10^5.

2. Description of the Models and the Experimental Set-up

The drag and pressure measurements have been carried out in the high pressure tunnel of the DLR in Goettingen. The test section has dimensions 0.6 m × 0.6 m. The turbulence intensity is between 0.1% and 0.3% and depends on the Reynolds number [5].

The used model is a vented sphere made of alloy of aluminium. It has a diameter of 200 mm and the vent channel has a diameter of 30 mm (Figure 1). This offers a vented area of 2.25% of the projected area of the sphere. There were 30 pressure ports on the surface of the sphere. They are used for the pressure measurements. The sphere was asymmetrically supported by a stem at the position 150° from the stagnation point of the basic sphere. The basic sphere was simulated by closing the vent channel using two plugs conserving the curvature of the sphere at the entry and at the rear.

The drag force was measured using a 6 component build-in balance (Task 0.75´´ MK XL AS/N 17330) which has a range of 111 N. It was located in the support stem 280 mm behind the sphere. Cross- bracing wires with a ring without connection with the stem were placed in order to avoid oscillations (Figure 2).

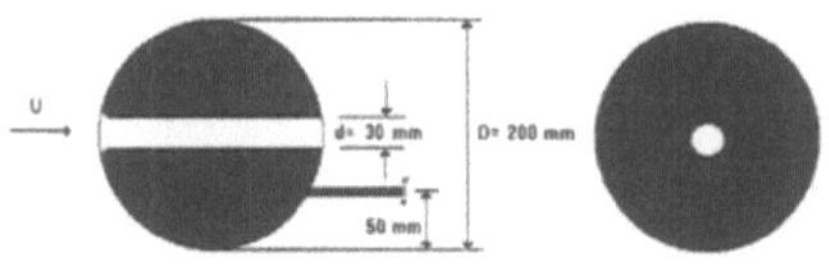

Figure 1. Schematic drawing showing vented sphere

Figure 2. A photograph of the sphere and the support

In water tunnel experiments a vented safety glass sphere of 180 mm diameter was used. It has a vent channel of 26 mm diameter. This provides a vented area of 2.1% of the projected area of the sphere. A basic sphere has the same diameter. The stem was placed at the same position as in the wind tunnel and has a diameter of 10 mm.

3. Results and Discussion

3.1 DRAG MEASUREMENTS

Figure 3 shows the drag coefficient versus Reynolds number. We see that the sphere with ventilation has the smaller drag coefficients. At subcritical Reynolds numbers the drag coefficient is reduced about 6% due to ventilation. In the supercritical range of Reynolds numbers the drag reduction is up to 63% until $Re \approx 6\times10^5$. In the range of $6\times10^5 \leq Re \leq 4.5 \times10^6$ the drag reduction is almost constant ca. 40%.

The Reynolds number range of transition does not seem to be influenced by ventilation.

A comparison between the present results and the results of Achenbach [1] and Bendig [6] is not meaningful since the latter measurements are all for spheres with symmetric supports. According to Raithby and Eckert [7], the support position plays a very important role beyond the critical Reynolds number.

Due to the strong oscillations the measurements have been stopped at the Reynolds number of 4.5×10^6.

3.2 PRESSURE MEASUREMENTS

There is little change of pressure distribution on the sphere by ventilation at the subcritical Reynolds number of 1.57×10^5 (Figure 4). Therefore there is a relative small drag reduction in the subcritical range of Reynolds numbers.

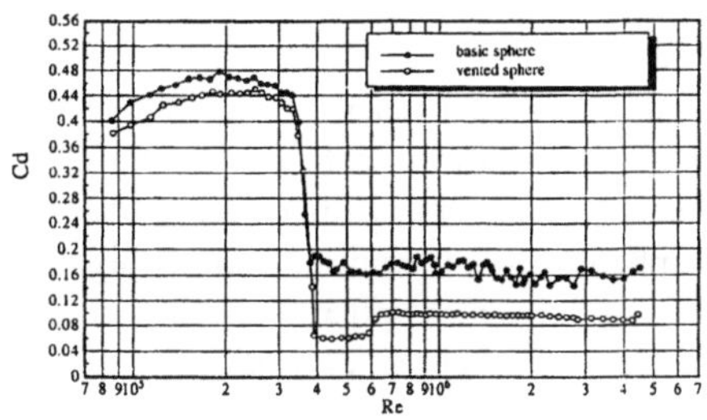

Figure 3. Drag coefficient versus Reynolds number of the vented and basic sphere

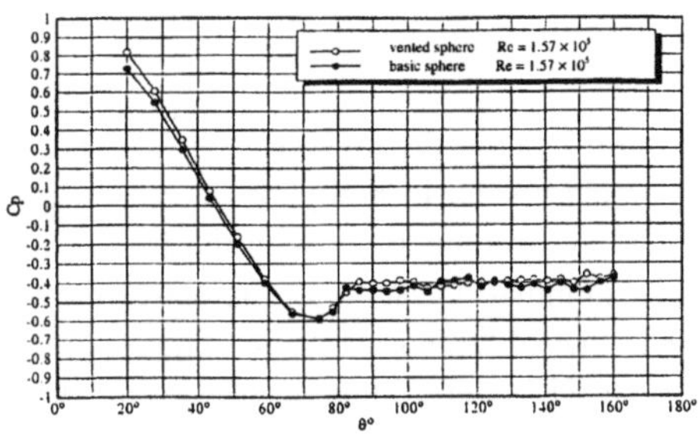

Figure 4. Pressure distribution around the vented and basic sphere at a subcritical Reynolds number ($Re = 1.57\times10^5$)

Figure 5 shows the pressure distribution on the vented and the basic sphere at the supercritical Reynolds number of about 5.1×10^5. Venting leads to slightly increased pressures on the windward side up to about 100°. Between 100°-110° the basic and the vented sphere show a laminar separation bubble. The reattachment line of turbulent boundary layer occurs at 110°. A rapid pressure recovery follows. The big difference of pressure distribution on the basic and the vented sphere concerns the base pressure. From 135° on the pressure coefficient of the basic sphere clearly decreases compared to that of the vented sphere. The turbulent boundary layer separation of the vented sphere stays at about 140°, and separation of the basic sphere stays below 135°. The present results are consistent with the measurements of G. K. Suryanarayana [8].

The pressure coefficients of the basic and vented sphere at Re $\approx 1\times10^6$ are shown in Figure 6. At all measurement positions on the sphere the pressure coefficient of the vented sphere lays above that of the basic sphere. The difference is small up to about 120°. The laminar separation bubble is hardly discernible. The difference of the base pressure between the basic and vented sphere is less than that at Re $\approx 5.1\times10^5$.

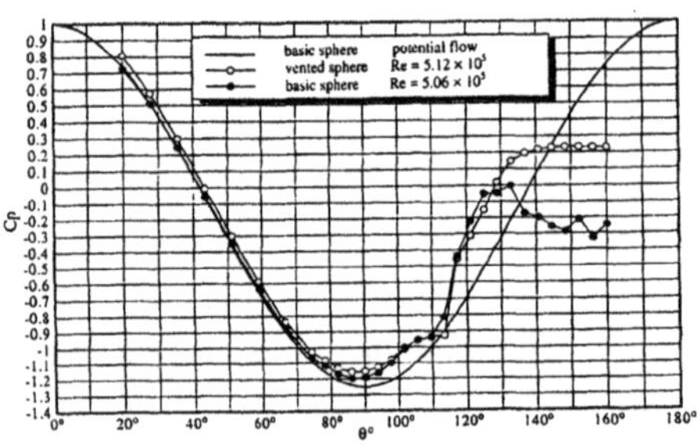

Figure 5. Pressure distribution around the vented and basic sphere at a supercritical Reynolds number (Re ≈ 5×10^5)

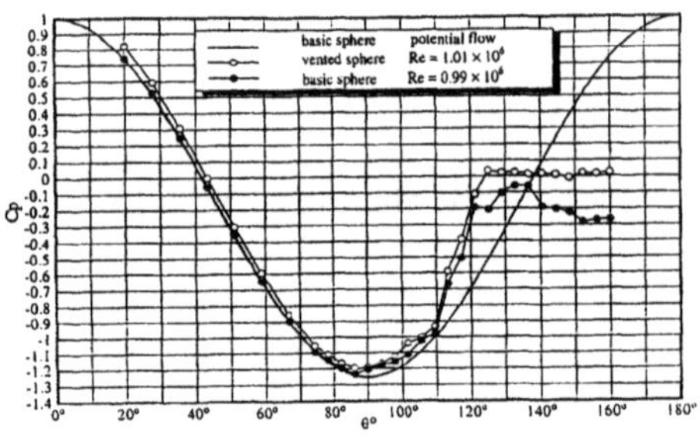

Figure 6. Pressure distribution around the vented and basic sphere at a supercritical Reynolds number (Re ≈ 10^6)

Figure 7 shows the pressure coefficients of the basic and vented sphere at Re $\approx 3.8\times10^6$. Venting results in a slightly increased pressure on the windward side up to 60°. The pressure difference between the basic and the vented sphere is large from 60° to 120°. The minimal pressure coefficient of the vented sphere stays at about 82° and is higher than that of the basic sphere. The separation point of the vented sphere is positioned at about 113°. The laminar separation bubble of the basic sphere vanishes. The base pressure on the basic and vented sphere is almost the same. Due to the higher pressure in the rear of the vented sphere (from 90° to 140°) the drag is smaller than that of the basic sphere, although the separation on the vented sphere occurs earlier.

Due to the wind tunnel interference effects all measurement results have been corrected [9]. The correction methods according to Holst [10] have been used.

3.3 VISUALISATION IN WATER TUNNEL

Figure 8 shows the hydrogen bubble visualisation of a basic sphere at a Reynolds number of 4.02×10^5 in a water tunnel. The wake region is asymmetrical, relatively large and unstable. Figure 9 shows the hydrogen bubble visualisation of a vented sphere at a

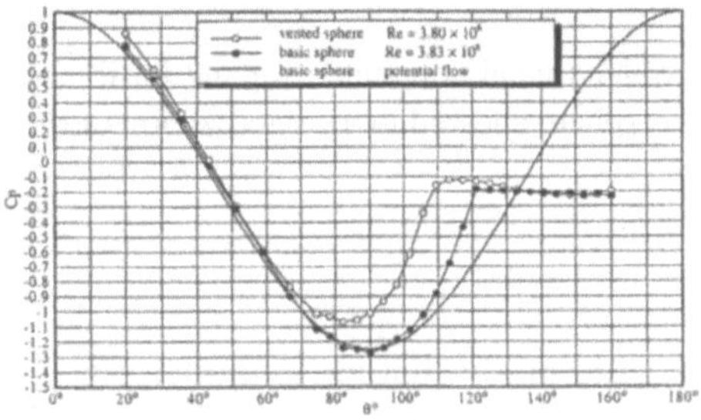

Figure 7. Pressure distribution around the vented and basic sphere at a supercritical Reynolds number ($Re \approx 3.8\times10^6$)

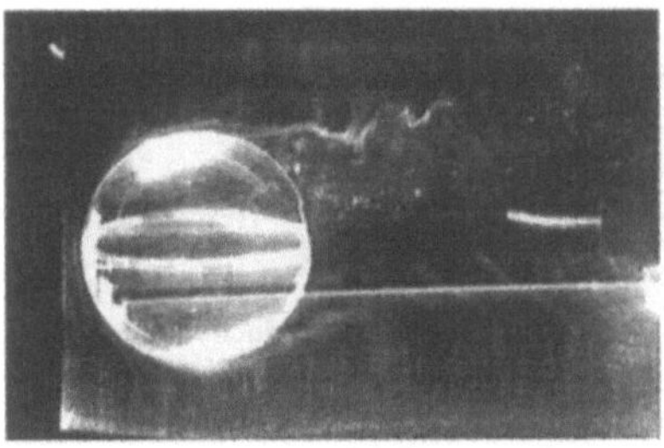

Figure 8. Hydrogen bubble visualisation of a basic sphere at $Re = 4.02\times10^5$ in water tunnel

Figure 9. Hydrogen bubble visualisation of a vented sphere at $Re = 3.57\times10^5$ in water tunnel

Reynolds number of 3.57×10^5 in the water tunnel. Compared to the last figure the wake region is more symmetrical, relatively small and more stable. The visualisations were recorded on a video tape.

4. Conclusions

This paper describes an experimental investigation of drag reduction of a sphere at supercritical Reynolds numbers up to $Re = 4.5\times10^6$ using passive ventilation. This ventilation was created by a channel connecting the stagnation region with the wake region. Passive ventilation is a very effective method of decreasing drag of smooth spheres in the named Reynolds numbers range. A drag reduction of about 40%-60% can be achieved with a cross-section area of the venting channel of about 2% of the sphere. Pressure distributions on the sphere show clear changes of base pressure and point of separation. In addition, the flow on the vented sphere showed smaller oscillations at supercritical Reynolds numbers than those on the basic sphere. The reduction in oscillation amplitude came basically from a stabilization of the wake.

Flow visualization in a water tunnel was made for the basic and vented spheres. The visualization of the vented sphere showed more stabilization of the wake than that of the basic sphere.

Acknowledgements

The work was done at DLR in Goettingen, Germany. The first author thanks the Friedrich-Ebert foundation in Bonn. The authors wish to thank the technicians Mr. Bendig, Mr. Mittmann and Mr. Mattner for their help during the work. Dr. Holst, Mr. Schneider, Mr. Denecke and colleagues at DLR are also acknowledged with thanks.

References

1. Achenbach, E. (1972): Experiments on the flow past spheres at very high Reynolds numbers. J. Fluid Mech., vol. 54, part 3, pp. 565-575

2. Meier, G.E.A.; Suryanarayana, G.K.; Pauer H. (1990): Widerstandsverminderung durch Ventilation. DLR Bericht 90-06, pp. 311-315

3. Meier, G.E.A.; Suryanarayana, G.K.; Pauer, H. (1991): Bluff-body drag reduction by passive ventilation. First European Conference on Fluid Mechanics EUROMECH, Cambridge, UK, 16-20 Sept.

4. Suryanarayana, G.K.; Pauer, H.; Meier, G.E.A. (1993): Bluff-body drag reduction by passive ventilation. Exp. in Fluids 16, pp. 73-81

5. Försching, H. (1981): Ein neuer Windkanal für gebäudeaerodynamische und aeroelastische Untersuchungen bei Reynoldszahlen bis 10^7. DLR Bericht 232-81 J 01

6. Bendig, H. -J. (1986): Widerstandsmessungen an einer Kugel im Hochdruckwindkanal HDG der DFVLR in Göttingen. DLR Bericht 29112-86 A 04

7. Raithby, G.D.; Eckert, E.R.G. (1968): The effect of support position and turbulence intensity on the flow near the surface of a sphere. Wärme und Stoffübertrag. 12, pp. 87-94

8. Suryanarayana, G.K.; Meier, G.E.A. (1995): Effect of ventilation on the flow field around a sphere. Exp. in Fluids 19, pp. 78-88

9. Lu, M.; Meier, G.E.A.; Grosche, F.-R. (1997): Verminderung des Widerstandes von Kugeln durch Ventilation des Nachlaufs bei überkritischen Reynoldszahlen. DLR Bericht 223-97 A 14

10. Holst, H. (1991): Procedure for determination of three-dimensional wind tunnel wall interferences and wall adaptation in compressible subsonic flow using measured wall pressures. DLR-FB 91-09

SEPARATION CONTROL OF A THIN AIRFOIL BY PROJECTILE VORTEX-RINGS

O. MOCHIZUKI, M. KIYA, AND N. SUZUKI
Division of Mechanical Science, Hokkaido university
Nishi 8, Kita 13, Kita-ku, Sapporo 060-8628, JAPAN

1. Introduction

Generally, thin airfoils in turbomachinery are not thick enough to install a device for controlling separation [1],[2],[3]. Our proposal for resolving this difficulty is to add rich momentum to the separation region through the separated shear layer from the outside flow region, away from the airfoil, by vortex rings.

In order to use vortex rings as a transmitter of momentum, the characteristics of vortex rings in a cross flow [4] to obtain the locus and position of transition from a laminar vortex ring to a turbulent one and the specification of actuator to generate vortex rings [5] have been investigated experimentally. To examine the momentum acceptance of the separation region, interactions between vortex rings and vortices of the separated shear layer have been simulated [6].

Separation control of a thin airfoil by using impinging vortex-rings on the separated shear layer is presented in this paper. Several combinations of circulation and generation frequency of the vortex rings were tested to obtain optimum suppression of separation. The momentum defect in the wake of the thin airfoil was examined for each combination to evaluate the effect of suppression of separation. Phase-averaged flow visualization and velocity measurement were carried out to clarify the mechanism of suppression due to interaction between the vortex rings and separated shear layer.

2. Experimental Setup and Technique

Experiments were performed in a low-speed and open-circuit wind tunnel with a test section of $400mm \times 200mm$. An orifice with a diameter d of $5mm$ to generate vortex rings was made in the wall of the test section at $25d$ upstream to a thin airfoil. The orifice was connected to a chamber to

G.E.A. Meier and P.R. Viswanath (eds.),
IUTAM Symposium on Mechanics of Passive and Active Flow Control, 133–138.

which a speaker was attached. The circulation Γ and generation frequency f were changed by a signal provided to the speaker by a computer. Γ of the vortex ring was calculated by the velocity $V_j(t)$ of ejection at the orifice as follows: $\Gamma = \int_0^{t_b} V_j(t)^2 dt$, where t_b is the time interval of the ejection. Since the maximum value V_{jmax} of $V_j(t)$ was fixed at $5.8m/s$ in this experiment, Γ was changed by changing t_b . The momentum of a vortex ring is presented by $\rho A_0 \Gamma$, where $A_0 (= \pi d^2/4)$ is the area of the orifice and ρ is the density of air. Thus, when vortex rings are produced at frequency f, the momentum supplied into the main flow from the orifice per unit of time, M_t, is given by $\rho A_0 \Gamma f$. The range of the Reynolds number of the vortex ring, Γ/ν, was $4.2 \times 10^3 - 1.2 \times 10^4$.

The thin airfoil was made by an acrylic plate of $2mm$ in thickness and $100mm$ in chord length. The leading edge was rounded in order to reduce the acceptance of disturbance at the leading edge. The reason for this was to eliminate the effect of rolling-up of the vortices due to shear layer instability from the results of suppression of the separated shear layer. The angle of attack was fixed at 10 degrees. The relative position of the thin airfoil to the orifice is shown in Fig.1.

The main flow velocity, U_∞ was $1.25m/s$. The velocity ratio V_{jmax}/U_∞ was 4.6. The Reynolds number based on the chord length , Re $(= cU_\infty/\nu)$, was 8.3×10^3.

The velocity field was measured by using both a hot-wire anemometer (HWA) and an LDV. The velocity measurements were synchronized with the generation signal of vortex rings. The time-mean value of velocity and rms velocity fluctuation are denoted by $\overline{q}$ and $\acute{q}$ for the HWA and $\overline{u}$ and $\acute{u}$ for LDV. The phase-averaged velocity is represented by $\tilde{q}$. The vortex rings and separated shear layer were visualized by the smoke wire technique, synchronizing with the generation signal of vortex rings. The exposure time was 2 seconds, and therefore the number of scenes at the same phase taken in one photo was $2f$.

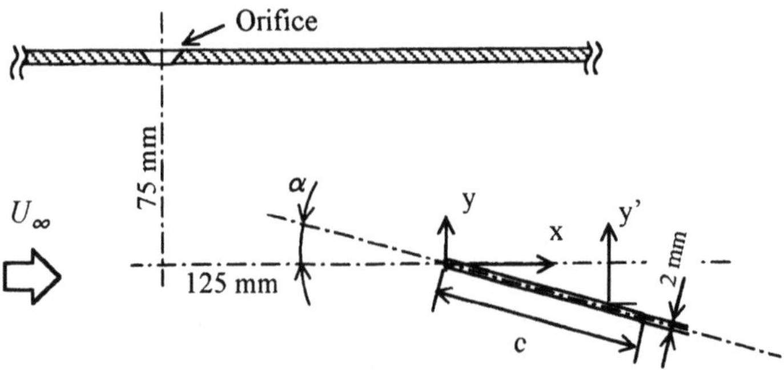

Figure 1. Schematic cross-sectional view of the thin airfoil and orifice on the wall of the test section.

3. Experimental Results and Discussion

Figure 2 shows the separation region visualized by smoke (Fig.2a) and velocity distributions on the suction side of the thin airfoil (Fig.2b). The inclined thin airfoil shows a complete stall at this angle of attack. The broken line in Fig.2b shows the position of $u/U_{\infty} = 0.5$, which is regarded as the center of the shear layer. The velocity distribution measured by the LDV in the separation region shows the existence of a reverse flow near the wall, as seen in Fig.2b. The velocity distribution measured by the LDV in the separation region shows the existence of a reverse flow near the wall, as seen in Fig.2b. The velocity distribution in the same part measured by the HWA is a plateau. Thus, we defined a plateau near the wall as indicating the existence of a reverse flow. The velocity distribution near the trailing edge remains a separation-type velocity distribution, and therefore the separated shear layer never reattaches to the wall at this angle of attack, as seen in Fig.2a.

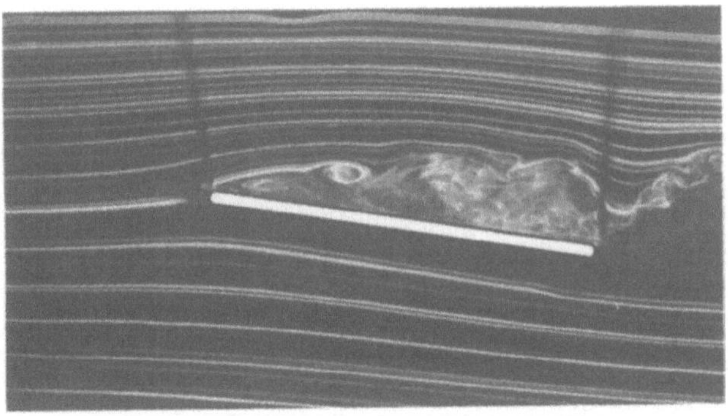

(a)

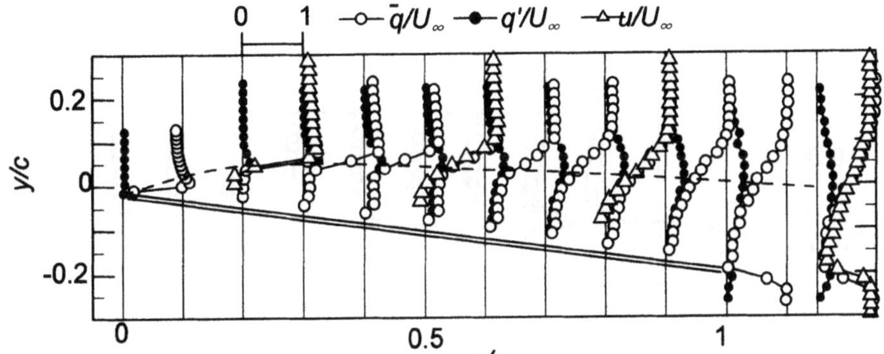

(b)

Figure 2. Separation region without impingement of vortex rings, (a) flow visualized by smoke, (b) velocity and rms velocity fluctuation distributions.

The interaction between the vortex rings and separated shear layer is shown in Fig.3. When the vortex ring arrives at the leading edge of the thin airfoil, the phase is defined as zero. The convection velocity of the vortex ring is $1.4m/s$, and is $1.12U_\infty$.

The vortex ring seems to stimulate the rolling-up of a large vortex in the separated shear layer near the leading edge, as seen in Fig.3. The separated shear layer seems to reattach at this phase. The phase-averaged velocity distribution is shown in Fig.4. This was obtained at the same phase of the photos shown in Fig.3. A comparison of velocity distribution with the visualized flow field showed that entrainment by the large vortex makes a dent in the solid line originating at the leading edge. Thus, the velocity at the center of the separated shear layer at the dent increases more than that at other positions. The velocity distribution is not the separation type at the dent. This position is regarded as the reattaching position of the separated shear layer.

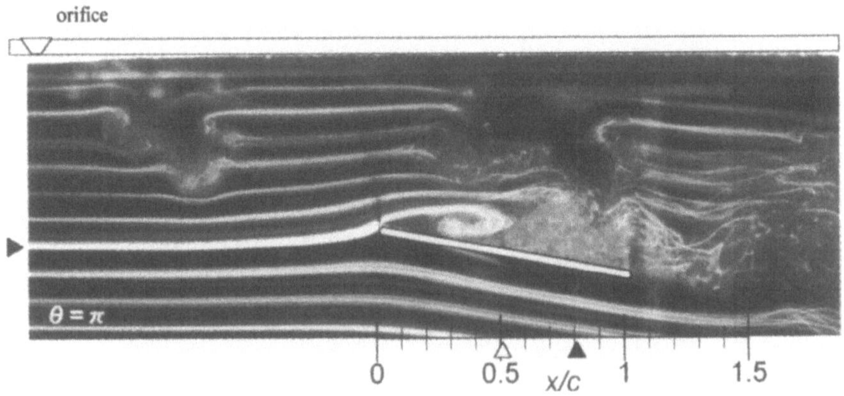

Figure 3. Change in the flow pattern by impingement of vortex rings. Phase π. $f = 10$Hz.

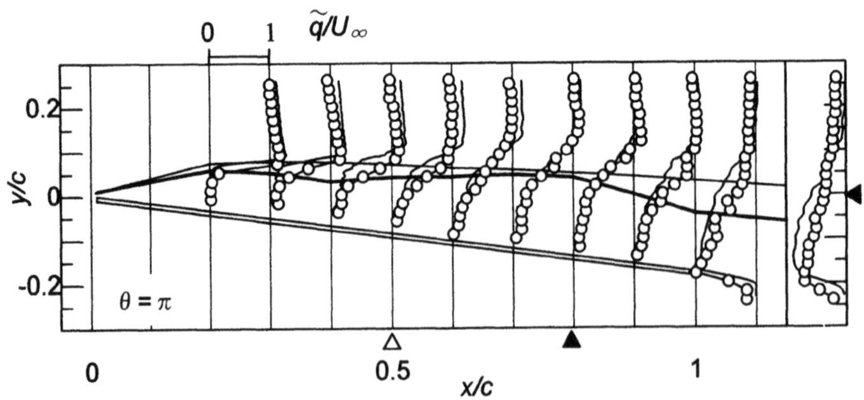

Figure 4. Instantaneous velocity distributions. Phase π. $f = 10$Hz.

To examine the effect of f on the suppression of separation, the momentum defect $M = \rho \int \overline{u}(U_\infty - \overline{u})dy$ per unit length in the spanwise directionin the wake of the thin airfoil was evaluated. Figure 5 shows the momentum defect ratio, M/M_0 for different values of f . Here, M_0 means the momentum defect of the thin airfoil without separation at zero angle of attack. Thus, $M/M_0 = 1$ means no separation. The momentum defect decreases

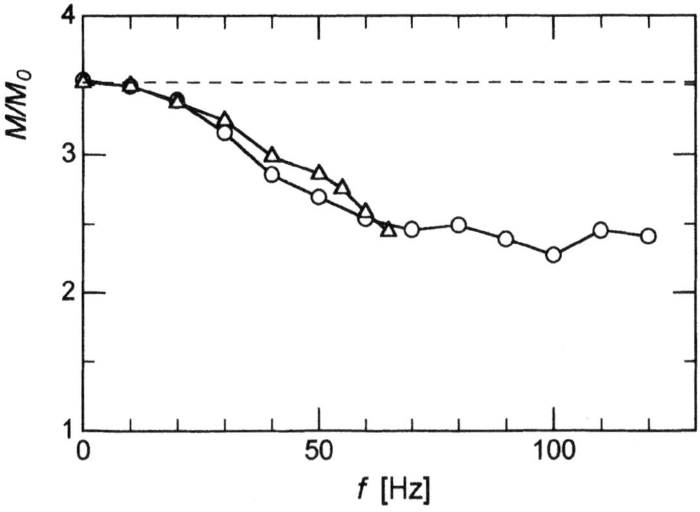

Figure 5. Momentum defect versus generation frequency of vortex rings. ○, $\Gamma/\nu = 6.6\times10^3$; △, $\Gamma/\nu = 1.2 \times 10^4$.

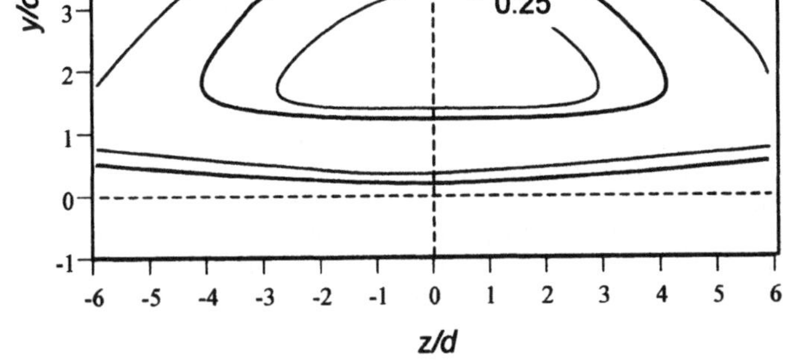

Figure 6. Gain distribution in velocity in a cross section behind the trailing edge of the thin airfoil. Contour lines show $(\overline{u} - \overline{u_0})/U_{infty}$. $\Gamma/\nu = 1.1 \times 10^4$.

with increases in f. However, when $f > 70$Hz, there is no difference. This means that $f = 70$Hz is sufficient for suppressing separation in this case. If we assume that the suppression of separation by vortex rings is related to instability of the separated shear layer, 50Hz should be the most effective frequency. However, since there was no typical decrease in the momentum defect at this frequency, the suppression mechanism in this study is thought to be independent of the instability.

It is necessary to know how extensive the influence of the vortex ring on the change of velocity in the separated shear layer is. A gain in velocity at the x-section just behind the trailing edge is shown in Fig.6. The region where $(\overline{u}-\overline{u_0})/U_\infty$ is larger than 0.2 is extended to $8d$ in the spanwise direction and $3d$ in the y direction. This indicates the distance between adjoining orifices arranged in the spanwise direction required to suppress wide separation in the spanwise direction.

4. Conclusions

This study has shown that the separation of a thin airfoil can be suppressed by impinging vortex-rings generated from outside the airfoil. The effects of several combinations of circulation and generation frequency of the vortex rings on suppression of separation were investigated. The main results obtained are as follows.

(1) Vortex rings stimulated the formation of a large vortex in the separated shear layer. This is independent of the instability of the shear layer. The large vortex enhanced entrainment of high-speed flow into the separation region.

(2) The frequency of impingement of the vortex rings affected the extent of suppression of separation. A higher frequency was more effective, but there was a limit.

(3) A relatively wide region was affected by the impingement of vortex rings. This indicates the possibility of suppression of separation in the spanwise direction by using multi-generators of vortex rings.

References

Gad-el-Hak, M., and Bushnell, D.M. (1991) Separation Control: Review, *J. Fluids Eng,* **113**, pp. 5-30.

Kiya, M., Shimizu, M., and Mochizuki, M. (1997) Sinusoidal forcing of a turbulent separation bubble, *J. Fluid Mech,* **342**, pp. 119-139.

McManus, K., Ducharme, A., Goldey, C., and Magill, J. (1996) Pulsed Jet Actuators for Suppressing Flow Separation, *AIAA paper,* **96-0442**, pp. 1-13.

Suzuki, N., Kiya, M., Mochizuki, O., and Jinzu, H. (1997) Sinusoidal forcing of a turbulent separation bubble, *JSME,* **63-605**, pp. 106-111 (in Japanese).

Rathnasingham, R., and Breuer, K. S. (1996) Characteristics of Resonant Actuators for Flow Control, *AIAA paper,* **96-0311**, pp. 1-10.

Kiya, M., Takeo, H., Mochizuki, M., and Kudo, D. (1998) Simulating Vortex Pairs Interacting with Mixing-Layer Vortices, *Fluid Dynamics Research,* (in press).

TURBULENCE REDUCTION IN THE INITIAL REGION OF A PLANE JET USING PASSIVE CONTROL

J.F. OLSEN, S. RAJAGOPALAN and R.A. ANTONIA
Department of Mechanical Engineering,
The University of Newcastle, N.S.W. 2308, Australia.

Abstract

The introduction of a thin cylinder in each mixing layer of a plane jet has been found to reduce both longitudinal and cross-stream rms velocity fluctuations (u' and v') by 47% and 50% respectively on the jet centre-line at x = 3H, where H is the nozzle width. The spread rate of the mixing layers was decreased by 27%. Normalised distributions of the mean velocity U and Reynolds stresses exhibit a higher degree of self-preservation than in the undisturbed jet. Such an improvement seems to be caused by an attenuation in the modified jet of the formation of organised motion in the mixing layer. Spectra of u support this observation as peaks associated with the natural roll-up frequency and pairing events are absent in the disturbed flow. There remains some organisation in the mixing layers since v spectra show a maximum at the natural instability frequency and a flat peak at the sub-harmonic frequency suggesting the possibility of vortex pairing.

1. Introduction

Since it was discovered in the early seventies that mixing layers in jets contained coherent structures (organised motion), it has been thought that control of turbulence, at least in the initial region may be achieved by manipulating these structures. Previous attempts to reduce turbulence in jets have typically involved periodic forcing of the flow using acoustic drivers or vibrating ribbons, (Zaman and Hussain [1]), at a frequency approximately 40% higher than the natural mixing layer instability frequency. The resulting reduction in turbulence intensity - as much as 80% [1] - was attributed to accelerated transition to turbulence due to the vortices being forced to roll-up rapidly and subsequently break down before any pairing process could occur.

An application of a reduction in turbulence could be in reducing the noise in the exhaust of jet engines. It is known that both the fine scale turbulence and coherent structures both have an effect on jet noise (Tam 1998 [2]). The effect of coherent structures on noise production only becomes apparent when the convection speed of the structures is above the ambient speed of sound. Hussain and Husain [3], showed that

G.E.A. Meier and P.R. Viswanath (eds.),
IUTAM Symposium on Mechanics of Passive and Active Flow Control, 139–144.

by using an acoustic excitation form of active control, the overall jet noise (i.e. jet noise plus speaker noise) measured in an anechoic chamber, was reduced. In their experiments, the convection velocity of the coherent structures was well below the ambient speed of sound suggesting that the major producer of sound would have been the fine-scale turbulence. This suggests that this method of turbulence control does not break down the initial structures into fine-scale turbulence but more likely into irregular substructures. Unfortunately this kind of control seems incompatible with the hostile environment in which a jet engine operates. This suggests that a passive means of turbulence reduction may prove to be more attractive.

Tong and Warhaft [4] placed a ring of slightly smaller diameter than their axisymmetric jet close to the jet exit plane to study the effects of shear on a scalar (temperature). They felt that the combination of the cross-sectional diameter of the ring and flow speed would inhibit vortex shedding. It is also known that incoming shear also inhibits vortex shedding (Kiya *et al* [5]). Preliminary measurements showed a reduction, relative to the undisturbed jet, in the longitudinal turbulence intensity level of approximately 30% on the jet centre-line. Tong and Warhaft claimed that spectra at $x/D = 2$ indicated an earlier transition to fully developed turbulence as the peaks associated with the natural roll-up frequency and vortex pairing were absent.

Recently, Rajagopalan and Antonia [6], modified a mixing layer of a plane jet by placing a thin cylinder in the mixing layer near the nozzle exit plane. Maximum values of u' in that mixing layer was reduced by approximately 20%. Spectra showed a reduction in amplitude of the fundamental and sub-harmonic frequencies and autocorrelation functions were less periodic with the addition of the cylinder. They suggested that an interaction between the cylinder vortex street and the mixing layer vortices inhibited the formation of large structures. Their cylinder, although thin, was not found to vibrate during these experiments. Vandsburger and Ding [7] found that self-excited vibrating wires had the effect of increasing the spread of the mixing layer.

The aim of the present work is to investigate the mechanism of passive control in a plane jet by modifying both mixing layers by the introduction of thin cylinders. Detailed measurements of u and v components of the fluctuating velocity were made with a cross-hot wire probe. The cross-stream velocity fluctuation and its spectrum are likely to indicate whether the formation of organised motion has been completely suppressed or only inhibited.

2. Experimental Conditions

In the present work, experiments were conducted in a plane jet facility with nozzle dimensions of 340 x 34mm and initially laminar boundary layers. Thin cylinders of diameter d = 0.2mm was placed in each mixing layer (see Figure 1 for experimental conditions). These cylinders were mounted on a separate frame to help isolate them from any tunnel induced vibrations. The jet exit velocity was 9.1m/s corresponding to $Re_H = 20500$. Measurements were made in the near field ($1.0 \leq x/H \leq 6.0$) of both the undisturbed (as a basis for reference) and modified jet with a 2.5μm cross-hot wire probe operated in constant temperature mode. The hot wire signals were digitised using a 12 bit a/d converter and were stored and processed on a PC.

Preliminary measurements were conducted with a single hot wire (5μm) to determine the optimum location of the cylinders for maximum reduction of turbulence intensity in the mixing layers. It was observed that placing the cylinders at x/H = 0.07 at a distance of 0.8mm (0.024H) from each nozzle lip towards the high speed side resulted in a reduction in u' of approximately 50% on the jet centre-line. The cylinders were continually monitored both visually and audibly for vibrations.

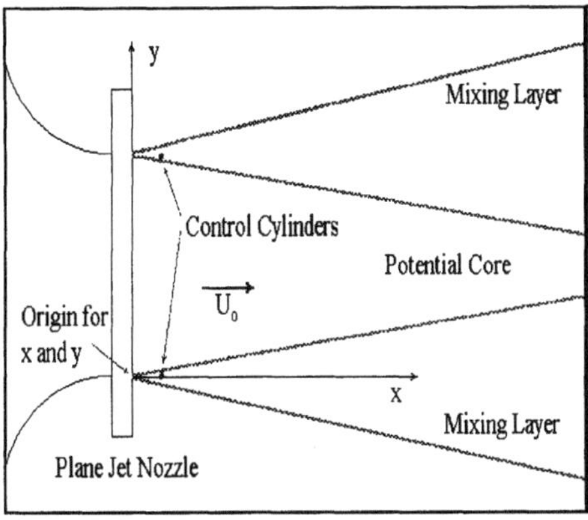

Figure 1. General layout of modified jet

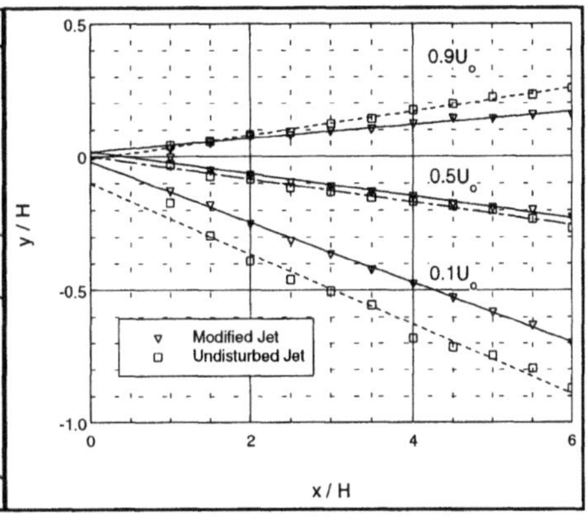

Figure 2. Growth of the mixing layer

3. Results and Discussion

From the mean velocity profiles of the undisturbed and modified mixing layers, cross-stream (y) locations where $U = 0.1U_0$, $0.5U_0$ and $0.9U_0$ were estimated for different x/H values and plotted in Figure 2 together with lines of best fit. This figure indicates that the introduction of the cylinders decreases the spread of the mixing layer. The location of the virtual origin can be more easily determined for the modified mixing layer as there is less "jitter" in the spread especially on the low speed side. In both undisturbed and modified flows, the virtual origin is zero ($x_0 \approx 0$).

Distributions for the undisturbed and modified flows of U/U_0, u'/U_0, v'/U_0 and $\overline{uv}/U_0^2$, plotted against $\eta = (y / (x - x_0))$ are shown in Figure 3 - a), b), c) and d). They indicate that the effect of the "control" cylinders is to reduce the spread of the mixing layers by 27% at x = 3H and to improve self-preservation. These distributions also show a decrease in the maxima of both u'/U_0 (≈15%) and v'/U_0 (≈30%) and a reduction in the magnitude of $\overline{uv}/U_0^2$ (≈20%) at x/H = 3. The reduction in v'/U_0 is consistent with the decreased growth rate of the mixing layer.

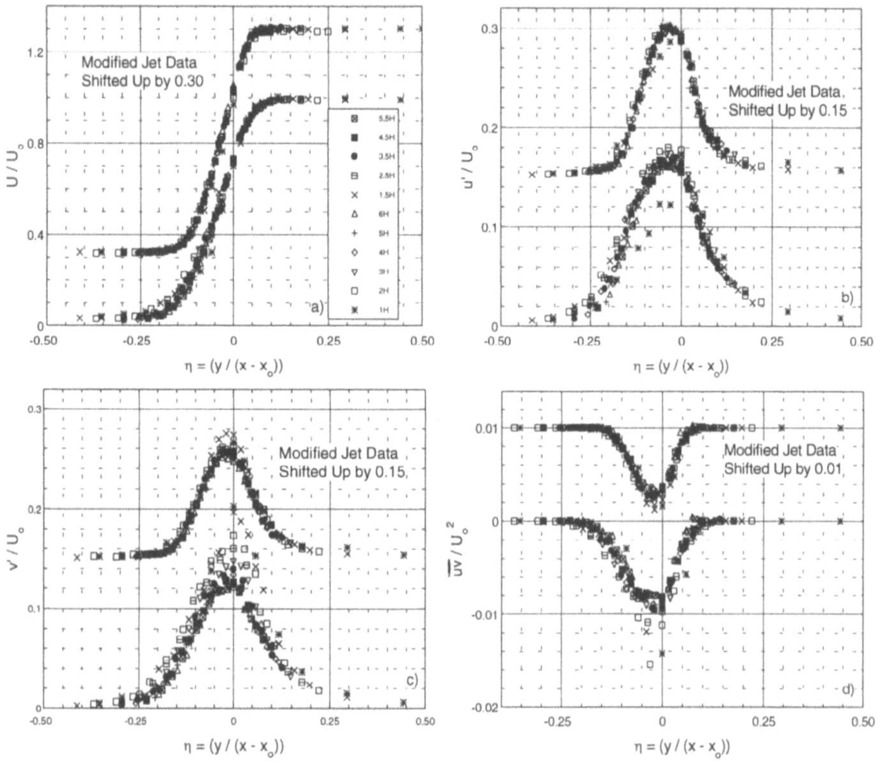

Figure 3. Distributions of a) U/U_o, b) u'/U_o, c) v'/U_o and d) $\overline{uv}/U_o^2$ in similarity coordinates.

Mixing layers that originate from turbulent boundary layers (Husain and Hussain [8]) also have an initially smaller spread rate than those originating from laminar boundary layers although the spread increases further downstream ($x/H > 2$). In an initially turbulent mixing layer, however, the mean flow achieves self-preservation significantly further downstream, in contrast to the initially laminar mixing layer [8]. It would appear that to achieve earlier self-preservation (or accelerate transition), the initial mixing layer vortices need to be modified, not by irregular turbulence but by the regular vortices shed from the cylinders. This is important since a turbulent boundary layer at the nozzle exit plane requires a trip located well upstream of the nozzle lip in the contraction for the boundary layer to become turbulent; this may not be possible in practice.

Figure 4 shows the distribution of U/U_o, u'/U_o and v'/U_o on the centre-line for the undisturbed and modified jets. For the modified jet, U/U_o is larger than that for the undisturbed jet at $x/H = 6$. This suggests that the potential core extends further downstream for the modified jet. This again is consistent with the suppression of the large scale structures in the mixing layer and may explain the lower turbulence intensities on the jet centre-line; for example at x/H = 3, a maximum reduction in u'/U_0 and v'/U_0 of 47% and 50% respectively was observed for the modified flow.

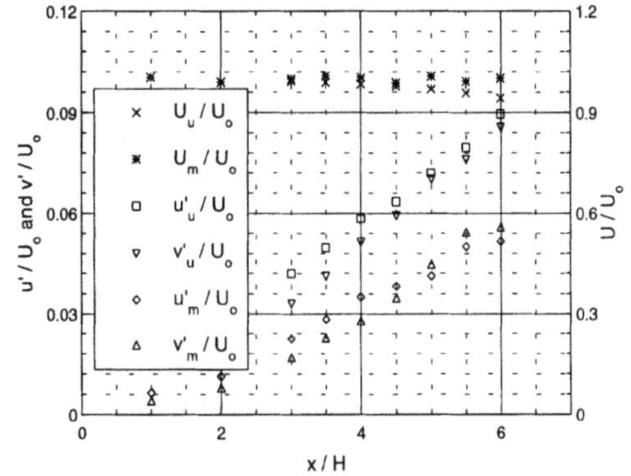

Figure 4. Centre-line mean velocity and rms velocity fluctuation distributions (subscripts: u:- undisturbed jet, m:- modified jet)

Spectra of u at x/H = 1 and 2 (Figures 5a and b) indicate that the natural roll-up frequency of the undisturbed shear layer is 610Hz. At x/H = 1, the first sub-harmonic was observed in the undisturbed jet in both u and v fluctuations whereas in the case of the modified flow, only the first harmonic was present. The instability frequency of the modified jet is also 610Hz which indicates that the control cylinder has little effect on the natural roll-up frequency. At x/H = 2, the first sub-harmonic for the undisturbed flow was the dominant frequency in the u spectrum whilst the v spectrum showed a broad sub-harmonic peak, both of which strongly suggest the occurrence of a pairing process. For the modified flow, a low flat peak was obtained at this frequency in v which suggests the possible pairing of vortices.

4. Conclusions

By modifying mixing layer vortices in the initial region of a plane jet by introducing a thin circular cylinder into each mixing layer, slightly downstream of the jet exit plane, the mean velocity and Reynolds stress distributions of the jet are altered. Turbulence intensities in both the x and y directions are reduced in the mixing layers and on the jet centre-line. Reynolds shear stresses are also attenuated in the mixing layers. The modifications have resulted in a reduction in the spread rate of the mixing layers. In particular the decrease in v' is consistent with the smaller growth rate of the mixing layers. The v spectra suggest that coherent structures in the mixing layers are not completely suppressed, and may possibly undergo pairing.

It appears that in the initial region of the jet, these modifications have an effect similar to that achieved by using various types of active control techniques, for example, excitation at a frequency approximately 40% higher that the initial instability frequency has yielded turbulence suppression [1]. It is likely that the present form of passive control will lead to a reduction in overall jet noise.

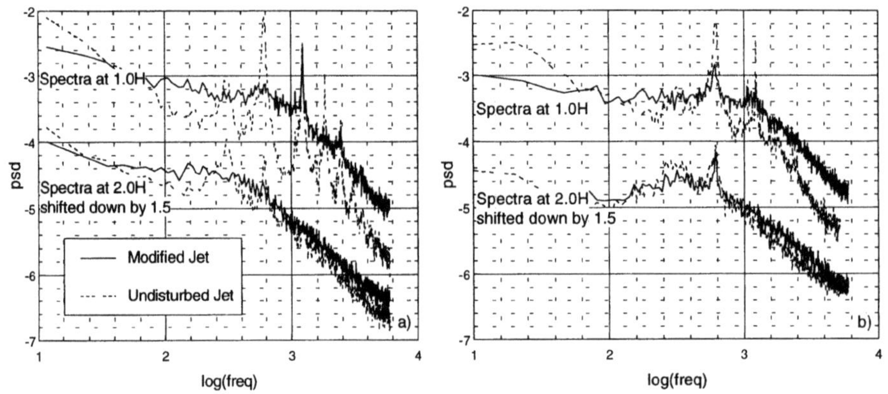

Figure 5a) u spectra and *b)* v spectra at different x/H (lower side of jet).

Acknowledgments

The authors are grateful to the Australian Research Council for their support.

References

1. Zaman, K.B.M.Q and Hussain, A.K.M.F.: Turbulence suppression in free shear flows by controlled excitation, *J. Fluid. Mech.* **103**, (1981) 133-159.
2. Tam, C. K. W.: Jet noise: since 1952, *Theoret. Comput. Fluid Dynamics*, **10**, (1998) 393-405.
3. Hussain, A. K. M. F. and Husain, M. A. Z.: Turbulence suppression in free turbulent shear flows under controlled excitation. Part 2. Jet-noise reduction. *J. Fluid Mech.* **150**, (1985) 159-168.
4. Tong, C and Warhaft, Z.: Turbulence suppression in a jet by means of a fine ring, *Phys. Fluids* **6**, (1994) 328-333.
5. Kiya, M., Tamura, H. and Arie, M.: Vortex shedding from a circular cylinder in moderate Reynolds number shear flow, *J. Fluid. Mech.* **141**, (1980) 721-735.
6. Rajagopalan, S and Antonia, R.A.: Turbulence reduction in the mixing layer of a plane jet using small cylinders, *Experiments in Fluids*, (1998) to appear.
7. Vandsburger, U and Ding, C.: Self-excited method for the control of turbulent mixing layers, *AIAA*, Vol **33**, No **6**, (1995) 1032-1037.
8. Husain, Z. D and Hussain, A. K. M. F.: Axisymmetric mixing layer: Influence of the initial and boundary conditions, *AIAA*, Vol **17**, No **1**, (1979) 48-55.

INFLUENCE OF RIBLETS ON A SUPERSONIC WAKE FLOW

E. AUGENSTEIN, F. LEOPOLD, F. CHRISTNACHER, E. BACHER
French-German Research Institute of Saint-Louis (ISL)
5 rue du Général Cassagnou, 68301 SAINT LOUIS, France

1. Motivation

In the late sixties biologists found that the skin of fast swimming sharks is structured. With the coming-up of the oil crisis a strong interest in drag reduction technology arose. One of the projects was to influence the structure of the boundary layer in the same manner as observed on the shark's skin. In spite of nearly 20 years of riblet research, there have been, up to now, only a few reports in literature concerning riblets at supersonic regime, e.g. Robinson (1988), Gaudet (1989) and Coustols & Cousteix (1994). The aim of this work is to determine the influence of riblets or of a rough surface on a supersonic wake flow, especially on the behaviour of the vortices.

Vortex shedding behind blunt bodies appears at supercritical Reynolds numbers in many flow regimes. Von Kàrmàn studied these structures at low velocities and was able to predict their frequencies. Nevertheless, the shedding of vortices from blunt bodies appears not only at low Mach numbers, as shown by Gold (1963). In recent years the concept of absolutely and convectively unstable wave propagation has been applied to many aerodynamic investigations, for example by Mack (1984), Huerre & Monkewitz (1985), Koch (1986) and Monkewitz (1988), to mention only a few names. In this work two different flow measurement techniques - a laser Doppler anemometer and a flow visualization method – and two different numerical simulation techniques are applied in order to describe the turbulent wake flow.

2. Experimental Set-Up

2.1. SUPERSONIC WIND TUNNEL

All measurements are carried out in the blow-down wind tunnel of ISL at a Mach number of 2.06. The Reynolds number based on the diameter of the cylinder (40 mm) is $1.9 \cdot 10^6$. The flow around a sting-mounted model (fig. 1) is considered. This model can be equipped with different surface roughnesses. In order to distinguish the different surfaces, the non-dimensionalized height h^+ was introduced; the corresponding values are 0.01 for a smooth surface, 23 for a surface with riblets and 190 for the rough sandpaper surface. The model is fixed in the settling chamber in order to suppress

G.E.A. Meier and P.R. Viswanath (eds.),
IUTAM Symposium on Mechanics of Passive and Active Flow Control, 145–150.

shocks which would be provoked by attaching the body in the supersonic part of the wind tunnel. The flow configuration is indicated in fig. 2, following Gaviglio et al. (1977). The initial boundary layer is fully turbulent, leaves the model at the edge, is accelerated through the expansion fan towards the axis of symmetry, then develops as a mixing layer which is subjected to recompression with the shock wave, and finally is transformed into the turbulent wake. The mixing layer encloses the recirculation zone.

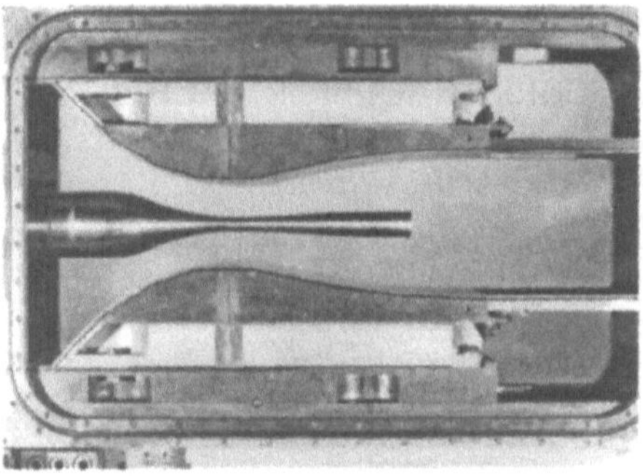

Figure 1. Sting-mounted model in the Laval nozzle for Mach 2.06

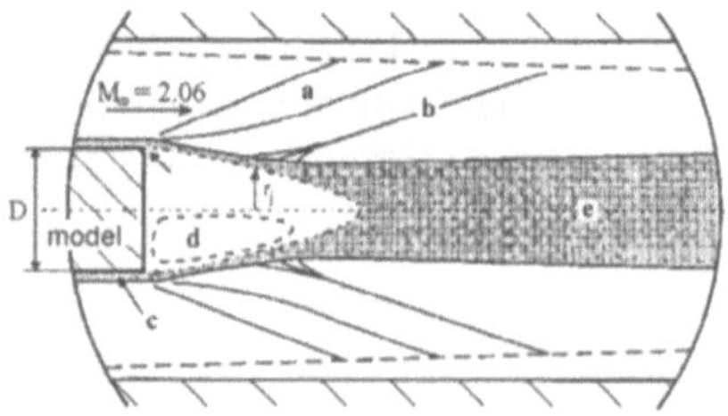

a: expansion fan, b: recompression shock, c: turbulent boundary layer, d: recirculation zone, e: wake

Figure 2. Schematic illustration of the wake flow

2.2. FLOW VISUALIZATION USING HOLOGRAPHIC FILTERS

The technique for flow visualization uses continuous laser illumination associated with holographic filters and a high-performance CCD camera. The technique is described in detail by Leopold et al. (1998). The camera is used to show the rapid phenomena of the flow, where even the turbulent structures are "frozen". A dynamic sequence to show the unsteady phenomena can also be recorded on a video tape.

The image in fig. 3 with a time of exposure of 100 μs represents the quasi-steady aspect of the flow. The image at 5 ns (fig. 4) shows the unsteady phenomena such as the coherent structures and the turbulence in the wake of the smooth cylinder. Figures 5 and 6 show the wake flow for a riblet-mounted cylinder and for a cylinder with rough surface at a time of exposure of 5 ns. The turbulent wake diameter of the rough cylinder (fig. 6) is bigger.

2.3. LASER DOPPLER ANEMOMETRY

2.3.1. *Boundary Layer Measurements*

The velocity measurements above the wind tunnel models are shown in figure 7. The velocities are normalized with the free-stream velocity. The distance to the wall of each point is normalized with the help of the boundary layer thickness of the smooth model (4 mm). The increase of the boundary layer thickness over the riblet-mounted body and the model covered with sandpaper can clearly be seen. The boundary layer thickness in the first case is 4.4 mm and in the rough case it is about 5 mm. The velocity profile for the riblet case is placed between the smooth surface and the sandpaper profile all over the investigated regime.

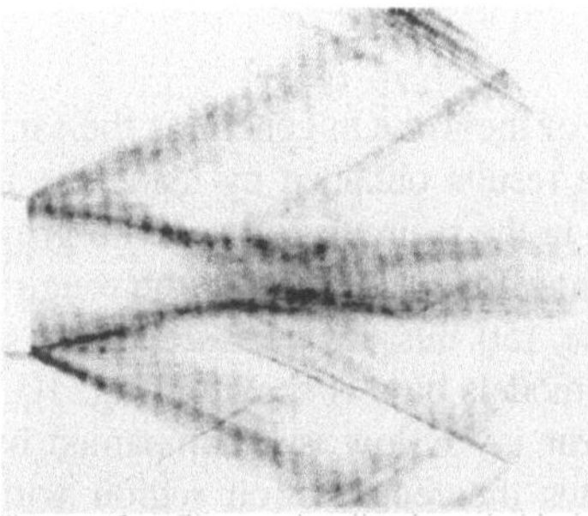

Figure 3. Flow visualization behind cylinder with smooth surface. Time of exposure: 100 μs.

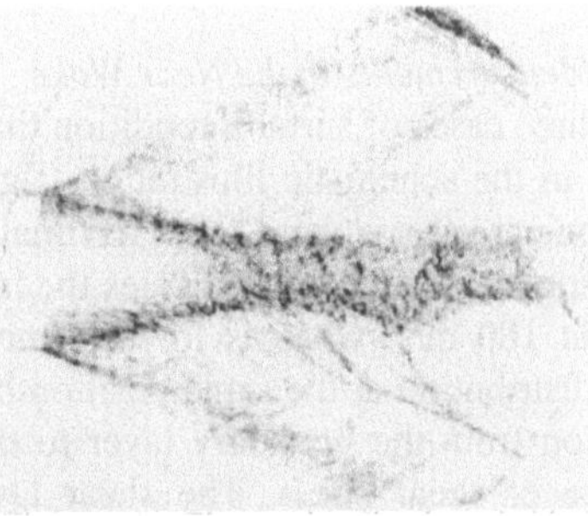

Figure 4. Flow visualization behind cylinder with smooth surface. Time of exposure: 5 ns.

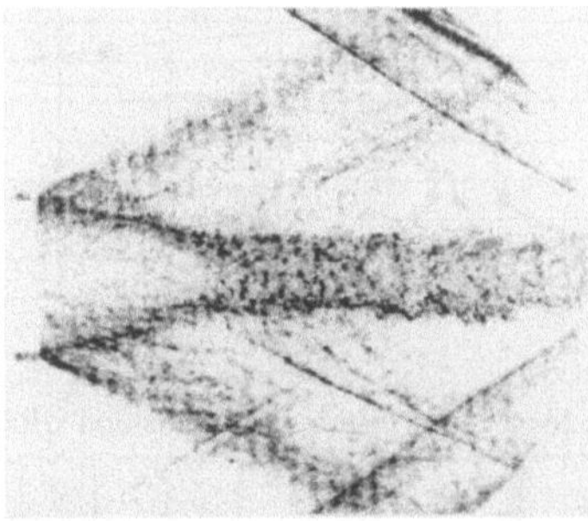

Figure 5. Flow visualization behind cylinder with riblet surface. Time of exposure: 5 ns.

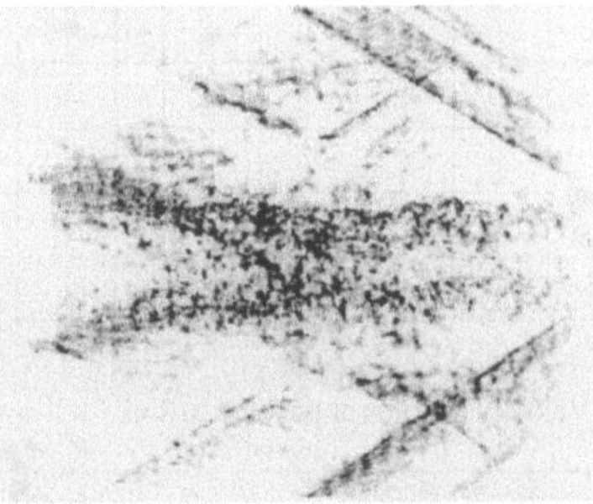

Figure 6. Flow visualization behind cylinder with rough surface. Time of exposure: 5 ns.

The distribution of the Reynolds shear stress throughout the boundary layer is represented in figure 8. The values observed for the sandpaper roughness are about the double of the values obtained for the smooth surface. For the riblet case we are able to separate clearly an inner from an outer layer. At the location y/δ about 0.5 there is a step in the profile towards the values near the smooth wall. The longitudinal riblets have a damping effect on turbulence near the wall. More details concerning these measurements can be found in Augenstein et al. (1998).

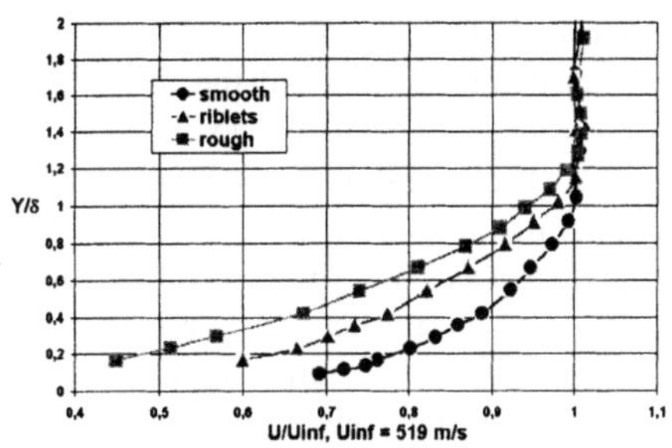

Figure 7. Boundary layer velocity profiles for different surface conditions

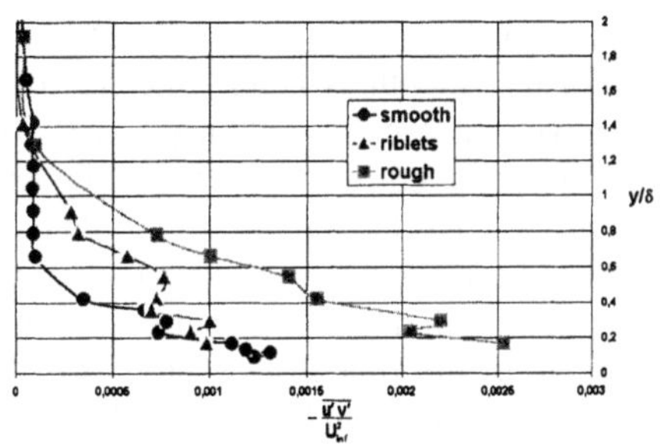

Figure 8. Shear stress distribution in the boundary layer

2.3.2. *Measurements in the Near Wake*

In all three cases of surface condition the structure of the wake in general is the same as shown in the schematic illustration (fig. 2). Some results obtained by using the LDA technique are shown here. The maximal velocities in the recirculation area are growing with roughness. Figure 9 indicates the highest counterflow velocities for the rough case at about 180 m/s, whereas for the smooth surface 150 m/s are measured. The shear stress distribution at the same position behind the models base is shown in fig. 10. The transition from the boundary layer to the free shear layer flow is accompanied by an increase of shear stress. The shear layer separates the recirculation region with its relatively low shear stress from the acceleration area with neglectable shear stress.

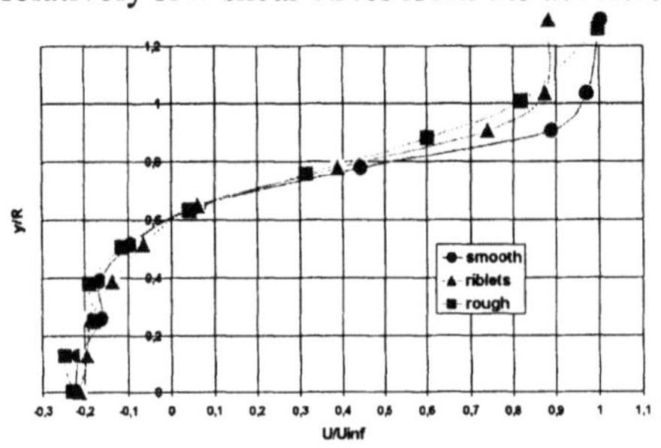

Fig. 9. Velocity profiles at position x/R=1

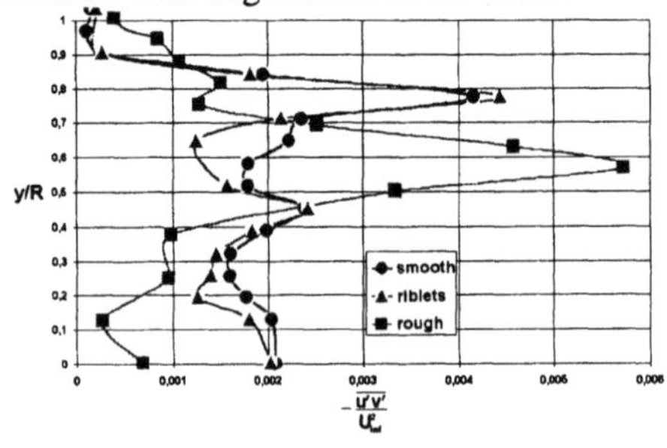

Fig. 10. Shear stress distribution at position x/R=1

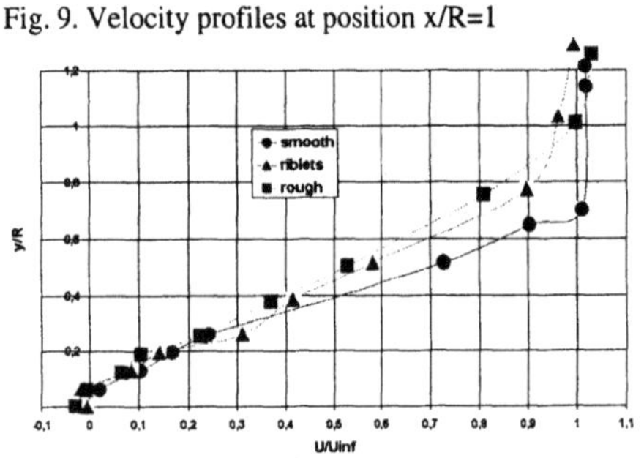

Fig. 11. Velocity profiles at position x/R=2.5

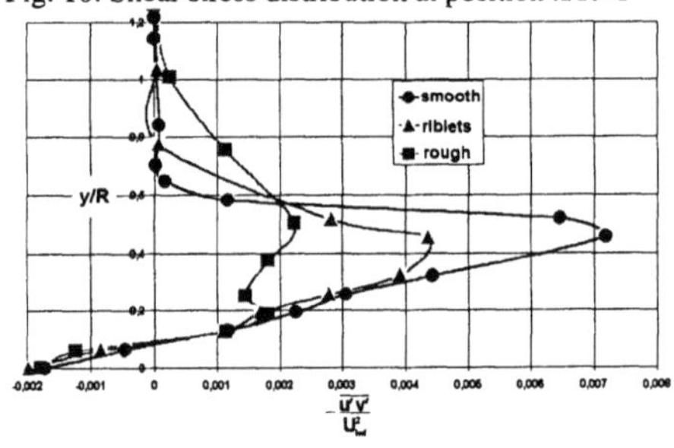

Fig. 12. Shear stress distribution at position x/R=2.5

The rear stagnation point is found to be 50 mm behind the model with smooth surface, whereas in the riblet-mounted case the distance is 51 mm and behind the rough cylinder it is 53 mm. In fig. 11 it can be seen that the velocity at the symmetry line is zero for the smooth cylinder, whereas for the riblet-mounted case and the rough one, respectively, a backflow already exists. At this position, the highest levels of shear stress are observed for the smooth cylinder (fig. 12). The profile in this case is sharply outlined, whereas the rough case profile has a smaller maximum, but a taller diameter of the wake flow structures. The profile for the sandpaper case still shows characteristics of a recirculation flow, because its stagnation point is further downstream. The maximum of the riblets shear stress distribution as well as the diameter of the wake structure can be found between the lines representing the rough and the smooth cylinders wake, respectively.

The velocity profiles in fig. 13 indicate that the flow is up to 30% slower in the wake of the sandpaper model, dependent on the distance from the axis of symmetry. The three shear stress profiles at this position (fig. 14) show in general the same tendencies. The highest stress values are reported for the smooth surface and the smallest for the rough surface case. The turbulent wake diameters of the smooth and the riblet-mounted models are almost the same, but the wake diameter is higher in the rough case. This is in good agreement with the results of the flow visualizations.

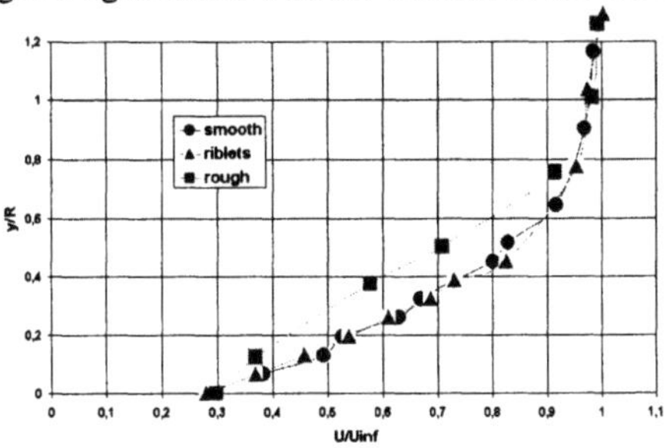

Fig. 13. Velocity profiles at position x/R=4

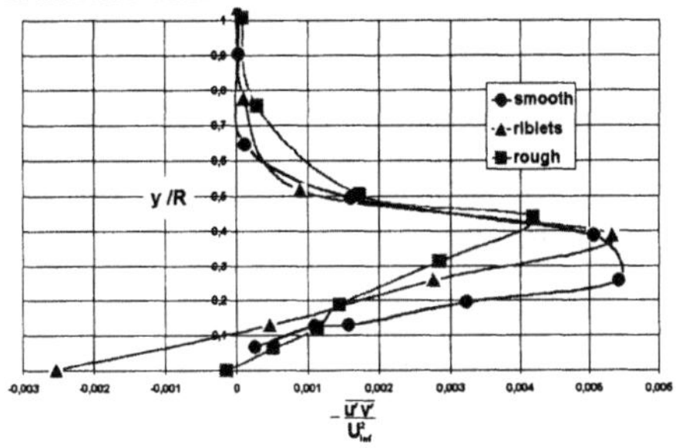

Fig. 14. Shear stress distribution at position x/R=4

3. Numerical Simulation of the Near Wake

In order to get a better insight into the origins of the unsteady flow phenomena, different numerical simulation methods are used. The quasi-steady aspect of the flow is simulated by solving the time-dependent Navier-Stokes equations written in conservative form for a compressible ideal gas. The approach used here is a finite volume technique coupled with the upwind flux splitting scheme. For time discretization a Runge Kutta scheme is applied. The comparison between the quasi-steady solution (fig. 15) and the flow visualization with a long time of exposure (fig. 3, 100 μs) shows a good agreement for the location of the expansion fan, the mixing layer and the recompression shock.

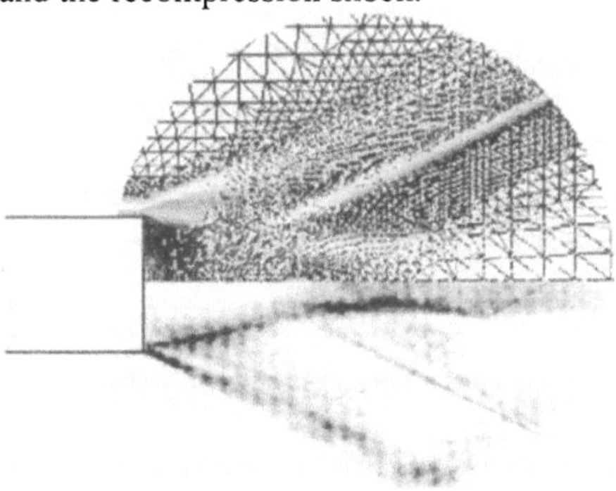

Figure 15. Comparison between the simulated density distribution for a smooth surface and the flow visualization at a time of exposure of 100 μs

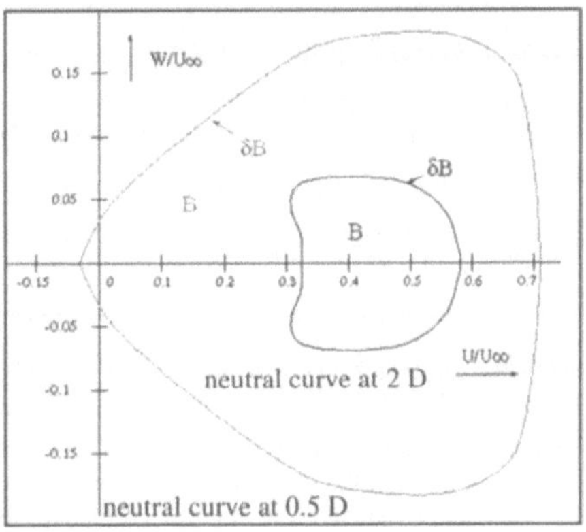

Figure16. Neutral curves at different positions in the wake

The stability characteristics of this flow can be determined by investigating its impulse response to small perturbations superimposed on the quasi-steady solution, as shown by Leopold (1993). The line between the amplification and the damping of these perturbations is called the neutral curve. The calculated neutral curves in figure 16 for a position in the recirculation zone show negative propagation velocities, which means that they travel upstream. The increase of surface roughness is directly related to higher velocities in the recirculation zone, to greater vortices and therefore, to a larger diameter of the far wake. The length scale of these vortices can be directly compared to the characteristic lengths of the shedding vortices visualized at a time of exposure of 5 ns (figures 4, 5 and 6).

4. Conclusions

The different surface roughnesses of the wind tunnel models have an important influence on the structure of the wake, the velocity distribution and therefore, on the shedding vortices. Shear stress near the wall is reduced by riblets. This leads finally to smaller coherent structures in the wake. For a rough surface the diameter of the turbulent wake is greater and higher velocities in the recirculation area are found

Acknowledgements

The authors wish to express their sincere thanks to the members of the Advanced Aerodynamics Department from AIRBUS INDUSTRIE for supplying us with the riblets and to Laurent Gross for his help throughout this work.

References

Augenstein, E., Bacher, E., Demeautis, C., Jagusinski, F. & Leopold, F. (1998) Influence de l'état de surface sur l'écoulement de culot en supersonique. Mesures par LDA 2D, 6e Congrès Francophone de Vélocimétrie Laser, Saint-Louis, France, September, 22-25.

Coustols, E. & Cousteix, E. (1994) Performances of riblets in the supersonic regime, *AIAA J.*, vol. 32, No. 2, pp. 431-433.

Gaudet, L. (1989) Properties of riblets at supersonic speed, *Appl. Scie. Research*, vol. 46, pp. 245-254.

Gaviglio, J., Dussauge, J.P., Debieve, J.F. & Favre, A. (1977) Behaviour of a turbulent flow, strongly out of equilibrium, at supersonic speeds, *Physics of Fluids*, vol. 20, No. 10, pp. 179-192.

Gold, H. (1963) Stability of Laminar Wakes, Ph.D. Thesis, California Institute of Technology, Pasadena.

Huerre, P. & Monkewitz, P. A. (1985) Absolute and convective instabilities in free shear layers, *J. Fluid Mech.*, vol. 159, pp. 151-156.

Koch, W. (1986) Direct Resonances in Orr - Sommerfeld Problems, *Acta Mechanica*, vol. 58, pp. 11 - 29.

Leopold, F. (1993) Simulation und stabilitätstheoretische Untersuchungen zum kompressiblen Nachlauf eines längsangeströmten Kreiszylinders, ZLR-Forschungsbericht 93-05.

Leopold, F., Augenstein, E., Damp, S., Christnacher, F. & Bacher, E. (1998) LDA measurements and visualization of the supersonic flow around a longitudinal cylinder with different surface roughnesses, 9th International Symposium of Laser Techniques to Fluid Mechanics, Lisbon, Portugal, July, 13-16.

Mack, L. M. (1984), Boundary Layer Stability Theory, Special Course on Stability and Transition of Laminar Flow, AGARD-R-709.

Monkewitz, P. A. (1988) A note on vortex shedding from axisymmetric bluff bodies, *J. Fluid. Mech.*, vol. 192, pp. 561-575.

Robinson, S.K. (1988) Effects of Riblets on Turbulence in a Supersonic Boundary Layer, AIAA-88-2628

RECEPTIVITY ISSUES FOR LAMINAR-FLOW CONTROL

J.D. CROUCH
Boeing Commercial Airplane Group
PO Box 3707, MC 67-LF, Seattle, WA 98124-2207, U.S.A.

Abstract

Boundary-layer receptivity is analyzed in the context of mean-flow-modification and wave-cancellation approaches to laminar-flow control. The adverse effects of receptivity for control via mean-flow modification are studied using linear amplitude methods to estimate the transition location. This includes both the receptivity and the linear growth of instabilities. For wave cancellation, the receptivity analysis provides the transfer function for a linear actuator. Distributed roughness and suction nonuniformities, suction strips, localized membrane actuators and suction/blowing actuators are considered.

1. Introduction

Laminar-flow control (LFC) offers the potential for very large reductions in airplane drag. The drag-reduction benefit, however, is partially offset by the weight and complexity of the LFC system. For LFC, or any other flow-control application, the ultimate measure of goodness depends on both the resulting change in value and the associated change in cost. Complexity of the flow-control system affects manufacturing costs, maintenance costs, and the degree of risks associated with application. A preferred flow-control system balances the performance improvement with the total system complexity. Thus the details of the control methodology are critical to the viability of flow control for practical applications.

The transition from laminar to turbulent flow in low disturbance environments is linked to the excitation, growth, and nonlinear interaction of instabilities. In general, LFC methods seek to delay transition by inhibiting the onset of the nonlinear interactions leading to breakdown. This can be achieved through a modification of the mean flow (basic state), or through linear wave cancellation. Mean-flow modifications (for example, due to suction) reduce the linear growth of the instabilities to delay transition. Wave cancellation affects the linear disturbances through the excitation of a wave form that is superposed on the initial disturbances.

The effectiveness of these LFC methods is influenced by the receptivity of the boundary layer to various source perturbations. Receptivity describes the process by which instability modes are excited. Significant levels of excitation occur only when there is a close coupling between the frequencies and length scales of the source perturbations and the instabilities. Depending on the method of LFC, the receptivity may be viewed as a hindrance or as a necessity. If the instabilities are being controlled through a modification of the mean flow, any receptivity resulting

G.E.A. Meier and P.R. Viswanath (eds.),
IUTAM Symposium on Mechanics of Passive and Active Flow Control, 151–158.

from the control can significantly degrade the system performance. For wave cancellation, on the other hand, the instabilities are controlled directly through the receptivity process. We consider receptivity associated with both the mean flow modification and wave cancellation approaches to LFC.

2. Receptivity formulation

For LFC applications, the most important instabilities are Tollmien-Schlichting (TS) instabilities and, for wing sweep greater than approximately 15 degrees, cross-flow (CF) instabilities. TS instabilities are traveling waves and CF instabilities can be stationary vortices or traveling waves. Receptivity describes the process by which these instabilities are excited. Stationary CF vortices can be directly excited by surface roughness or by variations in surface suction. Traveling CF instabilities and TS instabilities require an unsteady source with the correct length scales to match the instabilities at a given frequency; potential unsteady sources are surface modulations, suction/blowing, or free-stream disturbances that couple with steady surface-generated disturbances.

We consider the general case of a direct coupling between an instability and an unsteady source perturbation on the surface. The resulting disturbance decays to zero outside the boundary layer. The sources are modeled by the boundary conditions

$$u = v = w = 0 \quad \text{at} \quad y = \epsilon_r H(x,z)\exp(-i\omega t) \tag{1}$$

for surface modulation, and by

$$u = w = 0, \; v = -\epsilon_s H(x,z)\exp(-i\omega t) \quad \text{at} \quad y = 0 \tag{2}$$

for suction/blowing. The coordinates (x, y, z) and velocities (u, v, w) correspond to the chordwise, surface-normal, and spanwise directions, respectively. Steady perturbations are given by $\omega = 0$. The small parameter ϵ_r characterizes the modulation amplitude for the surface variation and ϵ_s the velocity amplitude for suction/blowing; the resulting velocity fields can be written as $\mathbf{v}_0 + \epsilon_r \mathbf{v}_r$ and $\mathbf{v}_0 + \epsilon_s \mathbf{v}_s$, respectively. The surface-modulation boundary condition is linearized using a Taylor expansion in terms of the amplitude ϵ_r to yield

$$u_r = -\frac{du_0}{dy}H(x,z)e^{-i\omega t}, w_r = -\frac{dw_0}{dy}H(x,z)e^{-i\omega t}, v_r = -i\omega H(x,z)e^{-i\omega t} \text{ at } y = 0. \tag{3}$$

Focusing on the modes which dominate the solution downstream, the modal disturbance amplitudes at the source location x_0 are given by

$$A(x_0;\beta,\omega) = \epsilon_s K_s(x_0;\beta,\omega)\tilde{\tilde{H}}(\alpha_e,\beta), \tag{4a}$$

and

$$A(x_0;\beta,\omega) = \epsilon_r K_r(x_0;\beta,\omega)\tilde{\tilde{H}}(\alpha_e,\beta), \tag{4a}$$

where A is defined as the disturbance-profile maximum (over y) – typically of the chordwise velocity component. K_s and K_r are the complex response functions (or receptivity coefficients). The geometry factor $\tilde{\bar{H}}$ is the Fourier transform of the surface variation evaluated at α_e, the instability chordwise wavenumber corresponding to the frequency ω and spanwise wavenumber β.

The excitation of traveling waves from steady suction or blowing is modeled by $\omega = 0$ in the surface boundary condition (2) with an unsteady free-stream condition. Here, the receptivity results from the interaction between the steady-suction generated disturbance, $\mathbf{v}_s$, and an unsteady disturbance, $\mathbf{v}_a$, that results from the free-stream modulation. In this case, the total velocity is given by $\mathbf{v}_0 + \epsilon_a \mathbf{v}_a + \epsilon_s \mathbf{v}_s + \epsilon_a \epsilon_s \mathbf{v}_{as}$. An acoustic disturbance, in the incompressible limit, is modeled by the boundary condition

$$u_a \to \exp(-i\omega t) \quad \text{as} \quad y \to \infty. \tag{5}$$

The acoustic receptivity due to the steady suction occurs at $O(\epsilon_a \epsilon_s)$ and is given by

$$A(x_0; \beta, \omega) = \epsilon_a \epsilon_s K_{as}(x_0; \beta, \omega) \tilde{\bar{H}}(\alpha_e, \beta). \tag{6}$$

The initial amplitude depends linearly on both the acoustic amplitude ϵ_a and the suction level ϵ_s. More details and results for the different receptivity mechanisms are given in [1].

3. Estimating the effects of mean-flow modification

Instability growth leading to transition can be reduced by modifying the boundary-layer thickness and profile shapes. Local stability analysis shows that "fuller" velocity profiles are more stable (or less unstable) to TS instabilities while the growth of CF instabilities is most strongly linked to the magnitude and shape of the cross-flow velocity profile. A fuller velocity profile can be achieved through favorable pressure gradient, through surface suction, or through surface cooling in air or heating in water. Control of cross flow, for a given wing sweep, usually requires suction. The combination of pressure-gradient control with an active means of control such as suction is referred to as hybrid LFC (HLFC). High Reynolds numbers, wing sweep, and other practical constraints motivate the use of HLFC for most applications to larger aircraft.

The use of suction introduces potential source perturbations that can partially offset the stabilizing benefits (see [2] for a discussion in the context of TS-instability receptivity). In the extreme case, oversuction can occur resulting in the complete loss of any suction benefit. Suction surfaces introduce velocity variations as a result of the flute-stringer layout, the suction-hole distribution, and nonuniformities in the hole porosity. Also, a suction surface is rougher than a typical solid LFC surface.

The effects of control can be studied using linear theory since the controller is acting on the instabilities during their linear evolution. The linear amplitude of a given mode, with frequency ω and spanwise wavenumber β, can be written as

$$A(x; \beta, \omega) = A(x_0; \beta, \omega) e^{N(x_0, x; \beta, \omega)} \quad , \tag{7}$$

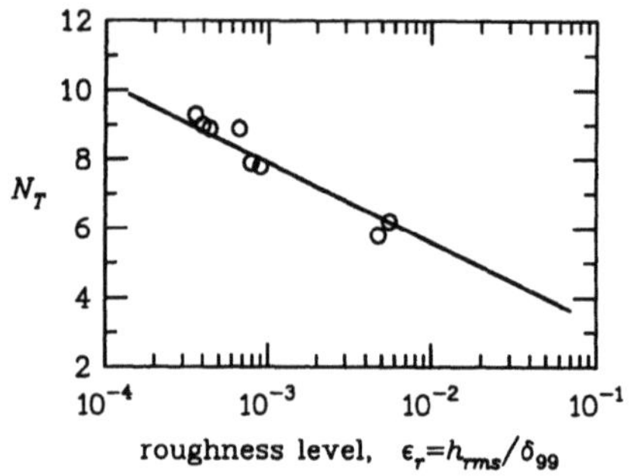

Figure 1: Stationary CF-instability N-factor variation with surface roughness level. Quasi-parallel calculations based on experiments of Radeztsky *et al.* [3].

with,

$$N(x_0, x; \beta, \omega) = \int_{x_0}^{x} \gamma(s; \beta, \omega) ds.$$

$A(x_0; \beta, \omega)$ is the initial amplitude at x_0, N is the amplification N-factor from x_0 to x, and γ is the linear growth rate.

The transition location is often estimated based on the e^N method. In this method, transition is taken to occur when the value of $N(x_I, x_0; \beta, \omega)$, referenced to the instability neutral point x_I for any given β or ω, exceeds an empirically defined value N_T. The value N_T depends on the type of instability (i.e. TS, stationary CF, or traveling CF instability). In its simplest form, the e^N method only accounts for the receptivity and nonlinear breakdown through the empirically defined value of N_T. The stabilizing effect of control is captured in the reduction of N as measured at a fixed downstream position.

A more physical approach, although still linear, is to link the transition onset with the instability amplitudes at transition. In this approach, transition is estimated to occur at x_T when the linear amplitude A exceeds an empirical transition-threshold value A_T,

$$\max_{\beta,\omega}[A(x_T; \beta, \omega)] = A_T \quad . \tag{8}$$

The value of A includes both the receptivity and the linear amplification, and the nonlinear physics of breakdown are embedded in the amplitude A_T. Linear amplitude methods, and the limiting case of a variable N-factor method, provide a more complete assessment of the effects of control by incorporating the receptivity.

3.1. DISTRIBUTED ROUGHNESS AND SUCTION NONUNIFORMITIES

We first consider the influence that roughness or suction nonuniformities have on transition due to stationary CF instabilities, $\omega = 0$. Assuming nonlocalized receptivity from a uniform source spectrum, the initial amplitude at the neutral point can be written as $A(x_I; \beta, 0) = \epsilon_r C_r$ for roughness and $A(x_I; \beta, 0) = \epsilon_s C_s$ for suction. In this case, the values of ϵ_r and ϵ_s are rms measures of the nondimensional roughness and suction nonuniformity, respectively.

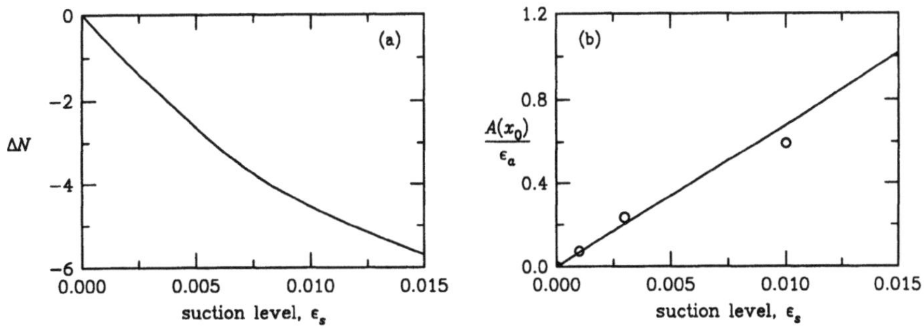

Figure 2: Effect of suction on: (a) the growth of a TS-wave as measured downstream of a suction strip and (b) the initial TS-wave amplitude resulting from acoustic receptivity due to the suction (normalized by the acoustic amplitude).

Combining the initial amplitudes with eq. (7), the transition criteria (8) can be written in terms of a variable N-factor. For example, in the roughness case this gives

$$N_T \equiv \max_{\beta}[N(x_I, x_T; \beta, 0)] = N_{Tr} - \ln(\epsilon_r), \tag{9}$$

where $N_{Tr} = \ln(A_T/C_r)$. Figure 1 shows the variation of N_T with roughness for the experiments of Radeztsky *et al.* [3]. The symbols show the values calculated from the experiment and the solid line is from eq.(9) with $N_{Tr} = 1$ chosen to provide the best fit. As the roughness level is increased, the enhanced receptivity reduces the the level of linear amplification needed to initiate transition. A suction surface would typically have a larger level of roughness resulting in a smaller value of N_T. In addition, the suction nonuniformities provide an excitation source that would show a variation of N_T with ϵ_s.

3.2. SUCTION STRIPS

We now consider a single suction strip used to control TS instabilities. In a two-dimensional flow, this is described by the boundary condition $v(x, y = 0) = -\epsilon_s H(x)$. The stabilizing effect of suction is shown in figure 2a in terms of the change in linear growth (N-factor) as measured at a fixed downstream position. The results are for a Gaussian suction distribution modeling a suction strip placed at the instability neutral point in a Blasius boundary layer [4].

In the presence of acoustic fluctuations in the free stream, the mean-flow distortion that stabilizes the boundary layer can also lead to the excitation of TS waves as described by eq.(6). For the conditions of figure 2a, the width of the suction strip is chosen to maximize the receptivity for a particular frequency $F = \omega\nu/U_\infty^2 = 56(10)^{-6}$. Calculating the values of K_{as} and $\tilde{H}$, eq.(6) gives the initial amplitude $A(x_0; 0, \omega) = 67.8\epsilon_a\epsilon_s$. This amplitude variation with ϵ_s is shown by the straight line in figure 2b. The symbols are numerical simulation results of Spalart taken from [4]. The theory and simulation are in good agreement – showing that the initial amplitude, normalized by the acoustic amplitude,

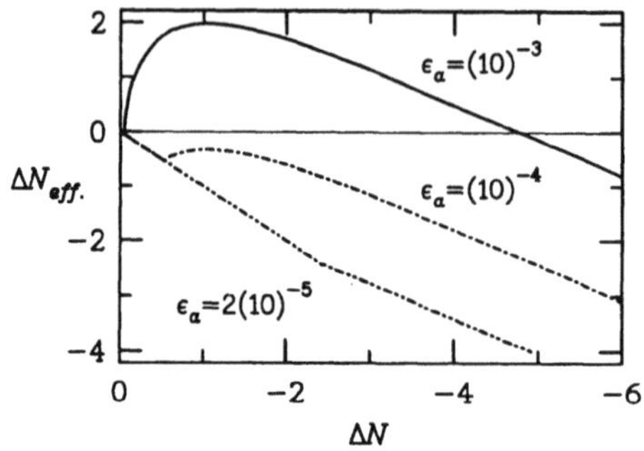

Figure 3: Effective N-factor changes (acounting for the enhanced receptivity) as a function of the change in the growth for three different values of the acoustic amplitude ϵ_a at the conditions of figure 2.

increases with the suction level.

To see the potential relevance of these results for LFC, we consider the effective change in the N-factor when accounting for the receptivity. In the absence of the suction slot, transition is assumed to occur at an N-factor value of $N_T \approx 9$. Experimental results for TS-wave transition give a transition amplitude of $A_T \approx 0.05$ [5]. The "background" initial amplitude for the critical mode causing transition can be backed out of eq.(7), $A_T e^{-N_T} \approx 6(10)^{-6}$; this value is consistent with the experimental results of Kosorygin [6]. If the suction-slot receptivity amplitude is less than the background level, the suction does not have an adverse effect and the effective change in growth is given by ΔN from figure 2a.

When the suction-slot receptivity amplitude $A(x_0; 0, \omega)$ for the critical mode exceeds the background level, the level of instability growth required to initiate transition is reduced by $\ln[A(x_0; 0, \omega)/\ (A_T e^{-N_T})]$. The stabilizing effect of the suction is partially offset by the increase in the initial amplitudes. Thus the total effect of the suction slot can be characterized by an *effective* change in the N-factor, $\Delta N_{eff.}$, which is greater than or equal to ΔN

$$\Delta N_{eff.} = \max \left\{ \begin{array}{l} \Delta N \\ \Delta N + \ln(A(x_0; 0, \omega)/A_T) + N_T, \end{array} \right. \tag{10}$$

where N_T is taken to be 9 and $A_T = 0.05$. Figure 3 shows an example of the effective N-factor variation versus the calculated N-factor change based on stability alone. This assumes that the dominant modes at transition are affected by the suction similar to the mode considered in figure 2. For the larger acoustic amplitude, $\epsilon_a = (10)^{-3}$, the effective change to the N-factor is positive (i.e. transition moves forward due to suction) unless the suction level is very large (i.e. where $\Delta N < -5$). For acoustic amplitudes of $\epsilon_a = (10)^{-4}$ or below, the net effect of the suction is always to delay transition. As the acoustic level is lowered, the adverse effects of the suction only occur at the higher suction levels. For example, at the smallest acoustic amplitude shown, the adverse effects due to receptivity are only significant for suction levels large enough to produce $\Delta N < -3$.

Although these results are based on a suction slot which maximizes the level of receptivity, a different suction slot would be similar to a change in the acoustic amplitude since A depends linearly on both ϵ_a and $\epsilon_s \bar{\bar{H}}(\alpha_e, \beta)$. The results of

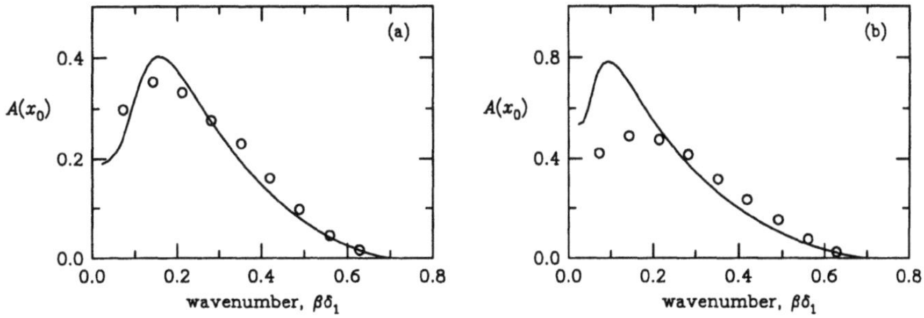

Figure 4: Spanwise wavenumber spectra resulting from a modulated surface membrane in a three-dimensional boundary layer, (a) $F = 22.2(10)^{-6}$, (b) $F = 36.5(10)^{-6}$. Symbols from the experiment of Gaponenko *et al.* [10].

figure 3 also illustrate the potential effects of localized regions of suction used to control steady CF instabilities. In this case, the receptivity depends only on the suction without the small parameter ϵ_a, as shown by eq.(4). Thus $\tilde{\bar{H}}(\alpha_e, \beta)$ is the key parameter controlling the level of CF-instability excitation.

4. Direct excitation from a localized source

For wave cancellation, we consider the receptivity from an unsteady (or steady) source on the model surface. The basic mechanisms for a modulated surface or unsteady suction/blowing are essentially the same. There have been a number of studies using wave cancellation as a means for LFC (see for example, [7, 8, 9]). These studies consider the entire control scheme. Here we focus on modeling the linear transfer function describing the dominant downstream modes from an actuator input.

We consider the general problem of the excitation of stationary and traveling instabilities in a three-dimensional boundary layer. The mean flow is described by the Falkner-Skan-Cooke velocity profiles as considered in the experiments of Gaponenko *et al.* [10]. In the experiments, traveling CF instabilities were excited by a well-documented modulated surface membrane. The resulting disturbance field was measured at a sequence of downstream positions and transformed into spanwise-wavenumber spectra. These spectra were then projected upstream to the source to provide the initial disturbance field. This process was repeated for different modulation frequencies, and the initial stationary-CF amplitudes were estimated by extrapolating to $\omega = 0$.

In the analysis, this problem is modeled by the linearized boundary condition (3). The initial amplitudes are given by eq.(4a). Figure 4 shows the comparison between the linear receptivity theory and the experiment for two values of the frequency $F = \omega\nu/U_0^2$ (where U_0 is the edge velocity at the source location). The Hartree parameter is $\beta_H = 0.42$ and the local Reynolds number is $R = U_0\delta_1/\nu = 498$ based on the displacement thickness δ_1. At the lower frequency, the agreement is very good. At the higher frequency, the theory overpredicts the experiment at wavenumbers below $\beta\delta_1 = 0.2$ – this difference increases with increasing frequency. The results are in good agreement over the most-unstable band of wavenumbers $\beta\delta_1 \approx 0.3$ to 0.6.

The results show that the receptivity theory provides a good model for the transfer function associated with unsteady and steady surface perturbations. Localized suction/blowing actuators, as described by eq.(2), should provide a comparable level of agreement.

5. Conclusions

The adverse effects of receptivity for mean-flow-modification LFC are studied using linear amplitude methods that incorporate both receptivity and linear growth. The effects of receptivity are characterized by a reduction in the growth required to initiate transition. These effects are quantified in terms of a reduced transition N-factor N_T (for distributed roughness and suction nonuniformities), and an effective change in N-factor $\Delta N_{eff.}$ (for localized suction strips). For wave-cancellation LFC, receptivity theory is used to model the transfer function for actuators embedded in the surface. Results show good agreement between the linear theory and experiments for predicting the initial modal content associated with a vibrating membrane actuator.

This work has benefited from collaborations with P.R. Spalart, L.L. Ng, V.S. Kosorygin and Y.S. Kachanov.

6. References

[1] Crouch, J.D. (1994) "Theoretical studies on the receptivity of boundary layers," AIAA Paper No. 94-2224.

[2] Kerschen, E.J. and Choudhari, M. (1985) "Boundary layer receptivity due to the interaction of free-stream acoustic waves with rapid variations in wall suction and admittance distributions," *Bull. Am. Phys. Soc.*, **30**, 1709.

[3] Radeztsky, R.H., Reibert, M.S., Saric, W.S. and Takagi, S. (1993) "Effect of micron-sized roughness on transition in swept-wing flows," AIAA Paper No. 93-0076.

[4] Crouch, J.D. and Spalart, P.R. (1995) "A study of non-parallel and nonlinear effects on the localized receptivity of boundary layers," *J. Fluid Mech.*, **290**, 29-37.

[5] Crouch, J.D. (1997) "Transition prediction and control for airplane applications," AIAA Paper No. 97-1907.

[6] Kosorygin, V.S. (1996) private communication.

[7] Liepmann, H.W. and Nosenchuck, D.M. (1982) "Control of laminar-instability waves using a new technique," *J. Fluid Mech.*, **118**, 187-200.

[8] Cohn, R.K. and Gaster, M. (1992) "Active control of a wavepacket in a Blasius boundary layer," *Bull. Am. Phys. Soc.*, **37**, 1812.

[9] Fan, X., Hofmann, L.M. and Herbert, Th. (1993) "Active flow control with neural networks," AIAA Paper No. 93-3272.

[10] Crouch, J.D., Gaponenko, V.R., Ivanov, A.V. and Kachanov, Y.S. (1996) "A method of experimental determination of the linear receptivity coefficients of a 3D boundary layer to microscopic surface non-uniformities. Verification of theory," in: *International Conference on Methods of Aerophysical Research. Proceedings, Part II.*, Novosibirsk, 30-35.

CONTROL OF BOUNDARY LAYER TRANSITION AT HIGH FST BY LOCALIZED SUCTION

A.A.BAKCHINOV [1,2], M. M.KATASONOV [2],
P.H.ALFREDSSON [1], V.V.KOZLOV [2]
[1] *Dept. Mechanics, Royal Institute of Technology, S-10044 Stockholm, Sweden.*
[2] *Institute of Theoretical and Applied Mechanics, Russian Academy of Sciences, Siberian Branch, 630090 Novosibirsk, Russia.*

1. Introduction

Recently great attention has been paid to streamwise oriented structures (streaky structures-"puff"-structures, longitudinal localized structures) which appear during laminar-turbulent transition in boundary layer flows subjected to high levels of free stream turbulence (FST). The free stream disturbances give rise to streamwise oriented streaky structures inside the boundary layer. These structures, which appear randomly in the spanwise direction, consists of regions of high and low streamwise velocity. Their amplitude increases in the downstream direction and the resulting rms-level of the streamwise fluctuations grows in linear proportion to the boundary layer thickness and may reach up to 10% of the free stream velocity before transition occurs. The breakdown of the streaks into turbulence seems to originate from the development of a high frequency secondary traveling wave instability which results in the formation of turbulent spots.

In many situations it would be valuable to be able to control transition for these types of flow and in a previous experiments [Ref. 1] at "natural" conditions the influence of suction on a boundary layer transition point was studied. It was shown that suction through a spanwise slot was able to change the transition point downstream.

G.E.A. Meier and P.R. Viswanath (eds.),
IUTAM Symposium on Mechanics of Passive and Active Flow Control, 159–164.

2. Experimental set-up

In the present investigations we have used the MTL wind tunnel at KTH to study the influence of localized suction on the development of the streaky structures. The measurements were made at a free stream velocity U_o=5.8 m/s using hot-wire anemometry. The streaky structures were modeled in a similar way as previously was used in Refs. [2,3], i.e. by injecting a short duration disturbance through a 7 mm wide slot which is located at x=90 mm (x=0 is at the plate leading edge). The resulting streaky structure consists of low velocity region of a spanwise extent of 5 mm if measured at x =189 mm, and with a positive region at each side. On top of this disturbance a high frequency wave disturbance (200 Hz) is generated which interacts with the streaky structures and eventually leads to breakdown. In this case both disturbances are generated from the same slot, by connecting two loudspeakers in parallel which are driven by separate amplifiers. The inputs to the amplifiers are computer controlled.

Suction is applied through a spanwise slot which is located at x =189 mm. In order to arrange for localized suction the slot is covered by a thin plastic tape and holes at the desired positions are made in the tape. The holes have a diameter of approximately 0.5 mm. The suction rate through the holes was very small and an estimate of the flow was made
by putting the hot wire just above the hole, with no flow in the tunnel. The measured velocity was around 0.5 m/s. Another way to estimate the flow through the hole is to measure the disturbance inside the boundary layer and it was found that at x =250 mm (i.e. 60 mm downstream the suction hole) the streamwise velocity increase was about 3% U_o and the affected spanwise region was less than 3 mm wide.

3. Results and discussion

The measurements were made by studying the hot-wire signal at different x-positions, where measurements with and without suction were made sequentially. A typical example is shown in figure 1, which shows (z,t)-planes of the disturbance, first at x =189 mm, i.e. just before suction was applied and then the corresponding (z,t)-planes at x =500 mm with and without suction.

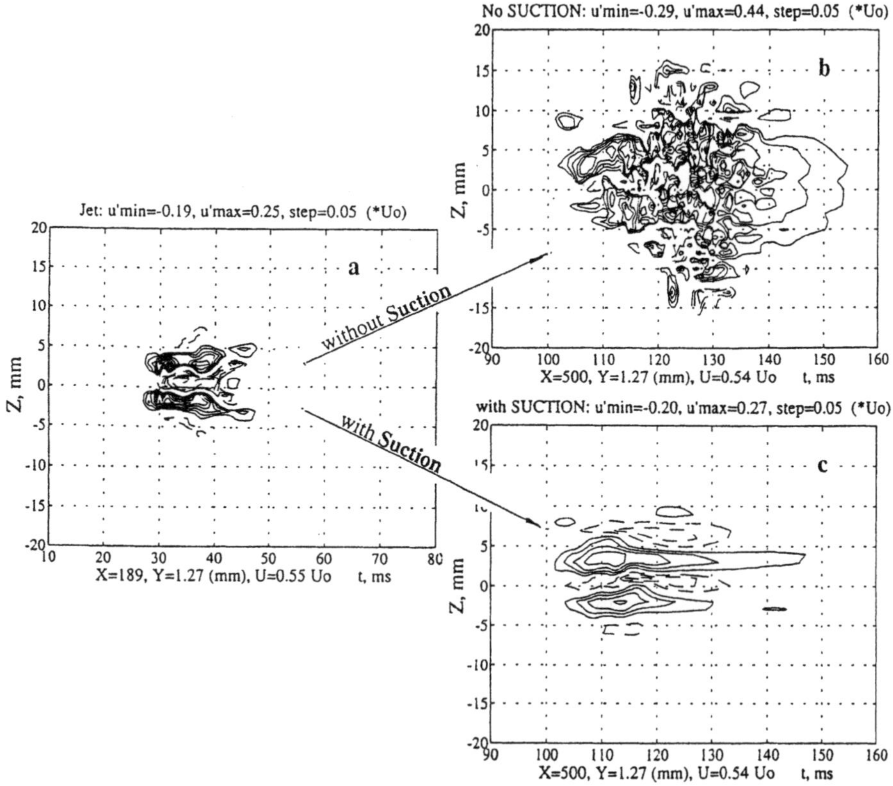

Figure 1. Contour lines of streamwise velocity disturbance, u', in (z,t)-plane at y =1.27 mm.a) x =189 mm, just before suction hole. b) x =500 mm, no suction, showing that the structure has broken down to a turbulent spot, c) x =500 mm, suction through one hole at x =190 mm, z =0, preventing breakdown.

Without suction the disturbance has already broken down to a turbulent spot (this figure is not averaged and only one realization is taken at each z-position which explains the irregularities of the contour lines). With suction on the other hand the streaky structure show no signs of breakdown, and has become elongated as compared to the structure at x =189 mm although its amplitude has only increased slightly. These figures clearly demonstrate the effect of minute localized suction, despite the fact that the spanwise size of the low velocity region is an order of magnitude larger than the size of the hole.

Contour lines of the streamwise velocity disturbance in (z,t) -plane at x =250 mm are shown in figure 2. In the figure 2a the disturbance without suction is plotted, whereas a result of subtraction of the disturbance with continuous suction from that without suction are presented (figure 2b). The spanwise distribution of ΔU_{mean} (see figure 2b) demonstrates the local in z-direction influence of the suction on the mean velocity profile. The suction was found to result in effective damping high frequency oscillations ridding on the low-frequency streaky structures.

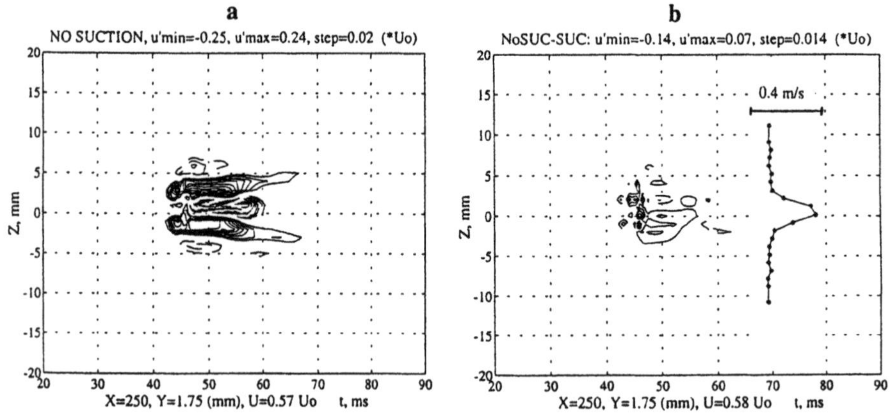

Figure 2. Contour lines of streamwise velocity disturbance, u', in (z,t)-plane at y =1.75 mm, x =250mm. without suction. b) the rest from subtraction the u'-flow field with suction from that without suction. In the right figure the ΔU_{mean} (z) distribution is also shown, where $\Delta U_{mean} = \Delta U_{mean}$ (NoSuc)- ΔU_{mean} (Suc).

Figure 3a and b show the structure of the streamwise velocity perturbations in (y,t)-plane at x =250 mm, z =0 with and without suction respectively. The distributions look similar at this x-position. By subtracting figure 3b from 3a we have obtained a distribution shown in figure 4a. The same as in (z,t)-plane the vertical distribution of u' revealed that the continuos suction reduced high frequency wave packet, which is observed in the middle of the boundary layer. Distortion of the mean velocity Y-profile due to the suction is shown in figure 4b. As it can be expected the influence of the suction on the boundary layer shear layer is located in the vicinity of the plate surface. But the maximum effect on the disturbance in suction case was found to occur at higher distance from the surface.

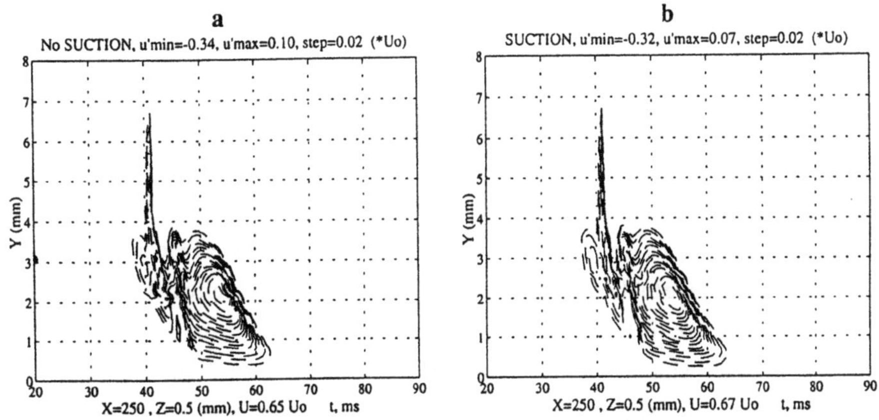

Figure 3. Contour lines of streamwise velocity disturbance, u', in (y,t)-plane at z=0, x=250 mm.

a) without suction. b) with suction.

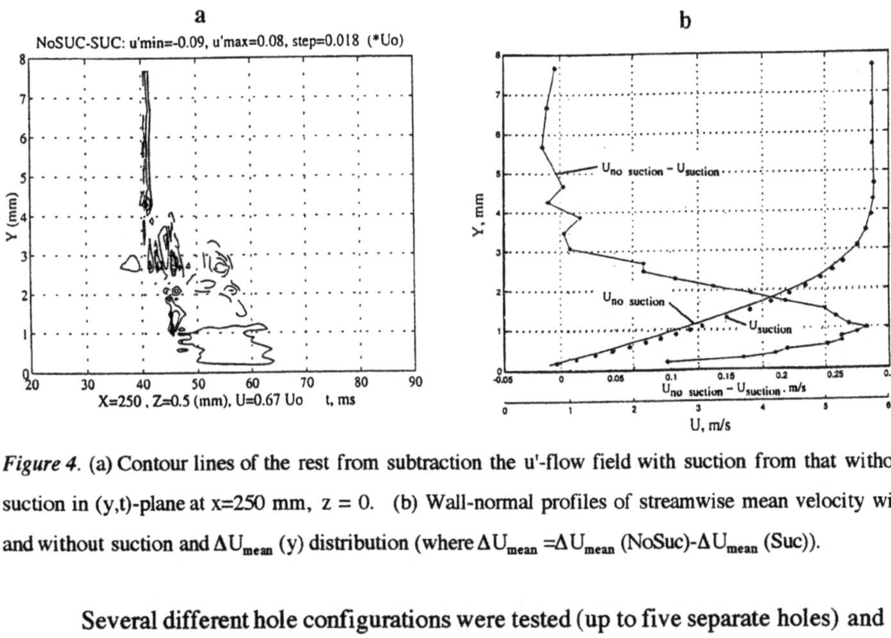

Figure 4. (a) Contour lines of the rest from subtraction the u'-flow field with suction from that without suction in (y,t)-plane at x=250 mm, z = 0. (b) Wall-normal profiles of streamwise mean velocity with and without suction and ΔU_{mean} (y) distribution (where $\Delta U_{mean} = \Delta U_{mean}$ (NoSuc)-ΔU_{mean} (Suc)).

Several different hole configurations were tested (up to five separate holes) and in all cases where suction was applied to the low velocity region there was a damping effect. On the other hand when suction was applied in the shear layer between the high and low velocity regions there was in fact an enhancement of the breakdown process. Also when suction was applied to the high velocity regions transition was enhanced.

Another interesting aspect of boundary layer control is attempts to control turbulent boundary layers in order to reduce skin friction. A prominent feature also of turbulent boundary layers is elongated regions of high and low streamwise velocity in the immediate vicinity of the wall. These streaks are believed to be strongly coupled to turbulence production and a possible way to control the turbulence generation process would be to apply suction at the wall below the low velocity streaks. The width of these streaks are of the order of 50 viscous units which for a [illegible]ow of 10 m/s corresponds to approximately 1.5mm, whereas for a velocity of 100 m/s it decreases to slightly above 100 microns.

In the present experiments we observed that suction could be very efficient if applied in the right position relative to the streaks, but could have a diametrical effect if applied at the wrong place. Therefore such devices in a turbulent boundary layer must both be small and be able to react with short time response. To apply a control strategy based on localized suction for high speed applications may at present seem to be futuristic, but may turn into a viable concept if the development of MEMS-technology will see a similar development as we have observed for miniaturization (of both size and cost) of electronic components.

Acknowledgment

The visits of Prof. Kozlov V.V. and Katasonov M.M. was financed through the Swedish-Russia Exchange program of The Royal Swedish Academy of Sciences (KVA) and under Grand RFFR 96-01-01892. Dr Bakchinov A. A. was supported through a Goran Gustafsson post-doc position.

References

1.Grek, G.R., Kozlov, V.V. & Ramasonov, M.P. Investigation of boundary layer stability in a gradient flow with a high degree of free stream turbulence. *Fluid Dynamics*, **25** (1990) 207-213.

2.Bakchinov,A.A., Westin,K.J.A., Kozlov,V.V. & Alfredsson,P.H. Experiments on localized disturbances in a flat plate boundary layer. Part 2. Interaction between localized disturbances and TS-waves. Eur. *J. Mech. B/Fluids* (1998) (accepted for publication).

3.Bakchinov, A.A., Grek, G.R., Katasonov, M.M. & Kozlov, V.V. Experimental study of the interaction between streaky structures and high frequency disturbances, in A.N.Anfimov (ed.) *International Conference on Experimental Fluid Mechanics*, Central Research Institute of Machine Building, Korolev (1997) 28-33.

EXPERIMENTS ON THE CONTROL OF CROSSFLOW INSTABILITY WITH THE AID OF SUCTION THROUGH PERFORATED WALLS.

H. BIPPES, M. WIEGEL AND F. BERTOLOTTI
DLR Institut für Strömungsmechanik
Bunsenstraße 10, D-37073 Göttingen, Germany

1. Introduction

In swept wing flow of large transport aircraft, transition occurs predominantly in the leading edge area due to leading edge contamination. Therefore, attempts to keep swept wing flow laminar have been concentrated on this wing portion. Recent experiments have demonstrated that suction enables the relaminarisation of the attachment line flow [1], [2]. This success inspired aircraft industry and research to study the feasibility of laminar swept wing flow by damping crossflow instability. In flight and wind-tunnel tests known so far much less success was achieved than predicted by stability analyses. This may partly be due to inadequacies in the suction techniques which do not allow to approximate satisfactorily uniform suction as assumed in stability analyses. On the other hand the stability problem in the leading edge area of swept wings especially where stagnation flow instability switches over to crossflow instability is not well understood. Moreover, the strong non-parallelism in this area leads to numerical difficulties. In order to provide a better physical insight into this problem basic experiments are needed. The present work contributes to this goal. It is specifically aimed at studying the stabilisation of crossflow instability by means of suction isolated from stagnation flow instability. The results should make obvious the influence of the effect of the shortcomings in the suction technique and give some hints on possible improvements and benefits.

2. The experiment

In favour of a realistic estimate of the effect of suction on crossflow instability, for the present study the same model was used as in previous experiments on polished surface which provide an extensive data base for comparison. It is a swept flat plate where the pressure gradient is induced by a wing-like body outside the boundary layer. Details of the model design and the pressure distribution are reported in [3]. For the suction experiment, the front part of the plate is equipped with a suction device (Fig. 1a). Along the chord, the whole suction area is subdivided in four chambers (Fig. 1b) not only to account for the chordwise pressure gradient but also to enable the study of the effect of suction limited to various states of the instability process. At the conditions chosen for the

G.E.A. Meier and P.R. Viswanath (eds.),
IUTAM Symposium on Mechanics of Passive and Active Flow Control, 165–170.

tests, the flow becomes unstable only well downstream of the stagnation area. Suction is applied through holes spaced 400 μm apart as in previous experiments and flight tests. The nominal diameter is 50 μm but due to manufacturing deficiencies, it varies in a wide range. Some of the holes are even closed. This results in a significant non-uniformity of the suction velocity.

The tests are performed in a wind tunnel with an open 1m × 0.7m test section and a turbulence level of $Tu = 100/Q_\infty \sqrt{1/3\ (\overline{u'^2} + \overline{v'^2} + \overline{w^2})} = 0.15\%$, in the frequency range 2 Hz < f < 2 kHz. The unsteady velocity field is measured with the use of hot-wire anemometry (for further details see [3]). Only for the determination of the suction velocity distribution, LDA is applied to extend the measurements closer to the wall.

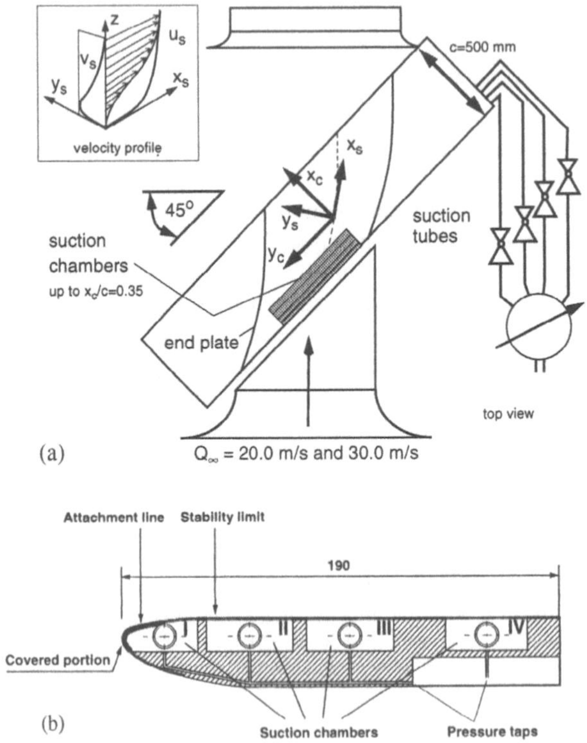

Figure 1. (a) Experimental set-up. (The displacement body which generates the pressure gradient on the flat plate is not shown. For details of the model design we refer to [3]). (b) The arrangement of the suction chambers in the front part of the model.

3. Results

Preliminary tests aimed at optimizing suction revealed that suction in the first chamber actually increases the disturbance content. Only if the suction surface was covered beyond the stagnation line (see Fig. 1b), the negative effects could be avoided. Not using this chamber altogether and applying suction only in chambers II, III, IV, denoted case (0, II, III, IV), turned out to be more effective.

In order to demonstrate the attenuation of the disturbances due to suction, in Fig. 2 the spanwise velocity distribution in the boundary layer is depicted. It was measured at a chord position in the nonlinear regime of the instability process at the wall distance z/δ of maximum disturbance amplitudes. The spanwise variation of the mean velocity $\overline{U}_s(y_c)/Q_e$ is due to the presence of stationary vortices (left column) whereas the size of the travelling modes is directly given by the r.m.s. fluctuations u_{rms}/Q_e (right column). The spanwise variation of the unsteady disturbances indicates the nonlinear state of the disturbance development. Comparing the measurements with and without suction, it becomes obvious that suction has considerably damped the growth of the unsteady modes whereas the amplitudes of the stationary modes are only marginally influenced. In previous experiments on the model without suction insert, i. e. on

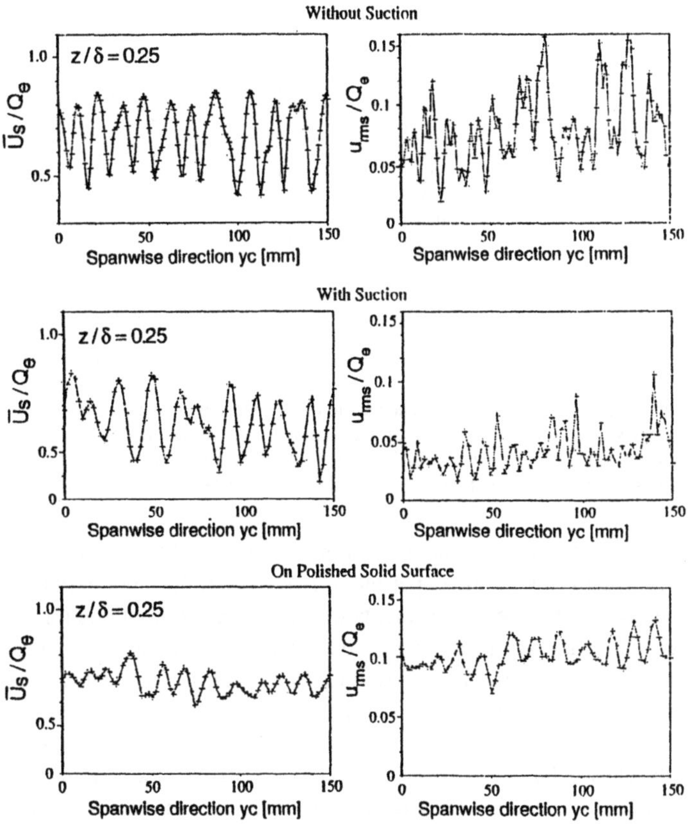

Figure 2. Spanwise variation of the mean velocity ($\overline{U}_s/Q_e$) identifying steady disturbances and r.m.s. amplitudes of the unsteady modes ($u_{s,rms}/Q_e$) developing with and without suction. Q_e is the velocity of the external flow, $c_q = \overline{\overline{W}}_{suct}/Q_\infty$ is the suction rate. Q_∞ = 20.4 m/s (free stream velocity), Φ_∞ = 43,5° (sweep angle), z/δ = 0.25, and x_c/c = 0.9.

polished surface but at the same flow conditions, even smaller amplitudes of the stationary vortices were measured than with suction (on the model with suction insert) whereas the travelling modes were found to be roughly of the same size. In order to also attenuate satisfactorily the stationary vortices which have been shown in [3] to be most dangerous for transition, the suction rates had to be increased up 0.5 %.

For the physical interpretation of these observations, two results of related stability calculations may be considered. Fig. 3a, shows the chordwise growth of the stationary vortices i. e., the disturbances most dangerous for transition, for different chambers activated. It turns out that in contrast to the experiment, the largest effect is predicted for case (I, II, III, IV), i. e. with all the suction chambers activated. Fig. 3b compares the decrease in the disturbance growth of the most amplified instability modes due to suction in case (0, II, III, IV). It becomes obvious that the stabilisation effect decreases

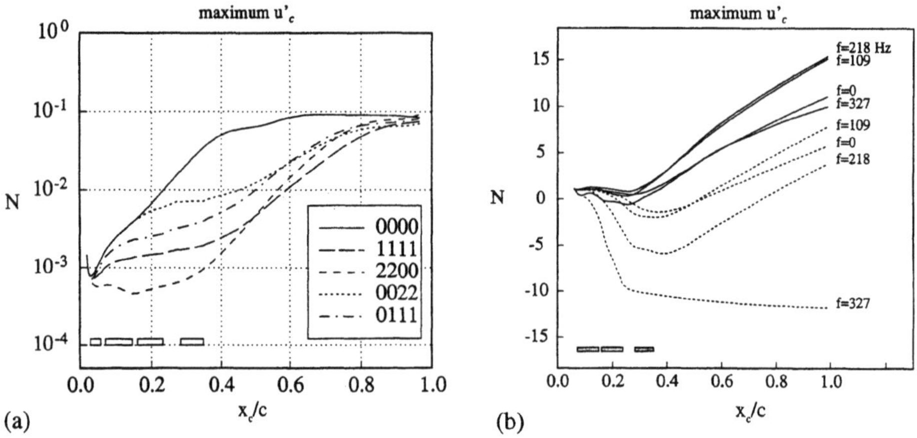

Figure 3. Stability calculations to the experiment. The location and extension of suction chambers is indicated by the shaded rectangles. (a) Nonlinear growth of the stationary crossflow modes for various suction dispositions. The numbers to the curves give the suction rates in the 4 chambers in ‰. (b) Comparison of the linear growth of the most unstable modes without (——) and with suction (- - - -), case (0, II, II, IV).

with the frequencies of the unstable modes and is smallest for the stationary vortices (f =0). In tendency, this agrees with the experiment, However, it does not explain why the stabilisation is negligible. Former experiments on the model without suction insert [3] have shown that stationary vortices become dominant even if they are linearly less amplified than the unsteady modes, in the presence of a rough surface or low free stream turbulence. This suggests that with the suction technique currently in use, stationary vortices are strongly excited at very large amplitudes. Comparative tests indicated that this is not mainly caused by the increase in roughness due to the perforation of the surface since the main difference in the excitation of stationary modes was found on the model with suction insert between the tests with and without suction. Another possible cause was seen in the non-uniformity of the suction velocity. LDA measurements revealed a significant variation along the surface which is due to tolerances in the hole diameter and sporadically appearing closed holes. This non-uniformity only appears closer to the wall than 0.5 to 1mm. They are similar to the perturbations due to surface

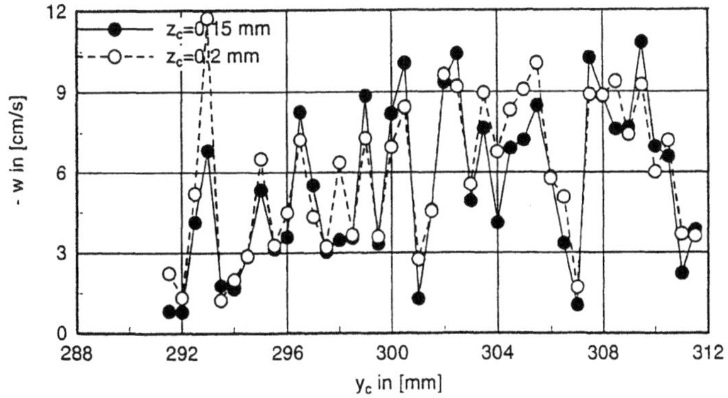

(a)

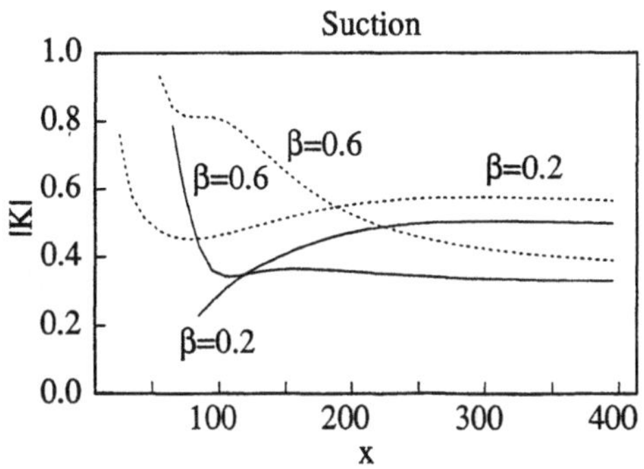

(b)

Figure 4. (a) LDA traces of the suction velocity across the surface at 2 wall distances z. (b) Receptivity coefficients calculated for swept Hiemenz flow based on parallel (- - - -) and non-parallel (——) approaches.

roughness, and therefore stimulate stationary vortices in the same way. Our receptivity calculations for swept Hiemenz flow demonstrate that such non-uniformity in the suction velocity leads to considerably larger receptivity coefficients than surface roughness and thus to a strong excitation of stationary vortices (Fig. 4a) which reduces the possible delay of transition. This consideration infers that an improvement in the suction technique would grant a more extensive use of the benefit of suction.

An additional interesting feature becoming obvious in Fig. 4a is the strong decrease of the receptivity coefficient with increasing distance from the attachment line. This could explain why the best results were obtained without suction in chamber I. In order to get an overview on the influence of the location of suction on the stabilisation, further tests were performed with different groupings of activated suction chambers. The result is summarized in Fig. 5 where the r.m.s. amplitudes of the unsteady disturbances, measured at $x_c/c = 0.9$, are plotted versus the total suction rate, i. e. the sum of the flow rates in the activated chambers. In addition, the total flow rate of suction is given to each measuring point. This way it becomes obvious that with equal suction rates in the individual chambers, case (I, II, III, IV), leads to a decrease of the unsteady modes which is slightly larger than for case (0, II, III, 0) but needs a much larger total suction rate. For this reason, an increase in the suction rates in the two chambers to have the same total rate as in case (I, II, III, IV) would increase the efficiency of suction. It should

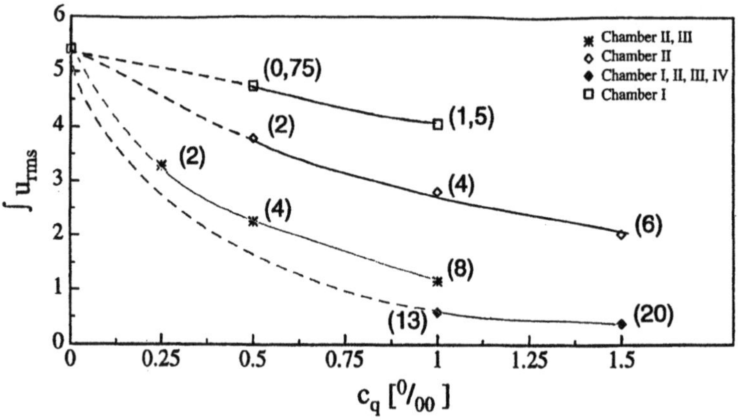

Figure 5. Reduction of unsteady disturbance content due to suction for various suction groupings. The number in the brackets give the total suction flow-rate.

be mentioned that in our experiments up to suction rates of 2 - 3% an over-suction could not be detected.

Detailed measurements of the disturbance growth showed all the features observed without suction except of saturation. This phenomenon is not yet understood and needs further experiments.

4. Conclusions

The experiments have shown that suction can be a powerful means to stabilise three-dimensional boundary layer flow to crossflow instability. In agreement with theory, travelling modes are more damped than stationary modes. In addition, especially in the proximity of the stagnation area, non-uniformity in the spatial distribution of suction is a strong agent of streamwise vortices. Therefore, suction rates > 0.5 % were needed to also reduce the stationary vortices and to make use of the full potential of suction. Especially in the case of swept wing flows when suction has to be applied already at the stagnation area because of leading edge contamination, economic use of suction requires an improvement in the suction technique. Laminar flow on the attachment line without suction would improve the stabilisation effect of suction in the subsequent boundary layer flow subject to crossflow instability.

5. References

1. Juillen, J., C. & Arnal, D.: Experimental study of boundary layer suction effects on leading edge contamination along the attachment line on swept wing. *In Laminar-Turbulent Transition, IUTAM Symp., Sendai, Japan, 1994* (ed. R. Kobayashi), pp. 137-144. Springer, 1995.
2. Poll, D. I. A & Danks, M. 1994 Relaminarisation of wept wing attachment-line by surface suction. *In Laminar-Turbulent Transition, IUTAM Symp., Sendai, Japan, 1994* (ed. R. Kobayashi), pp. 137-144. Springer, 1995.
3. Deyhle, H. & Bippes, H. Disturbance growth in an unstable three-dimensional boundary layer and ist dependence on environmental conditions. *J. Fluid Mech.* **316** (1996), 73-113.

Control of Crossflow Instability Field by Selective Suction System

EGAMI, Yasuhiro and KOHAMA, Yasuaki
Institute of Fluid Science, Tohoku University, Sendai 980-8577, Japan

1 Introduction

In order to obtain the overall aerodynamic drag reduction for next generation large aircraft development, it is necessary to control the Crossflow instability generated on the main wing. However, mainly because of its complexity, detailed transition process is not well understood. It can be said that at least there are two explanations for the transition process. One explanation is that the primary instability appears as streamwise stationary instability (crossflow vortices), and which drives unsteady secondary instability (inflectional instability) locally along primary instability [1, 2]. Other explanation is that the unsteady wave instability becomes absolutely unstable at certain Reynolds number and eventually brings the flow field into fully turbulent state [3]. Further investigations will be needed until detailed and exact transition process is detected, but we believe that at least in the case of a swept wing flow, transition is being occurred by former process. In our understanding, key process in the crossflow dominant turbulent transition is the generation of the secondary instability. Vortical motion of the crossflow vortices (streamwise vortices) pumps up low momentum boundary layer flow near the wall up to the main flow region where high momentum flow prevails. As a result, unstable flow condition, high shear with point of inflection is created along each streamwise vortex. After the appearance of 7 to 10 cycles of the secondary instability, full turbulent transition is reached. Therefore, it can be said that critical mechanism for turbulent transition is the pumping up of the low momentum flow by the crossflow vortices. Namely, crossflow vortices pump up low momentum boundary layer flow existing very close to the wall surface to much higher momentum flow region near the boundary layer edge, and as a result, unstable flow condition is created locally.

One of the important characteristics in the transition region is that crossflow vortices tend to appear quite stable at the position where they started to appear at first. Therefore, it is quite possible to groove streamwise suction slits along the crossflow vortices accurately. Taking such flow situation in mind, we got an idea that it may be effective if we could suck selectively such low momentum flow being gathered by the streamwise vortices under their foot regions before they are pumped up. In order to do so, selective suction holes are grooved to the swept flat plate of the present model.

2 Experimental set-ups

Experiment is performed using Low-Turbulence Wind Tunnel [4] equipped at the Institute of Fluid Science, Tohoku University, Japan. Figure 1 shows open jet type working section of the wind tunnel with the experimental model, a swept flat plate with upper displacement body. Sweep angle of the plate (Chord, C=800mm, Span, S=1520mm) is kept constant at λ=45°. NACA0015 wing profile is used for the displacement body in order to get constant favorable pressure gradient on the flat plate. From laminar to full turbulent transition state is created in three-dimensional (3D) boundary layer developed

G.E.A. Meier and P.R. Viswanath (eds.),
IUTAM Symposium on Mechanics of Passive and Active Flow Control, 171–176.

on the flat plate by using present system. The flat plate surface is polished to reduce the surface roughness as much as possible. The surface roughness value defined as averaged peak-to-peak value is 0.52μm and maximum peak-to-peak value is 1.36μm.

Details of the suction system are shown in Figure 2. In figure 2(d), Visualized secondary instability is shown. In order to control this secondary instability selectively, two types of suction surfaces, system [A]: uniform suction and [B]: selective suction, are used. Diameter of suction holes is 0.5mm and hole interval is 1mm. Suction holes are distributed along each crossflow vortex axis in the case [B]. In the case of selective suction [B], 4 lines (4mm) of suction holes are drilled for each crossflow vortex and this set of line suction holes is distributed at 11mm intervals in spanwise direction in accordance with the vortex interval. The ratio of selective suction holes to uniform suction is 4/11.

Suction is applied at x/C=0.65-0.75 at around transition region in the case of $Re = U_\infty C/\nu = 6.6\times10^5$. Air is sucked by centrifugal fan through the plenum chamber equipped under the flat plate. Suction rate is measured by flow meter. Hot-wire anemometory is used to measure unsteady motion of the flow in the boundary layer. Hot wire probe is traversed precisely by three-dimensional traversing system controlled by computer at 0.01mm step accuracy. Acquired output signal is stored in a hard disk of the computer and analyzed. Turbulence level of the freestream is 0.016% of U_∞ in the measured range.

3 Results and discussion

3.1 Comparison with selective suction and uniform suction

Figure 3 shows the effect of the suction to the transition. Flow condition is adjusted so that the secondary instability appears near the center of the suction surface. Applied suction rate is 0.03 m^3/m^2s (0.03m/s at the uniform suction case and 0.087m/s at the selective suction case.) It corresponds to $2.8\times10^{-5}m^3/s$ per one vortex. Figure 3(a) shows the results without suction. In the case of no suction at this stage, it seems that the flow field is at the later stage of the transition. One can see two vortices structure with high turbulence intensity region at right side of each vortex. On the other hand, flow field becomes much stable when suction is applied as shown in figures (b), (c). Selective suction case (figure(c)) shows the most stable condition among three. Therefore, it can be said that present idea is successfully certified from these results. Namely, occurrence of the secondary instability is restrained by the selective suction.

Figure 4 shows the transition characteristics for distributed suctions. Selective suction shows the effect that is as same as uniform suction with less than 1/3 suction volume rate as shown in (a). Comparing from the viewpoint of suction velocity, there isn't much difference between both results even though uniform suction's volume is 2.75 times at the same suction velocity as shown in (b). Therefore, it can be said that selective suction is more effective method than uniform suction to delay the turbulent transition in the case of a swept wing.

3.2 Optimum suction position

3.2.1 Spanwise direction

It is important to select the appropriate position to apply present selective suction. Crossflow component (low momentum flow) flows near the wall from left to right in figure 5. Low momentum flows are pumped up when they get over vortex A shown in the sketch of figure 5(a). Suction line holes are applied just right and left side of vortex A as shown in (b), (c). It is clear that selective suction applied at left side (figure b) is more effective as shown in figure 6(a). Considering from these results, to control secondary instability it was found out that to suck the low momentum flows before they are pumped up is most effective.

3.2.2 Chordwise direction

Suction position and area in streamwise direction is varied as shown in table 1. In order to find out optimum position and area, four different suction cases are selected. Figure 6(b) shows the transition characteristics for distributed suctions at different chord wise position.

Table 1: Suction positions

Case	Suction position	Transition stage
(a)	x/C=0.650 $\sim$ 0.750	Whole region of f_2
(b)	x/C=0.650 $\sim$ 0.700	Before occurrence of f_2
(c)	x/C=0.675 $\sim$ 0.725	Occurrence of $f2$
(d)	x/C=0.700 $\sim$ 0.750	After occurrence of f_2

It is clear from the results that case (b), in which suction was applied before onset of the secondary instability, is the most effective position. Same flow control effect with less than 1/2 or 1/3 suction velocity is attained compared to other cases. The results show that it is effective to control the crossflow vortices at its linear growth stage. By comparing the results of (a) and (c), one can see that the effect is almost same even though suction volume of (a) is about double against (c). From this, it can be said that higher the suction velocity is more important than suction volume. It is quite natural to consider that the higher suction velocity can effectively control the occurrence of the secondary instability that appears rather high in the boundary layer. However, too much suction again changed the condition worse. Excess of the suction brings not only high shear stress, but also creates new streamwise vortex structure in the different span positions. Therefore, present investigation clearly shows that the transition control by selective suction system is effective, and there exists adequate suction condition for present flow control technique. In all, by selecting the best suction rate, suction position, transition was delayed up to about 30%.

4 Conclusion

Control of the Crossflow instability generated on swept flat plate with displacement body is performed taking the transition mechanism into consideration. By introducing selective suction system, it is attained that the control effect is as same as uniform suction with less than 1/3 suction volume rate. It is also found out that suction velocity is more important than suction volume because secondary instability is generated at higher position in the boundary layer. From this point, selective suction system has an advantage to uniform suction. It is important to select the appropriate position to apply present selective suction. It is effective to control crossflow instability at linear growth stage in chordwise position.

References

[1] Kohama, Y. (1987) *Acta Mech*, **Vol.66**, pp.21–38.

[2] Kohama, Y., Saric, W.S. and Hoos, J. A. (1991) *Proc. Roy. Aero. Soc. Conf. On: Boundary Layer Transition and Control* , Cambridge Univ., England, 8-12, Apr, pp.4.1-4.13.

[3] Lingwood, R. J. (1995) *J. Fluid Mech.*, **Vol.299**, pp.17–33.

[4] Ito, H., Kobayashi, R. and Kohama, Y. (1992) *Aeron. J. of the Roy. Aeron. Soc.*, pp.141-151.

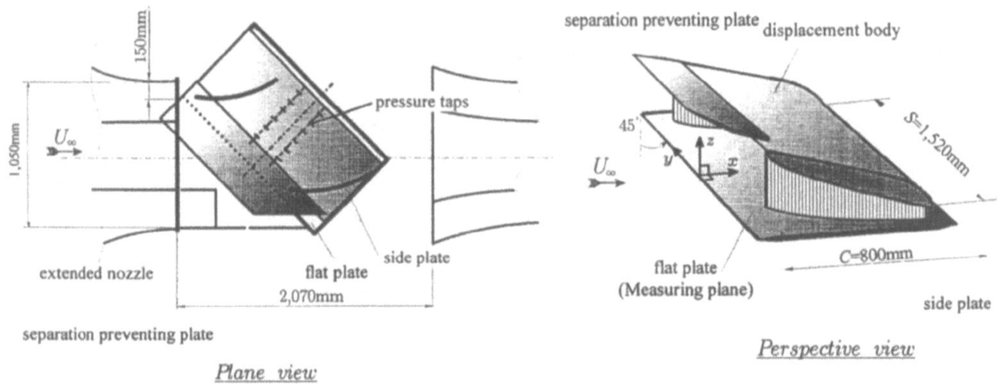

Low turbulence wind tunnel (Institute of Fluid Science, Tohoku University)
Wind tunnel velocity U_∞=10-15m/s, Turbulence intensity of the tunnel : U_{rms}= 0.07%

Figure 1. Experimental set-up and coordinate system

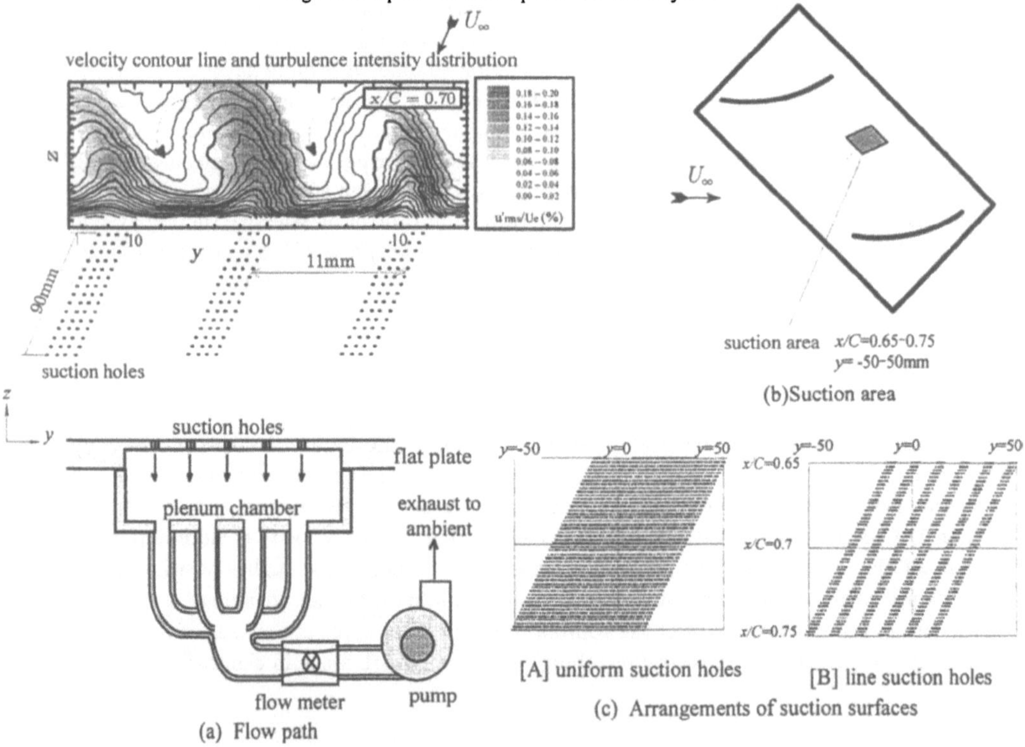

Figure 2. Details of the suction system

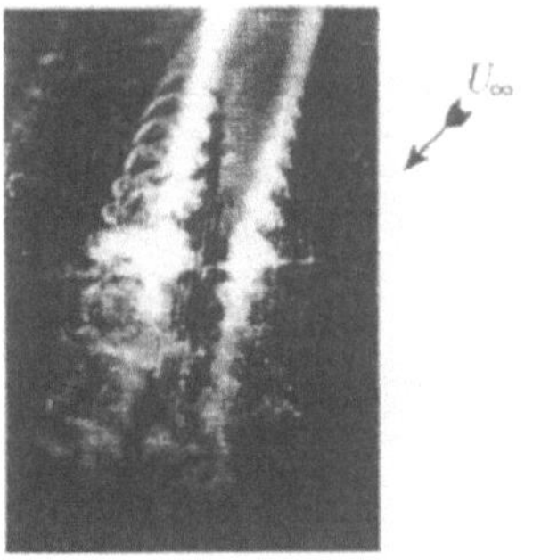

Flow visualization by Stage-smoke method
of high frequency secondary instability f_2.

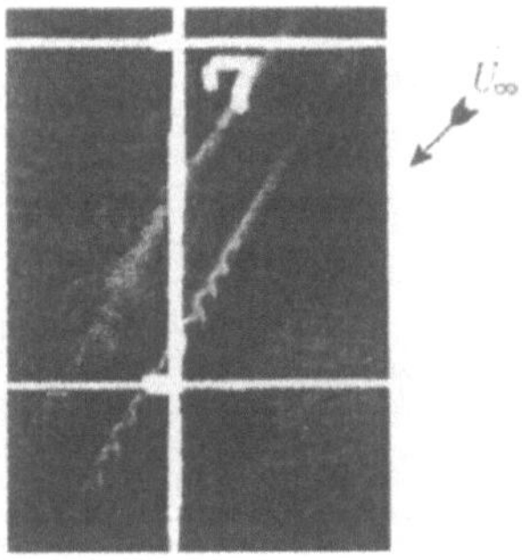

Flow visualization by Stage-smoke method.

(d) Flow visualizations

Figure 2. Details of the suction system

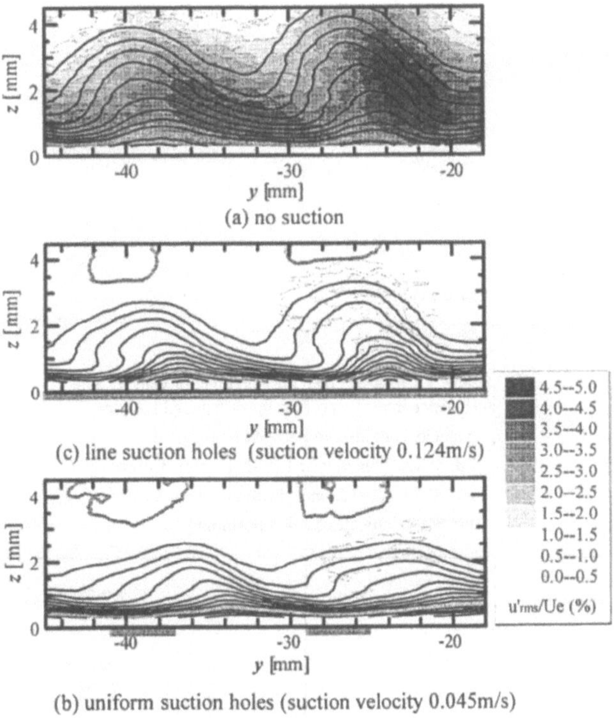

Figure 3. Cross sectional measurement of velocity and turbulence intensity profiles for different suction surfaces.
(velocity contour line step: 5% of U_e,
Bandpass filter range : 1.25-3kHz for disturbance f_2 ,
x/C=0.79, U_∞ =12.3m/s, Suction volume 0.045m^3/m^2·s,
▬ indicates suction positions)

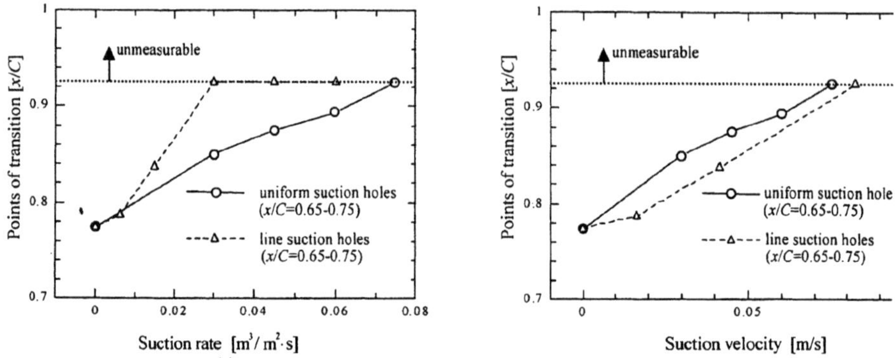

(a) (b)

Figure 4. Transition characteristics for distributed suctions
(U_∞=12.3m/s, $Re=U_\infty C/\nu=6.6\times10^5$)

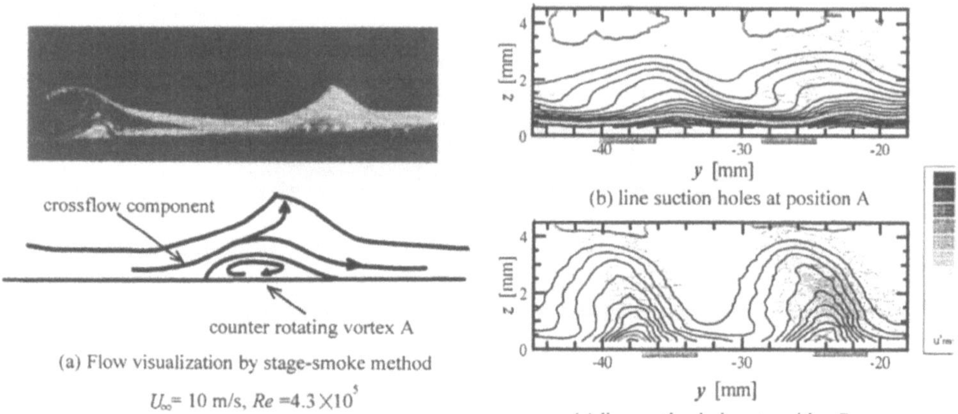

(a) Flow visualization by stage-smoke method

U_∞= 10 m/s, Re =4.3×10^5

(b) line suction holes at position A

(c) line suction holes at position B

Figure 5. Cross sectional measurement of velocity and turbulence intensity profiles for distributed suctions.
(velocity contour line step: 5% of U_e,
Bandpass filter range : 1.25-2kHz for disturbance f_2,
x/C=0.79, U_∞=12.3m/s, Suction volume 0.03m^3/m$^2\cdot$s,
▬ indicates suction positions)

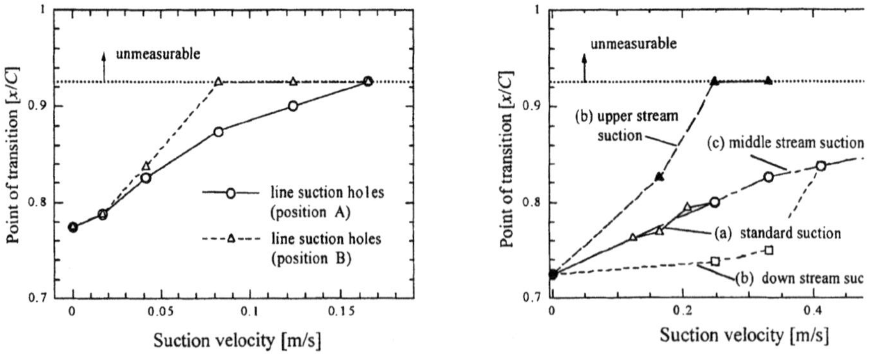

(a) distributed suction in spanwise direction
($Re=U_\infty C/\nu=6.6\times10^5$)

(b) line suction holes
(U_∞=13.5m/s, $Re=U_\infty C/\nu$ =7.:

Figure 6. Transition characteristics

CONTROL OF THE LAMINAR-TURBULENT TRANSITION BY SUCTION OF THE BOUNDARY LAYER IN 2D FLOW.

A. SERAUDIE[1], M. CORREGE[2], G. CASALIS[1], P. MOUYON[2]
ONERA-CERT, 1 : DMAE, 2 : DCSD
2 avenue Edouard Belin - BP 4025
31055 Toulouse Cedex 4 (FRANCE)

1- Introduction

The streamwise location of the boundary layer transition on a flat plate depends on many parameters, such as the unit Reynolds number, the pressure gradient, a wall suction, etc.

The present study takes place in the active flow control framework developed at ONERA. The goal of this work is to keep constant the abscissa of transition (Xt) in spite of low frequency perturbations which are superposed to the incoming flow. In a first part of the study, a theoretical approach has been performed in order to develop an efficient numerical code to simulate the transition displacement when the wall suction and/or the free stream velocity vary in time. In a second part, a control law has been proposed, and tested using the numerical simulation code for the transition location evolution. Finally, experiments have been performed to control the transition location in real time, using the control law developed with numerical tools. So, some parameters such as wall suction, "troublemaker", time response of the valve suction,.. have been analysed to define their action on the transition in open loop. Moreover the control law has been adjusted to the experimental conditions, and used in closed loop configuration, first to move the transition abscissa to a chosen location, secondly to reduce the transition motion in real time when the free stream velocity is perturbed periodically.

This paper describes essentially the experimental part of the study : the test conditions, the important parameters of the control law studied in open loop configuration, and finally the closed loop control of the transition performed in real time.

2- Control principle

The goal of this study is to develop theoretically and to apply, in a well known aerodynamic condition, a control system to keep constant the transition location of a flat plate placed in an unsteady low speed flow [1,2]. A suction panel is placed close to the leading edge and the suction mass-flow rate is adjusted in order to control the fixed transition location when the free stream velocity changes periodically during the time. The three important parameters used in this control system are the hot film transducers to determine the transition location, the "troublemaker" to modify the free stream

G.E.A. Meier and P.R. Viswanath (eds.),
IUTAM Symposium on Mechanics of Passive and Active Flow Control, 177–182.

conditions and the "actuator" to change the velocity suction (V_w) in order to compensate the transition motion (fig. 1).

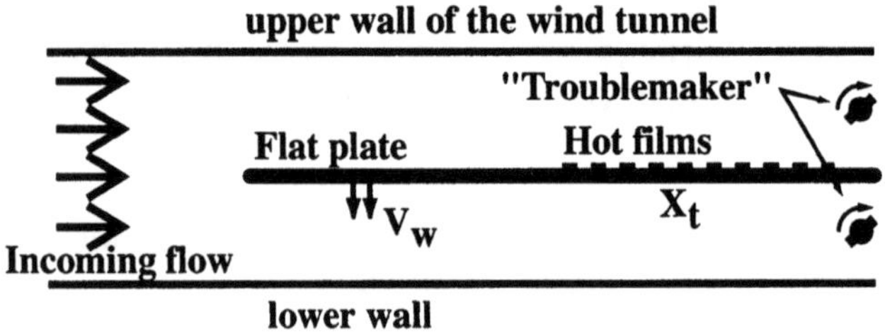

Figure 1 : Experimental set-up

3- Experimental devices and test conditions

3.1. MEASUREMENT TOOLS

The experimental part of the study takes place in a low velocity wind tunnel of DMAE [3]. The flat plate is set close to zero degree angle of attack (α) at mid height of the test section (300x400 mm^2). The plate, 1083 mm length, is extended downstream by a flap used to adjust the pressure gradient on the upper side. Twenty centimeters downstream of the leading edge, the suction section 5cm length, is covered by a thin perfored wall. The chamber is connected to a vacuum pump and a proportional valve allows to adjust the control mass-flow. Moreover the plate is fitted with static pressure taps (steady pressure distribution) and with twelve hot films, glued on the wall every 30 mm on the upper side. Downstream of the test section, two rotating bars are fixed on the lateral walls, above and below the flap. They act as the "troublemaker" to change periodically the wind tunnel section and so the flow velocity. The real time velocity conditions are given by a hot wire probe fixed at the beginning of the test section.

The data acquisition system has been developed on a P.C. computer coupled with a D.S.P. computer which has in charge the real time work (acquisition, transition determination, suction valve command). The data acquisition is performed at 2 KHz on 16 channels.

3.2. TEST CONFIGURATIONS

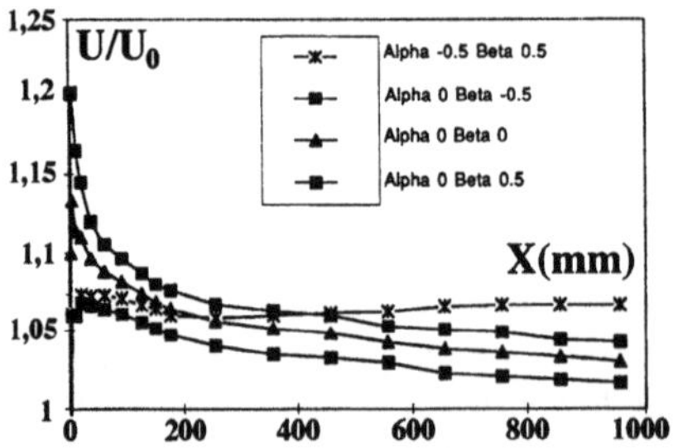

Figure 2 : Velocity distributions on the flat plate upper side

Different aerodynamic configurations were tested for different flat plate incidences (α) and different flap angles (β) for one velocity flow : 45 m/s ; 4 pressure distributions are plotted in figure 2. Only two of these cases were finally chosen ($\alpha = 0°$, $\beta = 0°$ and $\alpha = 0°$, $\beta = 0.5°$) to perform the tests with control. These configurations correspond to the most upstream mean transition location on the plate, with respect to the lowest pressure gradient.

4- Open loop parameters description

In the first part of the experimental study, it was necessary to qualify the effect of parameters such as the wall suction or the "troublemaker" on the transition location.

4.1. WALL SUCTION STUDY

First we measured the transition motion under the action of the wall suction (V_w) for the two aerodynamic configurations tested. The transition motion is detected by the hot film signals analysis and the curve plotted (fig. 3) corresponds to the middle of the transition location ; the total displacement is close to 120 mm for the two configurations. The slope of the curves is not quite constant, but a linear evolution is taken for the control operation. Moreover we observe that the transition length is higher for the mean transition locations the most upstream (for low suction rates).

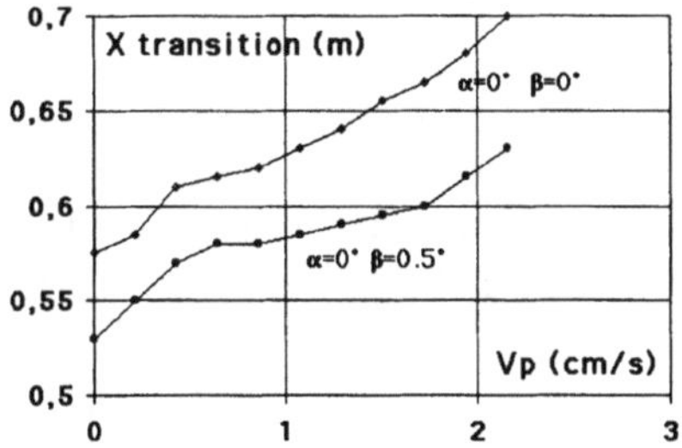

Figure 3 : Wall suction effect on transition location

Secondly we performed static and dynamic suction valve calibrations. For the most efficient valve, opening and closing duration close to 35 milliseconds were measured, and the valve works, quite linearly, between 5 and 8 volts supply for the total suction magnitude.

4.2. "TROUBLEMAKER" DESCRIPTION

The periodic change of the flow velocity is produced by two rotating bars. Each one is composed of a cylindrical metallic bar ($\phi = 30$ mm) on which a plate (35 and 40 mm wide) is screwed to change the wind tunnel section during the rotation. These two bars are connected together to change simultaneously the flow velocity above and below the plate. This device is coupled with an electric motor to adjust and maintain the rotating speed for the tests.

An example of the "troublemaker" effect is given in the aerodynamic configuration $\alpha = 0$ °, $\beta = 0.5°$ performed without wall suction. The hot film signals change periodically with the incoming velocity (fig. 4). During the perturbation period, we can follow for instance the evolution of the intermittent signal 6. The transition moves upstream when the velocity is close to the maximum ; a better accuracy of its evolution will be given by the analysis of the hot film signals.

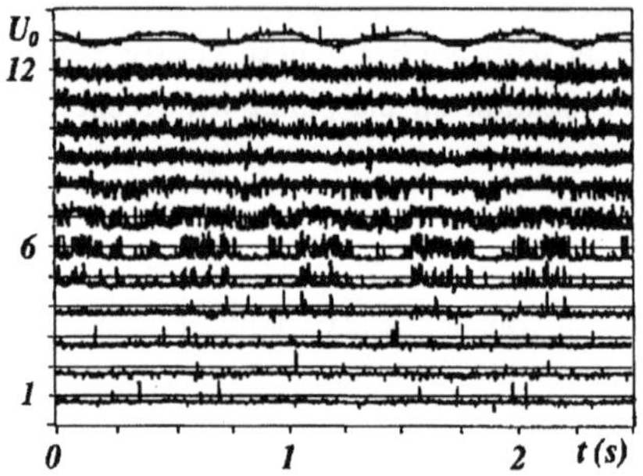

Figure 4: Hot film signals

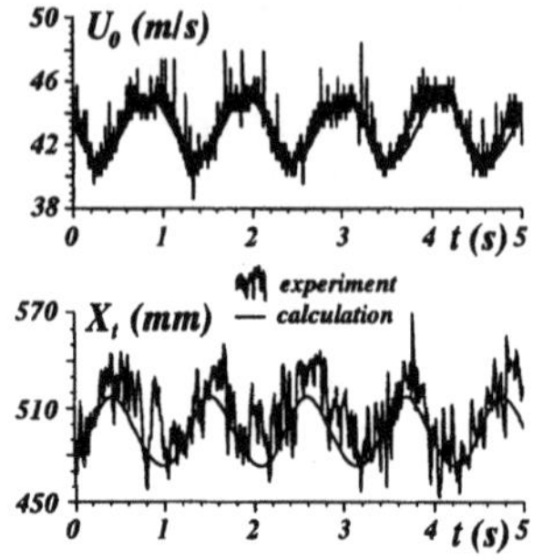

Figure5:Open loop calculation-experiment

4.3. ANALYSIS OF THE TRANSITION LOCATION

The transition location determination (Xt) is based on the analysis of the 12 hot film signals. For each signal we consider a low level for a laminar condition and a high level for the turbulent flow, between these two extreme levels, the signals are intermittent. The laminar and turbulent levels are determined with a calibration performed at the beginning of the tests. During each test, the signals are normalised between 0 (laminar) and 1 (turbulent), and an average is calculated inside a shifting window of 50 measurements ($\Delta t = 25$ milliseconds). So, at any time we can plot the curve of the normalised, filtered values of the films versus their abscissa (fig. 6). A linear approximation of the "intermittent part" of the curve is done and the transition is located at the abscissa corresponding to the Y axis equal to 0.5 ; so the middle of the transition location corresponding to an intermittence factor of 0.5 is considered.

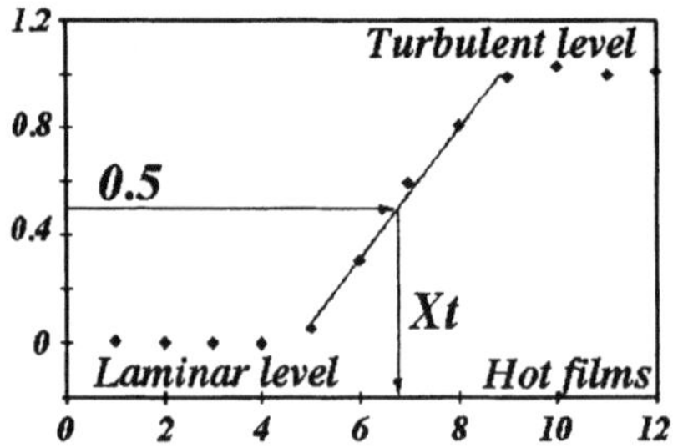

Figure 6 : Transition location determination

A subroutine has been written with this method for the real time control process, and using it, an analysis of the open loop tests has been performed, in batch mode. An example of this post data processing is plotted on figure 5 : the upper part gives the velocity U_0, the lower part the experimental transition location (Xt). In this analysis the transition motion can be compared to the evolution of the flow velocity : we can notice a phase difference of a quarter period between maximum/minimum of the two curves. The most upstream transition doesn't correspond to the maximum of incoming velocity. There is a combination between a modification of the pressure gradient on the plate and the Reynolds number effect produced by the velocity evolution.

4.4. CALCULATION-EXPERIMENT COMPARISON

The simulation code based on the unsteady boundary layer code [2,4], the linear stability analysis solved by a simplified approach (parabola method) and the e^n method [5] has been developed with some approximations physically justified. Among them we assume that the inviscid velocity is everywhere the same (U_0), and that the variations are low in terms of time scale. In fact the typical frequency is close to 1 Hz, which is small compared to the scale of the unstable Tollmien-Schlichting waves frequencies (1000 Hz). This assumption allows us to numerically integrate the unsteady boundary layer equations and, at each time step, a classical stability analysis is performed. An example of result is plotted in figure 5 and compared to the experimental result : the pulsation of the incoming flow is well adapted to the experimental condition, and there is a good agreement between the computation and the measurement of the transition amplitudes and motions.

5- Closed loop experimental results

The last implementation of the full control system was performed in two stages. In the first step the device controls the transition at a chosen location, without flow perturbation, in the second step the transition is controlled at a chosen location with incoming flow variations.

5.1. TESTS WITHOUT FLOW PERTURBATION

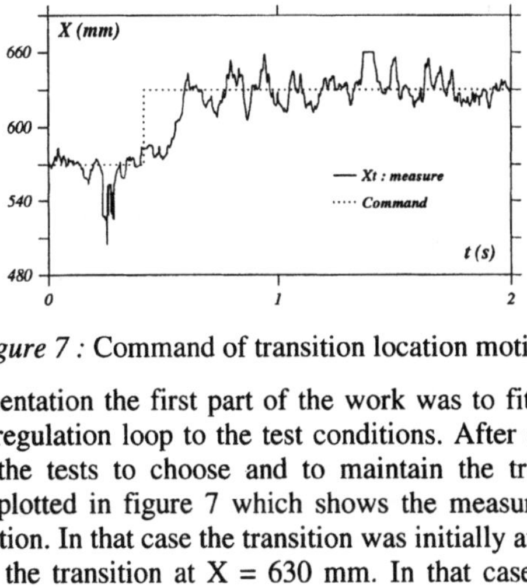

Figure 7 : Command of transition location motion

In this implementation the first part of the work was to fit the proportional and integral gains of the regulation loop to the test conditions. After some validation tests, we have performed the tests to choose and to maintain the transition location. An example of result is plotted in figure 7 which shows the measured transition and the command for the location. In that case the transition was initially at X = 570 mm and the command was to set the transition at X = 630 mm. In that case the time delay (150 milliseconds) is produced by the response time of the suction valve.

The same kind of study has been performed at the Southampton University [6], in which the transition was got through the analysis of 4 microphone signals ; in this experiment the transition location is fixed by the microphone abscissa.

5.2. TESTS FOR TRANSITION LOCATION CONTROL

The final tests performed in closed loop prove that it is possible to completely reduce the transition motion when the free stream velocity is perturbed by the two rotating bars, at least at low frequency. In that aerodynamic configuration ($\alpha = 0°$ and $\beta = 0°$) the incoming velocity pulsation is close to 4 m/s and without control the transition location moves on about 60 mm (fig. 8). When the control loop is working the transition is maintained at the chosen abscissa X = 570 mm with an accuracy better than 30 mm which is the distance between two consecutive hot films (fig.9).

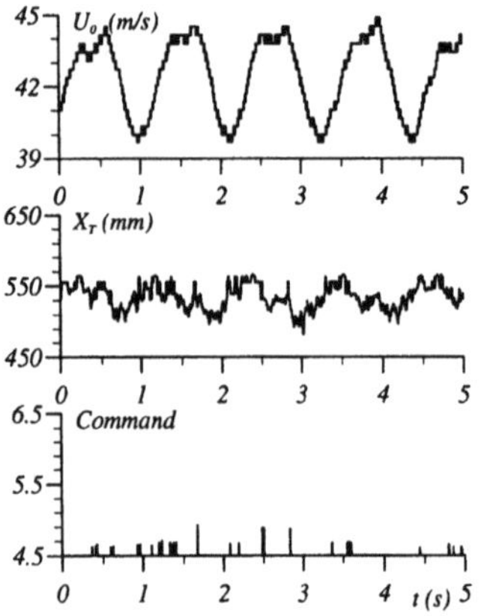

Figure 8 : Transition without control

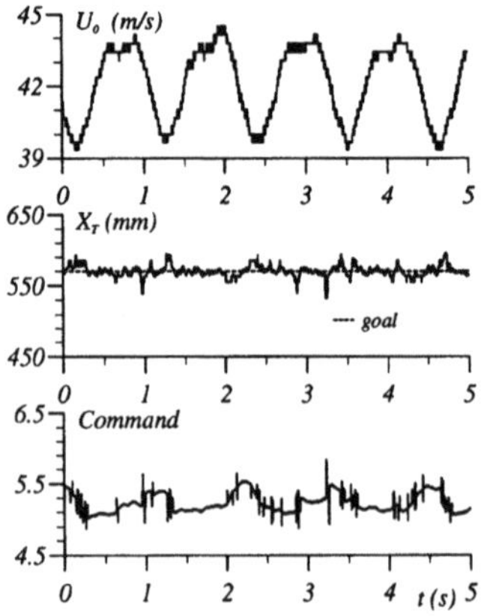

Figure 9 : Transition with control

6- Conclusion

After the development of a simulation code, and in parallel with the control law development, experimental tests were performed. We first choose the test configurations and the different systems able to change the incoming flow and able to adjust the transition location with wall suction. These devices were tested to know their features in static and dynamic conditions.

Finally, the developed control law was tested, adapted to the testing configurations and used with success to set the transition at a chosen location, first without, secondly with a perturbed incoming flow, at low frequency.

In the next future we would like to improve the transition location determination in order to increase the flow perturbation frequency, for testing the robustness of the control law in more severe condition.

References

[1] : Casalis G., Mouyon Ph., Prudhomme S., Séraudie A. *Contrôle actif des écoulements.* R.F. n°1/7600.29/DERA et n°104/5604.82/DERAT

[2] : Mouyon Ph., Casalis G., Séraudie A, Prudhomme S. *Feedback control of the laminar-turbulent transition in a boundary layer by suction.* Experimental Thermal and Fluid Science (1998)

[3] : Séraudie A., Corrège M., Casalis G. *Contrôle dynamique de la position de transition par aspiration de la couche limite..Phase 1 : Ralliement à une consigne.* R.T. 138/5618.46 DMAE Juillet 1998

[4] : Michel R., Cousteix J., Houdeville R. *Unsteady turbulent shear flows,* IUTAM Symposium Toulouse, Ed. Springer-Verlag, 1981

[5] : Arnal D. *Boundary layer transition : predictions based on linear theory.* Special course Agard on Progress in Transition Modeling. 1993

[6] : Rioual J.L., Nelson P.A., Fisher M.J. *Experiments on the automatic control of boundary layer transition.* J. Aircraft, vol. 31, n°6, 1995

CONTROL OF TRANSITION IN 3-D BOUNDARY LAYERS

W. S. SARIC, R. B. CARRILLO, M. S. REIBERT
Mechanical and Aerospace Engineering
Arizona State University
Tempe, AZ. 85287-6106

1. Introduction

Transition to turbulence in swept-wing flows has resisted correlation with linear theory because of its sensitivity to freestream conditions and 3-D roughness and because one of the principal instability modes quickly becomes nonlinear. In the face of such a formidable problem, two rather long-term fundamental efforts have been underway at DLR Göttingen and Arizona State University that address swept-wing transition. These efforts have been recently reviewed by Bippes (1997) and Reibert & Saric (1997). The present work is a continuation of studies on swept-wing boundary layers. In particular, we have taken advantage of the sensitivity to 3-D roughness and the modal nature of the instability in order to propose a particular control strategy.

Complementing the two aforementioned reviews, general reviews of the swept-wing problem are found in Arnal (1997) and Kachanov (1996). Other reviews include Reshotko (1997), Crouch (1997), and Herbert (1997a,b).

The basic idea is that the combination of sweep and chordwise pressure gradient within the boundary layer creates a velocity component perpendicular to the inviscid streamline. This *crossflow* profile is inflectional and exhibits both traveling and stationary unstable waves called crossflow vortices that are (approximately) aligned along the inviscid streamlines. Under conditions of low freestream turbulence levels, the dominant crossflow wave is stationary (Reibert & Saric 1997) while moderate to high turbulence levels initiate dominant traveling waves (Deyhle & Bippes 1996; Bippes 1997). The mechanism is relatively insensitive to sound and 2-D surface roughness (Radeztsky et al. 1993) but very sensitive to 3-D roughness near the attachment line.

We concentrate our work on low-turbulence freestream flows and stationary crossflow waves. Although the v' and w' components of the disturbances are very small, by convecting streamwise momentum in the wall-normal direction, they produce $O(1)$ changes in u'. Thus the mean flow is highly distorted with localized inflection points. Transition is then triggered by a high-frequency secondary instability of the distorted mean profile (Kohama et al. 1991; Malik et al. 1994).

This experiment is conducted in the ASU Unsteady Wind Tunnel, a low-turbulence facility in which the stability and transition of boundary layers are studied (Saric 1992a). The NLF (2)-0415 airfoil is mounted vertically in the 1.4 m x 1.4 m x 4.9 m test section. Floor and ceiling contours in the test section produce an infinite-span swept-wing flow (Reibert et al. 1996). With a 45° sweep and a -4° angle of attack, the pressure gradient produces large crossflow while suppressing T-S modes.

The aluminum surface of the model is hand polished to 0.2 μm rms in order to establish reference conditions. Roughness elements, of minimum height of 6 μm

G.E.A. Meier and P.R. Viswanath (eds.),
IUTAM Symposium on Mechanics of Passive and Active Flow Control, 183–188.

($Re_k \approx 0.1$), are introduced in a spanwise periodic distribution. These roughness elements linearly excite specific modes within the boundary layer at a level above the background level. Detailed hot-wire measurements within the boundary layer provide two-dimensional maps of the stationary disturbance structure, while spectral techniques are used to identify and follow specific stationary modes (see Reibert et al. 1996).

2. Nonlinear Interaction and Amplitude Saturation

In the absence of artificial surface roughness, the naturally occurring stationary crossflow waves are nonuniform in span due to submicron surface irregularities near the leading edge. However, a dominant feature at a 12 mm spanwise spacing appears, which is approximately the most amplified stationary wavelength according to linear theory. At the same time, the spectral content is rich and indicates nonlinear interaction among many modes. The random nature of the leading-edge roughness leads to difficulty because a computational model will decompose the unstable motion in modes and thus one needs the individual mode behavior from the experiment if a meaningful comparison can be made.

The initial conditions are controlled by applying a full-span array of 6 μm roughness elements at $x/c = 0.023$ where the displacement thickness is approximately 250 μm and $Re_k \approx 0.1$. Thus, we expect a weak linear input to the unstable waves. The roughness elements are circular and the diameter is 2 mm. Radeztsky et al. (1993) provide data on the effect of chord location and diameter of the roughness.

When the spanwise spacing of the elements is 12 mm, corresponding to the naturally occurring most-amplified wavelength, Figure 1 shows the streamwise velocity contours where the dominance of the 12 mm mode is striking. This allows a direct calculation of the stationary disturbance amplitude (see Reibert et al. 1996 for a description). Figure 2 compares the experimental amplification factor ("*N*-factor") for the 12 mm roughness forcing with the predictions of the Orr-Sommerfeld equation (OSE), the linear parabolized stability equations, and the full nonlinear parabolized stability equations. All computational results are provided by Haynes and Reed (1997). The saturation phenomenon is clearly evident, and can be quantified. The early growth shows excellent agreement with linear PSE, however strong nonlinear effects develop well before transition at $(x/c)_{tr} = 0.5$. The importance of nonparallel effects is indicated by the failure of traditional linear stability theory (OSE) to accurately predict the growth even in the linear range. When nonlinearity is added, the agreement is remarkable over the entire measurement region and all aspects of the growth are predicted. This is explained in more detail in Reibert et al. (1996) and Haynes & Reed (1997). The outstanding agreement shown in Figure 2 results from the inclusion of curvature in the computations, without which the disturbance growth is significantly overpredicted. The sensitivity to very weak curvature is due to the strong stabilizing Görtler effect with convex curvature (Benmalek & Saric 1994).

3. Role of Spanwise Spacing and the Control of Transition

Two observations concerning the data of Reibert et al. (1996) and Saric et al. (1998) are:

1. Unstable waves occur at integer multiples of the primary disturbance wavenumber.
2. No subharmonic disturbances are destabilized.

In other words, spacing the roughness elements 12 mm apart excites disturbances with spanwise wavelengths of 6 and 4 mm. Spacing the roughness elements 36 mm apart excites disturbances with spanwise wavelengths of 36, 18, 12, 9, 7.1, 6, 5.2, and 4 mm. The 18 mm spacing excites the 18, 9, 6, and 4.5 mm wavelengths, but *not* the most unstable wave at 12 mm. These spacings do not produce any unstable waves with "intermediate" wavelengths or with wavelengths *greater* than the imposed spacing.

Following this lead, we investigate the effects of distributed roughness whose primary disturbance wavelength is less than the most unstable wavelength according to linear theory $(\lambda_z < 12\ \text{mm})$. For this study, the 6 μm-high roughness elements are applied in a full-span array at $x/c = 0.023$ with a spanwise spacing of 8 mm.

3.1. FLOW VISUALIZATION

For the reference roughness case (k = 0.2 μm), surface-shear-stress patterns and transition locations are determined using naphthalene flow visualization. Figures 3 and 4 show the naphthalene pattern for the reference and distributed roughness cases, respectively at $Re_c = 2.4 \times 10^6$. The flow is from left to right and the leading edge and chord locations are marked. The stationary crossflow vortices transpose high-and low-momentum fluid within the boundary layer, yielding a spanwise modulation of the surface shear stress. This modulation is clearly indicated as alternating streaks in the naphthalene. Regions in which the boundary layer is turbulent are indicated by complete sublimation of the naphthalene (dark area). Note that this does not hold true in the region near the leading edge, where naphthalene is not sprayed to prevent roughness contamination. Transition appears as a series of turbulent wedges, which is characteristic for swept-wing flows. Note that the transition pattern is nonuniform under natural surface roughness conditions. This is consistent with past experiments and is indicative of submicron roughness irregularities near the leading edge.

The transition pattern of Figure 4 shows that distributed roughness has effectively delayed transition *past* the pressure minimum x/c = 0.71. In fact, transition has been moved back even past the transition location observed under natural surface roughness conditions (Figure 3). Transition occurs for the distributed roughness case beyond 80% chord, which is the location of the junction between the wing and trailing-edge flap. Thus, it is entirely possible that the junction actually causes transition.

Figure 4 shows definite vortex streaks with an 8 mm spacing for $0.3<x/c<0.45$. For $x/c>0.45$, the 8 mm structure "washes out" and the vortex streaks appear at larger spanwise spacings. Although the spacing between streaks becomes larger in general, there is definite spanwise nonuniformity in the spacing. Thus, it is difficult to make any conclusions about the longer wavelengths based solely on the flow visualization. Detailed boundary-layer measurements shed more light on the phenomenon.

3.2. DISTURBANCE AMPLITUDE

The normalized velocity contours for $Re_c = 2.4\text{x}10^6$ and distributed roughness are shown in Figures 5and 6 for chord locations *x/c* = 0.3 and 0.6, respectively. The disturbance is first measurable at *x/c* = 0.1, and a very uniform and dominant 8 mm mode develops for 0.1<*x/c*<0.25. At *x/c* = 0.3 (Figure 5), although the contour still shows a dominant 8 mm mode, there is noticeable development of some slight nonuniformity. This becomes more dramatic with increasing *x/c*, and the 8 mm structure fades out in favor of some longer wavelength disturbances. By *x/c* = 0.5, the dominance of the fundamental mode is gone, and *x/c* = 0.6 (Figure 6) the fundamental mode is indistinguishable in the contours. This is consistent with the results of the naphthalene flow visualization.

The total disturbance amplitude *N*-factors were computed from the mode shapes. Typically, the total disturbance amplitude grows rapidly from 0.1<*x/c*<0.3; at which point the amplitude saturates and then slows strong decay. At *x/c* = 0.3 strong nonlinear behavior is evident in the mode shape. The amplitude continues to decay for 0.3<*x/c*<0.45, after which, the amplitude levels off and then shows a second region of strong growth for 0.5<*x/c*<0.6. The computation of Haynes & Reed (1997) confirms the theory and experiment for the 8 mm excitation. The nonlinear PSE accurately predicts the maximum *N* reached by the first growth stage and even captures the dramatic amplitude decay. The linear computations do not predict this feature. The nonlinear PSE does not predict the second growth stage downstream. This is expected because the second growth region is due to longer-wavelength broadband disturbances, and not the fundamental 8 mm or superharmonic disturbances.

The modal content of the stationary vortex structure for distributed roughness is determined with spanwise scans and spectral analysis. The fundamental mode (8 mm) is first detected at *x/c* = 0.1 and grows rapidly to *x/c* = 0.3. The 4 mm mode becomes measurable at *x/c* = 0.2. For *x/c*>0.3, nonuniformities develop as longer wavelength modes grow and eventually become dominant. At *x/c* = 0.6, the spectrum shows no evidence of the original modes, rather, the spectrum is dominated by broadband energy contained in longer wavelengths $(\lambda_z \geq 10\,\text{mm})$. This is similar to the reference roughness results which show nonuniformities due to the excitation of longer wavelength disturbances. The fundamental mode shows excellent agreement with the total disturbance for 0.1<*x/c*<0.25. For *x/c*>0.3, the fundamental amplitude drops dramatically from the total disturbance amplitude. Unlike the results of distributed roughness and Reibert (1996), the total disturbance is *not* completely made up by the fundamental and its superharmonics. In other words, strong growth of the harmonic modes is not seen in the region where the fundamental drops from the total disturbance (0.3<*x/c*<0.6). Rather, the continuing growth of the total disturbance occurs because the longer wavelengths become unstable downstream. Note that the longer-wavelength disturbances are broadband and are not subharmonics of the fundamental disturbance.

In short, subcritical roughness delays transition past that of the natural roughness case. Strong early growth of the 8 mm mode effectively suppresses initial growth of the very unstable 9 mm and 12 mm modes near the leading edge, which linear stability theory predicts to be the region where these modes have the largest growth rates. The fundamental mode saturates and then decays rapidly. This decay allows for longer wavelength disturbances (due to submicron surface irregularities) to become unstable. The growth of these broadband disturbances eventually leads to transition.

4. Conclusions

Subcritical roughness spacing has a dramatic affect on transition location. In the absence of artificial roughness, transition occurs before the pressure minimum at $x/c = 0.71$ for $Re_c = 2.4 \times 10^6$. Adding 6 μm roughness with a spanwise spacing equal to (or a multiple of) the wavelength of the most unstable wave moves transition forward to $x/c<0.5$. However, the subcritical forcing at 8 mm spacing actually delays transition beyond the pressure minimum and onto the trailing-edge flap at $x/c = 0.80$.

This work was supported by NASA Langley Research Center Cooperative Agreement NCC-1-194 and Boeing Contract ZA0078.

5. References

Arnal, D. 1997 Laminar-turbulent transition: research and applications in France. *AIAA Paper 97-1905.*

Benmalek, A., Saric. W.S. 1994. Effects of curvature variations on the nonlinear evolution of Görtler vortices. *Phys. Fluids* **3**:3353-67.

Bippes, H. 1997 Environmental conditions and transition prediction in 3-D Boundary layers. *AIAA Paper 97-1906.*

Crouch, J. 1997 Transition prediction and control for airplane applications. *AIAA Paper 97-1907.*

Deyhle, H., Bippes, J. 1996 Disturbance growth in an unstable three-dimensional boundary layer and its dependence on environmental conditions. *J. Fluid Mech.* **316**:73-113.

Haynes, T.S., Reed, H.L. 1997 Numerical simulation of swept-wing vortices using nonlinear parabolized stability equations. *SAE Paper 97-1479.*

Herbert, T. 1997a Transition analysis of flow over aerodynamic bodies. *AIAA Paper 97-1908.*

Herbert, T. 1997b On the stability of 3-D boundary layers. *AIAA Paper 97-1961.*

Kachanov, Y.S. 1996 Experimental studies of 3-D instability of boundary layer. *AIAA Paper 96-1976.*

Kohama, Y., Saric, W.S., Hoos, J.A. 1991 A high-frequency secondary instability of crossflow vortices that leads to transition. In *Proc. R.A.S. Boundary Layer Transition and Control,* Cambridge, U.L.

Malik, M.R., Li, F., Chang, C.L. 1994 Crossflow disturbances in three-dimensional boundary layers: Nonlinear development, wave interaction and secondary instability. *J. Fluid Mech.* **268**:1-36.

Radeztsky, R.H. Jr., Reibert, M.S., Saric, W.S., Takagi, S. 1993 Effect of micron-sized roughness on transition in swept-wing flows. *AIAA Paper 93-0076.*

Radeztsky, R.H. Jr., Reibert, M.S., Saric, W. 1994 Development of stationary crossflow vortices on a swept wing. *AIAA Paper 94-2373.*

Reed, H.L., Saric, W.S., Arnal, D. 1996. Linear stability theory applied to boundary layers. *Ann. Rev. Fluid Mech.* **28**:389-428.

Reibert, M.S. 1996 *Nonlinear Stability, Saturation, and Transition in Crossflow-Dominated Boundary Layers.* Ph.D. diss., Arizona State University.

Reibert, M.S., Saric, W.S., Carrillo, R.B. Jr., Chapman, K.L. 1996 Experiments in nonlinear saturation of stationary crossflow vortices in a swept-wing boundary layer. *AIAA Paper 96-0184.*

Reibert, M.S., Saric, W.S. 1997 Review of swept-wing transition. *AIAA Paper 97-1816.*

Reshotko, E. 1997 Progress, accomplishments and issues in transition research. *AIAA Paper 97-1815.*

Saric, W.S. 1992. The ASU transition research facility. *AIAA Paper 92-3910.*

Saric, W.S., Carrillo, R.B., Reibert, M.S. 1998 Nonlinear stability and transition in 3-D boundary layers. *Meccanica* **33**: 471-89.

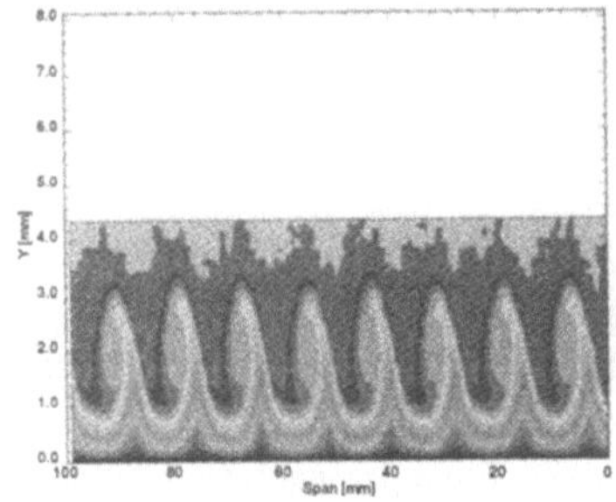

Figure 1. Streamwise velocity contours at $x/c = 0.45$, Re_c x 10^6. k = 6 μm at 12 mm spacing.

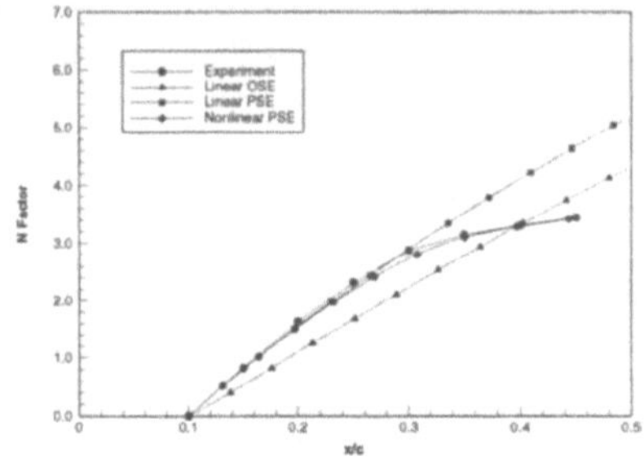

Figure 2. Measured and theoretical N-factors for the conditions of Figure 1. All theoretical calculations include curvature (Haynes and Reed 1996).

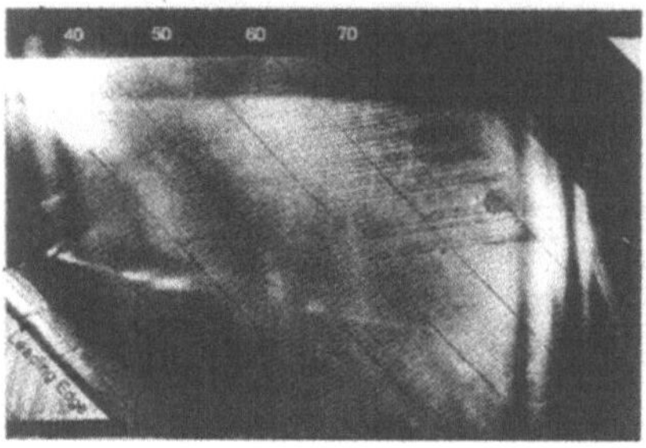

Figure 3. Flow visualization for $Re_c = 2.4$ x 10^6 and no artificial roughness

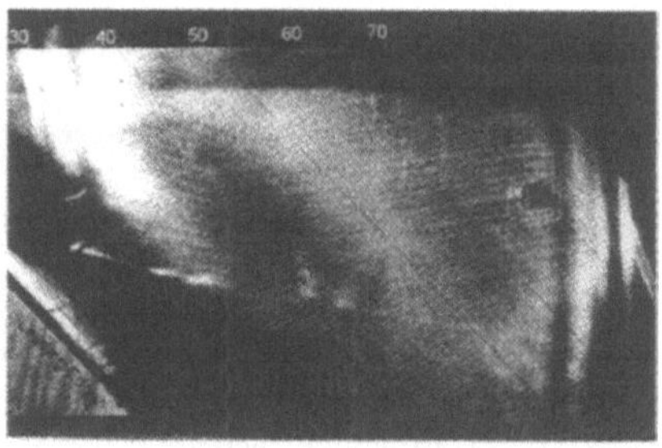

Figure 4. Flow visualization for $Re_c = 2.4$ x 10^6 with k = 6 μm at 8 mm spacing.

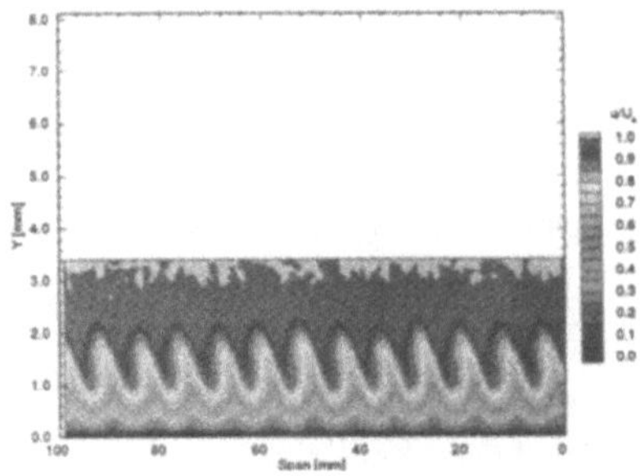

Figure 5. Streamwise velocity contours at x/c = 0.30. $Re_c = 2.4$ x 10^6. k = 6 μm at 8 mm spacing.

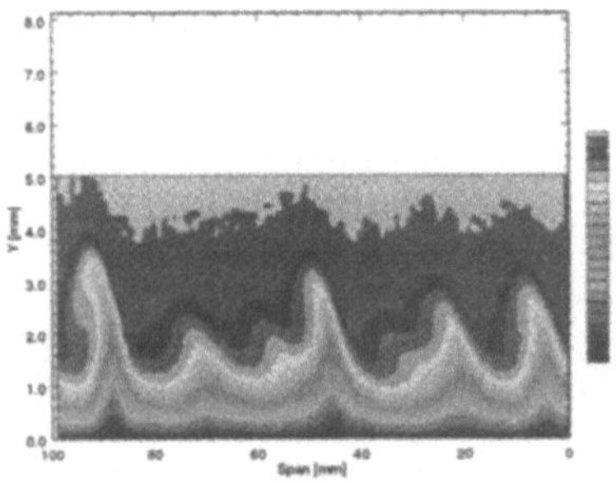

Figure 6. Streamwise velocity contours at x/c = 0.60. Conditions of Figure 5.

TRANSITION CONTROL USING LEADING EDGE ROUGHNESS FOR THE ONERA-D WING

E. JANKE AND P. BALAKUMAR
Old Dominion University†
Department of Aerospace Engineering
Norfolk, Virginia - 23529, U.S.A.

1. Introduction

Recent results by Saric et al. (1998) from experiments with an infinite swept wing document a passive transition control mechanism that originates from micron-sized roughness elements near the leading edge. In the experiments performed at Arizona State University (ASU), Saric et al. (1998) studied the natural transition case first. In that case, they determined the most amplified spanwise wavelength as λ_{x_2}=12mm and observed transition at $x_1/c \simeq 0.71$. Second, they forced disturbances of equal, or larger wavelength than the most amplified one by placing roughness elements at spanwise distances of 12mm, 18mm and 36mm near the leading edge. When forcing the 12mm disturbances, they found that transition moved forward to $x_1/c \simeq 0.52$. Forcing the 18mm disturbances, they observed that neither the 12mm, nor the 36mm modes were amplified in the downstream direction . Third and most interestingly, they were able to delay transition to $x_1/c \simeq 0.80$ by forcing disturbances with a wavelength smaller than the most amplified one. In that case, they placed the roughness elements at a spanwise distance of λ_{x_2}=8mm. The transition delay was attributed to the strong initial growth of the 8mm disturbances that suppressed the 12mm disturbances. The early saturation and decay of the 8mm disturbances then gave rise to the growth of longer wavelength disturbances that eventually lead to transition. However, this successful transition delay could be accomplished for the Reynolds number of Re=2,400,000 only. Increasing the Reynolds number slightly moved the transition location forward.

2. Mean Flow Computation and Linear Stability Analysis

For the present study, we consider an experiment performed by Arnal *et al.* (1984). The experimental conditions are given as follows: freestream ve-

† First author currently at the German Aerospace Center in Göttingen.

G.E.A. Meier and P.R. Viswanath (eds.),
IUTAM Symposium on Mechanics of Passive and Active Flow Control, 189–194.

locity Q^*_∞=80 m/s, Reynolds number $Re_\infty = Q^*_\infty \cdot c^*/\nu^*$=1,840,000, sweep angle ϕ=40°, angle of attack α=-8° and model chord length c^*=0.35 m. We compute the mean flow by solving the boundary layer equations based on fourth-order-accurate compact scheme formulation. The crossflow profiles remain positive over the entire chord length, and hence, the mean flow is assumed to be dominated by a pure crossflow instability. Seeking the most amplified wave number, we introduce steady disturbances before the neutral point at x_1/c=0.02 and compute their spatial evolution downstream using the linear PSE. Figure 1 shows the spatial growth rates based on the chordwise velocity component and the corresponding N-factors. The most amplified disturbance (λ_{max}=3mm) and its subharmonics with λ_{x_2}=1.5mm and λ_{x_2}=2mm behave according to the observations in the ASU experiment. A strong initial growth of the subharmonics is followed by a fast decay.

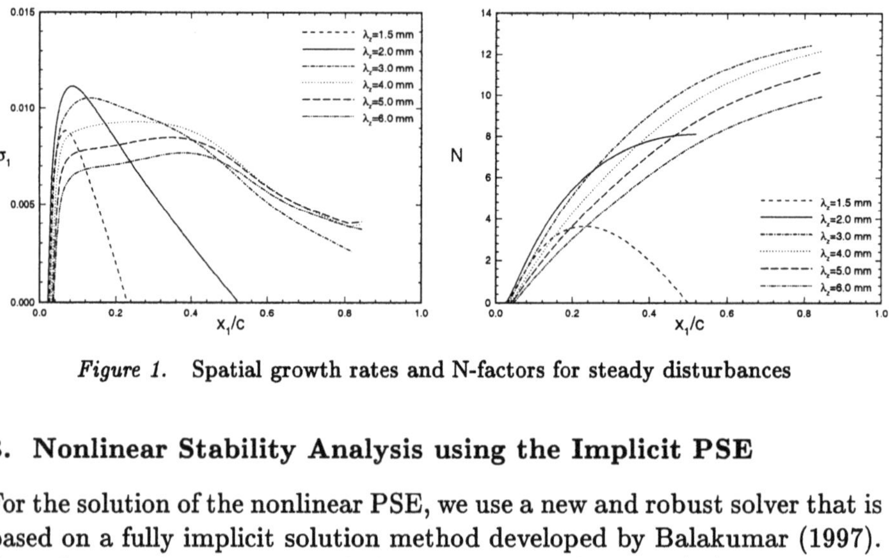

Figure 1. Spatial growth rates and N-factors for steady disturbances

3. Nonlinear Stability Analysis using the Implicit PSE

For the solution of the nonlinear PSE, we use a new and robust solver that is based on a fully implicit solution method developed by Balakumar (1997). According to the experimental observations, we distinguish four cases:

1. "Natural transition", dominated by the most amplified disturbance (λ_{max}=3mm).
2. Forcing of the fundamental disturbance (λ_{max}=3mm).
3. Forcing of a subharmonic wave number disturbance (λ_{x_2}=6mm).
4. Forcing of a superharmonic wave number disturbance ($\lambda_{x_2}=2/3\cdot\lambda_{max}$=2mm).

The initial disturbance amplitudes as defined in table 1 are obtained using a dimensional analysis neglecting the scaling influence of the receptivity coefficient. A value of $A=|u_1|_{max}/U_\infty$=0.2% is selected to model forced disturbances. For the other disturbances, an initial amplitude of A=0.008% is specified according to the ratio of k^*_{rough} and the distributed surface roughness height in the experiment ($k^*_{rough}/k^*_{distrib}$=24).

TABLE 1. Definition of the passive control cases

Case	λ_1 [mm]	λ_2 [mm]	λ_3 [mm]	A_1	A_2	A_3
1	3	-	-	0.008%	-	-
2	3	-	-	0.200%	-	-
3	6	3	2	0.200%	0.008%	0.008%
4	6	3	2	0.008%	0.008%	0.200%

3.1. NATURAL TRANSITION (CASE 1)

In order to establish a reference case that the forced disturbance cases can be compared to, the "natural transition" case is considered first. The influence of a small distributed roughness height is modeled by assuming the dominance of the most amplified disturbance. Based on disturbance energy, the disturbances do not saturate until $x_1/c \simeq 0.50$ in that case.

3.2. FUNDAMENTAL FORCING (CASE 2)

The presence of roughness elements at a spanwise spacing of 3mm is simulated by increasing the initial amplitude of the 3mm disturbance. Due to the higher initial amplitude, the regularly spaced 3mm disturbances saturate earlier than in Case 1 at $x_1/c \simeq 0.20$. Otherwise, there is no qualitative change in the evolution of the crossflow vortices compared to Case 1.

3.3. SUBHARMONIC FORCING (CASE 3)

A qualitatively different disturbance evolution is observed when disturbances with twice the most amplified wavelength are forced by placing roughness elements at a spanwise spacing of 6mm. Figure 2 shows the growth rates of the forced disturbance, of its first superharmonic, and

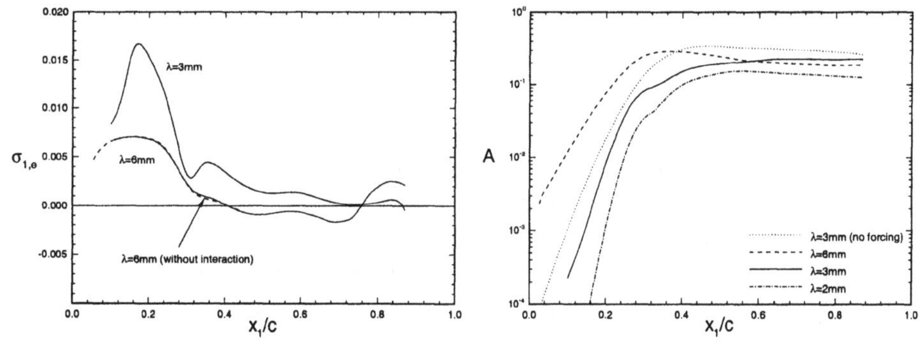

Figure 2. Growth rate based on disturbance energy and disturbance amplitudes

of the 6mm disturbance with the same initial amplitude when there is no nonlinear interaction. Also shown are the u_1-disturbance amplitudes of the forced 6mm mode and its first two superharmonics, where the amplitude evolution of the 3mm mode in Case 1 is plotted for comparison. Due to the forcing of the 6mm disturbance, the growth rate of the 3mm disturbance is largely modified compared with the previous two cases. The large initial amplitude of the (1,0) mode (λ=6mm) excites the (2,0) mode (λ=3mm) nonlinearly in contrast to its initially linear growth in Cases 1 and 2. Due to nonlinear interaction with the strong 6mm mode, the 3mm disturbances start saturating much earlier than in Case 1 near $x_1/c \simeq 0.30$. A secondary growth at $x_1/c \simeq 0.35$ delays the final saturation until $x_1/c \simeq 0.66$.

3.4. SUPERHARMONIC FORCING (CASE 4)

The last case simulates the most efficient means of controlling the transition onset, according to the ASU experiment. The fast growing $2/3{\cdot}\lambda_{max}$ mode is forced by placing roughness elements at a spanwise spacing of 2mm. Figure 3 shows the growth rates of the forced disturbance, of the fundamental disturbance, of the 2mm and the 3mm disturbances with the same initial amplitudes in the cases without nonlinear interaction, as well as the u_1-disturbance amplitudes. The growth of the fundamental mode is largely

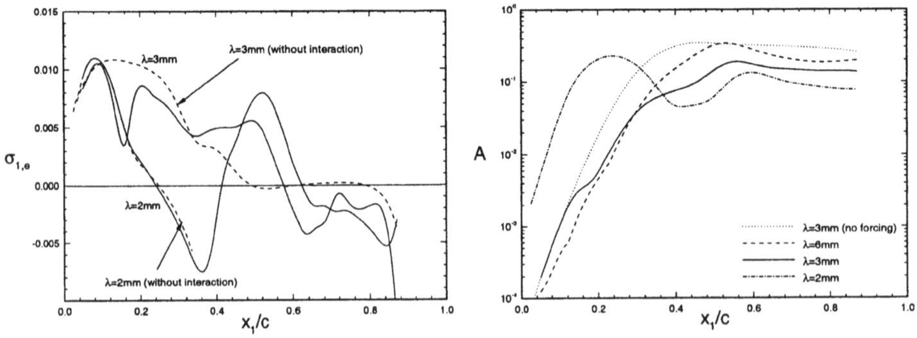

Figure 3. Growth rate based on disturbance energy and disturbance amplitudes

affected, and its development shows significant differences to the previously investigated Cases 1-3. First of all, the initial growth of the fundamental disturbance is suppressed. Second, the u_1-disturbance component saturates at the late location of $x_1/c \simeq 0.55$. It is the early tendency towards saturation that keeps the 3mm disturbance mode at low amplitude levels. The u_1-component of the 3mm mode does not reach its maximal amplitude of A=19.0% until x_1/c=0.56, and hence, the 3mm mode is considered successfully suppressed. Due to its strong initial growth, however, the forced 2mm mode reaches high amplitude levels at early chordwise positions. The interaction of the different disturbances is also seen in figure 4, where contours of the total u_1-disturbance are plotted at different streamwise stations. At $x_1/c=0.20$, the forced 2mm mode dominates the disturbance development and the crossflow vortices are spaced accordingly. At $x_1/c=0.35$, the 6mm mode with an amplitude of $A \simeq 10\%$ is also distinguished. This represents the onset of a subharmonic disturbance growth which agrees with the results by Saric et al. (1998). At $x_1/c=0.40$, the footprint of the 2mm mode has disappeared and the 3mm mode can be distinguished instead. Finally at $x_1/c=0.50$, the 6mm mode clearly dominates the crossflow vortex pattern.

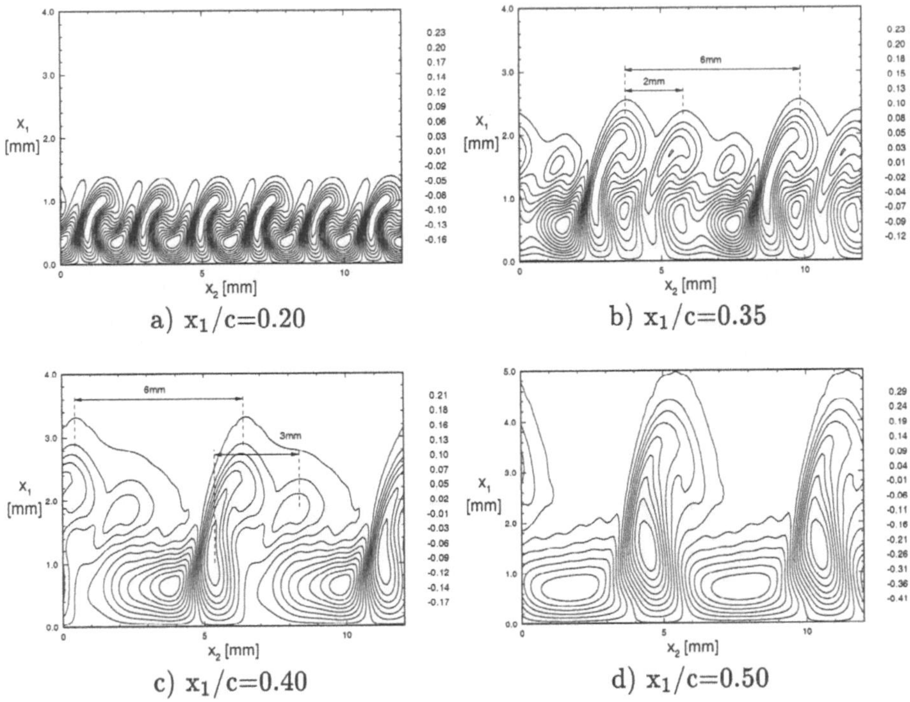

Figure 4. **Contours of the u_1-disturbance at different streamwise locations (Case 4)**

4. Conclusions

The study presented explains a nonlinear wave interaction mechanism that favorably modifies the evolution of the most amplified disturbances. By a proper placement of micron-sized roughness near the leading edge, the growth of the most amplified disturbance can be successfully delayed, if not suppressed. From our results, we conclude the following:

- Forcing the most amplified disturbance leads to an earlier disturbance saturation at the same high amplitude level as in the "natural transition" case (Case 2), and hence, to an earlier transition.

- Forcing a subharmonic disturbance with a wavelength larger than the most amplified one leads to an initial dominance of the forced disturbance. This results in a significant decrease in the growth of the most amplified mode. After the saturation of the forced disturbance at high amplitude levels, the nonlinear disturbance interaction causes a break-up of the forced disturbance into its superharmonics (Case 3).
- Forcing a superharmonic disturbance leads to a strong initial growth of the forced disturbance that entirely suppresses the most amplified disturbances over large regions of the chord. After its strong initial growth, the forced disturbance saturates early and its decay gives rise to the growth of subharmonic disturbances. These become dominant in the downstream direction and eventually lead to transition, if the transition was not caused earlier by the forced disturbance (Case 4).

However, for the prediction of the transition onset, the evolution of the total disturbance amplitude rather than the amplitudes of individual modes needs to be considered. Concluding this study, figure 5 presents the amplitude evolution of the total u_1-disturbance component for the four different cases. Plotted is an average of the total disturbance quantity in the spanwise direction. The results indicating a possibly useful technique for the passive control of laminar three-dimensional boundary layers are in qualitative agreement with the experimental observations by Saric et al. (1998).

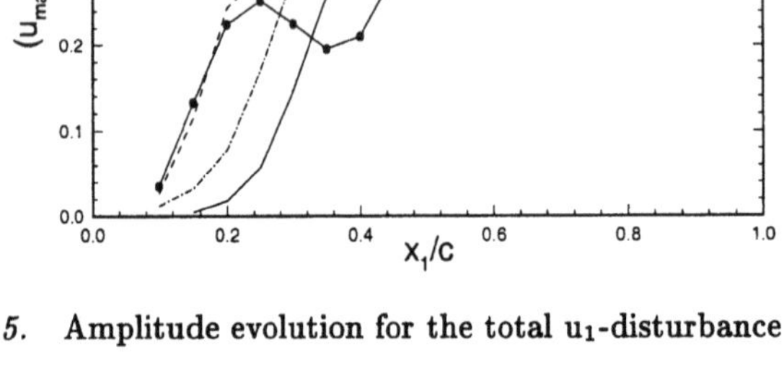

Figure 5. Amplitude evolution for the total u_1-disturbance component

References

Saric, W., Carillo, R. and Reibert, M. (1998) Leading edge roughness as a transition control mechanism, *AIAA-Paper 98-0781*.

Arnal, D., Coustols, E., Juillen, J. (1984) Experimental and theoretical study of transition phenomena on an infinite swept wing, *Recherche Aérospatiale*, **Vol. no. 4**, pp. 39–54.

Balakumar, P. (1997) Finite amplitude equilibrium solutions for plane Poiseuille-Couette flow, *Theoretical and Computational Fluid Dynamics*, **Vol. no. 6**, pp. 103–119.

Active Flow Control by Micro Systems

CHIH-MING HO, PO-HAO HUANG, JOON MO YANG, AND GWO-BIN LEE
Mechanical and Aerospace Engineering Department
University of California, Los Angeles, USA

YU-CHONG TAI
Department of Electrical Engineering,
California Institute of Technology, USA

Micro-Electro-Mechanical-Systems (MEMS) technology provides miniature sensors and actuators. These MEMS transducers enable the ability to control flows in a new domain. In this paper, flow control issues in length scales differing in over six orders of magnitude will be discussed. Within a hand-held device, we will detect biological agents through DNA/RNA identification using MEMS fabricated filters, pumps, and sensors. Methods for handling minute amount of fluid in an accurate and efficient manner are challenging problems. On the other extremity, micro-sensors and micro-actuators can be used to control macro aerodynamic devices. With the use of flexible MEMS shear stress sensors and "bubble" actuators, leading edge vortices can now be altered with near real-time response. This leads to full control of a delta wing aircraft and provides the possibility of a future agile, tailless, and stealthy aircraft.

1. Introduction

Due to length scale compatibility, MEMS technology is ideally suited for small-scale applications [1,2]. A bio-agent detection system consisting of airborne particle sampling, cell preparation, and DNA/RNA detector sub-systems have been developed by applying MEMS technology. In this system, minimizing the air flow resistance through micron-size holes as well as controlling the motion of micron sized particles on solid surfaces involve interesting surface-forces problems. In cell preparations, the nano-liter fluid flow-rate presents numerous challenges. On the other hand, it has also been found that through appropriate coupling of physical phenomena, we are able to control macro scale aircraft using micro scale transducers. This paper will begin by presenting results pertaining to small-scale flow manipulation for DNA detection. Then, we will move up the scale with a discussion of MEMS flow control on a macro scale object: a delta wing aircraft.

G.E.A. Meier and P.R. Viswanath (eds.),
IUTAM Symposium on Mechanics of Passive and Active Flow Control, 195–202.

2. DNA/RNA Identification

Figure 1 is an illustration of the DNA identification system. The objective of this system is to detect very small airborne particles (5~10 μm) or biological threat agents in the air. The system is designed so that it would be able to perform the complete task from sample collection to positive identification. Micro fluidic liquid delivery and circulation using thermopneumatic mechanisms optimized for liquid flow are being developed and integrated to enable on-chip analysis. In order to attain this goal, it is necessary to control the fluid flow and the particle motion in a very precise way. Understandings of the surface and viscous forces are critical.

The airborne particle collection system is composed of a suction fan and a micro-filter (Fig. 2(a)) on which an array of three-phase electrodes is deposited. The micro-filter is fabricated so that the hole size is smaller than that of the targeted particles (5~10 μm)[3]. Once the particles are captured by the micro-filter, they are transported to the liquid delivery system by the 3-phase electrode (Fig. 2(b)).

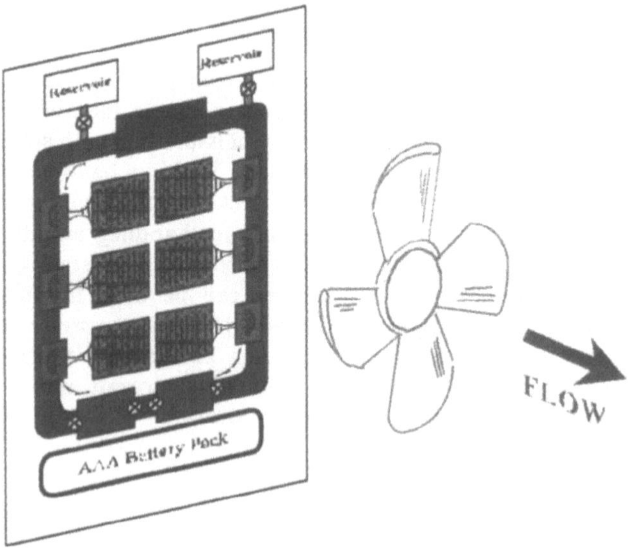

Fig. 1: Illustration of the system for biological agent detection.

In this case, the electrostatic force generated by the electrodes should be strong enough to overcome the surface force between the particle and the surface. Extensive testing of particle transportation has been done, and it is found to be feasible to electrostatically transport particles with sizes from 5~10 μm [4].

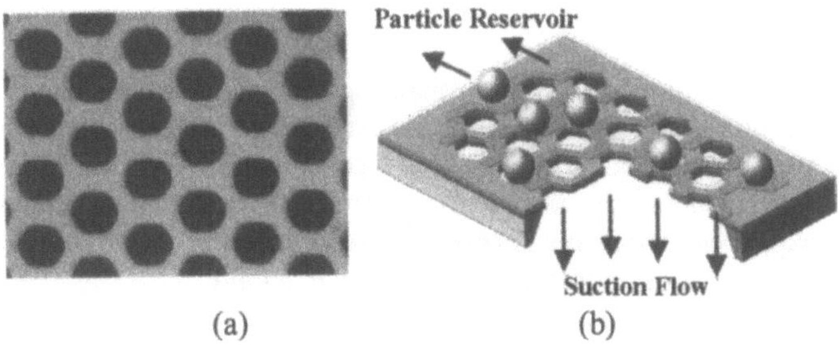

Fig. 2: Airborne particle collection system : (a) Micro-filter; (b) Particle transportation using three-phase electrode array.

After the particles or the biological agents are captured, DNA or RNA from the biological agents should be extracted for identification. A sample preparation system is developed to attain this goal. In this system, the cells attached on the captured particles will first be lysed and then the DNA/RNA will be extracted before sending to the detector. In the preparation process, several hydrodynamic components, such as pumps and valves, are necessary for moving the fluid. A micro-pump, micro-valve, and a cell lysis chip [5,6], as shown in figure 3, was designed and fabricated using an in house micromachining process. Basically, the micro-pump and the micro-valve utilize the deflection (actuation) of silicon rubber. The extracted DNA is then sent to the detector.

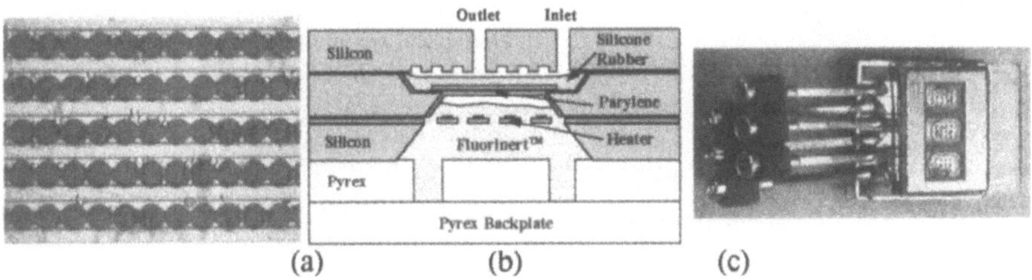

Fig.3: Micromachined components for sample preparation: (a) cell lysis chip; (b) schematic diagram of micro-valve; (c) micro-pump.

Since this system includes fluid flows in very small scales, very interesting phenomena are frequently observed. As an example, figure 4 shows the pressure drop through the micro-filter in an airflow and through the orifice in a liquid flow. Here, numerical calculations are carried out based on the Navier-Stokes equation with a no-slip boundary condition at the wall. The results indicate that a fluid flow in a very small scale is different from that in an ordinary size, and that the behavior in an airflow is different from that in liquid flow.

Here, numerical calculations are carried out based on the Navier-Stokes equation with a no-slip boundary condition at the wall. The measured pressure drops lower than calculated values in gas flow could be caused by the slip boundary condition. The opposite trend in liquid flow is due to the additional surface forces, which are not fully investigated. These features make micro flows, especially liquid flows, a very interesting area.

For particles moving in the micro screen, the surfaces involved are dielectrophoretic forces, viscous drag, and adhesive forces. We can obtain reasonable values of the first two forces. The adhesive forces include Van der Waals forces, electrostatic forces (image force), and meniscus forces. The calculation of these forces on micro-particles, even in vacuum, is approximate in orders of magnitude at best. In other words, extensive research efforts are needed in this field.

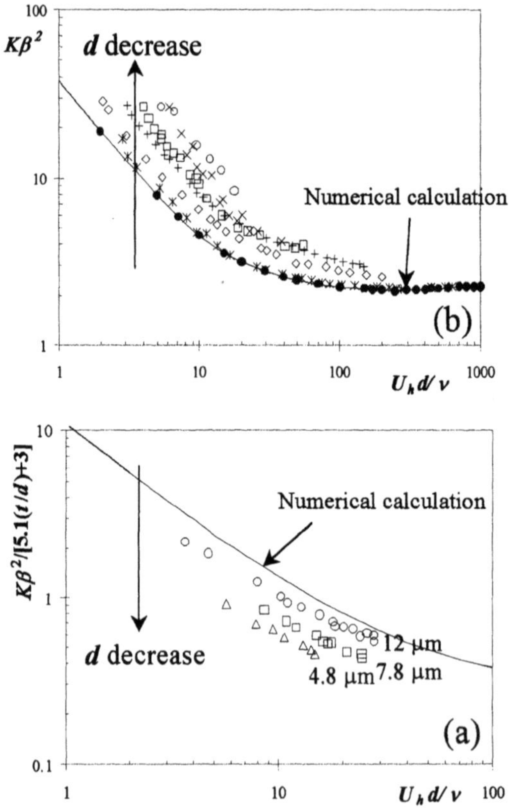

Fig. 4: Abnormal behavior commonly observed in a micro-fluid flow: (a) Lower pressure drop through micro-filters in airflow [1]; (b) Higher pressure drop through single orifices in liquid flow [5].

3. Delta Wing Vortex Control

A MEMS based technology for providing aerodynamic control is currently undergoing development at UCLA and Caltech. This seems to be daunting at first due to the fact that it is not apparent such a macro scale system can be controlled by micro devices. However, the fundamental link these two different scales used for aircraft control is the natural phenomena that the leading edge vortex is very sensitive to perturbation at its origin [7]. By using MEMS sensors to detect the separation line of the leading edge vortices and using MEMS actuators to alter it while interacting with the boundary layer, the resulting leading edge vortex pairs lose their symmetry. The ultimate result is the ability to create force imbalance of the high suction vortex pairs and provide independent flight control of pitching, rolling, and yawing. On an aircraft, conventional surfaces such as elevators, ailerons, and rudders are used to provide such controls.

In addition to its potential lift from the geometry of the airfoil cross section, a delta wing aircraft also has a pair of symmetrical vortices (Fig. 5), which provides up to 40% of total lift [8]. In order to utilize the vortex pairs for aircraft control, a detector must first be used to find the separation line where these vortices originated. This is where MEMS comes in at its best. A flexible shear stress skin of 80μm thick and 1x3 cm planar area containing 64 sensors was developed. Such a sensor has a frequency response of 10 kHz and can be wrapped around the leading edge curvature of a delta wing aircraft (Fig. 6).

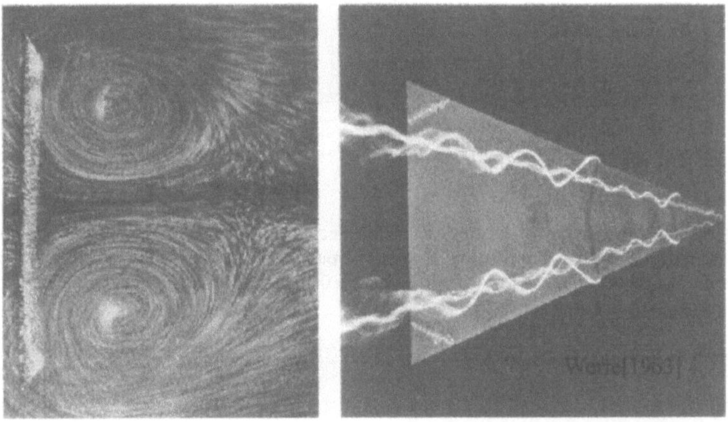

Fig. 5: Primary vortex pairs of a generic Delta Wing.

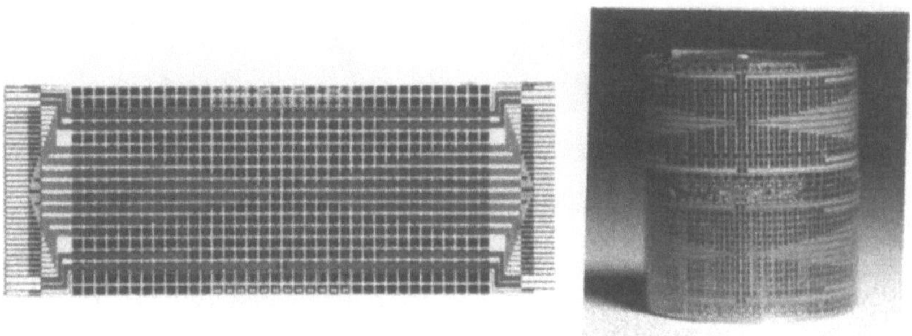

Fig. 6: Flexible MEMS Shear Stress Sensor [9]

After mapping the entire separation line on the leading edge (Fig. 7), an array of MEMS "Bubble" actuators can then be used to alter the separation line and break the vortex symmetry. Figure 8 shows the bubble used in the vortex flow control with actuation height being variable and based upon the pressure supplied. Furthermore, particle flow visualization was used to provide the physical evidence on how vortex flow control can be achieved. For example, figure 9 demonstrates that the starboard primary vortex is being shifted by the MEMS actuators. The resulting pressure field from such a vortex shift will create a rolling moment. Similarly, independent pitching and yawing moments can be created.

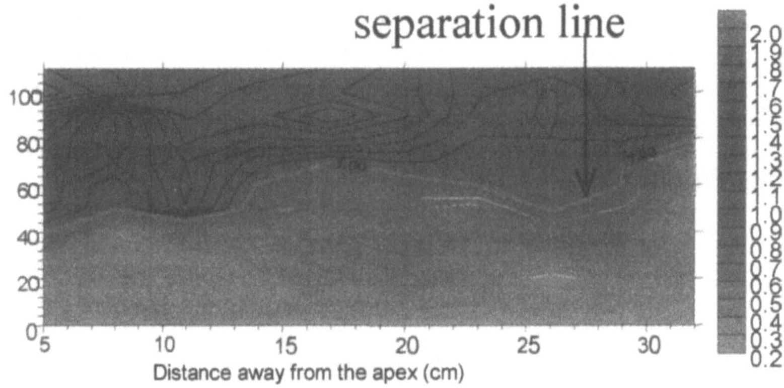

Fig. 7: Typical signal topography of leading-edge RMS fluctuation produced by MEMS shear stress sensors. The vertical axis shows the position in degrees around the leading edge radius, where 0 degrees is at the bottom side and 90 degrees at the centerline (using a 0.25 in leading edge radius cylinder).

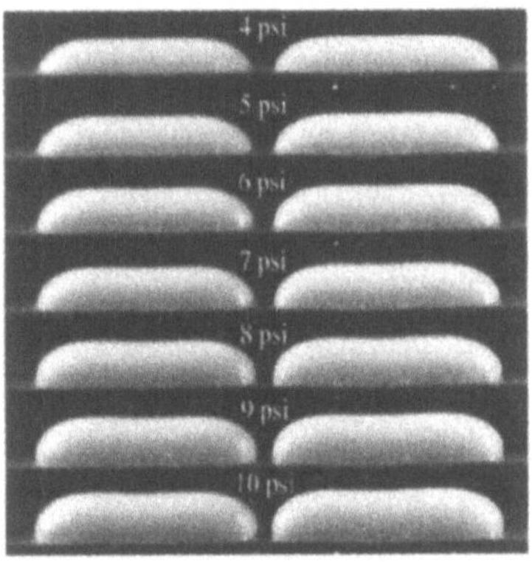

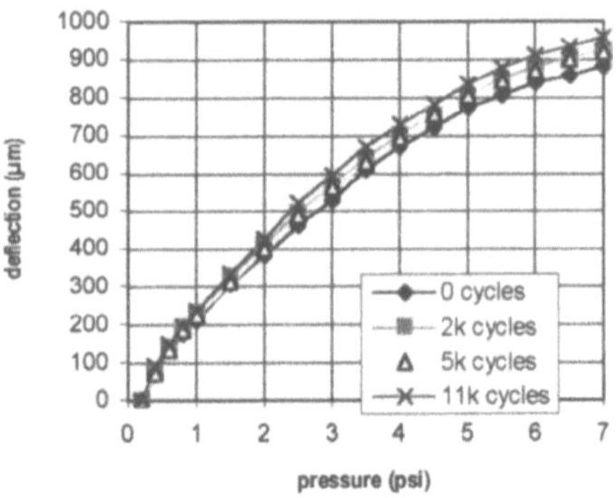

(a) (b)

Fig. 8: (a) Bubble actuation size versus pressure. (b) Actuator deflection versus pressure under various cycles [9].

This simply means that a tailless aircraft without large drag-inducing control surfaces is within the possibility with such MEMS flow control technology. Currently, a fully instrumented research UAV (Un-inhabited Aerial Vehicle) has been built to test out such a futuristic concept on actual flying machines. This first generation MEMS aircraft, co-developed by UCLA, Caltech, and AeroVironment, called "GRYPHON" will pave the way for future super-agile, tailless, and stealthy aircraft (Fig. 10).

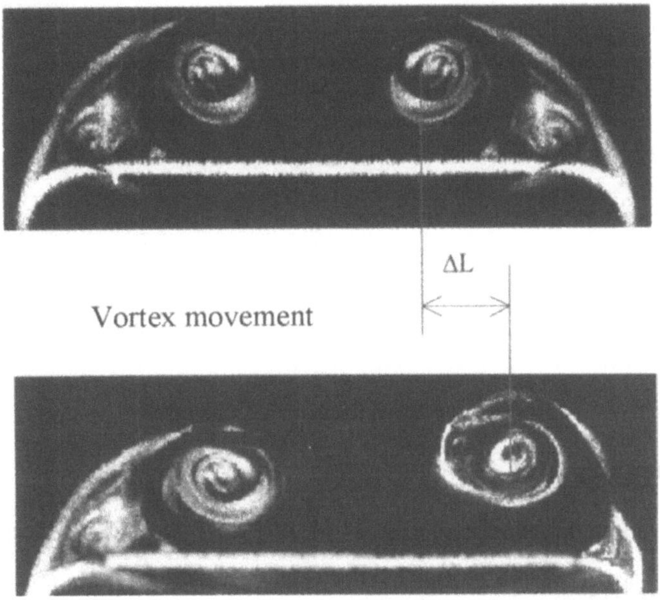

Fig. 9: Flow visualization showing shift of right primary vortex core.

4. Concluding Remarks

By applying MEMS based transducers, we demonstrated flow control in two extreme; one in manipulating nano-liter range fluid for bio-applications and another on controlling the motion of an airplane. In the first case, more research on the fundamental fluid-surface phenomena is critical for precise control of minute amount of fluid. For aerodynamic applications, micro transducers are able to control an aircraft by efficiently using the flow instability as an amplifier of their input.

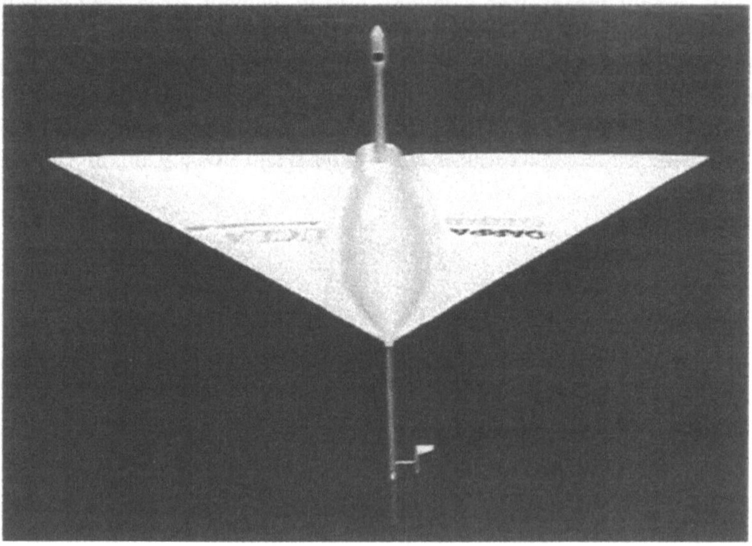

Fig. 10: MEMS research UAV, GRYPHON

5. Acknowledgement

The work in this paper is supported by DARPA ETO and TTO offices.

6. References

[1] Ho, C.M. and Tai Y.C., "MEMS and its applications for Flow Control". Journal of Fluid Engineering, Vol. 118, pp. 437-447, 1996.

[2] Ho, C.M. and Tai, Y.C., "Micro-Electro-Mechanical-Systems and Fluid Flows", Annual Review of Fluid Mechanics, Vol. 30, pp. 579-612, 1998.

[3] X.Yang, J.M. Yang, X.Q.Wang, E.Meng, C.M. Ho and Y.C. Tai, Proceedings of IEEE Workshop on Micro Electro Mechanical Systems, Heidelberg, Germany, Jan.(1998). 137-142

[4] A.Desai, S.W.Lee and Y.C.Tai, Proceedings of IEEE Workshop on Micro Electro Mechanical Systems, Heidelberg, Germany, Jan.(1998) 124-126

[5] S.W.Lee, H.Yowanto and Y.C.Tai, Proceedings of IEEE Workshop on Micro Electro Mechanical Systems, Heidelberg, Germany, Jan.(1998) 443-147

[6] T.Hasegawa M.Suganuma and H.Watanabe, Physics of Fluids, 9, No.1(1997)1-3.

[7] Ho, C.M. and Huang, L.S., "Subharmonics and Vortex Merging in Mixing Layers", Journal of Fluid Mechanics, Vol. 119, pp. 443-473, 1982.

[8] Werle, H "Flow Visualization Techniques for the study of High Incidence Aerodynamics", AGARD/VKI Lecture Series, Mar. 121-3(1982).

[9] G.B. Lee, F. Jiang, T. Tsao, Y.C. Tai, and C.M. Ho, IEEE *Aerospace Conference*, 1997

STRATEGY OF THE FLOW MEMS CONTROL AT LAMINAR-TURBULENT TRANSITION IN A BOUNDARY LAYER

A.V. BOIKO, V.V. KOZLOV
Institute of Theoretical and Applied Mechanics SB RAS
Novosibirsk, 630090, Russia

Introduction

Most of the techniques to control laminar-turbulent transition in near-wall flows are so-called passive approaches, being quite well developed for two-dimensional flows. Favourable pressure gradients, surface cooling and heating, as well as localised boundary layer suction [1] are most well known. Riblets placed in the boundary layers have also been used to suppress formation and interaction of organised three-dimensional structures, e.g., Λ- and stationary vortices in the transition region [2, 3]. They are passive, since they have no means for detecting and dynamically manipulating the flow structures, i.e. a possibility to adjust the control in time and space is essentially absent. A known alternative method is the direct suppression of the two-dimensional instability waves through their linear superposition (wave cancellation). To achieve this, it is necessary to place an actuator, which effectively generates the two-dimensional waves with the same amplitudes, but with opposite phases [4-7].

Meanwhile, the stage of a two-dimensional instability in certain flows is extremely short or overlapped by three-dimensional effects as in three-dimensional flows at swept wings, rotating disks, and concave surfaces. In these cases the instability development depends usually on local conditions in stationary vortices of different nature and the instability waves in the neighbour vortices can have different amplitudes [8]. Such inhomogenity of the flow field can lead to independent occurrence of the transition inside the individual vortices, that is observed as the characteristic saw-tooth shape of the transition line in visualisation pictures. The two-dimensional passive control techniques for such flows are usually difficult to apply, since they result in different effect in different points of the flows [9].

Nevertheless to control the transition, at least in certain three-dimensional situations, it is possible to develop active control methods. The general scheme of the control system in such cases can be formulated as follows [10]. Receiving data from flow sensors, analysers process the signals and reconstruct amplitudes, phases and other necessary characteristics of the disturbances and determine an optimal actuator response. Then the actuators excite, e.g., three-dimensional disturbances with the same

G.E.A. Meier and P.R. Viswanath (eds.),
IUTAM Symposium on Mechanics of Passive and Active Flow Control, 203–208.

characteristics, but opposite phase, produce a local suction or other spatially local effects on the flow. From the one side, a practical application of such technique needs special micro-electrical-mechanical-systems (MEMS) technology [10, 11], which allows to produce mechanical parts and the whole devices of extremely small sizes and large numbers combined with micro-chip analysers. From the other side, it is necessary to understand physical processes of the control to design the analysers and actuators. The methods of introduction of artificial disturbances in the flow are very well suitable for the investigation of the processes. Usually in this case, it is possible to introduce flexibly disturbances of different kinds and to achieve a good repeatability of the data comparing with a 'natural' flow development as well as to measure not only amplitude, but also phase disturbance characteristics in essentially any point of the flow.

Present work is a continuation of the investigations [2], where a passive method of the stationary vortex control by riblets in the transition region of a three-dimensional flow was tested using the method of the control disturbances. Below we propose in the same way two methods of the active control in such flow: the wave cancellation and a localised point-like suction.

Experimental set-up

The experiments were carried out in the subsonic low-turbulent wind tunnel of the Institute of Theoretical and Applied Mechanics SB RAS T-324. The free-stream turbulence level was less then 0.04% at the operating velocity U_0=8.5 m/s. A general outline of the experimental arrangement is shown in fig. 1. The test model consisted of profile C-12 with obstacles placed close to the leading edge, which were used as generators of strong vortices (see [2, §2.1] for detail). One of the wing surfaces was adjusted so that to achieve zero pressure gradient above it. Special tests [12, 13] showed that in the measurement region (X>130 mm) the streamwise gradient was essentially absent and the streamwise mean velocity profile was close to the Blasius one.

To generate control travelling waves from the instability region of the stationary vortices a flow acoustic forcing of the flow with the help of a loudspeaker placed in the diffuser part of the wind tunnel was applied. It radiated the sound against the flow direction. Sound amplitude and frequency were controlled by a function generator. To introduce the cancelling signal a flow forcing through a round orifice with diameter 0.8 mm at distance X=170 mm was used. The orifice was connected by means of a tube to the other loudspeaker, which produced the pressure pulsations and was controlled by the same function generator. The phase was adjusted by the tube length. The same orifice was used to produce continuos point-like suction. A constant temperature hot-wire anemometer DISA 55M01 with a single-wire probe served as the sensor to obtain the mean and fluctuating velocity components. Hot-wire data were digitised by an analog-to-digital converter and processed with a personal computer.

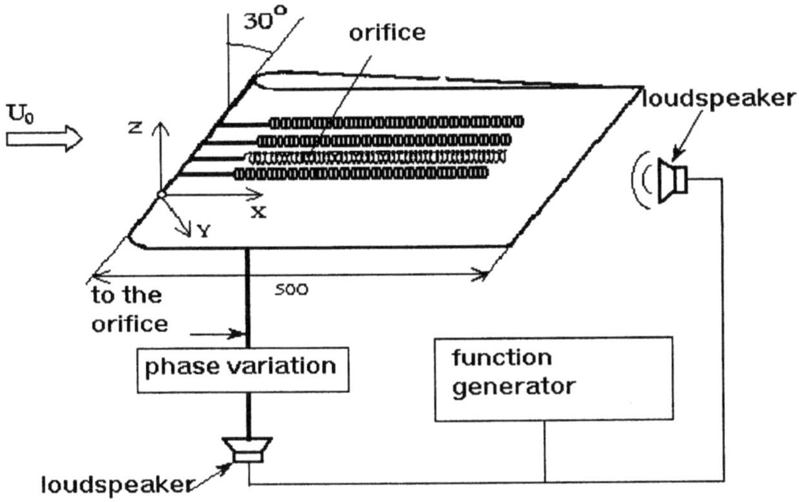

Fig. 1. Experimental arrangement.

The wave cancellation

A representative distribution of the streamwise velocity defect in respect to the undisturbed boundary layer at the absence of the vortex generation $\Delta U/U_0$ is given in fig. 2, *a* for one of the section in *YZ* plane. The measurements were carried out in section *X*=270 mm that is 100 mm downstream the orifice.

The wave excitation by means of the localised periodic blowing-suction and sound was applied. The r.m.s. velocity distributions in the vortex system at the frequency of excitation by the both means are given in fig. 2, *b* and *c*. It is seen that the instability wave at the suction-blowing appears only in the one vortex, under which the orifice is located, the disturbance distributions in the both cases being very similar and analogous to the single vortex case [13].

Under the conditions of the control excitation of the waves, the question on the linearity of the waves generated in the set of the vortices was studied and it was found that amplitude independence and superposition criteria are satisfied. Based on these results a method of a local active instability control in one of the vortex is proposed: at certain phase and amplitude of the disturbances excited by the blowing-suction it is possible to control the wave excited outside, e.g., by sound, as in our case. The picture of the wave superposition at their mixed excitation in the counter-phase is shown in fig. 2, *d*. It is seen that the amplitude of the waves excited by sound u'/U_0=0.6% (fig. 2, *b*) is reduced in all region of the vortex localisation in accordance with the amplitude of the control signal (fig. 2, *c*). It should be noted that the wave subtraction occurs in the most region of the vortex localisation. So, in the three-dimensional flow under investigation a possibility to control the development through the wave cancellation is shown. If MEMS-actuators will be able to excite the natural wave packet with opposite phase characteristics in the vortices, then almost full their cancellation and laminar regime prolongation will be achieved.

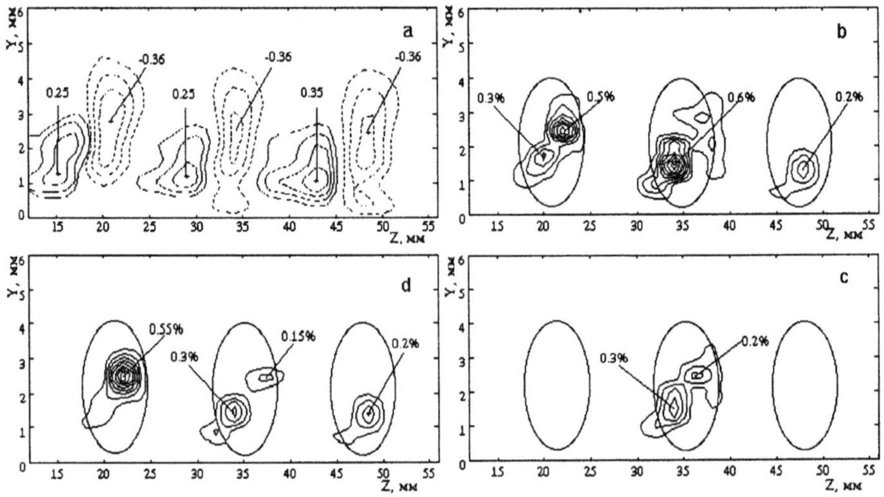

Fig. 2. Distributions of the streamwise mean velocity defects and r.m.s. disturbance values at forcing frequency f=320 Hz, X=270 mm. The spacing between the vortices is 13 mm. The solid elliptic lines used for a qualitative vortex 'representation'. *a* - mean velocity defect distribution $\Delta U/U_0$; *b* - disturbance distribution in the vortex set at the wave excitation by sound u'/U_0, %; *c* - the same at the suction-blowing; *d* - the same at the both waves counter-phase excitation.

Point-like continuos suction

In the experiments described below all the measurements were done, when the orifice was located directly below one of the vortex (see fig. 3). Spectral measurements inside the vortex showed that the suction affected the whole spectrum of the natural instability waves downstream the orifice. Comparative measurements of the mean velocity and sound excited disturbance distributions were carried out at X=270 mm (fig. 3). Comparison of figs. 2, *b* and 3, *b* shows that the local suction leads to approximately 30-fold decrease of the disturbance pulsations in the central vortex. Resulting mean velocity distortion generated by the central vortex is reduced approximately by 25% of its initial value. Meanwhile, the effect on the mean velocity distribution in the side vortices is not large, even at so quite strong point-like suction value and manifests itself not in changes of the peak amplitudes, but rather in a certain deformation of the contour lines.

It is seen, that the vortex structure undergoes changes on the stage of the linear development of the travelling waves, which lead, as in the two-dimensional case [1] to the flow stabilisation and to decrease of the disturbance level.

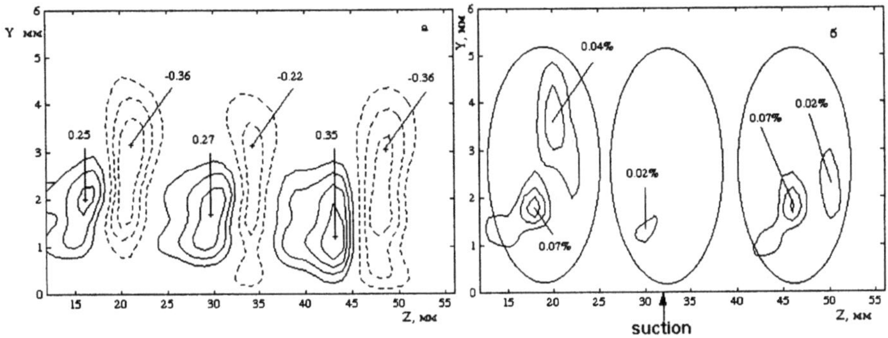

Fig. 3. Mean velocity defects $\Delta U/U_0$ distribution (a) and disturbance r.m.s. values u'/U_0, % (b) in the vortex set at the continuos suction; $U_{suction}/U_0$=1.2.

Some measurements were carried out for the suction position shifted a half of the spanwise vortex spacing (in the experiment the vortex set was shifted relating to the orifice). It was found that in this case in spite of also observed certain mean velocity changes in the vortex, the amplitudes of the travelling waves were modified only slightly. Probably, the suction in this case does not touch that part of the flow, which is responsible for the instability generation [13]. In particular the velocity shear inside the vortex leaves essentially the same, in spite of some observed flow distortions in the periphery.

Conclusions

Two methods of active control of instability and transition in stationary steady vortices [12, 13] in the three-dimensional boundary layer were proposed. It was found that by means of the control wave excited in the vortex due to periodic suction-blowing through the orifice it is possible to control the development of the linear instability waves and reduce their amplitudes. The strong localisation of this influence inside one of the vortex is observed.

It was found that the suction can significantly weaken the travelling instability by means of the local influence on the flow structure. It was shown that the influence depends on the suction location in respect to the vortices. The maximal effect is reached when the suction occurs below one of the vortices.

We suppose that in presence of secondary instabilities in other stationary vortex dominated flows it is possible to apply similar techniques, but some additional tests are necessary.

The work was carried out with the financial support of the Russian Foundation for Basic Researches (grants No 96-01-01182 and 96-15-96310) as well as the Siberian Branch of RAS.

1. Kozlov, V.V., Levchenko, V.Ya., Scherbakov, V.A.: Development of the disturbances in the boundary layer at a slit suction, *Uch. Zap. TsAGI* **IX**, No. 2 (1978) 99-105 (in Russian).
2. Boiko, A.V., Kozlov, V.V., Syzrantsev, V.V., Scherbakov, V.A.: Transition control by riblets in swept wing boundary layer with embedded streamwise vortex. *Eur. J. Mech.* **16**, No. 3 (1997).
3. Grek, G.R., Kozlov, V.V., Titarenko, S.V., Klingmann, B.G.B.: The influence of riblets on a boundary layer with embedded streamwise vortices, *Phys. Fluids A*, **7**, No. 10 (1995) 2504-2506.
4. Milling, R.W.: Tollmien-Schlichting wave cancellation, *Phys. Fluids* **24** (1981) 979-981.
5. Gilev, V.M.: Tollmien-Schlichting wave excitation on the vibrator and laminar-turbulent transition control. V.V. Kozlov (ed.) *Laminar-Turbulent Transition*.-Springer: Berlin, 1985, 243-248.
6. Kozlov, V.V., Levchenko, V.Ya.: Laminar-turbulent transition control by localized disturbances. *IUTAM Symposium on Turbulence Management and Relaminarization*, Bangalore, India, H.W. Liepmann, R. Narasimha (Eds.), 1987.
7. Gilev, V.M., Kozlov, V.V.: Influence of the periodic suction-blowing on the transition process in the boundary layer. *Preprint SB AS USSR, Inst. Theoret. and Applied Mech.*, 1985, No. 1, Novosibirsk (in Russian).
8. Swearinger, J.D., Blackwelder, R.F.: The growth and breakdown of streamwise vortices in the presence of the wall, *J. Fluid Mech*, **182** (1987) 255-290.
9. Bippes, H.: Basic experiments on transition in three-dimensional boundary layers dominated by cross-flow instability, to be published in *Prog. Aero. Sci.*
10. Ho, Ch.-M., Tau, Yu-Ch.: MEMS: Science and technology. *Application of Microfabrication to Fluid Mechanics*, 1994, ASME FED **197** 39-48.
11. Boiko, A.V., Kozlov, V.V.: Methods of controlled disturbances for laminar-turbulent transition research and control. *In Proc. of Intern. Conf. on the Methods of Aeroph. Research*, Novosibirsk, 1996, 25-33.
12. Boiko, A.V., Kozlov, V.V., Syzrantsev, V.V., Scherbakov, V.A.: Experimental investigation of the process of transition to turbulence in a single stationary disturbance in a swept wing boundary layer. *J. Applied Mechanics and Technical Physics*, No. 1 (1995) (in Russian, English translation is available).
13. Boiko, A.V., Kozlov, V.V., Syzrantsev, V.V., Scherbakov, V.A.: Investigation of influence of the internal structure of a streamwise vortex to the development of travelling disturbances inside it. *Thermophysics and Aeromechanics* No. 3 (1997) (in Russian, English translation is available).

MANIPULATING COHERENT STRUCTURES BY HEAT IN A BOUSSINESQ JET

R. NARASIMHA AND I.V.R. SIVAKUMAR
Jawaharlal Nehru Centre for Advanced Scientific Research,
Jakkur Campus, Bangalore 560 064, India

1. Introduction

Entrainment and mixing are problems of major concern in many fluid-dynamical applications, especially in the presence of heat release (e.g. combustion, clouds). A simple but interesting idealization here is to use the Boussinesq approximation, which is not directly applicable to most combustion problems, but happens to be quite realistic in many geophysical applications. With this objective in view, Basu & Narasimha [1] have performed a series of simulations using a pseudospectral method to compute the temporal evolution of a jet-like flow with volumetric heat injection over some region well away from the origin of the flow. Furthermore, recent experimental work on such flows [2-4] provides a basis for mutual validation and enhanced understanding. The experimental studies show that off-source heating accelerates the flow and arrests jet growth; absolute values of turbulence intensity increase but not as rapidly as the mean velocities, so normalized turbulence intensities are lower. It is gratifying that strong qualitative similarities with the experiments have been observed in the numerical simulations. This gives us considerable confidence in the results we shall present in this paper, where we describe a new study on how off-source volumetric heat injection can alter and manipulate the coherent vortical structure and entrainment characteristics of a jet.

2. Numerical method

The object of investigation is the *temporal* development of a cylindrical mixing layer with volumetric heating over a specified duration (Figure 1). Under the Boussinesq approximation the sole effect of density changes in the flow (because of heating) is to introduce a buoyancy term in the momentum equation, the changes in the inertial terms or the values of the transport parameters being negligible [5]. The heat injected into the flow appears as a source term in the energy equation. The governing equations for the problem are therefore [1]

G.E.A. Meier and P.R. Viswanath (eds.),
IUTAM Symposium on Mechanics of Passive and Active Flow Control, 209–216.

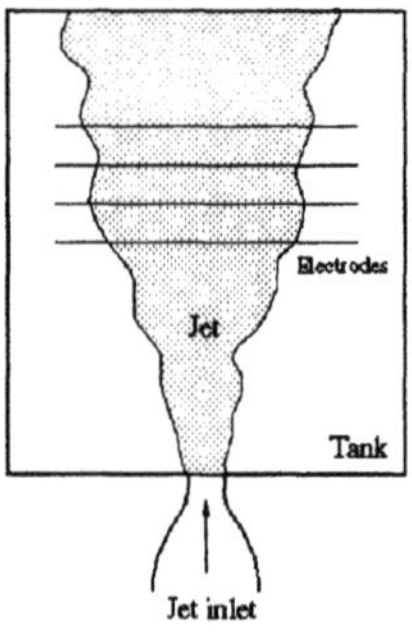

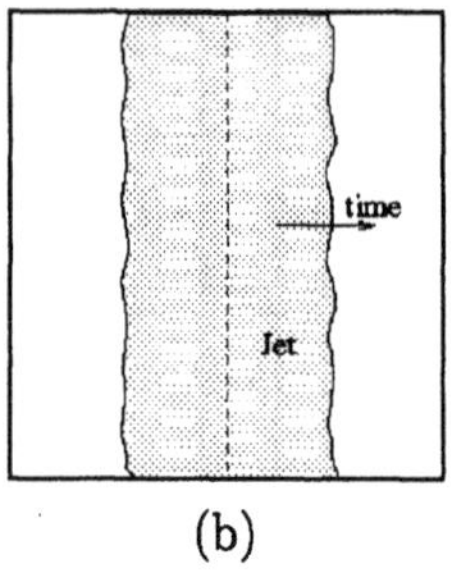

Figure 1. Flow situation considered. (a) Laboratory set-up, with voltage applied across the electrodes to produce heating in an electrically conducting fluid. (b) Temporal problem handled in numerical simulation.

$$\nabla \cdot \mathbf{u} = 0, \tag{1}$$

$$\frac{\partial \mathbf{u}}{\partial t} + (\mathbf{u} \cdot \nabla)\mathbf{u} = -\frac{1}{\rho}\nabla p + \nu \nabla^2 \mathbf{u} - \mathbf{g}\alpha T, \tag{2}$$

$$\frac{\partial T}{\partial t} + (\mathbf{u} \cdot \nabla)T = \kappa \nabla^2 T + \frac{J}{\rho c_p}, \tag{3}$$

where $\mathbf{g}$ is the acceleration due to gravity, α is the coefficient of thermal expansion, κ is the thermal diffusivity, J is the rate of head addition per unit volume, and T is the change in temperature above ambient; the rest of the notation is standard.

Equations (1-3) respectively express conservation of mass, momentum and energy. To non-dimensionalize these equations, we use as scales the initial diameter d_0 of the jet, the initial centreline velocity U_0 and a characteristic temperature difference T_0. If the heat is injected at a constant rate J for a total time duration t_h into the flow, we may define T_0 as the net temperature change that would result if this heat was injected *uniformly* over a unit volume: $T_0 \equiv J t_h / \rho c_p$. We then get

$$\frac{\partial T}{\partial t} + (\mathbf{u} \cdot \nabla)T = \frac{1}{Re\ Pr}\nabla^2 T + \frac{d_0}{U_0 t_h} g(r) \tag{4}$$

$$\frac{\partial \mathbf{u}}{\partial t} + (\mathbf{u} \cdot \nabla)\mathbf{u} = -\nabla p + \frac{1}{Re}\nabla^2 \mathbf{u} + G C_h T \tag{5}$$

where all variables are now non-dimensional, $Re = U_0 d_0 / \nu$ is the Reynolds number, $Pr = \nu / \kappa$ is the Prandtl number, $g(r)$ is a prescribed radial distribution function, and

$$G = \frac{g \alpha d_h^2}{\rho c_p} \frac{J}{U_h^3}, \tag{6}$$

d_h, U_h being respectively relevant length and velocity scales in the heating region. The heating parameter G is analogous to the non-dimensional heat release number introduced in [3], except that (6) incorporates a modification to take account of the fact that in the present simulation the heat release occurs over a finite duration

t_h in time, and not over a finite region in space as in [3]. Either version of G is a measure of the ratio of buoyancy to inertia forces, like the Richardson number. The other non-dimensional parameter in (5) is

$$C_h = \frac{U_h t_h}{d_h} \frac{d_0}{d_h} \left(\frac{U_h}{U_0}\right)^2,$$

whose value will remain constant during the present study. (This corresponds to fixing the spatial region over which heat is added in a real jet.) Since G and C_h appear only as a product, we use $G^* = GC_h$ as the relevant non-dimensional parameter governing heat release in the present study.

Equations (4-5) are solved, along with the continuity equation (which does not change upon non-dimensionalization), in a *Cartesian* coordinate system using the Fourier-Galerkin (spectral) technique under periodic boundary conditions in all three space dimensions. More details can be found in [1]. No dealiasing is used, so the scheme may be considered to be 'pseudospectral'.

The initial conditions for the velocity components are set up so that we have a tubular shear layer aligned along the z-direction (vertically upwards) at time $t = 0$. The w-velocity has a top-hat profile with a tan-hyperbolic shear layer. We put in streamwise and azimuthal perturbations to expedite roll-up and the development of the Widnall instability. The initial velocity field is made divergence-free using the Helmholtz decomposition. The size of the computational domain (one periodic cubical box) is $4d_0$ on each side.

We report here specifically on the results of three simulations based on a 128^3 grid. The first is the simple unheated jet. In the second case, heat is applied between the times $t = 25$ and $t = 32$ and the flow evolution is computed upto $t = 35$. In the third case, to be called "active" heating, heat is injected only in two quadrants in the cross-section of the jet, over bands whose position is related to that of the coherent structures in the flow in a manner that we shall describe later.

The relevant parameters for the computations are $Re = 1600, Pr = 7, G^* = 0.04$; $g(r)$ mimics the mean velocity distribution. The time-step used is $\Delta t = 0.0025$ for the unheated jet and $\Delta t = 0.00125$ for the heated jets.

The vorticity transport equation, obtained by taking the curl of equation (2), is

$$\frac{\partial \omega}{\partial t} + (\mathbf{u} \cdot \nabla)\omega - (\omega \cdot \nabla)\mathbf{u} - \nu \nabla^2 \omega = \alpha \mathbf{g} \times \nabla T \tag{7}$$

where the terms on the left are familiar from classical theory. The term on the right is a source of vorticity; it arises from the baroclinic torque $\nabla p \times \nabla \rho$ when the pressure gradient ∇p is replaced by its value in the hydrostatic approximation, and the density gradient $\nabla \rho$ by the temperature gradient, to which it is proportional in the approximation we are using.

3. Coherent Structures

There have been several experimental studies on coherent structures in jet flows [6-9]. It is not our objective to discuss in depth the question of coherent structures in jets in general (we shall do this elsewhere), but rather to note the effect of heating, taking advantage of the vorticity field that the present numerical simulations provide: coherent structures are probably best educed from the vorticity field. A detailed analysis of the vorticity field in the jet has recently been carried out [10,11]; it emerges from this analysis that the distribution of azimuthal vorticity ω_θ in axial sections of the jet is particularly revealing. Figure 2 shows, for all three cases, this component of the vorticity in the yz plane, from time $t = 25$ to 35 in steps of two time units. The contour levels start from 0.5 with increments of 0.5 .

At $t = 25$, the unheated jet is roughly cylindrical in shape. The jet boundaries are relatively sharp, more or less continuous and not too highly convoluted. Also, vorticity is nearly uniformly distributed throughout the jet. By $t = 27$, the boundaries begin to break up and "bulges" swell out on either side of the jet axis. At $t = 29$, the bulges take the form of an "arrowhead" shaped region. Another nascent arrowhead is seen developing at the bottom of the domain. The jet boundary is now highly convoluted; also, the vorticity distribution is much less uniform, as patches of irrotational fluid appear near the jet axis. By $t = 35$, one structure has moved downstream from its earlier nearly central position to the top of the box, and a second structure is at the centre. The distinct arrowhead shape of the coherent structure, most clearly seen in this frame (and on a larger scale in Figure 4), is characterized by a conical sheath of azimuthal vorticity resting on a toroidal base. A third structure (which is a wrap-around of the one now at the top) is seen developing at the bottom.

The contrast between the heated and unheated cases is noticeable only after $t = 31$. The heated jet has a vorticity field with both a larger magnitude and a more uniform distribution : while vorticity decays with time in the unheated jet, it is enhanced by buoyancy in the heated case. Also, the bulginess in the boundaries, observed in the unheated case, is less pronounced. The width of the jet is also slightly lower. By $t = 33$, the structure in the flow can barely be discerned - the only indication being the long filaments of vorticity seen near the edges. By $t = 35$, the heated jet bears little resemblance to its unheated counterpart. Boundaries show weaker convolutions. There is a clearer demarcation between the jet and the surrounding irrotational ambient fluid. Also, the jet is packed with vorticity. The highly distorted structures, barely identifiable, appear to be laterally compressed and longitudinally extended, and telescope into each other.

Results with active heating are shown in panel (c) of Figure 2. The lines here indicate the band over which heat is injected; there is no heating in the quadrant behind the dotted lines. The heating has clearly produced vorticity; the decay associated with the streamwise development of the jet has been largely suppressed. An interesting point to note is that, from $t = 31$ on, the two structures in the domain seem virtually disjoint; there is almost a rupture in the vorticity field from $t = 31$ to 34. At $t = 35$ the structure in the middle of the domain is rather similar to that in the unheated jet; the structure below has a relatively strong vorticity field, but it does not show the deep penetration that full heating results in.

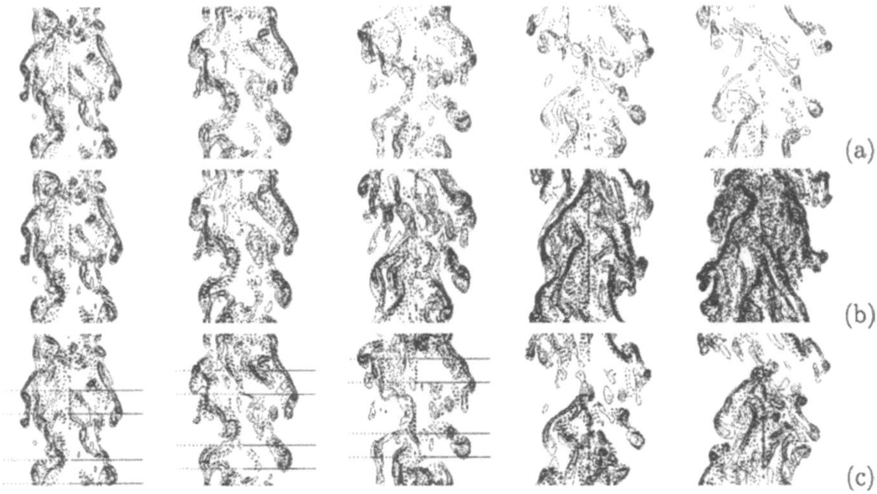

Figure 2. Evolution of azimuthal vorticity field, $t = 27$- 35 in steps of 2 units: (a) unheated, (b) fully heated, (c) with active heating. Solid (dashed) lines : positive (negative) values. Horizontal lines in bottom panel show band over which heating is applied; solid (dashed) lines indicate heating in quadrant out of (behind) plane of paper

It is clear therefore that heating alters the structures significantly, in ways that can be exploited to manipulate them.

4. The Entraining Velocity Field

Because of the periodic boundary conditions imposed on the computational domain in the present simulation, the net entrainment over the domain has to vanish. Nevertheless, considerable insight into the problem can be obtained by examining what we shall call the 'entraining velocity field', which displays velocity vectors in the ambient fluid in the *immediate* neighbourhood surrounding the jet. For lack of space, we content ourselves with a study of the flow at $t = 35$. Figure 3 shows the computed entraining velocity field at two representative sections each in the unheated, fully heated and actively heated cases. It is clear that heating substantially increases the entrainment velocities. This can be attributed to the higher vorticity generated due to by the baroclinic torque in (7), as is also evident from the contours of streamwise vorticity shown in the same figure. Secondly, at any given station, fluid is not uniformly entrained (or detrained) : outer fluid may be *inducted* into the core flow, but core fluid may also be *expelled* (as we shall say). Finally, as we go round the flow, these velocities vary appreciably, in both sign and magnitude; in fact, in all cases (most dramatically with full heating), the entraining velocity field is organized on a scale which is something like a fifth of the diameter of the jet (cf. [1]). This is also the scale on which the streamwise vorticity is organized : as positive and negative values of this vorticity often occur

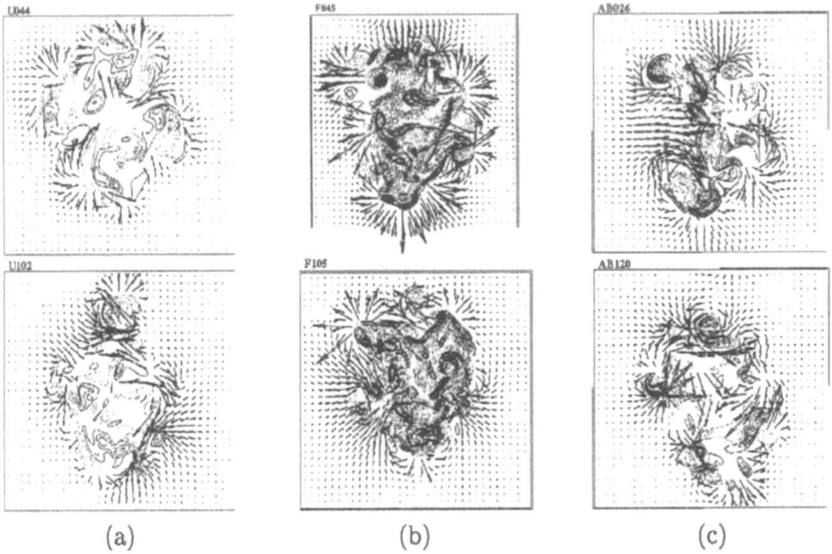

Figure 3. Entraining velocity field, against background of streamwise vorticity contours, in two representative sections for each of unheated (a), fully heated (b) and actively heated (c) cases. Outer lines in (b) and (c) indicate area of heating - full in (b), quadrantal in (c).

next to each other, there is a strong indication that streamwise vortex loops of some kind are responsible for this character of the entraining velocity. At the secton F045 in the heated jet (top, Figure 3b), five such expulsive "cells" can be identified; at F105 (bottom), there are three expulsive and three inductive cells. There are similar cells in the other two cases also.

5. The Volumetric Influx

Figures 4a, b, c show the variation of the volumetric influx into the jet with axial position along the jet at $t = 35$; to help correlation, the corresponding azimuthal vorticity contours in the mid plane are also shown. From the unheated case (Figure 4a), it is seen that there is an influx (considered positive) near the toroidal base of the coherent structure, and an outflux just above it. There is a further region of influx towards the vertex of the structure; the bigger chunk of influx above this is associated with the toroidal base of the next structure. In the fully heated case (Figure 4b) the vorticity field is far more intense, and so is the volumetric flux. The influx and outflux regions near the toroidal base of the primary structure can again be identified, but beyond $z/d \simeq 2$ there is another pair of similar regions; the secondary bump seen in Figure 4a seems to be absent. This may be due to the way the structures have telescoped, so that the fluxes of the two structures are mixed up. Figure 4c shows an interesting behaviour. Towards the base ($1 \leq z \leq 1.75$), and just above ($1.75 \leq z \leq 2.1$) the variations are similar to those in

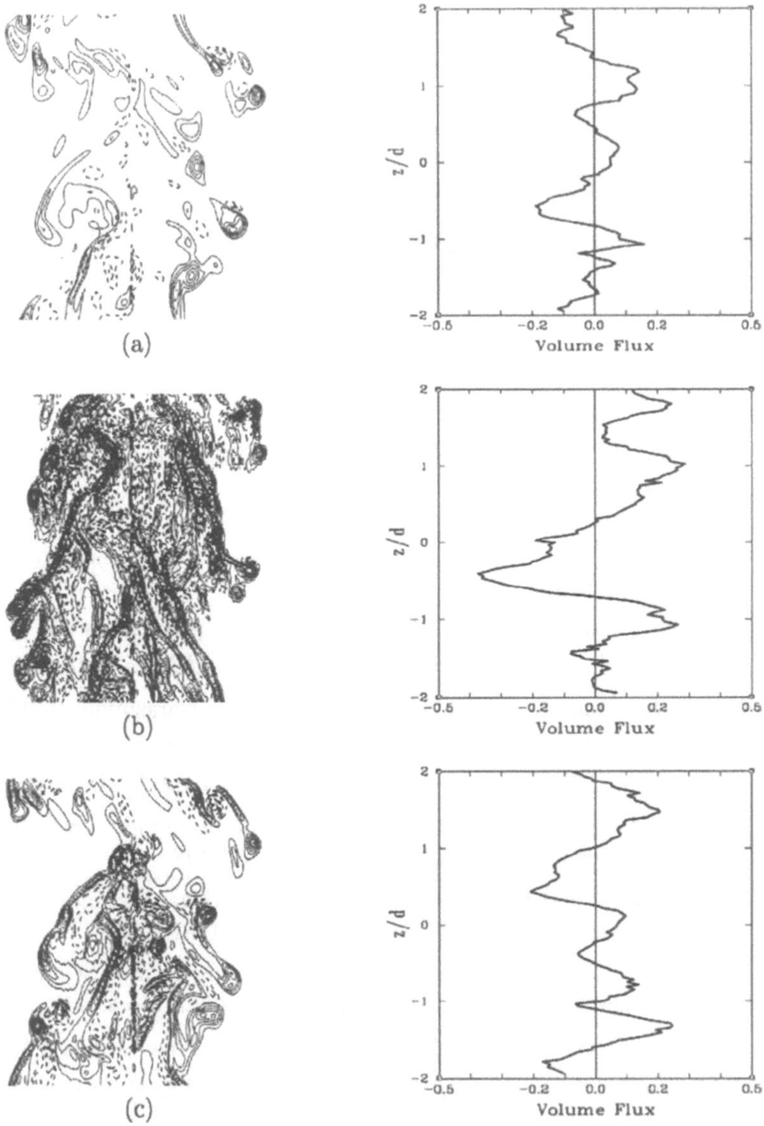

Figure 4. Azimuthal vorticity field with associated vertical distribution of volume influx into core of jet : (a) unheated, (b) fully heated and (c) actively heated cases. Full (dashed) lines positive (negative).

the unheated case. On the other hand, this is quickly followed first by a zone of outflux ($2.1 \leq z \leq 3.0$), and further downstream by a peak in the influx from the base of the upper structure. The fact that the upper structure is distinctly separated from the lower, with the telescoping completely prevented, appears to be responsible for this phenomenon.

6. Conclusion

The simulations reported here show how heating can exert a strong influence on the coherent motion in a jet, and consequently also on the entraining velocity field just outside the jet. By confining the heating to parts of the coherent structure in the jet, it is shown that it is feasible to manipulate these structures.

This work was supported in part by ONR contract No.N00014-94-1-1133; we are grateful to Dr. Gabriel Roy, ONR monitor, for his understanding support. We thank Dr. A.J. Basu and Mr. S.S. Siddhartha for their contributions to the work reported here.

References

1. Basu, A.J. and Narasimha, R. (1998) Direct numerical simulation of turbulent flows with cloud-like off-source heating, To appear in *J. Fluid Mech.*
2. Elavarasan, R., Bhat, G.S., Narasimha, R. and Prabhu, A. An experimental study of a jet with local buoyancy enhancement, *Fluid Dynamics Research* **16**,189–202 (1995).
3. Bhat, G. S. and Narasimha, R. 1996 A volumetrically heated jet: large-eddy structure and entrainment characteristics, *J. Fluid Mech.* **325**, 303–330.
4. Venkatakrishnan, L., Bhat, G.S., Prabhu, A. and Narasimha, R. 1998 Visualization studies of cloud-like flows, *Curr.Sci.* **74**, 597–606.
5. Turner, J.S. 1973 *Bouyancy Effects in Fluids.* Cambridge University Press.
6. Tso J. and Hussain, F. (1989) Organized motion in a fully developed turbulent axisymmetric jet, *J. Fluid. Mech.* **203**, 425-448
7. Mungal, M.G. and Hollingsworth, D.K. (1989) Organized motion in a very high Reynolds number jet, *Phys. Fluids* **A1**, 1615–1623
8. Dahm, W.J.A. and Dimotakis, P.E. (1990) Mixing at large Schmidt number in the self-similar far field of turbulent jets, *J. Fluid Mech.* **217**,299–330
9. Yoda, M., Hesselink, L. and Mungal, M.G. (1992) *Phys. Fluids* **A4**, 803-811.
10. Siddhartha, S.S., Narasimha, R. and Kailas, S.V. (1998) To be published.
11. Shivakumar, I., Narasimha, R. and Basu, A.J. (1998) To be published.

BOUNDARY LAYER CONTROL OVER AN ACTIVE RIBLETED SURFACE

N.F.YURCHENKO*, J.W.DELFS**

* *Institute of Hydromechanics, National Academy of Sciences*
8/4 Zheliabov St., 252057 Kiev
Ukraine
** *Institut für Entwurfsaerodynamik,*
Deutsches Zentrum für Luft- und Raumfahrt e.V.
Lilienthalplatz 7
D-38108 Braunschweig
Germany

A technique is proposed to control a boundary layer by means of preferred-scale counter-rotating streamwise vortices. Regular vortices were induced thermally with a given scale due to an imposed surface temperature gradient periodic in the spanwise direction z. Practically it was implemented using flush-mounted streamwise electrically heated strips regularly spaced along z. Physics and potential applications of the method were studied experimentally and numerically for boundary layers under centrifugal forces where streamwise vortices represent a naturally dominating flow structure governed by the Görtler instability mechanism.
Experimentally and numerically obtained spanwise and normal velocity profiles, as well as the developing vortical structure topology, were analyzed depending on the scale of the induced vortices correlated with the basic flow parameters. Good agreement of experimental and numerical results revealed physical mechanisms of the studied phenomenon. An active flow control was realized by switching on different temperature fields, thus providing a varying effect of longitudinal ribs or riblets while keeping the surface smooth.

1. Introduction

The role of coherent vortical structures in fluid transport near a wall is recognized both for laminar, and turbulent flows. In particular, this relates to streamwise counter-rotating vortices naturally developing in boundary layers affected by body forces. An exhaustive review of their studies relating to the Görtler instability mechanism was given in [1]; some practical aspects were considered, for instance, in [2] where a possibility of local heat transfer enhancement up to 360-400% depending on Prandtl number was reported.

Thus, the objective of the present work is (1) to demonstrate the practical benefit for boundary-layer control of regular longitudinal vortices and (2) to give a physical background for an engineering method of streamwise vortical structure generation and control over a smooth surface.

In this connection, riblets or ribs are proposed to be modeled by inducing thermally-driven flow inhomogeneities in the boundary layer flow. Heated flush-mounted strips

G.E.A. Meier and P.R. Viswanath (eds.),
IUTAM Symposium on Mechanics of Passive and Active Flow Control, 217–222.

oriented in the streamwise direction and separated from each other by thermally insulated strips create a temperature variation periodic in the spanwise direction, at the same time keeping the surface smooth. Thus imposing a surface temperature field, one can cause the development of a specially "designed" vortical structure in the boundary layer flow.

2. Experimental modeling

The experiments aimed
1) to demonstrate an effect of thermally induced disturbances/longitudinal vortices;
2) to investigate the boundary layer receptivity to the scale of induced vortices;
3) to give preliminary assessments for the numerical simulation.

2.1. EXPERIMENTAL FACILITY AND MEASUREMENT TECHNIQUES

Experiments were carried out in a low-turbulent water channel with an open test section of *10 x 25 x 300 cm* in the free-stream velocity range, U_0=*0.5-50.0 cm/s*, or typically at the length-based Reynolds number, $Re=(0.6\text{-}1.0)\ 10^5$ [3]. The studied boundary layer was developing on its changeable bottom, *25 x 300 cm*, which could be flat or contain a concave section (curvature radii, *R* = *1.0, 4.0* and *12 m*). A flow field was examined using LDV and tellurium-visualization method [3] similar to the well-known smoke-wire visualization technique in air.

Longitudinal vortices were induced using mechanical vortex-generators [3, 4] and flush-mounted heated longitudinal strips *z*-periodically spaced at λ_g=*1.2 cm* and *2.4 cm* from each other. There were tested two sets of the strips of different length: L_1=*2 cm* and L_2=*20 cm*. The first one was supposed to provide a concentrated effect on the flow as an initiating and scale-giving factor; the second one was to influence in a distributed manner creating a constant boundary condition.

2.2. EXPERIMENTAL RESULTS AND DISCUSSION

Selective response of a boundary layer to the scale of generated vortices was demonstrated in terms (1) of integral characteristics and (2) velocity field modifications downstream of the vortex-generator array. As regards the first task, considerable heat transfer enhancement [4] and the possibility of boundary-layer separation control were observed, both effects having been definitely dependent on the λ_g scale. Relating to the velocity fields, there was explicitly shown the boundary layer receptivity to the λ_g scale, or what happens in the flow if its basic parameters mismatch the scale of generated vortices: opposite to an adequate flow response, $\lambda_z=\lambda_g$, one can observe either fast attenuation of generated disturbances or a subharmonic boundary layer response, $\lambda_z=n\lambda_g$. The superharmonic response can be observed too as the modulation of a larger-scale, λ_{z1}, structure (close to the natural *U(z)* wave) with a small-scale forced structure, $\lambda_{z2}=\lambda_{g2}$. Visualized *U(z)* velocity profiles appeared to be very sensitive and convenient

characteristics for express-estimations of a boundary layer response to a certain excitation and, subsequently, to the choice of the control parameters.
Figure 1 shows visualized vortex patterns of the boundary layer over a plate with the flush-mounted strips modeling thermally a ribleted surface. Its effect is displayed in a

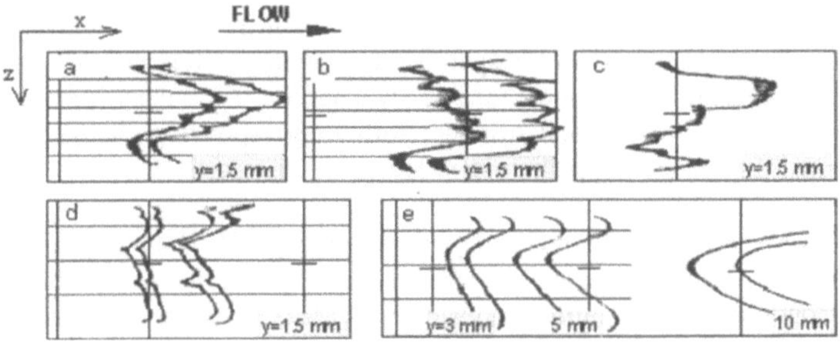

Figure 1. Visualized flow field (top view) influenced by 20cm heated strips, R=12 m;
a, b, c: λ_z=1.2cm; Re=10^5, $1.4{\cdot}10^5$, $1.5{\cdot}10^5$ (downstream development);
d, e: λ_z=2.4 cm; Re~10^5.

form of typical wavy *U(z)* distributions. Even such mild disturbances proved to be sufficient to modify the large-scale, large-amplitude, nonstationary and irregular natural velocity distributions into a smoothed picture observed both in the upper and downstream parts of the boundary layer, both under the "concentrated" and "distributed" conditions of the thermal excitation (Figure 1,e). At the same time, the boundary layer showed the tendency to support larger scale vortices (d-e) transforming the introduced small-scale vortices to the dominating vortical scale (a-c).

3. Numerical simulation

This first numerical study aims at revealing the dependency of the formation of the vortices due to different thermal excitation. Moreover, given a flow situation with relative body- and viscous forces comparable to the experiment, the simulation was to show that the described effect of "thermal ribs" is seen in the case of compressible air flows as well.
The numerical approach is based on a code developed for the direct numerical simulation of the laminar-turbulent transition in compressible subsonic boundary layers [5]. It was successfully applied to study disturbances of various nature like vorticity- , temperature- and density near-turbulent fluctuations [5, 6]. A formally 6th order accurate compact FD-scheme [7, 5] is applied in the wall normal direction while a spectral formulation is used in the other directions. The solution is advanced in time employing the Crank-Nicolson scheme in conjunction with a 3-step iteration method.

3.1. FORMULATION OF THE SIMULATION PROBLEM

The present application of the code to the boundary layer control by thermally modeled ribs is enabled by the specification of a spanwise periodic temperature boundary condition and the inclusion of body force terms. The numerical procedure can be described in the following way. For times $t<0$, the disturbance-free flow is subject to (constant, uniform) adiabatic temperature wall boundary conditions. At $t=0$, a z-periodic wall temperature field with a specified temperature difference ΔT with respect to the laminar adiabatic wall temperature is switched on. As a consequence, the flow responds by establishing a vortical structure with a corresponding change in the mean momentum distribution across the boundary layer.

Simulations were carried out for a Görtler number of $Gö_d = 15$, and a Reynolds number of $Re_d=595$, where both parameters are based on the Blasius reference length $d=(\nu_o\, x / U_o)^{1/2}$. The free-stream Mach number considered is $M=0.8$, the free-stream temperature is $T_o=290\ K$, the fluid is air with constant isentropic exponent $\gamma=1.4$ and Prandtl number $P=0.71$. The undisturbed flow is the 2D-zero-heat-flux similarity solution to the compressible boundary layer equations. The momentum thickness ϑ, based on this profile is $\vartheta = 0.66\ d$. The fundamental wave number in the spanwise direction, $\alpha_z=1$, i.e. was taken from the range of most amplified disturbance wavelengths, corresponding to linear stability analysis. Given this fundamental mode wavelength, the second harmonic is still in the domain of amplified disturbance wavelengths, while all further harmonics are linearly non-amplified. Twelve de-aliased harmonics were found to resolve the flow patterns sufficiently for all simulations.

From the set of investigated excitation cases two main cases will be considered here. For case 1, the strips were placed so as to ideally excite the second harmonic and harmonics thereof. The wall temperature was fixed at $\Delta T=30$ K above the recovery temperature of the initial, disturbance-free flow at $0<z<1.54$ and $3.08<z<4.62$. Also, a small initial wall temperature variation of about 0.3 K along the spanwise direction was introduced, including components of the fundamental mode, in order to account for small imperfections. For case 2, the first strip of case 1 was slightly shifted in the spanwise direction, thus introducing a permanent direct forcing component on the fundamental mode. So the strips were placed at $0.13<z<1.67$ and $3.08<z<4.62$.

3.2. NUMERICAL RESULTS AND DISCUSSION

The contour-lines of the streamwise velocity component $U(y,z)$ in the y,z-plane, as well as V-W-stream-traces were obtained to analyze the flow topology and the vortical structure evolution for the two cases (an intermediate stage see Figure 2 below). It was shown that in case 1, a transformation occurs of the generated second-harmonic vortices to larger scales corresponding to the fundamental wavelength as found in the experiments. Besides, the rate of the vortex formation strongly depends on the arrangement of the thermal strips. The same conclusions can be drawn from the observed evolution of the $U(z)$ velocity profiles (see upper part of Figure 2).

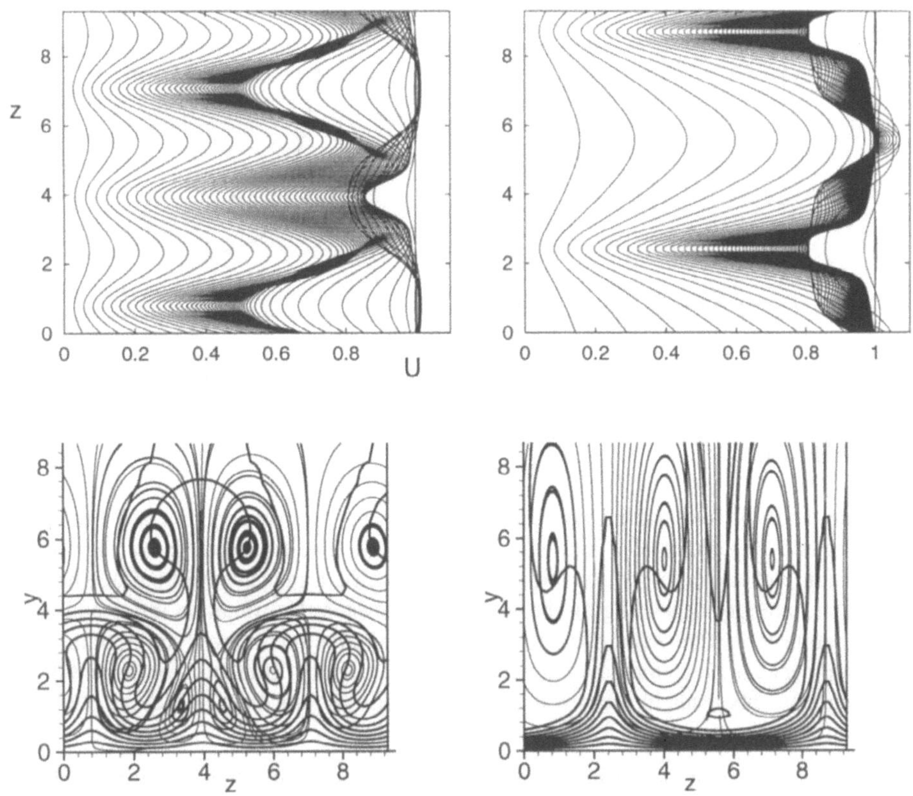

Figure 2. Above: Profiles $U(z)$ at various heights in the boundary layer in an intermediate stage of vortex formation; Below: Corresponding contour plots of $U(y,z)$ (thick lines, $\Delta U=0.1$) and V-W-streamtraces (thin lines). Left: case 1, Right: case 2.

Figure 3 shows $U(y)$ and $T(y)$ averaged in the spanwise direction for cases 1 and 2, as well as the unforced case (uniform, adiabatic wall), designated as case 3 in late stages of the vortex formation. Indicative for this late stage is the fact that further changes to the profiles remain small. It is seen that there is a remarkable change in both, velocity- and temperature profiles due to the thermal strips, compared with the unforced case. While the overall change in profile shapes is certainly to a considerable extent an effect of the mean heat added to the flow, the differences of the mean profiles between case 1 and 2 still remain characteristic. Interestingly, the long-term effect of the different wall heating is felt mainly well off the wall, in an upper shear layer seen at $3<y<5$ in the velocity profiles. The effectiveness of the flow manipulation is anticipated to become more pronounced as the heating (ΔT) is increased and the strips are narrowed.

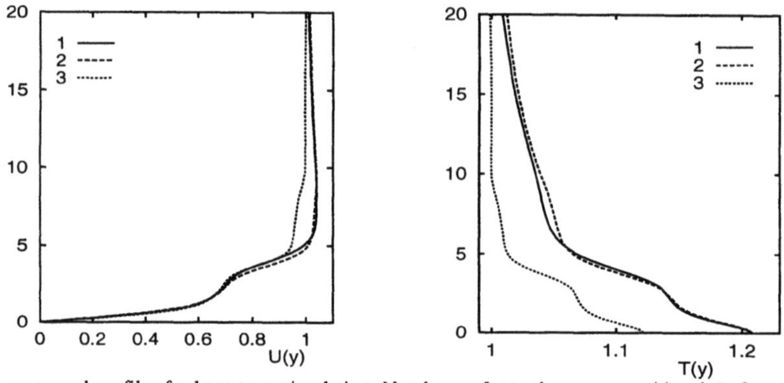

Figure 3. z-averaged profiles for long term simulation. Numbers refer to the cases considered. Left: velocity $U(y)$, Right: Temperature $T(y)$.

4. Conclusions

The initiation of longitudinal vortices of Görtler-type by structured surface heating is investigated experimentally and numerically. Both investigations show the vortex formation process to be controllable due to the specific arrangement of surface heating elements, strongly depending on the correlation between the excitation mode and basic flow parameters. The scale of the resultant vortical structure, as well as the rate of its development can be manipulated as needed. Only a slightly different arrangement of the heating elements leads to a characteristic change in the flow even in the late stages of the vortex evolution.

It is conjectured, that the thermal strips, embedded in an aerodynamic surface may be an effective mechanism of flow control if tailored to a given flow situation. Moreover, this mechanism may be equally applied for the flow control over convex surfaces given (locally) appropriate centrifugal force conditions, e.g. due to blowing through a 2D slot in the surface.

5. References

1. Saric, W. S. (1994) Görtler vortices, *Annu. Rev. Fluid Mech.*, **26**, 379-409.
2. Liu, J.T.C., Lee, K. (1995) Heat transfer in a strongly nonlinear spatially developing longitudinal vorticity system, *Phys. Fluids*, **7** (3),559-599.
3. Yurchenko, N.F. (1981) Experimental technique to study longitudinal vortices in a boundary layer, *Engineering-Physical J.*, **41**, 6, 996-1002.
4. Yurchenko, N.F. (1998) Optimization of heat transfer control based on a receptivity approach, in B. Launder (ed.), *Proc. of Turbulent Heat Transfer II Conference*, Manchester, 1998.
5. Delfs, J. (1994) Numerische Simulation der transitionellen schallnahen Plattengrenzschichtstroemung, *Dissertation TU Braunschweig*.
6. Delfs, J. (1995) Sound generation in the transitional flat plate boundary layer flow, *ZAMM 75*, S1, S379-S380.
7. Lele, S.K. (1992) Compact finite difference schemes with spectral-like resolution, *J. Comput. Phys.*, **103**, 16-42.

CONTROL OF COHERENT STRUCTURES SHEDDING AND GENERATION OF SOUND IN BLUFF-BODY WAKE FLOW

S.P. BARDAKHANOV, O.V. BELAY
Institute of Theoretical and Applied Mechanics
(Siberian Branch of Russian Academy of Sciences)
Novosibirsk 630090 RUSSIA

1. Introduction

The boundary layer has been a subject of investigation since the beginning of the XXth century because it does constitute in significant extent forces affecting a body placed to a stream of liquid or gas. Conditions in which the boundary layer detaches the surface of the body can change the drag catastrophically - from extremely small value in a flow without separation over an airfoil to quite large one when global separation occurs. If flowed body is bluff, then, with the Reynolds number getting great enough, unsteady flow regime set in. The flow pattern is constituted in this case by interaction between vortical disturbances of boundary layer and acoustical disturbances of external flow. Vortical-acoustical interaction consists of the receptivity of the boundary layer (and, in particular, of the separation region) [1] and of the sound generation by separating boundary layer. To comprehend the mechanism of this interaction, it is necessary to study the wake flow past the body, where disturbances develop introduced by sound in the separation region.

Turbulent wake behind bluff bodies contains large-scale regular structures resulting from interaction between vortical disturbances of the boundary layer and external flow, when the boundary layer separates from the trailing edge of the body. These structures, which can be called as *coherent*, shed alternately from one side and from the other, forming an almost periodical sequence. The general flow pattern for the semicircular trailing edge is presented in Figure 1.

G.E.A. Meier and P.R. Viswanath (eds.),
IUTAM Symposium on Mechanics of Passive and Active Flow Control, 223–228.

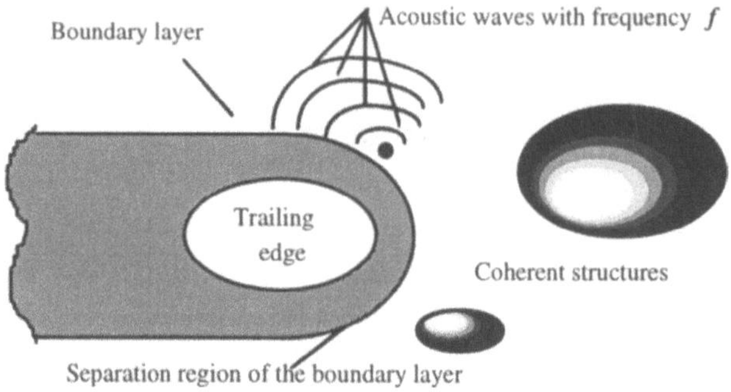

Figure 1. Flow pattern near the trailing edge.

The interaction of the vortical disturbances with the disturbances of the external flow can lead in favorable conditions to appearance of the aeroacoustic resonance [2]. The objective of this work was the study of the possibilities of flow control in resonance regime. Both *passive* (through changes of geometrical parameters) and *active* (by applying external disturbances) kinds of control are considered. The delimitation on *active* and *passive* methods is conventional to a significant extent - our study deals with the cases where the sound generator is the flow itself, but this sound can be considered as external in some sense, with the different parts of flow generating sound *almost independently*.

2. Experimental setup

The experiments were carried out in the subsonic closed-circuit wind tunnel MT-324 with a closed working section of square cross-section, 0.2 x 0.2 m with the length 0.8 m at the Institute of Theoretical and Applied Mechanics of Siberian Division of Russian Academy of Sciences. Following experimental models were used: rectangular flat plate 200 mm length and 8 mm thick; rectangular flat plate 300 mm length, 40 mm thick; *sliding* plate 150 mm length (in downstream direction), 10 mm thick angled by 60°; and *crosses* - the models constructed from rectangular flat plates joined by their side edges. Three kinds of *crosses* were tested.

The first kind below referred to as *model 1* consisted of three rectangular plates 8 mm thick joined by their side edges on a single axis so, that angles between plates were all 120°. Two models of this type were tested: 15 (model 1a) and 30 cm (model 1b) length (in downstream direction).

The second kind of experimental models - *model 2*, it consisted of a plate 20 cm wide (referred to as *horizontal*) and another plate joined to the axis of symmetry of the first one so, that the latter could be rotated around the hint. The model of this type 15 cm length is below referred to as *model 2a*, 30 cm length - *model 2b*.

The second kind - *model 3* - was assembled from a plate 20 cm wide (referred to as *horizontal*) and two plates of the same length (referred to as *vertical*) joined to the *horizontal* plate's axis of symmetry, the one of the two was mounted below, and the other - above. The angles between *horizontal* and *vertical* plates were 90°. The model length was 15 cm.

The leading and trailing edges of all the models were rounded. The models were installed parallel to the flow. The scheme of the experimental setup is presented in Figure 2.

Mean flow speed *U* was measured using pitot static probe. Measurements were done in flow speed range from 5 to 45 m/s. Streamwise components of mean velocity and velocity fluctuations were measured by means of constant-temperature hot-wire anemometer with a single-wire probe. The signal of the hot-wire anemometer was passed through analog-digital converter to a personal computer, where it was then processed in digital form. In particular, narrow-band (4 Hz) spectral analysis was done.

A microphone was used for qualitative observation of sound pressure level, sound wave spectral analysis and as a reference when synchronized hot-wire readings were done.

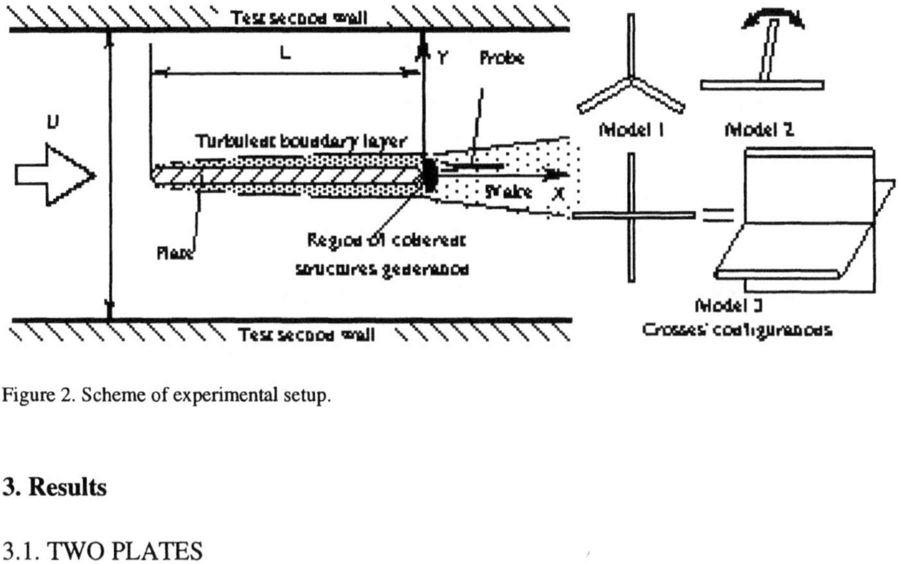

Figure 2. Scheme of experimental setup.

3. Results

3.1. TWO PLATES

When loudspeaker is used to introduce a disturbance into the boundary layer, the amplitude and direction of propagation of incident sound wave depend on location and orientation of the loudspeaker as well on its amplitude-frequency characteristic. If channel flow is considered, resonant properties of the channel are important too. It leads to non uniform sound field, which also depends on sound frequency. The factors enumerated result in practical unpredictability of sound velocity magnitude and direction in the region where sound transforms to boundary layer disturbance. By the reasons mentioned above, a question appears to be interesting about effectiveness of aeroacoustic resonance as an origin of disturbances of boundary layer on a body, that is not involved in resonance generation. An effort was made to introduce a disturbance with the frequency quite different from the natural one to the wake behind the plate 40 mm thick (it is below referred to as *thick*). Resonant sound generated by additional (*thin*) plate 8 mm thick was used as an origin of the disturbance. The *thin* plate was installed between the *thick* plate and test section wall. It was shown by means of phase measurements in the wake behind the *thick* plate, that there was no sound transformation to disturbance of the wake.

3.2. CROSSES

Model 1a was installed in square test section of the wind tunnel. This model did not present clearly heard resonance. Model 1b when installed in square channel generated louder sound. Mounting it in the round test section (the round insertable unit was set into square test section) resulted in greater sound level and allowed to observe dependence of resonant oscillations from orientation of the model (when it was being turned around its longitudinal axis). As it was found, the plates of the model equally contributed in resonant sound generation with some orientations chosen (it damaged the resonance when pivot of half-span length was placed in the region of coherent structures shedding on the trailing edge of any plate). When the other orientations were chosen the only two of plates were important. Not only trailing edges but also the regions near the leading edges of the models had significance for generation of the resonance observed. Beating was heard in resonant sound at some flow speed.

Model 2a was installed in square test section and like model 1a did not generate intensive sound. Shortening to 15 cm (model 2b) and then setting the round insertable unit led to the increase of sound intensity. When upper cover of the test section was removed, the additional growth in sound level was observed (the insertable unit was shorter than the test section, and with the cover removed the flow was rebuilt). Unlike model 1 resonance could be broken by the pivot placed in separation region only at the *horizontal* plate.

Model 3 was installed in square channel and generated loud sound. Beating was heard in the sound, it is shown schematically in Figure 3. The model modifications (removal of one of the *vertical* plates, the changes of the angles between the plates) did not result in decrease of sound level.

3.3. SLIDING PLATE

An aeroacoustic resonance was observed in tests with the *sliding* plate. This resonance showed some features distinguishing it from an *ordinary* two-dimensional one on a square plate. The *shelf* on "frequency-flow velocity" plot has a marked slope, i.e., the resonant frequency gradually increases with overall change being about 30 Hz. Wake velocity fluctuation spectra make this resonance similar to a two-dimensional one with hysteresis occurred.

An attempt was done to *visualize* wake behind the model. To do this, velocity field was measured in the wake by means of hot-wire anemometer. Phase locked ensemble average of ten realizations of the signal was made for each location of measurements using microphone signal as a reference. Mean values of ten were calculated for each measuring location (realization readings were initiated by reference microphone signal). Mean velocities and streamwise velocity fluctuations were calculated from those *mean realizations*. Local values of velocity fluctuations gotten in some instant (the same for all measuring region) formed streamwise fluctuation *field.* Then the transversal flucıuations *field* was calculated by means of continuity equation, and then - the field of fluctuation amplitudes. Isocontours of this field are presented in Figure 4. One can see at the figure that the isocontours are grouped into several formations spaced almost periodically. The formations correspond to vortices forming almost periodical street.

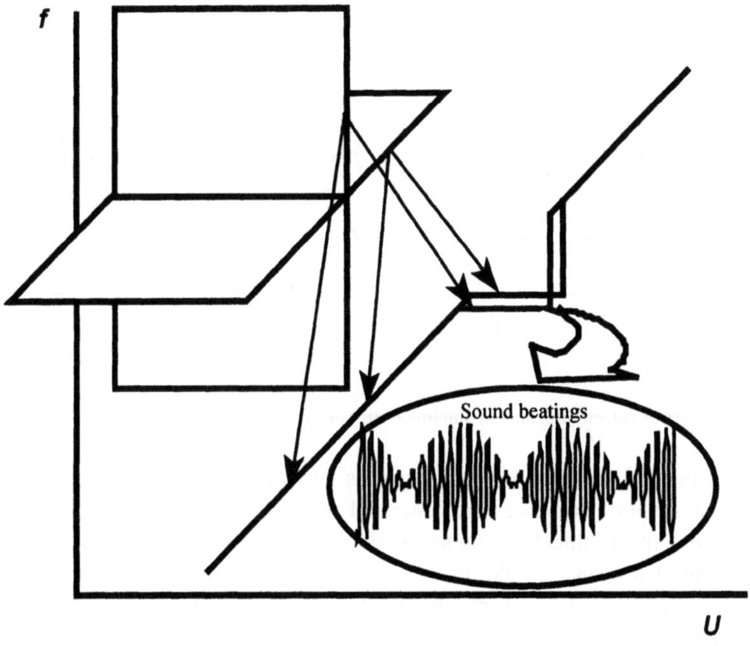

Figure 3. Dominant frequency of the wake behind model 3 versus flow speed (schematic illustration of sound beatings).

4. Conclusions

It was shown in the paper for all considered configurations the existence of resonant sound amplification regimes conditioned by coherent structures generation.

There was no possibility to introduce a disturbance into the wake behind the *thick* plate using resonant sound generated by the other (thin) plate, when the resonant sound frequency was quite different from the natural wake frequency of the *thick* plate.

Separation region modification either on leading or trailing edges can break the process of sound-wake synchronization. As a result the resonance can be removed.

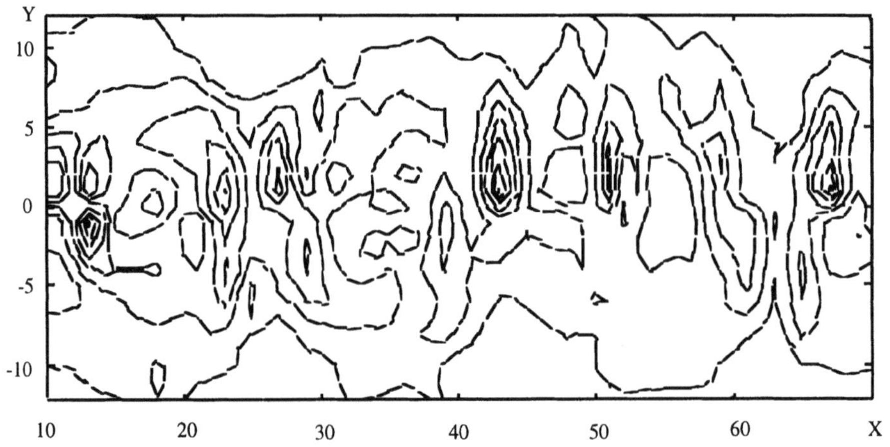

Figure 4. Isocontours of velocity fluctuation amplitudes in wake behind *sliding* plate.

For the "crosses" sound beating was observed in some flow speed range witnessing that several resonances are generated at the same flow speed. Each of these is generated by one of the plate forming the model.

Essential influence of the factors described above on the trailing edge region and on the flow as a whole confirms the importance of the conditions in the separation region for control of the integral flow characteristics.

5. References

1. Bardakhanov· S.P. and Kozlov, V.V.: Onset and Development of Coherent Structures in Turbulent Shear Flows, in H.-U.Meier, P.Bradshaw (Eds.), *Perspectives in Turbulence Studies*, Springer-Verlag, Berlin, (1987), pp. 154-187.

2. Parker, R.: Resonance Effects in Wake Shedding from Parallel Plates: Some Experimental Observations, *Journal of Sound and Vibration* **4** (1966), 62-72.

OSCILLATIONS OF SOLID SURFACES VERSUS SOUND AND OSCILLATORY SUCTION IN FLOW CONTROL

D. RONNEBERGER
Drittes Physikalisches Institut, Universität Göttingen
Bürgerstraße 42-44, D-37073 Göttingen

1. Introduction

The active dynamic control of flow fields often makes use of some convective instability of the considered flow (e.g., the Tollmien Schlichting (TS) or the Kelvin Helmholtz (KH) instability). Then it is sufficient to force the flow at a location where the instability is receptive to external perturbations while the choice of the forcing mechanism, e.g. oscillations of solid surfaces, sound, or oscillatory suction and blowing is not critical in most cases.

More subtle requirements have to be considered if unwanted perturbations which for some reason have inevitably been introduced into the flow shall be compensated by means of a proper forcing device. This demand may arise, e.g., in the abatement of aerodynamic noise or in the stabilisation of laminar boundary layers aiming at the delay of the turbulent transition, to mention only a few applications.

The effect that the forcing device has on the flow may be crucial in such situations because more than one mode may be excited by the actuator. So the stabilisation of one mode may lead to the excitation or the amplification of other modes that had been uncritical so far. This may occur, e.g. if sound is used as a forcing mechanism or is emitted by the forcing device: then unwanted effects may be triggered at distant locations, but also the near field of the actuator in sensor-actuator-sensor control systems may have implications on the efficiency of the system.

The paper will focus on the unsteady flow fields that are generated by various types of forcing devices and on the effect that these forcing fields have on the excitation of the KH and the TS instability. Experiments which have been performed in order to investigate the excitation and the suppression of KH instability waves in the free shear layer behind a splitter plate have provided the information to be dicussed in the following.

G.E.A. Meier and P.R. Viswanath (eds.),
IUTAM Symposium on Mechanics of Passive and Active Flow Control, 229–236.

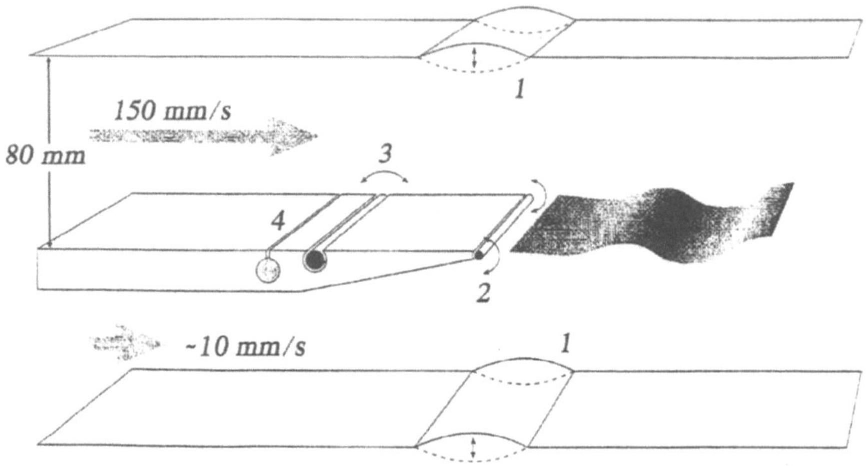

Figure 1. Experimental set-up: The free shear layer can be forced by oscillation of the membranes (1) and the rotating cylinder (2), while the wall boundary layer can be excited by the rotating cylinder (3) and blowing/suction through the slot (4)

2. The mean flow, the forcing devices, and the experiments

Figure 1 depicts schematically the trailing section of a splitter plate and some of the mechanical devices that have been used for the excitation of TS and KH instability waves. The splitter plate divides the channel into two parallel ducts, and nominally the fluid (water) is moving only in the upper channel. The laminar wall boundary layer separating from the trailing edge of the plate develops into a 2D laminar free shear layer. The slow fluid motion in the lower channel is to wash away the vorticity that otherwise would accumulate below the edge of the plate; the effect of this flow on the dynamics of the shear layer is small.

Various forcing methods like unsteady rotation of a cylinder that forms the edge of the plate and "acoustic" sources (oscillating membranes) in the wall of the channel have primarily been provided for the study of the unsteady flow separation at the edge of the plate (Rebel, 1992; Grützmacher, 1993; Dierken, 1995; Enghardt, 1996; Graf, 1998). Also TS waves have been used to model fluctuating vorticity that is propagated through the wall boundary layer of the plate and which contributes to the unsteadiness of the flow at the end of the plate; the TS waves have been excited by the rotating cylinder that is built into the surface of the plate some distance upstream of the edge as well as by oscillatory suction and blowing through a narrow spanwise slot. Finally, since the axis of the cylinder in the edge of the plate is not perfectly staight, small transversal oscillations appear together with the rotation of the cylinder. So also some information on the strong effect of transversal oscillations of the trailing edge has been obtained.

One of the major tools in the investigation of the various forcing methods is the mutual compensation of the the KH instability wave that is excited by different forcing devices. Implicitely, use has been made of the fact that the KH instability is receptive to unsteady forcing only in the close vicinity of the location where the shear layer separates from the wall. In most of these experiments a sinewave has been fed into two of these devices and the ratio of the amplitudes as well as the phase difference have been adjusted to compensate the far-field amplitude. Then an unsteady near-field is left which, however, does not trigger the KH instability.

Also broad-band disturbances have successfully been compensated by means of a sensor-actuator-sensor system which contains adaptive filters in order to drive the actuator in accordance with the incident disturbances. Two hot film sensors somewhat upstream of the trailing edge capture the disturbances. So the system is able to automaticly discriminate between pressure and vorticity fluctuations (Enghardt, 1996).

3. Receptivity of the Kelvin Helmholtz instability

Remarkable insight into the receptivity of the KH instability has been obtained by the comparison between the effects of the "acoustical" pressure gradient and of the rotating cylinder at the edge of the plate (Dierken, 1995). Figure 2 shows the complex ratio of the KH-amplitudes that are generated by the two devices if they are driven at the same forcing amplitude. The action of the cylinder (radius 1 mm) is described by the velocity of its surface (ca $0.01\,U_\infty$, $U_\infty = 150\,\mathrm{mm/s}$), and the acoustic pressure gradient is characterized by the normal velocity that would result 1 mm downstream of the trailing edge in the hypothetical (inviscid) potential field at zero mean flow.

The KH waves that are excited by the two forcing mechanisms have opposite phase while the respective forcing velocities are in phase. This is in contrast to the supposition that the oscillatory displacement of the point of separation is of primary importance in the excitation of the KH wave; at least in the quasi steady case the point of separation is shifted in the direction of a streamwise motion of the wall, respectively in the direction of the flow induced by a superimposed pressure gradient (see Gad-el-Hak & Bushnell, 1991 for a review of flow separation control). So the oscillation of the point of separation can be excluded from the true sources of the KH wave.

An alternative hypothesis on the source of the KH wave can be derived from a theoretical investigation of the inviscid dynamics of an infinitely thin

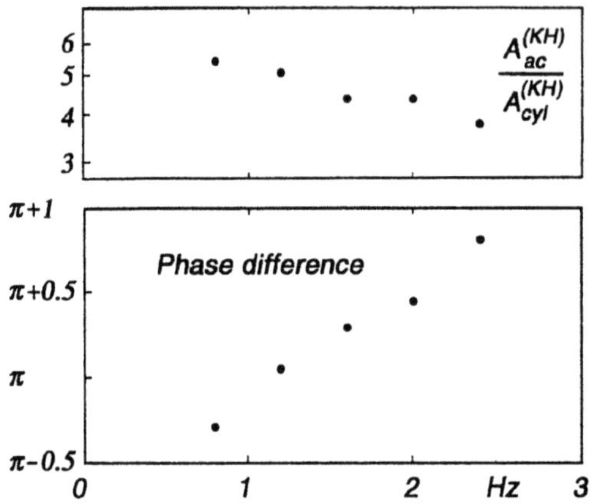

Figure 2. Amplitude ratio and phase difference between KH waves excited by acoustic pressure gradient and rotation of the cylinder, respectively, as functions of the frequency

vortex sheet leaving the sharp trailing edge of a thin semi-infinite splitter plate (Bechert & Michel, 1975). They calculate the source term from the potential field generated by the forcing mechanism at zero mean flow.

While the potential flow caused by the acoustical pressure gradient is well known, the flow due to the rotation of the cylinder needs some explanation. The normal velocity $\tilde{v}(x, y = \text{const})$ generated by the oscillatory rotation of the cylinder is shown on figure 3a as a function of the streamwise distance x from the trailing edge: the dashed and the dotted lines indicate that the flow is a superposition (solid lines) of two parts, namely the viscous Stokes layer which is driven by the wall shear stress at the moving surface of the cylinder, and a potential flow which balances the flux of mass transported through the Stokes layer – the Stokes layer is peeled off the cylinder by the plate, so to speak. In the limiting case of an half-infinite thin plate the potential flow caused by the cylinder decays in proportion to $x^{-3/2}$, in contrast to the acoustically driven flow which decays in proportion to $x^{-1/2}$ (on figure 3a the finite height of the channel and the virtual edge of the equivalent sharp edged thin plate have been taken into account).

Further insight into the excitation of the KH wave is provided by the unsteady flow field closely behind the edge in the case with superimposed mean flow; again only $\tilde{v}(x, y = \text{const})$ is shown on figure 3b. Three cases are depicted: forcing by rotation of the cylinder, forcing by oscillation of the membranes, combined forcing with compensation of the KH wave. In the latter case the quantity that acts as the intrinsic source of the KH wave has been adjusted to zero. Since $\tilde{v}(x)$ obviously has a finite amplitude in the non-exciting near-field it is rather unlikely that the normal velocity itself is

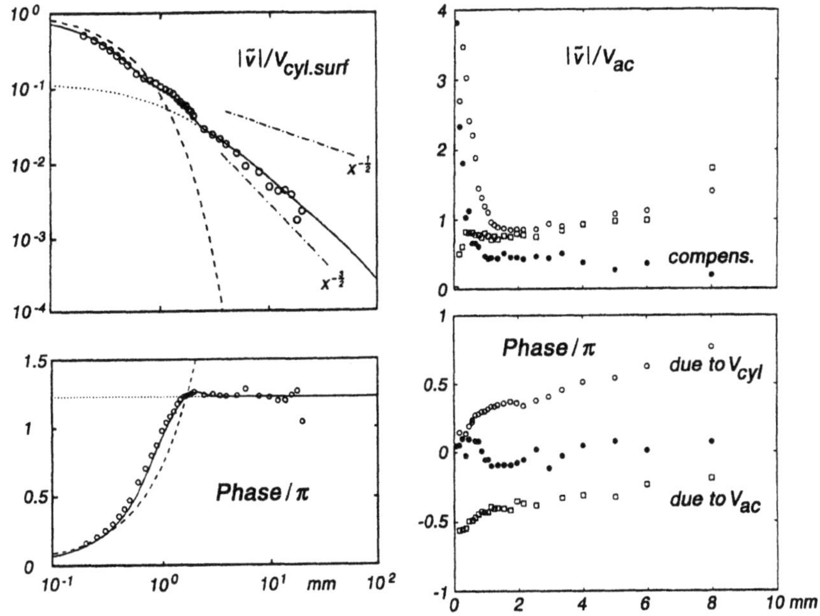

Figure 3. Normal velocity (magnitude and phase) as a function of the streamwise distance from the edge of the plate. A) Forcing due to rotation of the cylinder ($U_\infty = 0$): the field is composed of the Stokes flow (dashed) and recirculating potential flow (dotted); the experimental data are well reproduced by the superpostion of the two fields (solid line). B) $U_\infty \neq 0$; forcing by acoustic pressure gradient (□), by rotation of cylinder (○), combined with compensated KH wave (•)

intimately related to the source term. However, $\partial\tilde{v}/\partial x$ turns out to be very small in the range $1\,\mathrm{mm} \leq x \leq 3\,\mathrm{mm}$ which is outside the Stokes layer but still within the near-field of the edge. Also the non-exciting forcing field, i.e. the unsteady field which persists if the mean flow has been switched off exhibits only small $\partial\tilde{v}/\partial x$ (not shown here).

A high sensitivity of the KH instability with respect to $\partial\tilde{v}/\partial x$ would explain the high receptivity to the normal displacement of the cylinder that happened to be associated with the rotation of the cylinder due to a small misalignement (ca. $30\,\mu$m) between the axis of rotation and the cylinder axis. In fact the potential field due to a translation of the cylinder would decay in proportion to $x^{-3/2}$ like the potential field due to the rotation of the cylinder, and the ratio of the amplitudes of these fields explains the effect that the translation of the cylinder has on the KH wave amplitude.

One may even speculate that in all possible situations the interaction of the pressure field with the edge of the plate is the important process in the excitation of the KH instability wave, also if the KH wave develops out of a TS wave which comes along with a vorticity distribution that strongly resembles the KH wave: nevertheless, the development of the KH wave extends over the same length (ca half a KH wavelength behind the edge) as in the excitation by the acoustic pressure gradient or the rotating cylinder (Rebel, 1992).

The phase difference between the KH waves that are excited by the pressure gradient and the rotating cylinder, respectively, still depends on the frequency (figure 2). The evaluation of this dependency shows that the effect of the cylinder lags somewhat (ca 0.1 sec) behind the effect of the pressure gradient. This time lag may be interpreted as the time that is needed for the communication between the Stokes layer and the potential flow field. In fact a time lag in the same order has been found also in the opposite direction namely between the response of the hot film sensors and the outer potential flow in the aforementioned broad-band compensation experiment (Enghardt, 1996). The hot films are sensitive mainly to the wall shear stress the unsteady part of which is closely related to the dynamics of the Stokes layer. This time lag somewhat reduces the success of the flow stabilisation by means of the aforementioned sensor-actuator-sensor system.

4. Excitation of Tollmien Schlichting waves

Even though a great variety of actuators have been used in the past for the excitation of TS waves, and although these actuators have been applied in quite a few investigations both to the artificial excitation as well as the compensation of TS waves, not much is known about the receptivity of the TS instabilitity with respect to the different forcing mechnisms associated with these actuators. Besides the well established suction/blowing method, a rotating cylinder in the surface of the plate (figure 1) has been used in the present experiments for the excitation of TS waves.

Quite in contrast to the appreciable receptivity of the KH instability to the rotation of the cylinder at the trailing edge of the plate, the TS instability was only weakly excited by the first configuration of the cylinder in the surface of the plate; the gap between the cylinder and the spanwise boring in the plate was very narrow (ca 0.05 mm) compared to the thickness of the Stokes layer (ca 0.5 mm), and the slit in the surface of the plate through which the moving surface of the cylinder forces the flow had a width of 2 mm which is in the order of the distance of the critical layer from the wall. In a second configuration the gap around the cylinder as well as the slit were widened (to ca 0.25 mm and 3.3 mm, resp.) which considerably increased the amplitude of the excited TS wave. In any case, though, the efficiency of the cylinder was significantly lower than suction and blowing through the slot in the plate surface.

The attempt to interpret these observations is based on arguments that

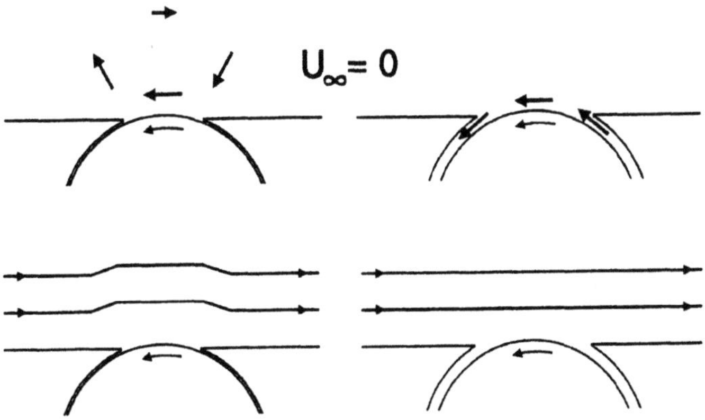

Figure 4. Caricature of the flow over the cylinder in the wall

have been given by Nishioka and Morkovin (1986) in their study of the receptivity of the TS instability. As a source of TS waves they consider the generation of unsteady vorticity by the imposed forcing field. They highlight two terms namely (1) the interaction of the imposed pressure gradient with the wall and (2) the redistribution of the mean vorticity due to the imposed normal velocity component, and they conjecture that only the spatial Fourier component that fits the TS wavelength influences the amplitude of the TS wave. The second term is large when fluid is injected or sucked through a narrow slot. So the large efficiency of this type of forcing seems to emphasize the role of the second term. The first term feeds the Stokes layer which can be considered to consist of the vorticity that has been generated at the wall by the oscillating pressure gradient and has diffused away from the wall.

Normal fluid motion and thus redistribution of mean vorticity is associated also with the Stokes flow namely with its inhomogeneous part. This particularly pertains to the strongly confined flow that is generated by the cylinder. As discussed with the cylinder at the edge of the plate, the Stokes flow is peeled of by the plate which gives rise to a recirculating potential flow. In the second configuration of the wall cylinder part of the Stokes flow "escapes" through the gap between the cylinder and the boring, and thus the recirculating flow is diminished. A scetch of this situation is shown on figure 4 (upper diagrams): actually the two extreme cases are depicted, namely (a) the whole flux of the Stokes layer contributes to the recirculation, and (b) there is no recirculation at all. The two lower diagrams show the streamlines of the superimposed mean and forcing flow. At configuration (a) the mean shear is "diluted" (in the depicted phase of the forcing cycle), and this counteracts the increase of the wall shear stress due to the rotation of the cylinder. Thus an appreciable modulation of the wall shear stress can be expected only with configuration (b), and one may conclude that the the wall shear stress is an essential ingredient in the excitation of TS waves by the rotating cylinder.

5. Conclusions

Various forcing devices designed for the controlled excitation of Kelvin Helmholtz (KH) and Tollmien Schlichting (TS) instability waves have been studied. The respective forcing fields have been decomposed into the Stokes flow which is dominated by viscous forces like the wall shear stress and into the outer (potential) flow which is governed by the pressure gradient. It clearly turns out that the excitation of the KH instability is related to the outer part of the forcing field, however, in the case of the TS instability the unsteady part of the wall shear stress contributes to the excitation of the instability wave as well. Indeed with regard to the fact that the wall shear stress is modulated also in the case of suction/blowing through a slot (due to the interaction with the mean flow), the relative importance of the Stokes layer and the outer flow remains an open question in the excitation of TS waves.

In addition it has turned out that the mechanical nature of the forcing device (oscillation of a solid surface, sound, suction/blowing) is not important with regard to the physical mechanisms that are involved in the excitation of the flow. From a technical point of view, suction/blowing is a more flexible tool in the design of a specific forcing field rather than moving surfaces. In any case this forcing field should be as confined as possible in order not to force the flow at unwanted locations; so sound radiation should be avoided as far as possible.

References

Bechert, D.W., Michel, U (1975) The control of a thin free shear layer with and without a semi-infinite plate by a pulsating flow field, *Acustica* **31** pp. 287–307

Dierken, C. (1995) *Receptivity of a separating shear layer at high frequencies/*. Diploma thesis (in German), III. Phys. Inst. Univ. Göttingen

Enghardt, L. (1996) *Adaptive suppression of the Kelvin-Helmholtz instability of a laminar 2D shear layer*. PhD thesis (in German), Univ. Göttingen

Gad-el-Hak, M., Bushnell, D. (1991) Status and Outlook of Flow Separation Control, *AIAA paper* **91-0037**

Graf, R. (1998) *Unsteady flow separation and vorticity transport at the trailing edge of a splitter plate*. Diploma thesis (in German), III. Phys. Inst. Univ. Göttingen

Grützmacher, V. (1993) *Suppression of instability waves in a separated 3D shear layer*. PhD thesis (in German), Univ. Göttingen

Nishioka, M., Morkovin, M.V. (1986) Baundary-layer receptivity to unsteady pressure gradients: experiments and overview, *J. Fluid Mech.* **171** pp. 219–261

Rebel, J., (1992) *Active control of a separating 3D wall boundary layer by means of Tollmien-Schlichting waves*. PhD thesis (in German), Univ. Göttingen

THEORETICAL AND COMPUTATIONAL STUDY OF THE WAKE CONTROL PROBLEM

B. PROTAS AND A. STYCZEK
Department of Aerodynamics,
Institute of Aeronautics and Applied Mechanics
Warsaw University of Technology,
Nowowiejska 24, 00-665 Warsaw, Poland

1. Introduction

In the present paper we are concerned with the problem of active drag minimization in the incompressible wake flow generated by a circular cylinder with turbulent Reynolds number. The objectives of the work are twofold: first, we show evidence coming from numerical simulations that a properly designed *open-loop* algorithm is indeed capable of performing effective flow control resulting in significant drag reduction, and then, with this motivation, we outline a rigorous *feedback* algorithm based on Optimal Control Theory. In both cases control is performed by rotary motion of the circular cylinder whose boundary we denote Γ_0. The flow domain Ω extends to infinity. By an open-loop algorithm we mean a scheme in which the control is determined a priori, without any reference to the instantaneous flow field information and/or its history. Because of computational limitations, throughout the paper we will study two-dimensional (2D) flows only.

2. Hydrodynamic Forces in Wake Flows

In this section we will consider the relation between the structure of the von Karman vortex street and the resulting hydrodynamic force. The latter is equal to

$$\vec{F} = -\oint_{\partial\Omega} \left(-p\vec{n} + \mu\bar{\bar{D}}\vec{n}\right) d\sigma = \oint_{\partial\Omega} \left(p\vec{n} + \mu\vec{n} \times \vec{\omega}\right) d\sigma, \tag{1}$$

where p is pressure, $\vec{n}$ is the unit normal vector directed into the body, μ is the fluid viscosity, $\bar{\bar{D}} = \frac{1}{2}\left[\nabla\vec{V} + \left(\nabla\vec{V}\right)^T\right]$ is the rate-of-deformation tensor

G.E.A. Meier and P.R. Viswanath (eds.),
IUTAM Symposium on Mechanics of Passive and Active Flow Control, 237–242.

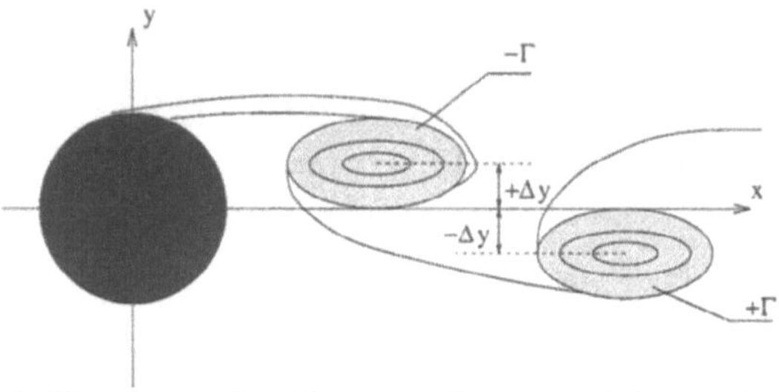

Figure 1. Sketch illustrating the distinctive features of the vorticity distribution in the von Karman vortex street behind an obstacle. $\pm\Gamma$ are the circulations of the vortices.

and ω denotes vorticity $\omega = \frac{\partial v}{\partial x} - \frac{\partial u}{\partial y}$. Applying the well-known *impulse formula* [1] one can express the drag force F_D using the time derivative of the vorticity moment with respect to the X-axis, i.e. $F_D = -\frac{d}{dt}\int_\Omega \omega y d\Omega$. This has a clear physical interpretation in terms of the vorticity structure in the wake. Let $\Gamma_i's$ be the circulations of the counter-rotating vortices that are consecutively shed and $\Delta y's$ denote the corresponding off-axis separations which are related to the sense of the vortex circulation (Fig. 1). The expression for the integral vorticity moment can be therefore roughly estimated as

$$F_D \simeq -\frac{d}{dt}\sum_i \Gamma_i \Delta y_i = -\frac{d}{dt}\left[(+\Gamma_1)(-\Delta y_1) + (-\Gamma_2)(+\Delta y_2) + \ldots\right]. \quad (2)$$

The above gives an idea of what could be done in order to reduce the drag: (i) decrease the off-axis separation Δy_i at which the eddies are advected downstream, (ii) reduce the strength of the vortices, thus effectively killing the von Karman vortex street.

3. Open-loop Algorithm - Results of the Numerical Simulations

The paper [2] reports on a successful laboratory attempt to implement this strategy in a turbulent wake flow at $Re = 15000$. The flow control was accomplished by performing rotary oscillations of the circular cylinder. Appropriate choice of the control parameters (i.e. the amplitude and frequency of the rotary oscillations) resulted in dramatic drag reduction, by as much as 80% in some cases. Drag reduction was accompanied by suppression of the vortex shedding. We have reproduced these results using numerical simulation. The angular velocity of the circular cylinder was given by $\dot{\varphi}(t) = A_0 \sin(\omega_0 t)$, where the amplitude A_0 and the frequency ω_0 are the two adjustable parameters. They can be normalized as $\tilde{A} = \frac{A_0 \cdot R}{U_\infty}$ and $St = \frac{\omega_0 \cdot 2R}{U_\infty}$, where R is the radius of the cylinder and U_∞ denotes the free stream velocity at infinity. The parameter St is called the *Strouhal*

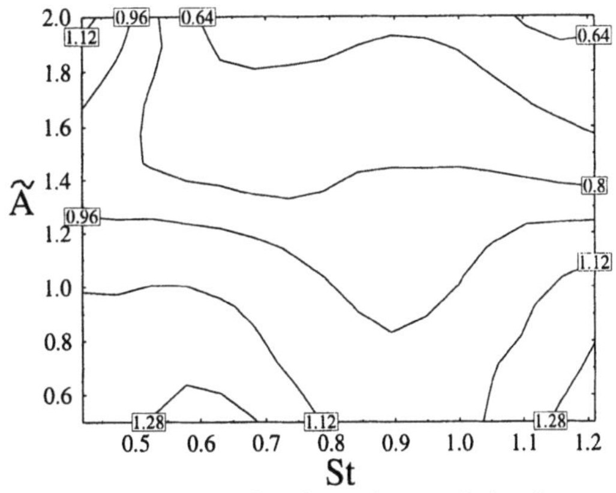

Figure 2. Contour plot presenting the dependence of the drag coefficient c_D on the control parameters $\tilde{A}$ and St. The iso-lines of c_D corresponding to the values of 0.64, 0.8, 0.96, 1.12 and 1.28 are indicated.

number. We performed a parametric study of the phenomenon exploring the ranges of $\tilde{A}$ and St roughly corresponding to those studied in [2]. In the numerical simulations we used the *Random Vortex Blob Method* [3]. In Fig.2 we present the contour plot of the drag coefficient $c_D = \frac{F_D}{\frac{1}{2}\rho U_\infty^2 (2R)}$ as a function of the two control parameters $c_D = f(\tilde{A}, St)$ that was obtained in our investigation. Remarkable reduction of the drag coefficient can be observed for properly chosen values of the control parameters $\left\{\tilde{A}, St\right\}$ which are close to those indicated in [2].

4. Towards a Rigorous Feedback Control Algorithm

In the present section we briefly outline the derivation of the gradient algorithm which can be used to determine the optimal control $\dot{\varphi}_{opt}(t)$ in the time interval $[0;T]$, where T is some characteristic time scale of the phenomenon. The control results in the minimum average drag in the interval $[0;T]$. The present method is an extension of the pioneering work [4] and its application to fluid dynamical problems [5]. The starting point is the formulation of the functional that will be minimized. It our case it represents the balance of the work done by the drag force and the work needed to control the flow

$$J(\dot{\varphi}) = \frac{1}{2}\int_T \left\{ \left[\begin{array}{c} \text{power needed to} \\ \text{control the flow} \end{array} \right] + \left[\begin{array}{c} \text{power related to} \\ \text{the drag force} \end{array} \right] \right\} dt. \quad (3)$$

The rotational velocity $\dot{\varphi}(t)$ of the cylinder is the control, therefore the energy needed to control the flow is introduced in the form of a moment of

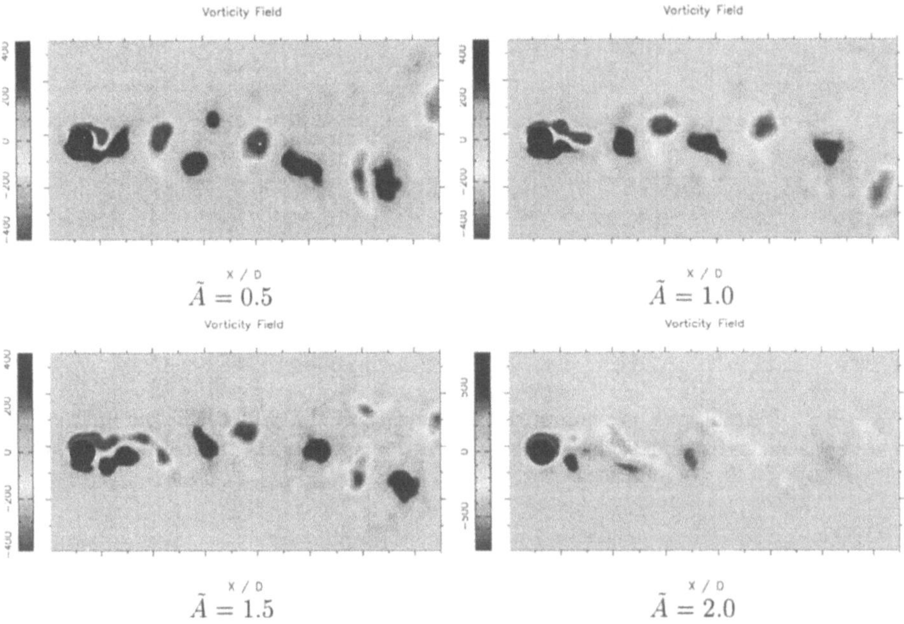

Figure 3. Instantaneous vorticity distributions in the turbulent wake flows past the rotating circular cylinder at $Re = 5000$. The Strouhal number is equal to $St = 0.91$. Values of the normalized amplitude of the rotary oscillations are indicated on the insets.

forces applied to the cylinder. Consequently, using the surface force density $\vec{f}(\dot{\varphi}) = -p(\dot{\varphi})\vec{n} + \mu\bar{\bar{D}}(u(\dot{\varphi}))\vec{n}$, the above relation can be expressed as

$$J(\dot{\varphi}) = \frac{1}{2}\int_T \oint_{\Gamma_0} \left\{ \left[\vec{f}(\dot{\varphi}) \times \vec{r}\right] \cdot \left(\dot{\varphi}\vec{k}\right) + \vec{f}(\dot{\varphi}) \cdot \vec{U}_\infty \right\} d\sigma dt, \tag{4}$$

where all the hydrodynamic quantities depend on the control $\dot{\varphi}$. Next we use the fact proved in [5] that the functional of the type (4) is Gâteaux differentiable and that the relation $\langle J'(\dot{\varphi}_{opt}), h\rangle = \langle \frac{DJ(\dot{\varphi}_{opt})}{D\dot{\varphi}}, h\rangle = 0$ is a necessary condition characterizing the optimal control $\dot{\varphi}_{opt}$ and the corresponding optimal state $\{u(\dot{\varphi}_{opt}); p(\dot{\varphi}_{opt})\}$. In this expression h denotes the *direction* in which the Gâteaux derivative (i.e. the functional gradient) is taken. The Gâteaux derivative of the functional (4) is equal to

$$\langle J'(\dot{\varphi}), h\rangle = \frac{1}{2}\int_T \oint_{\Gamma_0} \left\{ \left[-q(h)\vec{n} + \mu\bar{\bar{D}}(w(h))\vec{n}\right] \cdot \left[\vec{r} \times \left(\dot{\varphi}\vec{k}\right) + \vec{U}_\infty\right] \right.$$
$$\left. + \left[\left(-p(\dot{\varphi})\vec{n} + \mu\bar{\bar{D}}(u(\dot{\varphi}))\vec{n}\right) \times \vec{r}\right] \cdot \left(h\vec{k}\right) \right\} d\sigma dt, \tag{5}$$

where we used the denotations $\langle \frac{D\vec{f}(\dot{\varphi})}{D\dot{\varphi}}, h\rangle = -q(h)\vec{n} + \mu\bar{\bar{D}}(w(h))\vec{n}$ and $\langle \frac{D\dot{\varphi}}{D\dot{\varphi}}, h\rangle = h$. As was shown in [5], the pair $\{w(h); q(h)\}$ is related to the

Fréchet derivative of the mapping $\dot{\varphi} \rightarrow \{u(\dot{\varphi}); p(\dot{\varphi})\}$. Thus $\{w(h); q(h)\}$ can be obtained as the solution of the linearized form of the Navier-Stokes system

$$\begin{cases} Aw = \begin{bmatrix} \frac{\partial w}{\partial t} + (u_0 \cdot \nabla) w + (w \cdot \nabla) u_0 - \nu \Delta w + \nabla q \\ \nabla \cdot w \end{bmatrix} = 0 \\ w|_{\Gamma_0} = \vec{h}, \qquad w(x, y, 0) = 0 \end{cases} . \tag{6}$$

In the above, the field u_0 represents the basic state around which the linearization is carried out. The control h does not explicitly appear in the formulation (5). In order to factor it out we will use the *adjoint operator* A^* and the *adjoint state* $\{\tilde{w}; \tilde{q}\}$ defined by the following relation

$$(Aw, \tilde{w})_{L^2} = (w, A^* \tilde{w})_{L^2} + B, \tag{7}$$

where $(\cdot, \cdot)_{L^2}$ denotes the scalar product in the Hilbert space L^2 and B stands for the sum of the boundary terms

$$\begin{aligned} B = &- \int_T \oint_{\Gamma_0} \left[(u_0 w) + (u_0 w)^T \right] : (\tilde{w} \vec{n})\, d\sigma dt \\ &- \nu \int_T \oint_{\Gamma_0} \left\{ \tilde{w} \left[\nabla w + (\nabla w)^T \right] - w \left[\nabla \tilde{w} + (\nabla \tilde{w})^T \right] \right\} \vec{n} d\sigma dt \\ &+ \int_T \oint_{\Gamma_0} (q \tilde{w} - \tilde{q} w) \cdot \vec{n} d\sigma dt + \left[\int_\Omega w \cdot \tilde{w} d\Omega \right]_{t=0}^{t=T} . \end{aligned} \tag{8}$$

The adjoint state $\{\tilde{w}; \tilde{q}\}$ is the solution of the problem *adjoint* to the linearized Navier-Stokes system (6)

$$\begin{cases} A^* \tilde{w} = \begin{bmatrix} -\frac{\partial \tilde{w}}{\partial t} - u_0 \left[\nabla \tilde{w} + (\nabla \tilde{w})^T \right] - \nu \Delta \tilde{w} + \nabla \tilde{q} \\ \nabla \cdot \tilde{w} \end{bmatrix} = 0 \\ \tilde{w}|_{\Gamma_0} = -\vec{g}, \qquad \tilde{w}(x, y, T) = 0, \qquad dt < 0 \end{cases} . \tag{9}$$

It should be observed that (9) is a well-posed *terminal value* problem, i.e. it has to be marched backwards in time.

In the case of the rotating circular cylinder the control is limited to the tangential boundary velocity. This implies that $(u_0 \cdot \vec{n})\,|_{\Gamma_0} = 0$ and $(w \cdot \vec{n})\,|_{\Gamma_0} = 0$ in (8). The crucial problem is the determination of the boundary condition for the adjoint state $\tilde{w}$, i.e. the function $\vec{g}$ in (9). It is straightforward to verify that for the particular choice $\vec{g} = \vec{r} \times \left(\dot{\varphi} \vec{k} \right) + \vec{U}_\infty$ the adjoint state introduced above can be used to re-express the Gâteaux derivative (5) in such a way that the control h explicitly appears in all the terms. It can be therefore factored out and we obtain a close formula characterizing the necessary, though not sufficient, condition for optimality

$$\nabla J(t) = \mu \vec{t} \bar{\bar{D}}(\tilde{w}) \vec{n} R - \left[\left(-p(\dot{\varphi}) \vec{n} + \mu \bar{\bar{D}}(u(\dot{\varphi})) \vec{n} \right) \times \vec{r} \right] \cdot \vec{k} = 0. \tag{10}$$

Following [5], an iterative procedure will be presented which determines the optimal control $\dot{\varphi}_{opt}$ and the corresponding optimal state $\{u(\dot{\varphi}_{opt}); p(\dot{\varphi}_{opt})\}$ in the time interval $[0; T]$. First, we choose some initial control $\dot{\varphi}^1$ (for instance $\dot{\varphi}^1 \equiv 0$) and then:

1. solve the full nonlinear Navier-Stokes system in order to determine $\{u^i; p^i\}$,
2. solve (9) for the adjoint state $\{\tilde{w}^i; \tilde{q}^i\}$,
3. use $\{u^i; p^i\}$ and $\{\tilde{w}^i; \tilde{q}^i\}$ to determine the instantaneous values of $\nabla J^i(t)$,
4. use $\nabla J^i(t)$ to update the control $\dot{\varphi}^{i+1}(t) = \dot{\varphi}^i(t) - \lambda \nabla J^i(t)$, where λ is some properly tuned descent parameter,
5. iterate 2. through 5. until convergence, i.e. until (10) attains.

5. Final Remarks and Further Perspectives

In the present paper we have discussed the problem of active wake control. First we validated an open-loop strategy for the case of drag minimization. Good agreement was found with the available data from the laboratory experiments in a similar set-up. In the second part we derived a rigorous feedback algorithm using the tools of the Optimal Control Theory. Further developments of the algorithm may consist in its generalization for non-circular geometries of the obstacle. In such case the rotary control will involve not only the tangential, but also the normal velocity component. Such control is likely to be more effective, as it involves not only dynamic, but also kinematic forcing, the latter related to the change of geometry.

6. Acknowledgments

The first author was supported by a Fellowship of the Foundation for Polish Science. Calculations were performed at the Supercomputing Center of the University of Warsaw (ICMM UW).

References

1. P. E. Saffman, "Vortex Dynamics", Cambridge University Press, Cambridge, Great Britain (1993).
2. P. Tokumaru and P. Dimotakis, "Rotary Oscillation Control of a Cylinder Wake", J. Fluid Mech. **224**, 77, (1991).
3. A. Styczek and P. Wald, "Fast and efficient vortex blob simulation of the flow past the circular cylinder",Arch. Mech. Eng. **XLII**, 281 (1995).
4. J.L. Lions,"Contrôle Optimal des Systèmes Gouvernés par des Equations aux Dérivées Partialles", Dunod, Paris (1969).
5. F. Abergel and R. Temam,"On Some Control Problems in Fluid Mechanics", Theoret. Comput Fluid Dyn. **1**, 303-325, (1990).

MOVING SURFACE SEPARATION CONTROL FOR AIRFOIL AT HIGH ANGLES OF ATTACK

Tapan Sengupta, V. K. Gupta & G. Gopalan
Department of Aerospace Engineering, IIT Kanpur
Kanpur 208 016, India.

Abstract

Moving surface separation control problem is being numerically investigated for a NACA 0012 airfoil at 20° angle of attack. Here slip velocity is imposed on the portions of top and bottom surface of the airfoil to simulate the experiments of Modi et al [1] where rotating cylindrical portion at the leading edge of the airfoil exercised the control. We have studied the problem for Re = 3000, 35000 and 46200 by solving the unsteady 2D Navier- Stokes equation that utilises an orthogonal grid and higher order upwinding scheme for the nonlinear convection terms.

1. Introduction

Flow control has been practised for quite some time - but in recent times it has been a major activity with the adevent of newer technologies providing better sensing and actuating control signals and the ability to use more sophisticated fluid dynamical models for analysis that may involve thousands of degrees of freedom to the traditional practise of involving very few degrees of freedom.

There have been various methods, such as suction, blowing, vortex generators, turbulence manipulators that have been used for flow control with varying degrees of success. The use of moving surface to control flow separation has been experimented by Favre [4] who provided slip velocity on the complete upper surface of aerofoil to delay separation upto 55° and a $C_{L_{max}}$ of 3.5 was achieved. Continuing on the same line Cook et al [5], Johnson et al [6] and Modi et al [1] have experimentally used this idea for circulation control and separation delay for flow past airfoil, where a rotating cylinder in the nose portion of airfoil provided the desired effects. Theoretically this problem has received very little attention. Some theoretical calculations based on panel method coupled with boundary layer solution has been attempted earlier [1]. Only recently Chew et al [7] have simulated the flow past a symmetric Joukowski airfoil based on generalized conservation of circulation (GCC) method for Re = 143000.

Here numerical solution of 2D Navier- Stokes equation for flow past a NACA 0012 airfoil is reported when various stretches of the lower and upper surface were provided with the slip velocity.

G.E.A. Meier and P.R. Viswanath (eds.),
IUTAM Symposium on Mechanics of Passive and Active Flow Control, 243–248.

2. Method of Analysis

The 2D Navier- Stokes equation is solved by finite difference method in stream function (ψ) - vorticity (ω) formulation- the details are given in Sengupta & Sengupta [3]. This method was directly used for the case of Re = 3000. For $Re = 35000$ and 46200, in addition to using third order upwinding we have used skew symmetric differencing of the convection terms to remove aliasing errors which will be present at high Reynolds numbers. In skew symmetric differencing the convection terms are represented as an average of the conservative and nonconservative form. An orthogonal grid suitable for this purpose has been generated by an hyperbolic method developed here and given in Nair & Sengupta [2].

The boundary conditions for the tangential slip velocity for the control is given by,

$$u = U_s; \quad \& \quad v = 0 \tag{1}$$

so the condition for the wall vorticity is,

$$\omega_b = -\frac{U_s}{h_1}\frac{\partial}{\partial \eta}(\frac{h_1}{h_2}) - \frac{1}{h_2^2}\psi_{\eta\eta} \tag{2}$$

In these equations u and v are the contravariant component of velocities in the azimuthal (ξ) and normal (η) direction. The scale factors h_1 and h_2 are as defined in [3]. U_s is the prescribed non dimensional slip velocity on the surface of the aerofoil and shown in Figure 1. The pressure field is obtained by solving the pressure Poisson equation because of desired higher accuracy of load calculation.

3. Results & Discussion

All the calculations are performed using ($233X301$) grid. Various Reynolds number (Re) and U_s have been considered along with different extent of the top and bottom surface over which slip velocity is provided and are given below.

Table 1:

Case	Reynolds No.	U_s	Point1	Point2
I	3000	4	5%	5%
II	35000	3	7.78%	7.78%
III (Model 2)	46200	4	10.5%	40.6%
III (Model 4)	46200	4	40.6%	61%
III (Model 5)	46200	4	61 %	86%

Case I: This low Re case was considered to test the effectiveness of the method at low Reynolds numbers. All the experimental results, including the case studied by Modi et al [1] are for Reynolds numbers that are higher than this. Also unlike the experiments of Modi et al [1] a shorter extent is provided with U_s. For $U_s = 1$

and extent equal to 0.6% showed no aerodynamic benefit as compared to the base line case of no zero slip velocity. When U_s was increased to 4 and the extent was increased to 2.45% the drag was seen to decrease significantly. In Figure 2 we show the C_l and C_d for the case of $U_s = 4$ when the extent was further increased to 5%. This is an interesting case as C_d is always negative. This can be explained by noticing that the forward stagnation point is always on the top surface and there is a vortex near the leading edge which provides considerable suction on the aerofoil. In Figure 2 we show some representative C_p plots which support such an observation.

Case II: The C_l and C_d for this case is shown in Figure 3 and the results reveal that while the lift coefficient increases as compared to case I, the drag coefficient becomes positive. In Figure 4 we show the streamline plots at $t = 20$ and $t = 98$ when the drag attains local minimum. This case is characterised by repeated formation of leading edge bubble which are unstable and on bursting they are convected downstream.

Case III: This Reynolds number correspond to the experimental value used in reference [1] for the case of rotating cylinder as the nose portion of the symmetric Jukowski airfoil that has an extent of 10%. Modi et al [1] states that the experimental conditions were ***essentially two-dimensional*** and apart from the geometric discontinuity between the cylinder and the airfoil there was also flow along the gap between them. To avoid complexity of the resultant geometry we have considered an airfoil without surface discontinuity. Also to model the effect of gap flow the extent on the top and bottom surface over which slip velocity is provided is different. This is done by noticing that the strong favourable pressure gradient on the lower surface has no effect on the extent there and hence kept fixed at 10%. Based on calculations for Model 2 it was concluded that the combined effect of rotating nose and gap flow creates a wall jet after the gap. Thus in all the calculation the slip velocity is not ended abruptly, but is brought to zero over a longer stretch -as shown in Figure 1.

For the cases considered, the results are shown in Figure 4 where C_l and C_d are compared. It appears that the Model 4 performs the best in terms of drag reduction. However the Model 5 represents least unsteadiness and higher average lift generation. The optimum configuration and U_s for $(C_l/C_d)_{max}$ is different to that of $(C_d)_{min}$. The high frequency fluctuations on load are due to the unsteadiness even for those cases where the separation is delayed significantly and is due to the small vortices near the trailing edge. The low frequency oscillations are due to instability of the leading edge bubbles. In Figure 5 some instantenous streamlines and vorticity contours are shown for Model 4 and Model 5 cases. These correspond to maxima and minima of C_l and C_d respectively.

References

[1] Modi V.J., Mokhtarian F. & Fernando M.S.U.K. *J. Aircraft*, **28,2**, 104-111 (1991)

[2] Nair, M.T., & Sengupta T.K. Int. J. for Num. Methods in Fluids, (in press) (1998)

[3] Sengupta T.K. & Sengupta R. *J. Comp. Mech.*, **14,4**, 298-310 (1994)

[4] Favre A. *Thesis presented to Univ. of Paris* (1938)

[5] Cook W.L., Hickey D.H. & Quigley H.C. *AGARD Fluid Dynamics Panel on V/STOL Aerodynamics* Paper 10, Delft, The Netherlands (1974)
[6] Johnson W.S., Tennant J.S. & Stamps R.E. *J. of Hydronautics*, **9,2** 76-78 (1975).
[7] Chew Y.T., Pan L.S. & Lee T.S. Proc. Inst. Mech. Engrs. Part A *J. Power and Energy*, **212**, 69-77 (1998)

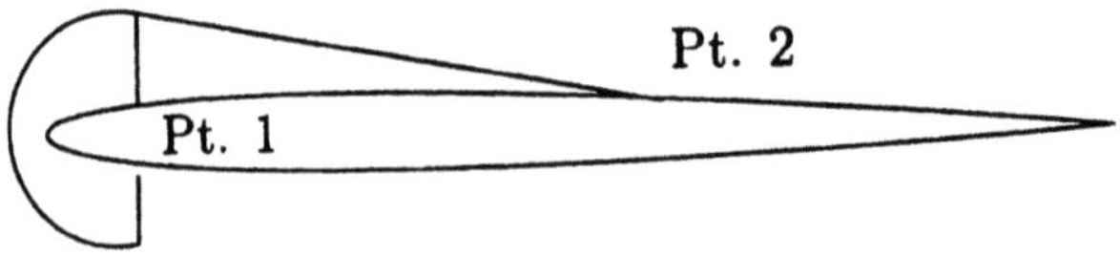

Figure 1. Schematic of the airfoil indicating the slip velocity distribution. Upto point 1 the slip velocity is U_s and it decays linearly upto point 2.

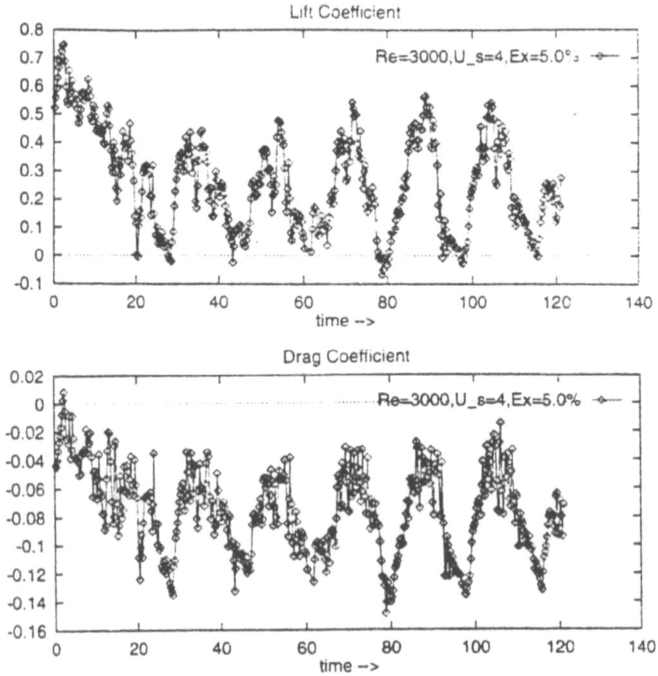

Figure 2a. C_l & C_d as a function of time for $Re = 3,000$; $U_s = 4$ & Pt. 1 at 5% chord. Pt. 1 & 2 are coincident.

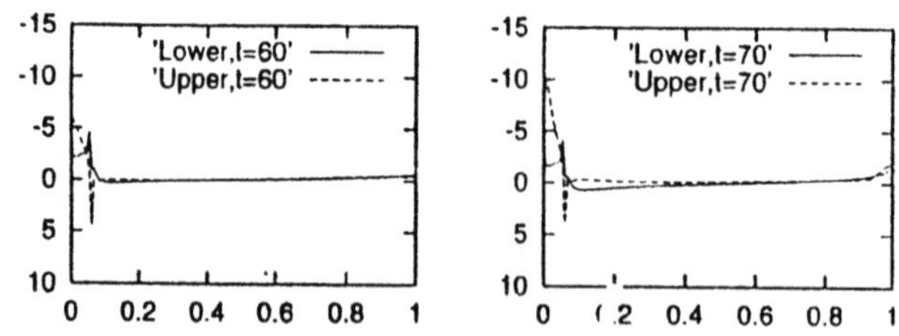

Figure 2b. Pressure Coefficient at indicated times for the case of Fig 2a.

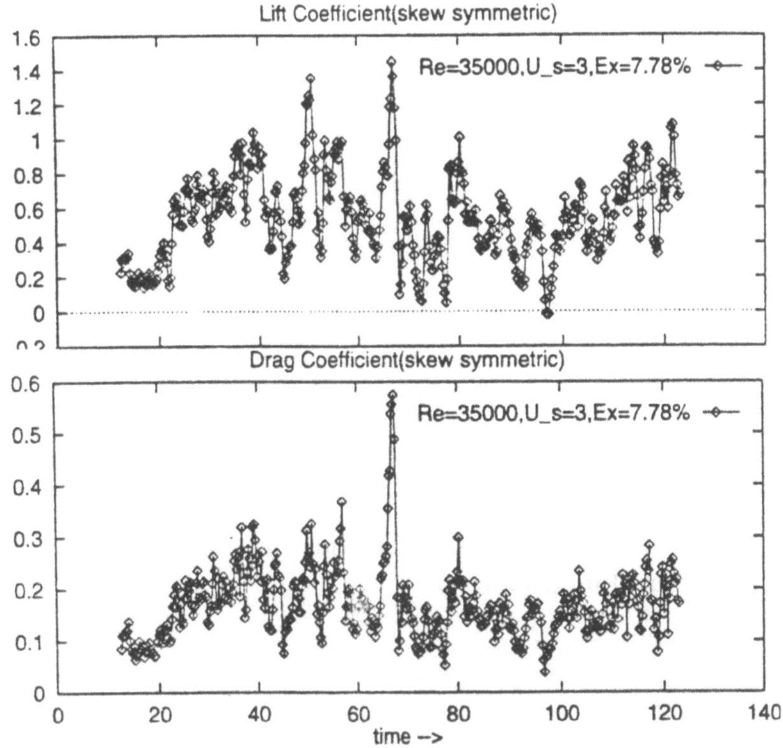

Figure 3. C_l & C_d as a function of time for $Re = 35,000$; $U_s = 3$ & Pt. 1 at 7.78% chord. Pt. 1 & 2 are coincident.

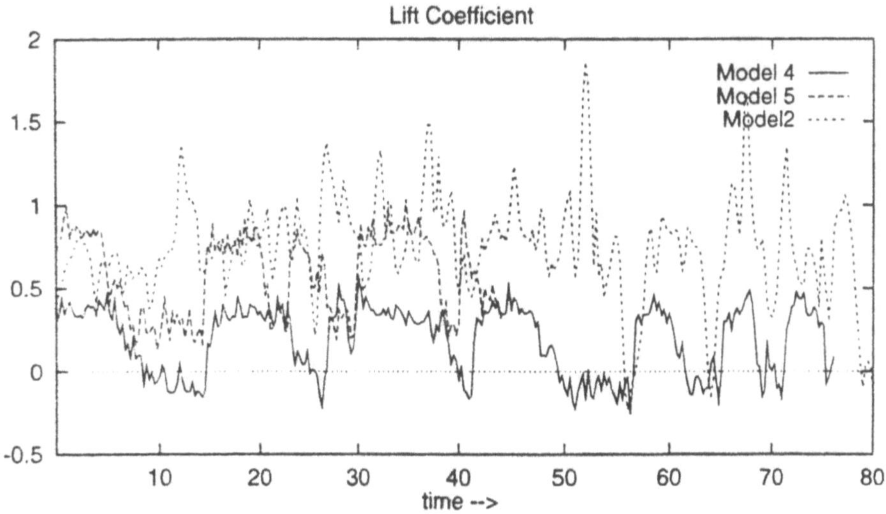

Figure 4. C_l & C_d as a function of time for $Re = 46,200$; $U_s = 4$ for Models as indicated in Table 1. (cont.)

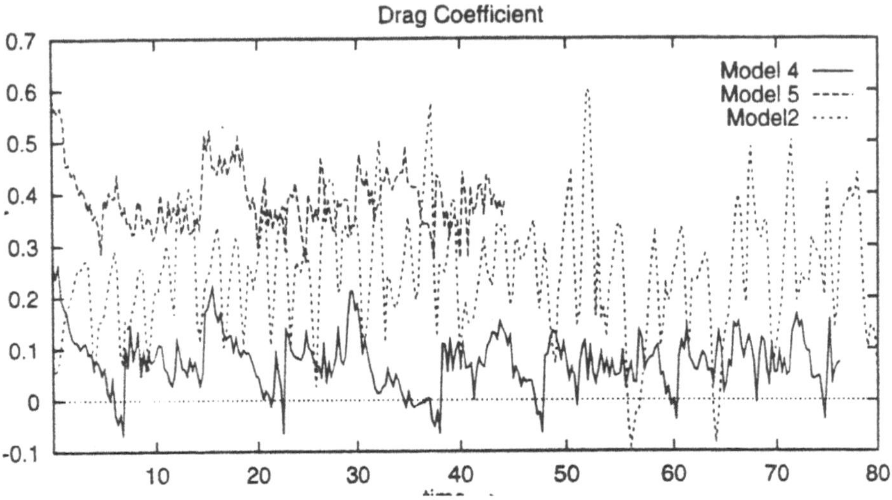

Figure 4. C_l & C_d as a function of time for $Re = 46,200$; $U_s = 4$ for Models as indicated in Table 1.

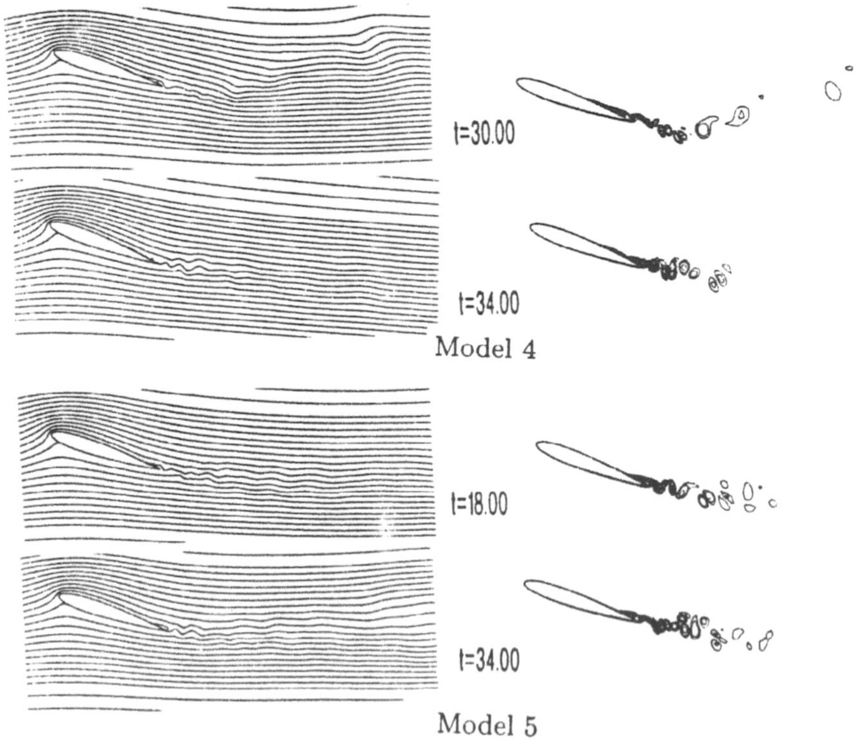

Figure 5. Instantenous streamline and vorticity contours for the cases of Model 4 and Model 5 at indicated times.

Separation control on an airfoil under post-stall conditions by mechanical excitation

F. Urzynicok (urzy@pi.tu-berlin.de) and H.H. Fernholz
Hermann-Föttinger-Institut für Strömungsmechanik
Technische Universität Berlin,
Straße des 17. Juni 135, 10623 Berlin, Germany

Abstract. The influence of mechanically generated perturbations upon the separated flow field on the suction side of an FX 61-184 airfoil under post-stall conditions (angle of attack $\alpha > 20°$) has been investigated. As the flow separated immediately at the leading edge, an oscillating fence located just downstream of the separation line was used to introduce spanwise vorticity into the flow field. Resulting from the induced downwash motion of the vortices, a portion of the initially separated flow reattached, and the separation line moved downstream to a relative streamwise location of up to 15%. Thus, the suction peak in the pressure distribution was partly restored leading to a lift augmentation by as much as 40%.

Keywords: active flow control, mechanical separation control, stalled airfoil, experiments

1. Introduction

Many technological applications, especially in the field of aeronautics and turbomachinery, suffer from the fact that the maximum angle of attack of an airfoil is limited by the onset of separation. If this angle is exceeded, a rapid decay of the performance results. For the improvement of airplane safety, as well as for possible benefits in many other disciplines where high lift is necessary, it is, therefore, desirable to enlarge the operational range of airfoils beyond their natural stall angle by preventing, or at least delaying, separation.

The present experiment employed active flow control by means of mechanical excitation to investigate its aptitude to delay leading-edge stall and, thus, to maintain high-lift conditions in the off-design range of an airfoil. The project followed previous studies using acoustic perturbations on the same wing model [1] and studies using mechanical perturbations on a different airfoil section [3].

2. Experiments

A series of measurements was carried out on a Wortmann FX 61-184 laminar glider profile with chord length $c = 450$ mm and span 900 mm.

G.E.A. Meier and P.R. Viswanath (eds.),
IUTAM Symposium on Mechanics of Passive and Active Flow Control, 249–254.

The experiments were conducted in a low-speed wind tunnel that allowed for velocities of up to 35 m/s in the test section. The angle of attack was held constant at $\alpha = 26°$, where the flow separated from the leading edge at all Reynolds numbers investigated. Immediately downstream of the separation line, the perturbations were introduced into the flow at a streamwise location of $x/c = 1.9\%$. They were generated by an oscillating fence along the entire span moving perpendicularly to the airfoil contour guided by a thin slit (figure 1).

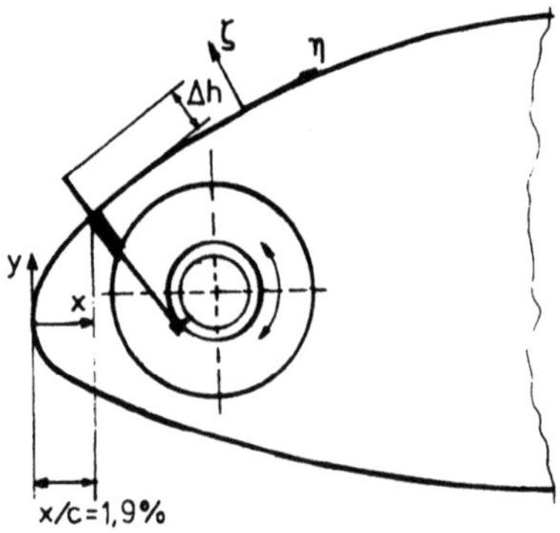

Figure 1. Cross section of the airfoil nose housing the mechanical perturbation source: via exchangeable eccentrics, a direct-current motor drives a shaft mounted in a spanwise duct. The shaft performs oscillations and, thus, extends and retracts the fence which is guided by a thin slit. Depending on the eccentric, three amplitudes Δh are feasible.

Neglecting all quantities that are held constant throughout the investigation, a parametric study provided that the lift coefficient is a function of the non-dimensional extension height of the oscillating fence $\Delta h/c$, the reduced excitation frequency St_c, and the Reynolds number Re_c. The latter two quantities are both based on the chord length of the airfoil model. The resulting relationship can be written as

$$c_l = \mathcal{F}\left(\mathrm{St}_c, \frac{\Delta h}{c}, \mathrm{Re}_c\right).$$

The height of the oscillating fence could be varied in discrete steps between $\Delta h = 2$ mm, 4 mm, and 6 mm with perturbation frequencies ranging from 5 to 70 Hz, 50 Hz, and 33 Hz, respectively. Reynolds numbers were from $\mathrm{Re}_c = 2.5 \times 10^5$ to 1×10^6.

A six-component wind-tunnel balance was used to systematically measure the lift force depending on the parameter field. Flow visualizations by oil-film and tuft techniques rendered an overall image of the flow topology at the airfoil surface and in a streamwise plane normal to it at midspan. In addition, pressure distributions around the airfoil were determined. Finally, data on the reverse-flow coefficient χ and the wall shear-stress τ_w along the centerline of the wing were measured by a traversable wall pulsed-wire probe.

3. Results

Visualizations of the surface flow demonstrated that, with mechanical excitation, the flow can be made attached until far downstream of the oscillating fence. The separation line could be shifted as far downstream as $x/c \approx 15\%$ in the mean sense. Its location tends to gradually move upstream with increasing excitation frequency St_c. At the same time, the reverse-flow region following downstream of the attached flow portion became more pronounced.

Visualizations in the plane normal to the airfoil surface showed that spanwise vorticity shed from the edge of the oscillating fence has a strong influence deep into the flow field. Consequently, the flow reattaches due to the downwash motion induced by the vortical structures originating from the oscillating fence, when the flow grazes over its edge [2].

As a result, considerable lift gains up to $\Delta c_l/c_l \approx 40\%$ were achieved with fence heights of $\Delta h = 4$ mm and 6 mm ($\mathrm{Re}_c = 2.5 \times 10^5$, $\mathrm{St}_c = 1.05$ and 0.6, respectively) as shown in figure 2. In contrast, the maximum lift gain resulting from the fence with height $\Delta h = 2$ mm was substantially smaller ($\Delta c_l/c_l \approx 28\%$ at $\mathrm{Re}_c = 5 \times 10^5$ and $\mathrm{St}_c = 0.2$). With rising Strouhal number, the lift rapidly increases in the range $0 \leq \mathrm{St}_c \leq 0.2$. The value $\mathrm{St}_c \approx 0.2$ seems to represent some sort of threshold above which the effect responsible for the lift gain is fully developed. An augmentation of the Reynolds number diminishes the possible lift gain.

A phenomenon denoted as saturation was observed: Although for small fence heights an augmentation of Δh results in an increased lift gain, there is a threshold ($\Delta h = 4$ mm in this case) above which a further increase of the height of the fence does not lead to a further improvement. As the thickness of the separation region at the location of the perturbation source was measured to be in the range of 2.5 to 3.0 mm, this implies that the fence has to extend into the outer flow to obtain a maximum effect. The occurrence of saturation with greater amplitudes of the fence appears to result from a blockage of the approaching boundary layer, thus promoting separation.

Pressure scans showed that the mechanical excitation causes the partial restoration of the suction peak which is mainly responsible for

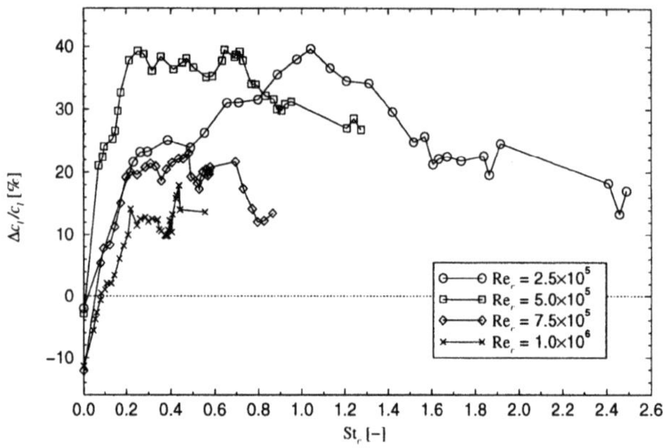

Figure 2. Lift gain due to mechanical perturbation by an oscillating fence with extension height $\Delta h = 4$ mm. The dashed line indicates the lift produced by the airfoil in the absence of flow control. For comparison, values at $\mathrm{St}_c = 0$ were taken with the fence permanently extended to act as a passive control tool.

the lift of an airfoil at high angles of attack (figure 3). The pressure distributions are frequency-dependent, as they exhibit extended regions of enhanced suction near the leading edge at low Strouhal numbers and shorter but more pronounced suction peaks with increasing St_c. This effect is likely to be caused by the resulting size of the vortices shed at the oscillating fence that scale with the excitation frequency St_c.

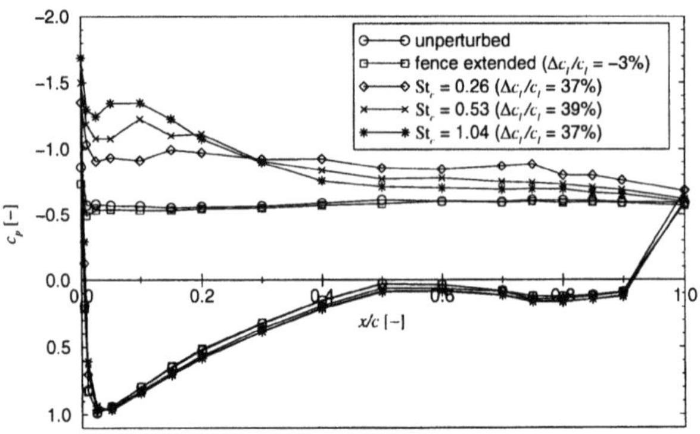

Figure 3. Pressure distribution at $\mathrm{Re}_c = 5 \times 10^5$ due to excitation by an oscillating fence with extension height $\Delta h = 4$ mm. Again, the results of the fence in its extended position are shown for comparison.

Due to the limits of the traversing gear, the data of the wall pulsed-wire probe cover the range of streamwise locations from $x/c \approx 0.2$ to ≈ 0.75. Therefore, all measurements were taken downstream of the separation line. Even in this region, the reverse-flow coefficient was largely reduced by the excitation with fence heights of $\Delta h = 4$ mm and 6 mm (figure 4) indicating that the amount of intermittently attached flow significantly increased up to a streamwise location of $x/c \approx 0.7$. The negligible difference between the data of the $\Delta h = 4$ mm and the 6 mm case is the result of the saturation effect. In contrast to this, the $\Delta h = 2$ mm case only differs slightly from the unperturbed flow scenario.

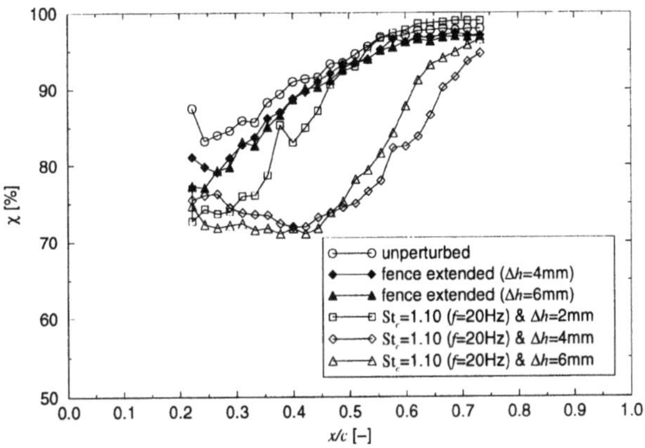

Figure 4. Reverse-flow coefficient for mechanical excitation with $St_c = 1.10$ and varying fence heights ($Re_c = 2.5 \times 10^5$). Data of the permanently extended fence are included for comparison.

The wall pulsed-wire probe also provided the wall-shear stress distribution along the centerline of the airfoil (figure 5). Surprisingly, with fences of heights 4 and 6 mm, the wall-shear stress became even more negative between $x/c \approx 0.2$ and 0.45, compared to the unperturbed case. This corresponds to observations during flow visualizations and indicates that the reverse-flow portion became much stronger, although with respect to time, there is less reverse flow. Again, the effect of the fence with amplitude $\Delta h = 2$ mm is small compared to the case without flow control.

For comparison, the impact of passive control tools such as tripping devices and the fence in its extended position was also looked at (included in figures 2 to 5). However, no beneficial effects could be observed in these cases. Under certain conditions, the lift is even reduced by as much as 15%. This fact illustrates that the flow does not simply reattach because of a transition of the boundary layer, but that the unsteady nature of the flow control is necessary, in this case, to create spanwise vortices which increase the momentum close to the wall and, thus, delay separation.

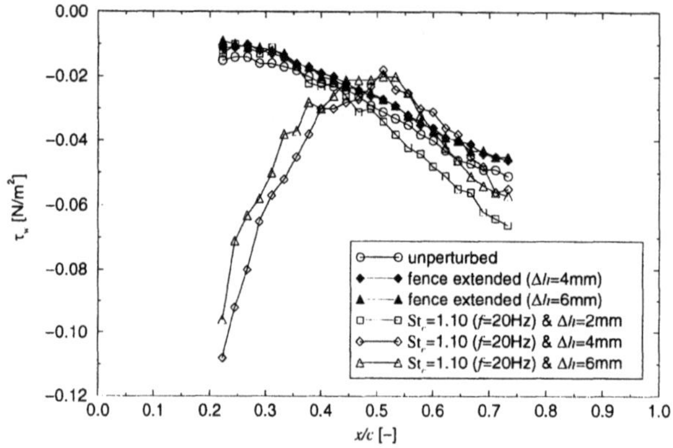

Figure 5. Wall-shear stress for mechanical excitation with $\mathrm{St}_c = 1.10$ and varying fence heights ($\mathrm{Re}_c = 2.5 \times 10^5$). As before, data of the permanently extended fence are included for comparison.

4. Conclusions

By means of mechanical excitation using an oscillating fence, leading-edge stall on an airfoil can be delayed, so that high-lift conditions can be maintained beyond the natural stall angle. In the present case, at an angle of attack of $\alpha = 26°$, the lift could be increased by as much as 40% over the original value.

To minimize the energy input needed for the active control, it seems possible to improve the location of the perturbation source. Smaller fences closer to the separation line should produce similar results, as long as they protrude far enough into the energetic outer flow field. In addition, detailed studies are still required on how the drag is affected by the flow control.

References

1. Erk, P.: 1997, *Separation Control on a Post-Stall Airfoil Using Acoustically Generated Perturbations*, No. 328 in Fortschr.-Ber. VDI Reihe 7. Düsseldorf: VDI Verlag.
2. Miau, J., M. Chen, and J. Chou: 1991, 'Frequency Effect of an Oscillating Plate Immersed in a Turbulent Boundary Layer'. *AIAA Journal* **29**(7), 1068–1074.
3. Zhou, M., H. Fernholz, H. Ma, J. Wu, and J. Wu: 1993, 'Vortex Capture by a Two-Dimensional Airfoil with a Small Oscillating Leading-Edge Flap'. *AIAA Paper 93-3266.*

EFFECT OF MICROBUBBLE CLUSTER ON TURBULENT FLOW STRUCTURE

H. Kato, T. Iwashina, M. Miyanaga
and H. Yamaguchi

Department of Environmental and Ocean Engineering,
University of Tokyo,
Hongo, Bunkyo, Tokyo Japan

1. Introduction

Wall shearing stress is decreased by the existence of microbubbles in a turbulent boundary layer as discovered by McCormick and Bhattacharyya [1] in 1973. Since that time experimental research has been conducted on air-water bubble flow [2]-[7]. It has been demonstrated that shearing stress is decreased to a value one order of magnitude lower than the original value under optimum conditions [2].

This reduction is caused by the modification of the turbulent flow by microbubbles. One possible mechanism is the reduction of turbulence by microbubbles in the buffer as well as the inner turbulent region. Many experimental results have shown that the void ratio of microbubbles gave a good correlation with the shearing stress reduction.

Theoretical analysis [8] has shown that a linear decrease of wall shearing stress against void ratio of microbubbles near the wall, when the size of microbubbles is assumed to be much smaller than the turbulence scale of the flow. However, many experimental results have shown that the wall shearing stress does not decrease when the microbubble injection rate is low. In some cases the wall shearing stress was increased slightly by the microbubble injection [2], [4]-[6]. There is a contradiction between theory and experiment especially at a low injection rate.

The authors measured the turbulence intensity with the mean velocity profile in a turbulent boundary layer filled with microbubbles, and propose a hypothetical mechanism of shearing stress reduction by microbubbles.

2. Experimental

The experiment was conducted in a recirculating water tunnel at the University of Tokyo. The test section was 120 mm in height, 50 mm in width, and 680 mm in length. It was made from Plexiglas, to enable the measurement of turbulent intensity using a laser Doppler velocimeter (LDV). The top wall of the test section was used as a flat surface for measurement, where a turbulent boundary layer developed (Fig. 1).

Microbubbles were generated through a porous plate (40 mm$\times$40 mm) of 15 μm in pore size, mounted flush with the top wall of the test section. Three shearing stress transducers were mounted on the plate at 120 mm (section 1), 260 mm (section 2) and 400 mm (section 3) from the air injection point. Measurement was conducted mainly at the two downstream transducers (sections 2 and 3). The transducer was a floating disk, 5 mm in diameter, fitted on a cantilever bar and the maximum capacity was a shearing stress of 500 Pa. Discussions on the accuracy was given in Ref. [9]. The velocity and turbulence intensity in the boundary layer were measured by a one-component LDV in forward scatter mode. Although no seeding particle was added, the signal rate was

G.E.A. Meier and P.R. Viswanath (eds.),
IUTAM Symposium on Mechanics of Passive and Active Flow Control, 255–260.

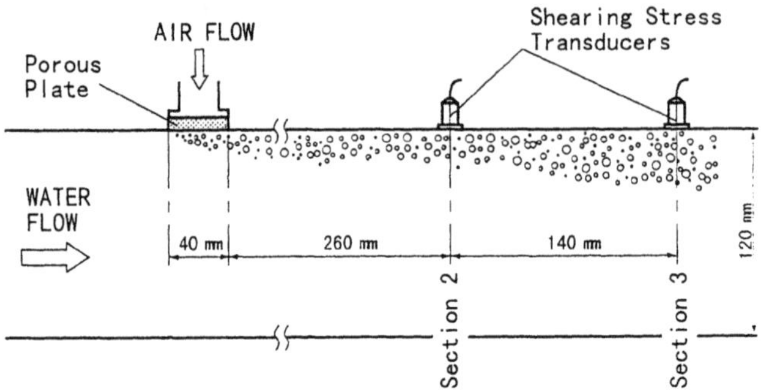

Fig. 1 Test Section

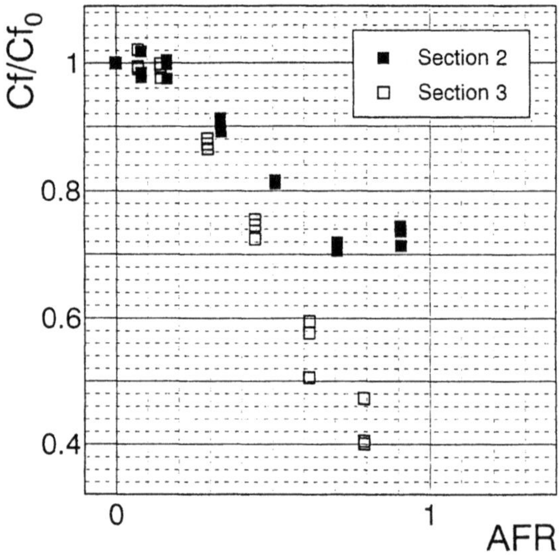

Fig. 2 Reduction of wall shearing stress against non-dimensional air flow rate (AFR) U = 8 m/s

enough for the measurement. When an air bubble interupted the LDV beams, the signal was holded and the effect of air bubble was easily discriminated. These are big advantages of using the LDV in forward scatter mode. The air flow rate through the porous plate was measured using a Rotameter.

The water flow rate was varied between 6 and 10 m/s. The Reynolds number was about 2 million at section 2 at U = 8 m/s, where the reference length is 260 mm.

3. Experimental Results and Discussions

Figure 2 shows the reduction of wall shearing stress against non-dimensional air (microbubble) flow rate (AFR) at section 2 and 3, where Cf_o is the wall shearing stress without microbubbles. The AFR is defined as,

$$AFR = \frac{Q}{UL\delta^*} \tag{1}$$

where, Q = air flow rate, U = main flow velocity, L = width of the porous plate (40 mm), and δ^* = displacement thickness without microbubble. The value of δ^* was 1.0 mm at

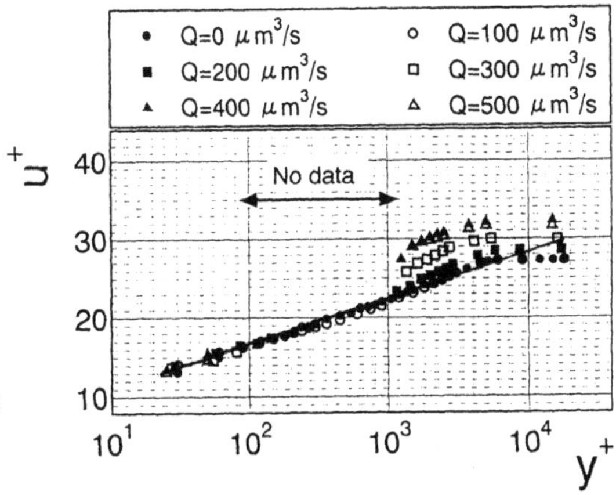

Fig. 3 Time-averaged velocity distributions in boundary layer Section 2, U = 8 m/s

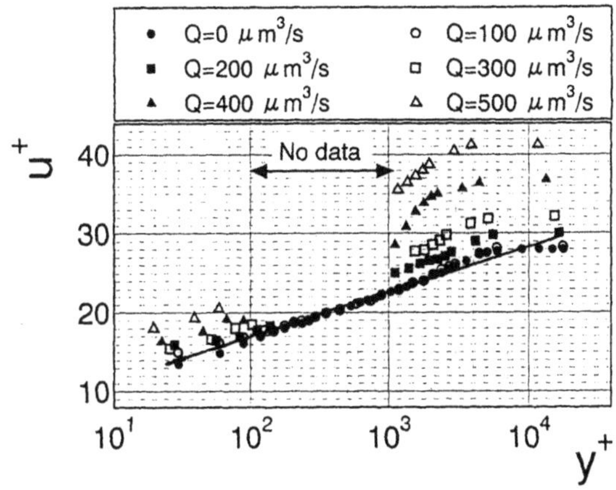

Fig. 4 Time-averaged velocity distributions in boundary layer Section 3, U = 8 m/s

section 2, at U = 8 m/s. The Cf/Cf_o does not decrease at the low AFR, then begins to decrease with the AFR. This tendency was observed in many previous experiments [3]-[6].

Figures 3 and 4 are the time-averaged velocity distributions in the boundary layer at sections 2 and 3, respectively. The solid lines in the figures are the empirical formulae for the turbulent region. The wall shearing stress decreases with the air flow rate Q. Therefore, the velocity and the distance were non-dimensionalized by the corresponding measured shear velocity. Velocities between about 100 and 1000 wall units could not be measured by the present LDV in forward scatter mode at higher microbubble density, because the light beams were interrupted by the microbubbles.

The turbulent flow structure in the buffer and inner turbulent regions was not changed at section 2 (Fig. 3). The microbubbles caused the thickening of the viscous sublayer, which became most pronounced at section 3 (Fig. 4), so that the universal velocity profile did not hold at higher air flow rates. This fact agrees well with the observed tendency of a reduction in wall shearing stress at section 3, U = 8 m/s (open squares in Fig. 2).

The turbulence intensities $\sqrt{u'^2}$ under the same conditions are shown in Figs. 5 and 6. Measurement was possible only in the region very near to the wall at a high

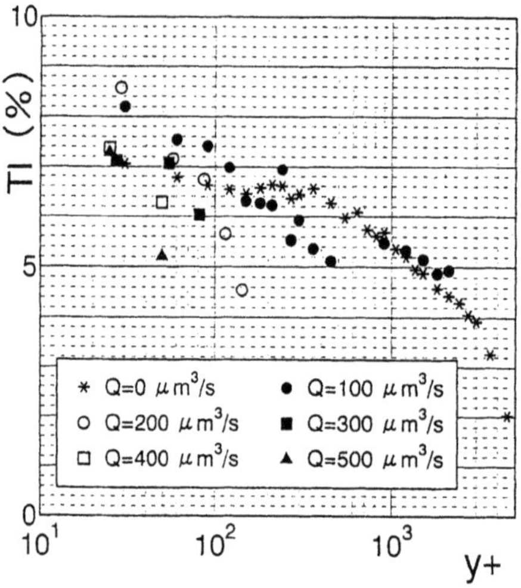

Fig. 5 Turbulence intensities TI ($=\sqrt{u'^2}$) in boundary layer, Section 2, U = 8 m/s

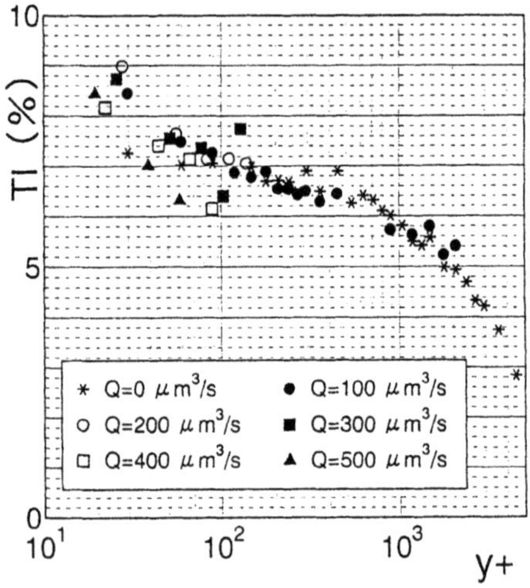

Fig. 6 Turbulence Intensities TI ($=\sqrt{u'^2}$) in boundary layer, Section 3, U = 8 m/s

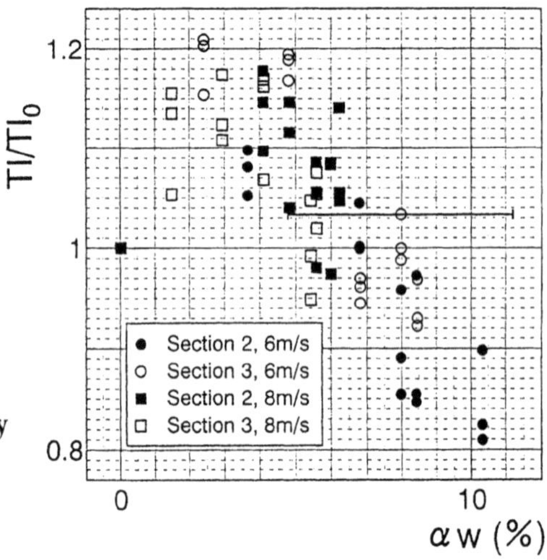

Fig. 7 Effect of microbubble density on turbulence intensity

microbubble ratio. For reference, data in the absence of microbubbles are also shown.

The turbulence intensity decreased at higher air flow rates. To the contrary, the intensity increased near the wall at lower air flow rates. This tendency was the same at section 2 (Fig. 5) and section 3 (Fig. 6).

The effect of particle size on turbulence intensity was analyzed by Gore and Crowe [10]. They gathered experimental data of two phase flow including gas-solid, gas-liquid and liquid-solid from 15 different papers, and concluded that the ratio dp/le $\simeq$ 0.1

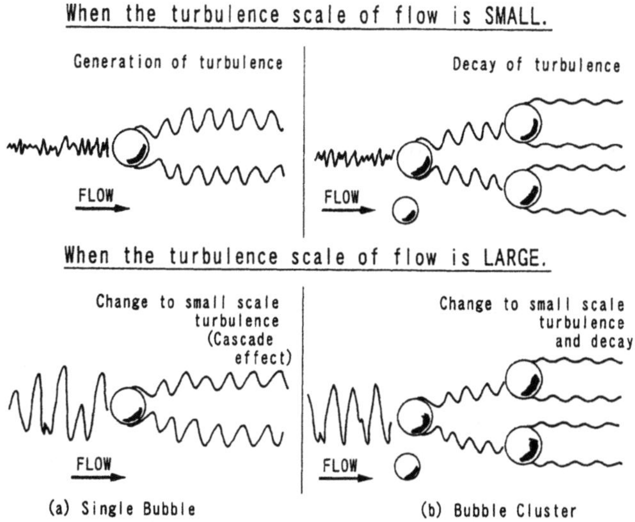

Fig .8 Hypothetical mechanism of microbubbles on turbulence structure

(dp : particle diameter, le : characteristic length of turbulence) provided an estimate of whether relative turbulence intensity of the carrier phase would be increased or decreased. Though their conclusion was for jet and pipe flows, the same effect should exist in a boundary layer flow.

In the present experiment, microbubble diameter did not change much by the air flow rate. The average measured diameter of microbubble was 0.47 mm at Qa = 100 μ m^3/s (0.1 liter/s). The diameter increased to 0.56 mm at 500 μ m^3/s. Therefore the present tendency, shown in Figs. 5 and 6, cannot be explained by a change in bubble size.

The effect of microbubbles on the turbulence intensity can be clearly seen in Fig. 7, where the turbulence intensity at 0.1 mm from the wall (about 30 wall units), which was the closest measurement point to the wall surface in the present experiment, was measured in detail. The abscissa is the averaged void ratio (microbubble density) near the wall α_w, where the averaging was carried out in the range 0-1.0 mm. Measurement of void ratio distribution in the boundary layer was made using a fiber-optic void meter. The details are presented in [11] and [7]. The ordinate is the non-dimensional turbulence intensities, TI/TIo, where TI and TIo are the turbulence intensities with and without microbubbles, respectively. The non-dimensional turbulence intensity first increase at the low void ratio, then begin to decrease on increasing the void ratio.

The existing models of turbulent flow with microbubbles consider only the decrease of turbulence. However, the effect of microbubbles on turbulent flow is more complicated. Figure 8 is a hypothetical mechanism of microbubbles on the turbulent structure. The microbubbles decay large scale turbulence and also generate new turbulence of bubble-size scale. Energy of large scale turbulence in the main flow is transferred into the smaller scale turbulence by the microbubbles, a process termed the "cascade effect". A single microbubble in a flow with small scale turbulence increases the turbulence intensity of the flow, whereas a cluster of microbubbles decays the turbulence of scale smaller than bubble size because of the interaction between bubbles. In other words, a cluster of microbubbles is more effective at decreasing the turbulent intensity in the boundary layer, resulting in a larger reduction in wall shearing stress at high microbubble densities. This mechanism explains well the tendencies shown in Fig. 2 as well as Fig. 7.

4. Conclusions

The effect of microbubbles on the turbulent flow structure was examined experimentally. The following results were obtained :

(1)Microbubbles are effective in reducing wall shearing stress, which is caused by the thickening of the viscous sublayer.

(2)The turbulence intensity is increased near the wall at a low microbubble density. On increasing the microbubble density the turbulence intensity reached a peak, then decreased.

(3)A cluster of microbubbles decays the turbulence of scale smaller than the bubble size. This is due to interaction between microbubbles.

References

1. McCormick M.E., Bhattacharyya R., : Drag Reduction of a Submersible Hull by Electrolysis, *Naval Engineers Journal* **85** (1973) 11-16
2. Bogdevich V.G., Evseev A.R., Malyuga A.G. et al. : Gas-Saturation Effect on Near-Wall Turbulence Characteristics. *Second International Conference on Drag Reduction*, Cambridge, England, BHRA (1997) 25-37
3. Madavan N.K., Deutsch S., and Merkle C.L., : Reduction of Turbulent Skin Friction in Microbubbles *Phys Fluids* **27**(1984) 356-363
4. Madavan N.K., Deutsch S., and Merkle C.L., : Measurements of Local Skin Friction in a Microbubble-modified Turbulent Boundary Layer, *Journal of Fluid Mechanics* **156**(1985) 237-256
5. Kato H., Miyanaga M., Haramoto Y., et. al : Frictional Drag Reduction by Injecting Bubbly Water into a Turbulent Boundary Layer. *Cavitation and Gas-Liquid Flow in Fluid Machinery and Devices.*, FED **190** (1994) ASME, 185-194
6. Kato H., Miyanaga M., Yamaguchi H., et. al : Frictional Drag Reduction by Injecting Bubbly Water into a Turbulent Boundary Layer and the Effect of Plate Orientation. In Serizawa A., Fukano T., and Bataille J. (eds) *Advances in Multiphase Flow*, Kyoto, Japan (1994) Elsevier, Amsterdam 85-96
7. Guin M.M., Kato H, Yamaguchi H., Maeda M., and Miyanaga M : Reduction of Skin Friction by Microbubbles and Its Relation with Near-wall Bubble Concentration in a Channel. *Journal of Marine Science and Technology*, **1** (1996) 241-254
8. Madavan N.K., Deutsch S., and Merkle C.L. : Numerical Investigation into the Mechanisms of Microbubble Drag Reduction, *Journal of Fluid Engineering*, **107**, (1985) 370-377
9. Kato H., Fujii Y., Yamaguchi H., and Miyanaga M. : Frictional Drag Reduction by Injecting High-viscousity Fluid into Turbulent Boundary Layer, *Journal of Fluids Engineering*, **115** (1993) 206-212.
10 Gore, R.A. and Crowe, C.T. : Effect of Particle Size on Modulating Turbulent Intensity, *Int. J. Multiphase Flow*, **15** (1989) 2, 279-285.
11. Kato, H., Iwashina, T., Miyanaga, M., and Yamaguchi, H. : Effect of Microbubbles on the Structure of Turbulence in a Turbulent Boundary Layer, (under reviewing).

STABILITY ANALYSIS OF THE WAKE CONTROL PROBLEM

MAREK MORZYŃSKI
Division of Basics of Machine Design, Poznań University of Technology, Piotrowo 3, 60-965 Poznań, Poland

AND

KONSTANTIN AFANASIEV AND FRANK THIELE
Hermann Föttinger-Institut für Strömungsmechanik, TU Berlin, Str. des 17 Juni 135, 10623 Berlin, Germany

1. Introduction. This study focuses on the stability analysis and low-dimensional modeling of the wake control problem. The influence of diverse symmetric and unsymmetric steady boundary conditions (blowing and suction) and the location of two additional control cylinders on the critical Reynolds number for a circular cylinder flow is numerically investigated with a 2D FEM code. The eigenfunctions of the linearized disturbance equation and Karhunen-Loeve (snapshot POD) modes of the numerical solution for the unsteady Navier-Stokes equations are used to obtain the low-dimensional dynamical systems. Using just a few of the eigenmodes yields the reconstruction of the unsteady behavior of the complex wake flow.

2. General concepts. In global linear stability analysis of the non-parallel flow the critical values of Reynolds number are determined by analysis of a basic steady solution of the Navier–Stokes equations. We assume that the unsteady solution of the Navier-Stokes equations can be expressed as the sum of the steady solution and a disturbance. This assumption leads us to the disturbance equation and the continuity equation for the disturbance. We linearize the disturbance equation and separate the time and space dependence. After the discretisation of the generalized differential eigenvalue problem we get an unsymmetric algebraic eigenvalue problem. An efficient algorithm based on the subspace iteration method (SIM) with an inverse Cayley transformation is developed for the solution of the eigenvalue problems resulting from non-parallel flow stability analysis. Details of the procedure can be found in Ref. [2].

G.E.A. Meier and P.R. Viswanath (eds.),
IUTAM Symposium on Mechanics of Passive and Active Flow Control, 261–266.

Most unsteady models used in industry either do not adequately model the essential unsteady flow bifurcations or are computationally expensive. For example, time-marching Navier-Stokes solvers have many thousands of degrees of freedom and are computationally too expensive for use in routine design studies. For many applications, it will be convenient to construct low-dimensional models of the unsteady flow, which are represented by a small set of ODE. To form such models, we use the information either from the proper orthogonal decomposition (POD) of the numerical unsteady solution or from the eigenvalue problem for the disturbance equations. The former approach needs an expensive data base describing the unsteady flow. This data base can be obtained from experiments or from unsteady simulations of the Navier-Stokes equations. We produce the unsteady solution numerically and use it for the POD. The latter approach needs no previous knowledge of the time-dependent solution of the Navier-Stokes equations. We compute the dominant leftmost eigenvalues and corresponding eigenmodes of the linearized disturbance equations and form the orthonormal basis to project the Navier-Stokes equations.

3. Computational Results. The control of circular cylinder flow is realized here by blowing and suction on the surface of the cylinder and the addition of two small control cylinders

In this studies, we investigated the situation using a few sources of steady blowing and suction on the surface of the cylinder and few additional small control cylinders. The goal of this investigation is to understand the influence of the position and intensity of blowing and suction and the position of control cylinders on the stability property of the flow.

a) Blowing and suction. The blowing and suction source was directed perpendicularly to the surface of the cylinder in the slot of length $l = 3/80 * L$, with $L = 2\pi R$ being the length of the cylinder surface. The position of blowing or suction is determined by the angle with respect to the stream direction and intensity is defined as $I = u * l$.

In Fig.1, the dependence of the critical Reynolds number on the angle of a single or two symmetrically positioned suction sources is shown. The intensity of the single source and the sum of the intensity of the two sources was kept constant ($I = 0.036$). The maximum effect is observed at a suction angle of 90^0 in both situations. The maximal critical Reynolds number is $Re_{cr} = 67.85$ in the case of a single source and $Re_{cr} = 64.77$ in the case of two sources.

In the analogous cases of blowing with the single and two symmetrically positioned sources (Fig.1.) the maximum effect is encountered at the angle 0°. The critical Reynolds number for this angle is equal to $Re_{cr} = 76.92$. The effect of stabilization of the flowa using sources of the same intensity is stronger for blowing than for suction, when both are positioned optimally.

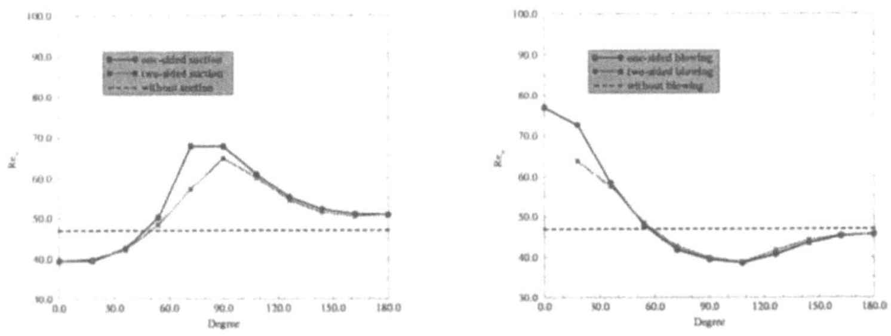

Figure 1. Critical Reynolds number by single and two symmetrically positioned suction sources for several positions, $I_{suc} = 0.036$ (left) and blowing sources, $I_{blow} = 0.036$ (right)

In a next step, we change the intensity of the suction and blowing while keeping positions fixed for most effective suppression of the instabilities. In Fig. 2, the dependence of the critical Reynolds number on the intensity of suction and blowing for a single and two symmetrically positioned sources is shown.

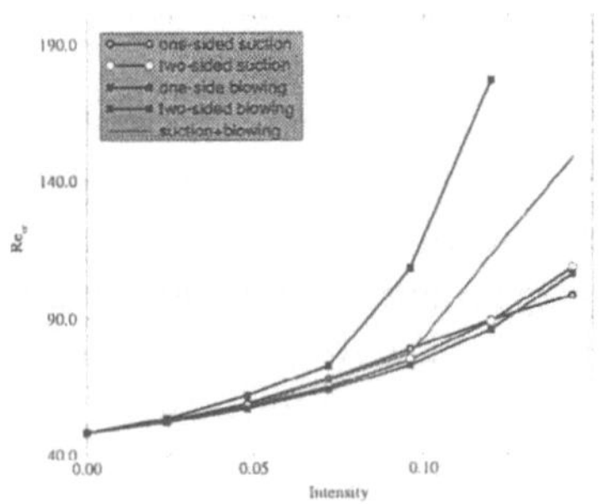

Figure 2. Critical Reynolds number for a single and two symmetrically positioned suction and blowing sources for several source intensities. $\alpha_{suc} = 90^0$

b) Two unsymmetrically positioned control cylinders. Next we consider the cylinder flow with two additional unsymmetrically positioned control cylinders. In our last investigation [1] of two symmetrically positioned control cylinders it was shown that the maximal critical Reynolds number was reached if the control cylinders were located in the point $x = 2.5, y = \pm 1.0$. The goal is to identify the importance of the symmetry in the mean flow for its stability. We fix the position of the first control cylinder in the point $x = 2.5, y = 1.0$ and change the position of the second control cylinder along the line $y = -1.0$. The critical Reynolds number drops monotonously with the distance of the second cylinder from the symmetry point. This

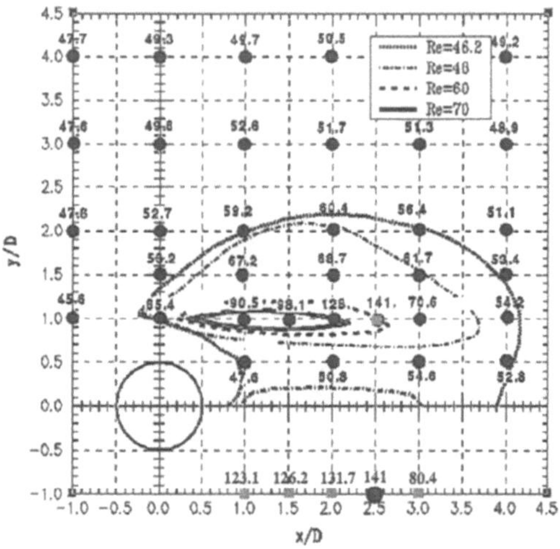

Figure 3. Critical Reynolds number for different positions of the control cylinders (single and two control cylinders cases)

feature is shown in Fig. 3. The isolines denote constant critical Reynolds numbers for the single control cylinder. The numbers above the dots in the upper half of the coordinate plain indicate the critical Reynolds number for the case with the two symmetrically positioned cylinders. The numbers above the squares in the lower half of the coordinate plain indicate the critical Reynolds number for the case with two unsymmetrically positioned cylinders.

c) Stability of the time-averaged solution. The stability analysis for the time-averaged solution for the circular cylinder flow without and with a single small control cylinder is performed. The small cylinder was positioned in the point $x = 1.2, y = 1.0$. An experimental investigation of these flows was described by Strykowski & Sreenivasan in [3]. The amplification rate of the time-averaged solution is negative for the Reynolds number range which we investigate. The time-averaged solution is stable.

The eigenvalue analysis of the time-averaged solution yields a frequency very close to the experimental values in contrary to the analysis based on the steady flow solution. It is shown in Fig. 4. However the calculation of the time-averaged solution needs usually a very expensive unsteady calculation.

d) Low-dimensional models of the unsteady flow. Traditionally, a reduced order model for unsteady flow is based on the projection of the Navier-Stokes equatioasn onto the Karhunen-Loeve space. In Fig. 4, the

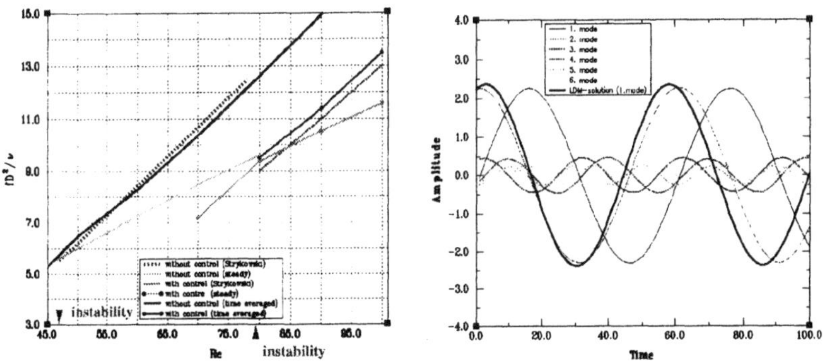

Figure 4. Non-dimensional frequency for cylinder flow with and without control cylinder in the point $x = 1.2, y = 1$ (left). Projections of unsteady solution in Karhunen-Loeve subspace (right).

projection of the unsteady solution on 6 dominant Karhunen-Loeve modes and the solution of a six-equation model for the single cylinder flow at $Re = 100$ is shown. The first two Karhunen-Loeve modes cover about 97% of the total energy of the unsteady solution. The amplitude of the first KL-Mode is very close to the projection of the unsteady solution on this mode.

The next calculation was performed for the cylinder flow with simultaneous blowing and suction on the cylinder surface for a Reynolds number $Re = 60$. The two sources of suction and the single source of blowing lie in the optimal position for stabilization of the flow: suction at $\pm 90^0$ to the mean flow direction and blowing at the angle of 0^0. The corresponding intensities are $I_{suc_1} = I_{suc_2} = 0.012$ and $I_{blow} = 0.024$.

The new 3-equation model was created by projection of the Navier-Stokes equations on the basis formed by the eigenfunctions of the linearized disturbance equation. The 3 dominant eigenmodes cover only about 60% of the total energy of the unsteady flow. Numerical integration of this model gives the spurious steady solution. The analysis of the single terms used to construct the dynamical system shows that the solution is very sensitive to the quality of the numerical differentiation. The FEM method with quadratical elements is not able to produce the necessary accuracy of the derivatives of the velocity field. The analogous problems arise by the low-dimensional model based on the Karhunen-Loeve eigenmodes. We have obtained the periodical solution, but the differences in the amplitudes of the eigenpairs are unacceptable. The only way to eliminate this problem in further computations is the use of high order discretization.

4. Conclusions. The calculations carried out here allow to clear some correlations in the onset of the instability of the wake flow and the mechanism of suppression of this instability.

The primary instability in the cylinder flow excite a form of a Karman vortex street behind the cylinder. This instability arises as a result of the collision of two shear layers. The suppression of the vortex street can be obtained by two measures: weakening or reduction of the energy of the shear layers or elimination of the collision between these layers.

The location of control cylinders in the area of the shear layers causes the loss of the energy for the collision between the shear layers and stabilizes the flow. The shear layers are symmetric to the primary axis of the flow, and the stabilization effect is more pronounced for symmetric positioning of the control cylinders. In the present work, the positions of the control cylinders, where the suppression of the shear layers causes the maximum stabilization of the flow, is determined.

The most effective weakening of the shear layers without change of the geometry of the flow is resulting from the suction in the area of developing boundary layer.

The collision between both shear layers can be prevented by locating a splitter plate or blowing in the wake. The linear increase of the intensity of blowing leads to the quadratical growth of the critical Reynolds number. Simultaneous application of blowing in the wake and suction in the boundary layer area is most efficient as no additional energy is needed.

The instabilities, which appear for laminar-turbulent transition in unsteady flow cannot be predicted with the techniques of the global linear stability analysis alone. In this work, the calculation technique based on the low-dimensional modeling of the unsteady flow is described. The new approach to construct such models uses the eigenfunctions of linearized disturbance equations. Preliminary results show the perspective of these models and the problems with its realization.

References

1. M.Morzyński, K.Afanasiev, F.Thiele, Eigensolutions of the global stability of non-parallel flow problem, XIII Polish Conference Computer Methods in Mechanics, 5-8 May 1997, Poznań, Poland, Proceedings of the XIII PCCMM, Vol.3, 925-932.
2. M.Morzyński, K.Afanasiev, F.Thiele, Solution of the eigenvalue problems resulting from global non-parallel flow stability analysis, in print in Comp. Meth. App. Mech. Eng.
3. P.J. Strykowski, K.R. Sreenivasan, On the formation and suppression of vortex "shedding" at low Reynolds numbers, J. Fluid Mech. 218, (1987), 71–107.

CONTROL OF THE PERIODIC WAKE BEHIND A PLANE BLUNT BASE

S.D. SHARMA and R.K. SAHOO
Aerospace Engineering Department
Indian Institute of Technology
Mumbai - 400 076, India.

The paper describes a new technique of active control of periodic near wake behind an aerofoil with blunt base. Tangential blowing, employed only on one of the surfaces at the trailing edge, is found effective in suppressing the alternate vortex shedding, and ultimately results in a nearly symmetric and stable wake with reduced turbulence.

1. Introduction

Injecting an air mass at varying rate into the wake from the base of a blunt trailing edged aerofoil has been used in the past as a technique to control a two-dimensional periodic wake. Wood (1964, 1967), Bearman (1967) and Schumm et al. (1994) have shown that a low speed base bleed can effectively suppress the regular Karman vortex shedding . A concept of momentumless wake wherein wake drag is balanced by the jet thrust has been tested by Cimbala and Park (1990). Apart from complete suppression of the vortex shedding, the experiments showed a rapid decay of turbulence in the wake and a strong dependence on the initial conditions (Park and Cimbala, 1991).

In the present work, an innovative technique of tangential blowing into a shear layer at separation is tried, instead of blowing from the model base straight into the wake. Since formation of vortex and its subsequent shedding is known to depend on the interaction between the two opposite shear layers, it was believed that stabilizing any one of the two shear layers would be effective in interfering with the process of vortex formation. Therefore, a high velocity jet with low mass flow rate was blown into one of the shear layers and the wake flow was examined to assess the efficacy of the technique.

2. Experimental Set-up

The experiments were carried out in a suction type low turbulence wind tunnel driven by a pair of contra-rotating axial flow fans. The test section is 305 mm high, 220 mm wide and 1830 mm long with its floor slightly diverged at an angle of about 0.8°. The

G.E.A. Meier and P.R. Viswanath (eds.),
IUTAM Symposium on Mechanics of Passive and Active Flow Control, 267–272.

free-stream turbulence level measures better than 0.1 % over a range of smoothly varying air velocity up to 26 m/s.

Figure 1 shows the sectional details of the test model with a semi-elliptic forebody and parallel sided rearbody having a blunt base. The model, constructed out of perspex with glass finish, is hollow and has a knife edged steel cover plate on the rearbody leaving a spanwise slit on the surface along one of the trailing edges. The cavity in the model rearbody is shaped into a smooth half contraction to form a two-dimensional nozzle with an opening of 0.22 mm from the knife edge of the cover plate for tangential blowing. A 500 mesh phosphorus-bronze screen is placed as a resistance upstream of the contraction for spanwise uniformity of the flow through the slit.

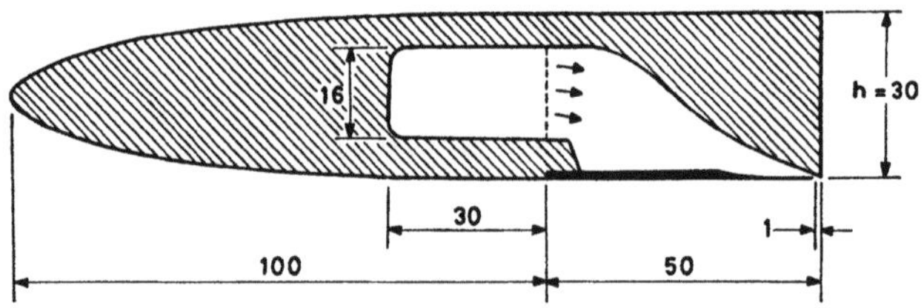

Fig. 1. Test model with the provision for tangential blowing.

The model was mounted at zero incidence between the test section side walls with an arrangement for regulated air supply from outside through an orificemeter into both the opposite open ends of the model. Hot-wire measurements were made using a 5 μ tungsten normal sensor probe that was traversed in the wake with the model slit facing down. The spanwise uniformity of the tangential jet was ascertained by measuring the velocity at a fixed blowing rate over a 70 % of the span in the middle. The jet was two-dimensional and uniform within ± 1 %. A velocity of 7 m/s was chosen as the free-stream velocity for the experiments and was monitored throughout. This gave the test Reynolds number of about 14000 based on the model base height. Full details of the experiments and acquisition of the data are described in Sahoo (1996).

A flow visualization study of the wake structures was also performed using the color dye injection technique on the same model in a water tunnel at the Reynolds number of about 400. The jet was produced by a regulated water supply from an overhead tank.

3. Results and Discussion

Figure 2 illustrates frequency spectra obtained by performing an FFT of the fluctuating component of the hot-wire signal acquired at several transverse locations on either side of the streamwise axis at x/h=4 for various blowing conditions denoted by C_q. The blowing coefficient, C_q, is obtained by normalizing the volumetric discharge rate through the slit by the free-stream velocity and the model base area. The frequency spectra for the case of C_q=0 in Fig. 2(a) show strong and sharp peaks registered at about 54 Hz in the outer wake regions suggesting occurrence of regular vortex shedding at this primary frequency. Further, an attenuation in the peak amplitude at the center (y/h=0) along with the appearance of its harmonic with comparable strength indicates that the vortex shedding is alternate. These results are consistent with the hot-wire traces obtained in a periodic wake, reminiscent of the Karman vortex street, at various

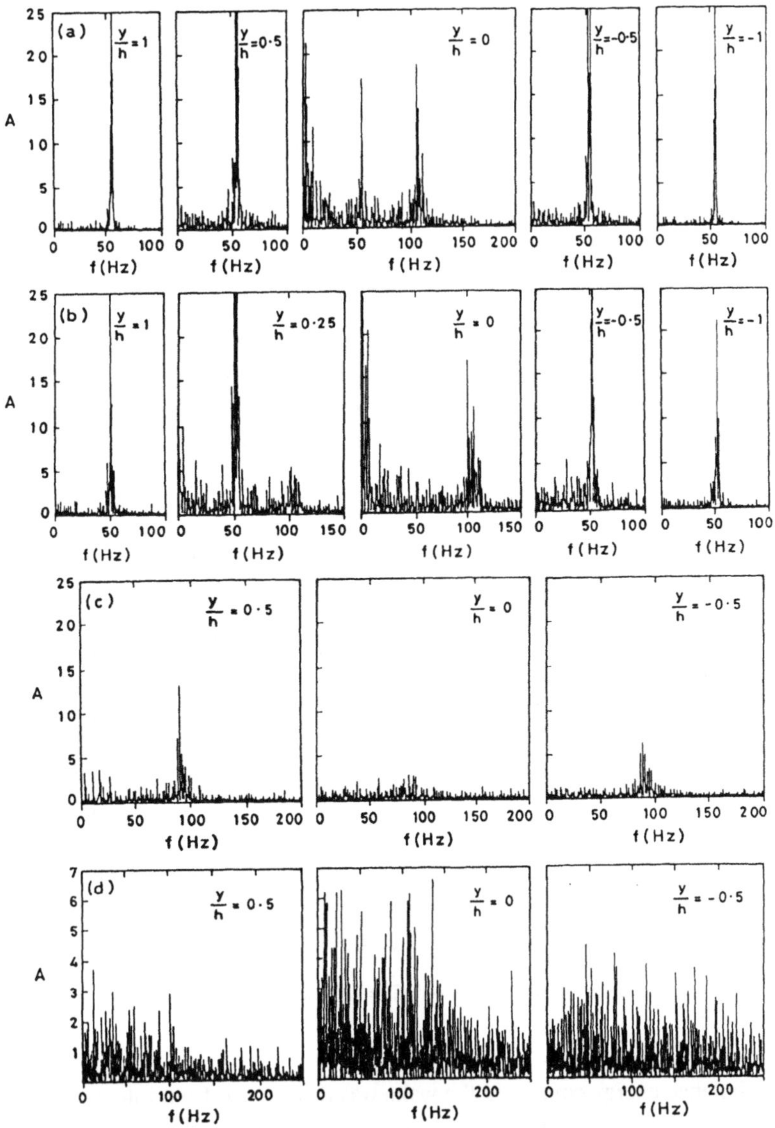

Fig. 2. Frequency spectra at several transverse locations in the wake at x/h = 4.
(a) Without blowing, $C_q = 0.0$, (b) $C_q = 0.058$, (c) $C_q = 0.129$, (d) $C_q = 0.2$.

transverse positions (Matsui, 1982). With blowing at C_q=0.058, Fig. 2(b) suggests that the vortex shedding process remains almost unchanged; however, frequency is reduced marginally to about 50 Hz and the vortex strength on the jet side is also attenuated. When the blowing is increased to C_q=0.129, Fig. 2(c), a remarkable attenuation of the spectral peaks with an increase in the frequency to about 90 Hz is achieved. Moreover, the absence of the distinct peak in the center (y/h=0) at twice the primary frequency suggests effective suppression of the vortex shedding except some weak oscillations in the wake. It is interesting to note that Wood (1964) and Bearman (1967) both have reported suppression of vortex shedding with base bleed in the range of Cq from 0.14 to 0.16. It may be noted that in the present case, because of the thin slit, the blowing of air mass carries higher momentum and therefore, an adequate increase in the blowing is expected to result in a surplus of momentum injection in the wake. Consequently, the frequency spectra in Fig. 2(d) with enlarged scale at C_q=0.2 exhibit the turbulence of the jet having no preferred frequency.

Mean velocity profiles, shown in Fig. 3 for C_q=0.129, reflect a process of entrainment and strong mixing in the near wake due to high velocity tangential jet - a substantial wake deficit is quickly dispersed by the distance of only two base heights. Further downstream, the profiles develop into a typical shape similar to that of a self-propelled body wake profile, reminiscent of the momentumless wake profile (Cimbala and Park, 1990). A rapid approach to symmetry in the wake, despite the initial conditions being strongly asymmetric, is rather surprising and contrary to findings of Park and Cimbala (1991).

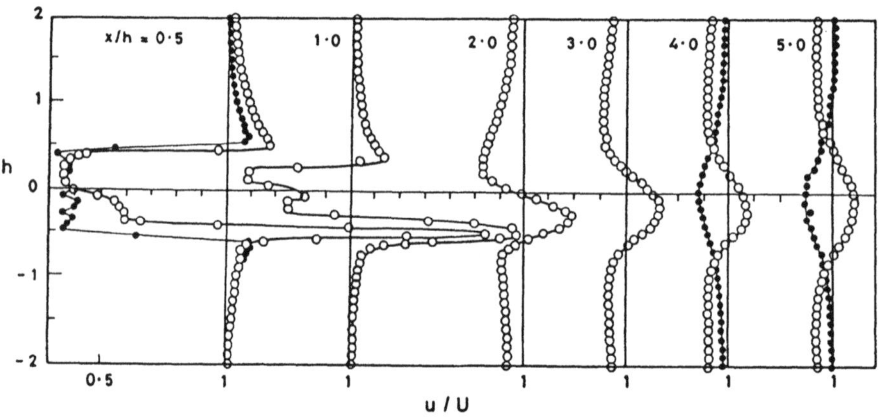

Fig. 3. Mean velocity profiles with C_q = 0.129 and C_q = 0.0 (filled symbols).

The turbulence energy content in the wake was determined from integration of mean square of velocity fluctuation. Figure 4 depicts distribution of the wake turbulence energy content, E, in non-dimensional form. The maximum turbulence energy exists in the wake at about base height for the case of no blowing, C_q=0. The blowing apparently results in the decay of wake turbulence as the peak in E is seen to subside along with an

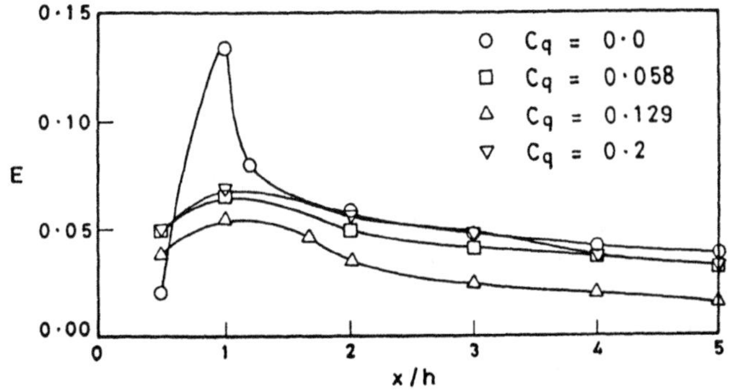

Fig. 4. Normalized turbulence kinetic energy in the wake.

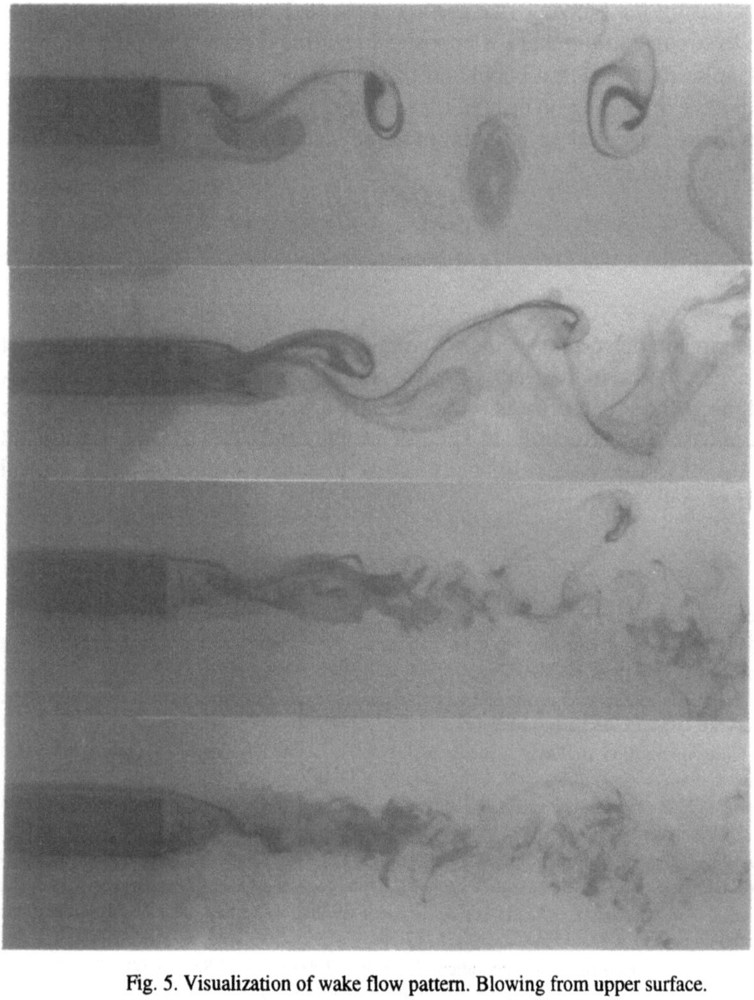

Fig. 5. Visualization of wake flow pattern. Blowing from upper surface.

overall reduction in the energy content. The turbulence energy first reaches its minimum for C_q=0.129 and then reverts with further increase in the blowing to C_q=0.2. The minimum level of turbulence energy is presumably an indication of a condition close to that of a momentumless wake (Cimbala and Park, 1990). It is estimated from their results that Cimbala and Park (1990) achieved such a wake condition at $C_q \approx 0.55$.

Flow visualization pictures illustrating changes in the wake features due to blowing are shown in Figure 5. The upper blue streakline marks the flow from the jet side. At C_q=0.05, the vortices are seen to be elongated with reduced lateral spacing. A tendency for the wake to become three-dimensional with blowing was observed. With an increase in the blowing rate to C_q=0.13, the dye filaments from opposite sides are seen to roll up to form a nearly close-region of about one base height size. The dye filaments were seen to undergo some mild but alternate flapping about this close-region and visible twisting at about two base height, however, no distinct vortex structures were seen further downstream. Increasing the blowing rate further to C_q=0.2 rendered an almost stable near wake. Due to strong entrainment the opposite streakline is seen to be pulled toward the jet and curl up to form a closed bubble on the model base. The jet is seen to graze over the closed bubble producing no discernible structure further downstream. Thus, the effect of a high speed tangential jet is found to aerodynamically change the blunt base into a smoothly tapered shape.

4. Conclusions

The present experiments have proved the technique of tangential blowing into one of the separated shear layers to be very effective in suppressing the periodic vortex shedding. A strong entrainment process renders nearly a steady and less turbulent wake by modifying the base flow structure. In spite of initial conditions bearing significant asymmetry for C_q>0.1, the wake ultimately tends to attain symmetry with typical features of a self-propelled body wake.

5. References

Bearman, P.W. (1967) The effect of base bleed on the flow behind a two-dimensional model with a blunt trailing edge, The Aeronautical Quarterly, **18**, 207-224.

Cimbala, J.M. and Park, W.J. (1990) An experimental investigation of the turbulent structure in a two-dimensional momentumless wake, J. Fluid Mechanics, **213**, 479-509.

Matsui, T. (1982) Flow visualization studies of vortices, in R. Narasimha and S.M. Deshpande (eds.), Surveys in Fluid Mechanics, Indian Academy of Sciences, Bangalore, pp. 145-163.

Park, W.J. and Cimbala, J.M. (1991) The effect of jet injection geometry on two-dimensional momentumless wakes, J. Fluid Mechanics, **224**, 29-47.

Sahoo, R.K. (1996) Plane Base Flow Structures of a Blunt Trailing Edge with Boundary Layer Control, Ph.D. Thesis, Aerospace Engineering Department, I.I.T. Bombay, India.

Schumm, M., Berger, E. And Monkewitz, P.A. (1994) Self-excited oscillations in the wake of two-dimensional bluff bodies and their control, J. Fluid Mechanics, **271**, 17-53.

Wood, C.J. (1964) The effect of base bleed on a periodic wake, J. Royal Aeronautical Society, **68**, 477-482.

Wood, C.J. (1967) Visualization of an incompressible wake with base bleed, J. Fluid Mechanics, **29**, 259-272.

AERODYNAMIC FLOW CONTROL OF BLUFF BODIES USING SYNTHETIC JET ACTUATORS

M. AMITAY[1], D. SMITH[2], V. KIBENS[3], D. PAREKH[1] AND A. GLEZER[1]
[1] *Georgia Institute of Technology, Atlanta, GA 30332*
[2] *University of Wyoming, Laramie, WY 82071-3295*
[3] *The Boeing Corporation, St. Louis, MO 63166*

Modification of mean aerodynamic forces on a 2-D cylinder model and on a 2-D thick airfoil using synthetic jet actuators is investigated experimentally. Surface pressure measurements on both models demonstrate that the control jets effect substantial increases in lift and reductions in drag. The transient mechanisms of the lift force as a result of a step change in the control input is investigated using phase-locked velocity measurements in the wake.

1. Introduction

The interest in separation control for aerodynamic applications on 2- and 3-D airfoils has led to the investigation of the efficiency of some control strategies in the nominally two-dimensional flow around circular cylinders (e.g., Williams et al. [1], Hsiao et al. [2] and Pal & Sinha [3]). This simple geometry is particularly attractive because the base flow has been extensively studied and documented over a broad range of Reynolds numbers. In the earlier experiments of Amitay et al. [4], the forces acting on a circular cylinder were significantly altered by the creation, with synthetic (i.e., zero mass flux) jet actuators (Smith and Glezer [5]), of small, closed, recirculating flow regions near the surface. Depending on the azimuthal location of the actuators, the lift and drag coefficients of the cylinder increased and decreased, respectively. Williams et al. [1], Hsiao et al. [2], and Pal & Sinha [3] reported similar modifications in the surface pressure distribution around a cylinder. In particular, the work of Hsiao et al. [2], and more recently of Bader and Grosche [6] (using internal acoustic excitation) demonstrated that these effects are not limited to a narrow spectral band around an unstable frequency (i.e., dimensionless frequency $f^+ \approx O(1)$) but persist through $f^+ \approx O(100)$. In a related paper, Chang et al. [7] have demonstrated a substantial improvement of the stall characteristics on a NACA 63-018 airfoil using internal acoustic excitation at Strouhal numbers up to 20. In the present work, the strategy of using synthetic jet actuators to modify a separated flow has been extended to an unconventional airfoil section for which flow control is necessary to maintain aerodynamic performance.

G.E.A. Meier and P.R. Viswanath (eds.),
IUTAM Symposium on Mechanics of Passive and Active Flow Control, 273–278.

2. Experimental Apparatus and Procedures

The present experiments are conducted in an open return, low-speed wind tunnel having a square test section measuring 0.91 m on the side, a maximum speed of 32 m/sec and turbulence level less than 0.25%. The cylinder model is machined from aluminum with a diameter of D = 6.22 cm and spans the entire test section. A pair of adjacent plane synthetic jet actuators each having an orifice width of 0.5 mm are mounted along the center section of the cylinder separated by 2.5 mm along the long side (140 mm) of their orifices. The jet orifices are flush with the external surface of the cylinder and colinear with respect to its axis. The performance of the two jets is quantified in terms of the conventional momentum coefficient, C_μ. The center section of the cylinder is instrumented with 47 pressure ports equally spaced around the circumference.

As noted in Section 1, synthetic jet control strategy was extended to an unconventional airfoil by fitting an aerodynamic fairing to the circular cylinder. In this configuration, the cylinder can be rotated about its axis relative to the fairing and the angle of attack of the airfoil, α, can be independently varied between -10° and $+25^\circ$. The fairing is instrumented with pressure taps at midspan along its top and bottom surfaces. In this paper, experimental results are presented for a chord Reynolds number of 310,000.

The transient response of the flow to pulsed excitation is measured across the wakes of the cylinder and the airfoil using phase-locked x-wire anemometry. These data are used to assess the evolution of the vortical structures in the wake, and the corresponding time-dependent lift force on each model. The present paper presents measurements of the streamwise and cross-stream velocity components, taken phase-locked to the modulating pulse, at x/D=3 for the cylinder and x/c=2 for the airfoil. The time rate of change in circulation is computed from the phase-averaged vorticity flux $d\Gamma/dt = \int_{-\infty}^{\infty} U\Omega_z dy$ and thus the incremental lift coefficient of the cylinder (computed relative to the lift at t = 0) is computed from $C_l = -2\Gamma/U_0 D$.

3. Results

3.1 CYLINDER

The variation of the lift coefficient C_l and of the normalized increment in (pressure) drag $\hat{C}_d$ ($\hat{C}_d = C_{d\,forced}/C_{d\,unforced} - 1$) of the cylinder with azimuthal jet position are shown in Figures 1a and 1b, respectively. These data are taken at Re_D = 75,000 for actuation frequencies 740 Hz (St = 2.6) and 1300 Hz (St = 4.5), and at Re_D = 131,000 for 740 Hz (St = 4.5). As is evident from these data, the distributions of C_l and $\hat{C}_d$ are qualitatively independent of the actuation frequency. At both frequencies, $\hat{C}_d$ decreases (i.e., a decrease in drag) while C_l increases with γ. At Re_D = 75,000, the maximum lift coefficient is approximately 0.54, while at Re_D = 131,000, the lift coefficient reaches a maximum of 0.93. For $\gamma > 100^\circ$, the lift reverses direction and $\hat{C}_d$ increases (i.e., an increase in drag). The largest reversed lift coefficient is -0.38 for Re_D = 131,000.

Finally at $\gamma = 135^0$, for both Reynolds numbers and actuator frequencies, the lift vanishes. It is also noted that for the nominal range $60^0 < \gamma < 110^0$, the local peak in the drag is probably associated with the protrusion of the jets into the free stream and an increase in frontal blockage of the cylinder.

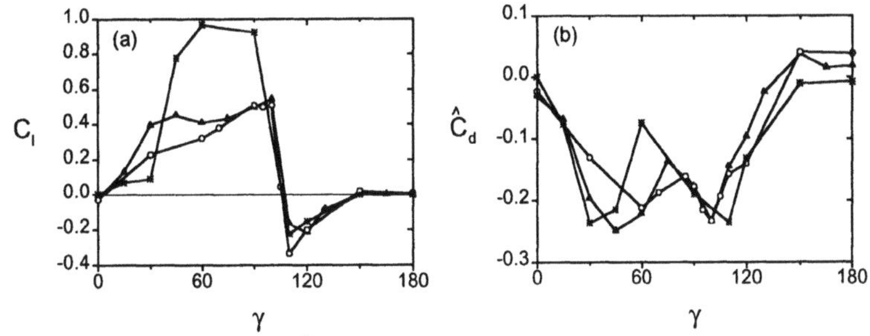

Figure 1. Variations of C_l and $\hat{C}_d$ with jet angle. (—o—) $Re_D = 75{,}500$, St = 2.6, (—▲—), $Re_D = 75{,}500$, St = 4.5 and (—✶—) $Re_D = 131{,}000$, St = 1.49.

Phase-averaged cross stream concentration of the spanwise vorticity is computed from the velocity distributions (not shown) and is shown in Figure 2 using a gray-scale raster plot (the time-averaged cross stream distributions in the presence and absence of the jets are also plotted on the right hand side). When the flow is unforced, the vorticity distribution in the wake is comprised of a train of vortical structures of alternating sense in the upper and lower cross stream halves of the wake (counter-clockwise vorticity as in the top half is taken to be negative). Note that during one period of the (unforced) shedding frequency, the total vorticity across the wake is approximately zero. The velocity measurements (not shown here) indicate that the arrival of the leading and trailing edges of the excitation pulse are associated with strong oscillations in the wake which corresponded to the passage of coherent vortical structures at the natural shedding frequency. Following the application of the pulse, a strong clockwise vortex is advected past the measurement station. This vortex is followed by a series of four counter-rotating vortices. When the wake reaches its final (limit) state, the cross-section of the vortices is substantially smaller (commensurate with the decrease in the cross-stream width of the wake), and the shedding frequency increases from 50 to 80 Hz. Similarly, following the trailing edge of the pulse, a longer train of counter-rotating vortices is advected past the measurement station at the natural shedding frequency. The decrease in lift is accompanied by a net change in the vorticity flux and a decrease in circulation.

The phase-averaged lift coefficient is estimated (not accounting for contributions of the fluctuating components) using the phase averaged cross-stream distributions of the streamwise velocity and spanwise vorticity and is shown in Figure 3. The steady lift coefficient of the unforced flow is nominally zero. When the pulsed excitation is applied, C_l increased to a value of 0.59 before subsequently relaxing, within 20T (T is the time period of the actuation frequency), to the steady value of 0.39. When the jets are turned off, C_l oscillated around a zero mean between 0.6 and –0.38 and within 250T decayed back to zero.

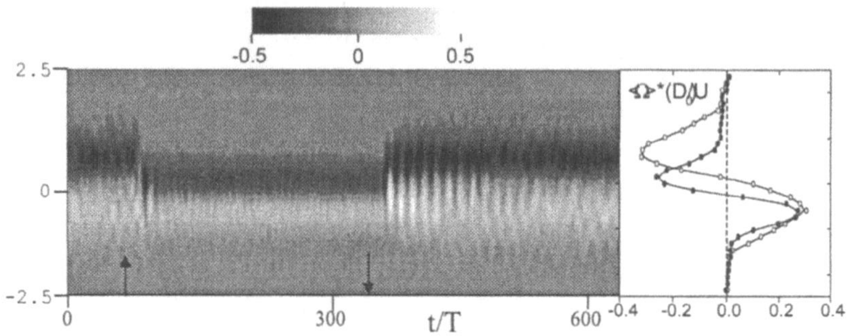

Figure 2. The transient response of the normalized spanwise vorticity, $\gamma = 60^0$.

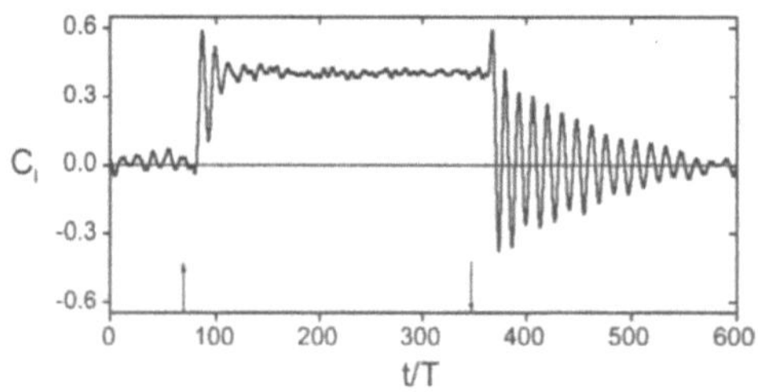

Figure 3. The time response of the phase-averaged lift coefficient, $\gamma = 60^0$.

3.2 AIRFOIL

In the absence of control, the thick airfoil stalls (for $\alpha \geq 5^o$). Figure 4 shows the pressure distribution around the airfoil at $\alpha = 15^o$. The unforced flow separates from the upper surface of the airfoil (open symbols) and is marked by region of constant C_p in this figure. With control, fully reattached flow is obtained for $\alpha \leq 15^o$ and partial reattachment is observed up to the maximum angle of attack tested, $\alpha = 25^o$. In fact, with control, the lift coefficient increases linearly with α as one might expect for a thin airfoil (see figure 5). At all angles of attack, dramatic increases in lift and decreases in pressure drag are observed as a result of the reattached flow. For example at $\alpha = 15^0$, forcing increased the lift by 160% and reduced the pressure drag by 45%.

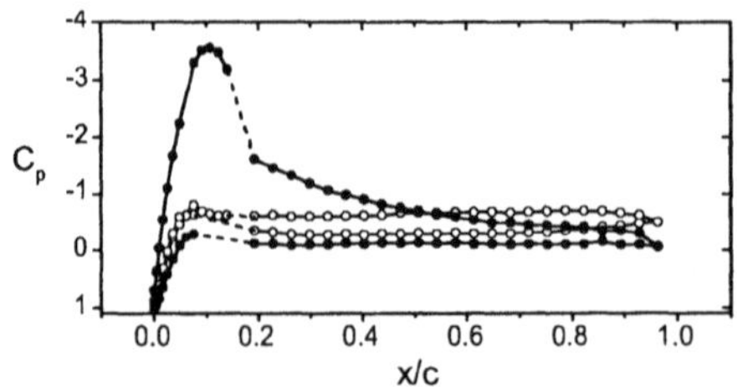

Figure 4. Distributions C_p for $\alpha = 15^o$ and $\gamma = 60^o$ (unforced open symbols).

Using the velocity measurements in the wake, the phase-averaged cross-stream distribution of spanwise vorticity is calculated and is shown in figure 6 using a gray-scale raster plot (the time-averaged cross-stream vorticity distribution for the separated and attached flows are shown on the right hand side of the figure). When the flow is separated, the airfoil sheds vortices at a frequency of 55 Hz. Note also that, without forcing, the time-averaged vorticity flux through a cross-stream plane in the wake is zero. The application of control leads to flow reattachment and to the establishment of a higher (positive) lift on the airfoil which must be accompanied by a change in the vorticity flux and a net increase in circulation associated with negative (clockwise) vorticity. During the reattachment, a strong clockwise vortex, indicating *a reduction in lift*, is initially shed into the wake followed closely by a stronger counter-clockwise vortex indicating the re-establishment of lift. These two large vortices are followed by a series of smaller vortices of alternating signs and diminishing strength (possibly as a result of some loss of phase locking to the modulating pulse).

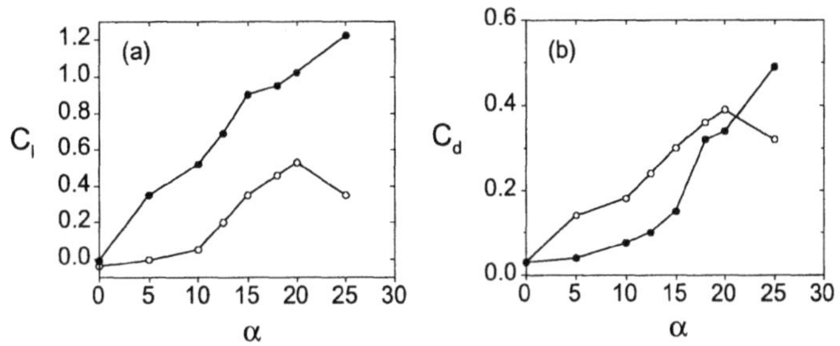

Figure 5. C_l and C_d vs. α for $\gamma = 60°$ (unforced open symbols).

Using the procedure described in Section 2, the increment in the lift coefficient ΔC_l is calculated relative to the separated flow (figure 7). As the flow reattached, ΔC_l initially drops to a value of –1.06 before recovering to a value of 0.61 with the shedding of the starting vortex. It appears that the shedding of the starting vortex causes partial separation at the trailing edge. This separation leads to the shedding of another (weaker) clockwise vortex followed by a train of vortices of alternating signs. The lift coefficient ultimately relaxes to a value, which is in good agreement with the lift coefficient obtained from the pressure measurements.

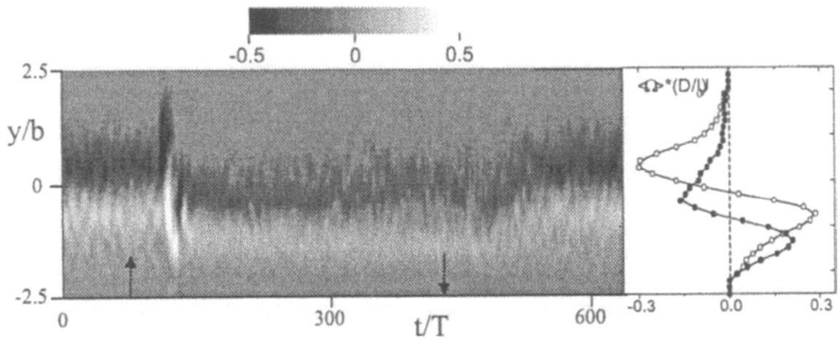

Figure 6. The transient response of the spanwise vorticity, $\alpha = 15^0$ and $\gamma = 60^0$.

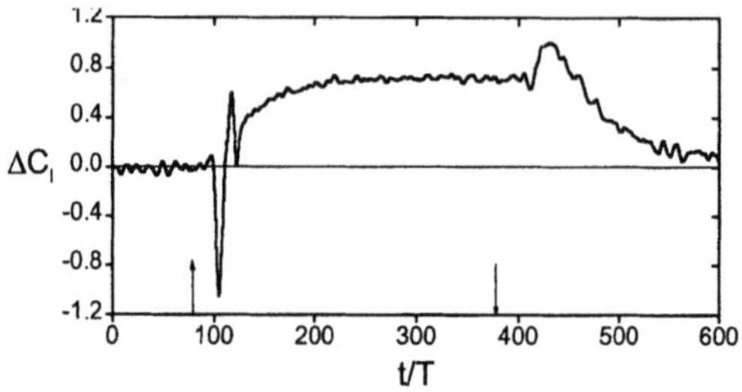

Figure 7. Transient variation of the phase-averaged lift coefficient, $\alpha = 15^0$ and $\gamma = 60^0$.

4. Conclusions

Modification of steady aerodynamic forces on a 2-D cylinder model and on a 2-D thick airfoil using synthetic jet actuators was investigated in wind tunnel experiments. The azimuthal position of the actuators was shown to affect the pressure distributions around the models resulting in an increase in the lift and a decrease in the drag. The transient flow response to the initiation of control was also investigated using a pulsed actuation.

On the cylinder, the response to the rising edge of the pulse, as gauged from the flow oscillations in the wake, was quicker than the response to the falling edge. We believe that this discrepancy can be explained in terms of variable damping. When the control is terminated (falling edge), the oscillations in the flow about the cylinder (and thus in the wake) decay to the unforced state through "natural" damping which comes from wall shear stress and the synchronous motion of the separation regions on the top and bottom surfaces. However, when control is initiated (rising edge), the interaction of the jets with the cross flow and the introduction of small scale motions by the actuators into the boundary layer (which increase the Reynolds stresses) leads to increased damping.

On the airfoil, a short train of large vortices associated with the readjustment of circulation around the airfoil was observed in the wake immediately after control was initiated. The shedding of these vortices produced a strong transient in the lift coefficient: first leading to a large reduction in lift but followed closely by a large increase in lift.

5. References

1. Williams, D. R., Acharya, M., Bernhardt, J. & Yang, P-M "The Mechanism of Flow Control on a Cylinder with the Unsteady Bleed Technique", AIAA Paper, 29th Aerospace Sciences meeting, Reno, NV, 1991.
2. Hsiao, F. B., Liu, C. F. & Shyu, J. Y. "Control of Wall-Separated Flow by internal Acoustic Excitation", AIAA Journal, Vol. 28, No. 8, 1990.
3. Pal, D. & Sinha, K. "Controlling an Unsteady Separating Boundary Layer on a Cylinder with an active Compliant Wall", AIAA Paper 97-0212, 35th Aerospace Sciences meeting, Reno, NV, 1997.
4. Amitay, M., Honohan, A., Trautman, M. & Glezer, A. "Modification of the Aerodynamic Characteristics of Bluff Bodies Using Fluidic Actuators", AIAA Paper 97-2004, 4th AIAA Shear Flow control Conference, Snowmass, CO, 1997.
5. Smith, B. L. & Glezer, A. "Vectoring and Small-Scale Motions Effected in Free Shear Flows Using Synthetic Jet Actuators", AIAA Paper 97-0213, 35th Aerospace Sciences meeting, Reno, NV, 1997.
6. Bader, V. & Grosche, F.-R. "Control of the Leading Edge Separation from an Airfoil by Internal Acoustic Excitation and Self-Induced Resonances," FLOWCON IUTAM Symposium, DLR-Gottingen, Germany, 1998
6. Chang, R. C., Hsiao, F. B. and Shyu, R. N. "Forcing Level Effects of Internal Acoustic Excitation on the Improvement of Airfoil Performance", Journal of Aircraft, Vol. 29, No. 5, 1992.

REDUCTION OF UNSTEADY FORCES ON BLUFF BODIES USING ROTATING CONTROL CYLINDERS

Sanjay Mittal
Department of Aerospace Engineering, IIT Kanpur 208 016, India

1. Introduction

Wind induced forces play an important role in the design of a variety of engineering structures. Control of vortex shedding leads to reduction in the unsteady forces acting on the bluff body and can significantly reduce its vibrations. Strykowski and Sreenivasan [1] have reported that the vortex shedding behind circular cylinders can in some cases be suppressed, over a limited range of Reynolds numbers, by proper placement of a second, much smaller, 'control' cylinder in the near wake of the main cylinder. Our computations using stabilized finite element formulations are in excellent agreement with their observations [2]. Modi et al. [3] have reported the application of the concept of Moving Surface Boundary-layer Control (MSBC) to flows past two-dimensional airfoils and bluff bodies via wind-tunnel testing. In this technique a moving surface, placed in the vicinity of the body of interest, delays the boundary-layer separation by injecting momentum into the already existing boundary-layer and by retarding its growth. Using small rotating cylinders as the moving surfaces they observed that this technique results in a significant reduction in the drag on bluff bodies and their flow induced vibrations. In the present work we report our results for the numerical simulation of the MSBC concept applied to flow past a circular cylinder (of diameter D) using two small rotating cylinders (of diameter $D/20$) as the moving surface.

2. Formulation

Stabilized finite-element method [4] is used to solve the incompressible Navier-Stokes equations in the primitive variables formulation. The SUPG (streamline-upwind/Petrov-Galerkin) and PSPG (pressure-stabilizing/Petrov-Galerkin) stabilization technique is employed to stabilize the computations against spurious numerical oscillations and to enable one to use equal-order-interpolation velocity-pressure elements. This flow involves the interaction of wakes of the main and control cylinders. To resolve all the flow structures adequately, a structured finite-element mesh with 3-noded triangular elements is used around the three cylinders while an unstructured mesh is generated using Delaunay technique in the rest of the domain. A typical mesh used in the computations (55,228 nodes and 108,172 triangular elements) is shown in Figure 1. The large-scale equation systems, resulting from the finite element discretization, are solved iteratively by using the preconditioned $GMRES$ technique.

G.E.A. Meier and P.R. Viswanath (eds.),
IUTAM Symposium on Mechanics of Passive and Active Flow Control, 279–284.

3. Results & Discussion

Computations are carried out for two sets of Reynolds numbers (100 and 10,000), for various spin rates of the rotating control cylinders. The effect of the gap between the main and control cylinders is also investigated. Reynolds number is based on the diameter of the main cylinder (D), free-stream velocity (U)and the viscosity of the fluid while the spin rate is characterized by the ratio of the tip speed of the control cylinder and the free-stream speed ($\alpha = U_c/U$). The flow is from left to right (Figure 1) and the upper control cylinder spins in the clockwise while the lower one in the anti-clockwise direction. The drag and lift coefficients are non-dimensionalized with respect to the free-stream flow speed and the diameter of the corresponding cylinder. Figure 2 shows the solutions for $Re = 100$ and $gap = .01D$ for various rotation rates of the control cylinder. Vortex shedding is completely suppressed for $\alpha = 5$ and is accompanied by a significant reduction in the drag coefficient. As expected, the lift coefficient for the steady-state solution is 0. The wake behind the cylinder also becomes narrower as a result of this flow control. At this low Reynolds number there is no vortex shedding for the control cylinders even when they are not spinning. Figures 3 and 4 show the solutions for $Re = 100$, $gap = 0.075D$ flow. The effect of the rotation rate can be clearly observed. The reduction of drag coefficient due to rotating control cylinders increases as the gap between the control and main cylinders decreases. Shown in Figure 5 are the results for $Re = 10^4$ and $gap = 0.01$. In this case for $\alpha = 0$ vortex shedding can be observed for, both, the control and main cylinders. Interesting interaction between the wakes of the three cylinders can be noticed. For $\alpha = 5$ the vortex shedding for the control cylinders is suppressed completely while that for the main cylinder is significantly reduced. The results for the $Re = 10^4$ and $gap = 0.01$ flow are shown in Figure 6. As observed for the $Re = 100$ flow, better flow control is achieved by a higher rotation rate of the control cylinders and by reducing the gap between the control and main cylinders.

4. Acknowledgments

Partial support for this work came from a research grant by Department of Science and Technology, India via project number DST/AE/95279.

5. References

1. Strykowski, P.J. & Sreenivasan, K.R. (1990) On the formation and suppression of vortex shedding at low Reynolds numbers, *J. Fluid Mech.*, **218**, 71–107.
2. Mittal, S. & Raghuvanshi, A. (1997) Suppression of vortex shedding using control cylinder, *Proc.* 7th *ACFM*, IIT Madras, India, 315–318.
3. Modi, V.J. & Shih, E. & Ying, B. & Yokomizo T. (1992) Drag reduction of bluff bodies through momentum injection, *J. Aircraft*, **29**, 429–436.
4. Tezduyar, T.E. & Mittal, S. & Ray, S.E. & Shih, R. (1992) Incompresible flow computations with stabilized bilinear and linear equal-order-interpolation velocity-pressure elements *Comput. Meth. in App. Mech. & Engg.*, **95**, 221–242, 1992.

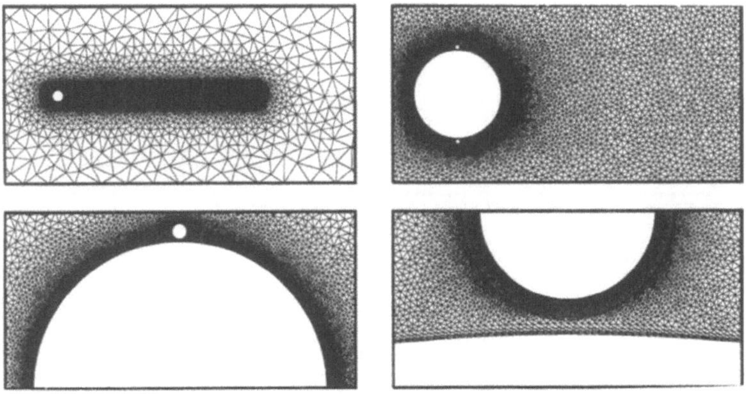

Figure 1. Successive close-ups of a typical finite element mesh with 55, 228 nodes and 108, 172 triangular elements.

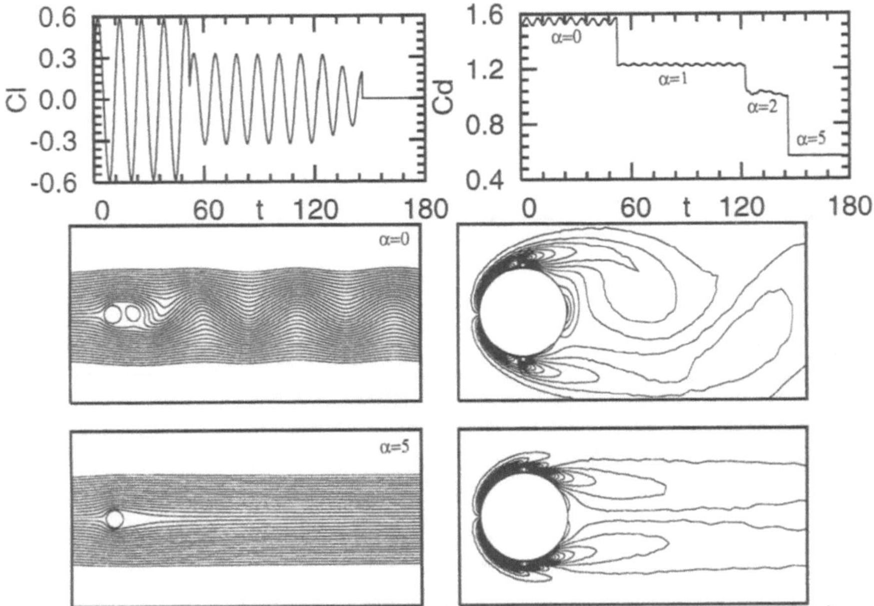

Figure 2. Time histories of the lift and drag coefficients for the main cylinder and stream-function and vorticity fields for $Re = 100$, $gap = .01D$ flow.

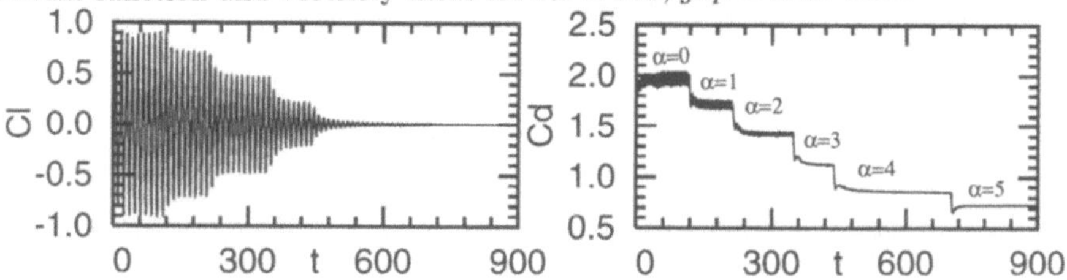

Figure 3. Time histories of the lift and drag coefficients for the main cylinder for $Re = 100$, $gap = .075D$ flow.

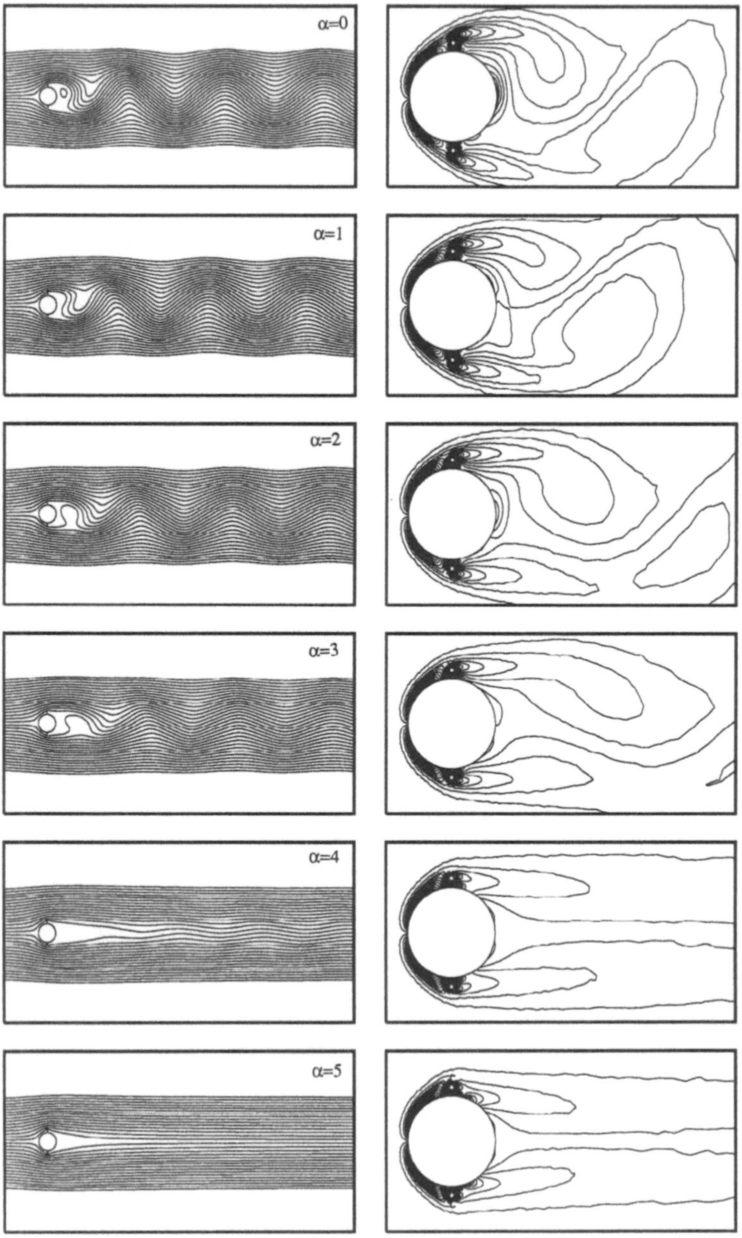

Figure 4. Stream-function and vorticity fields for $Re = 100$, $gap = .075D$ flow.

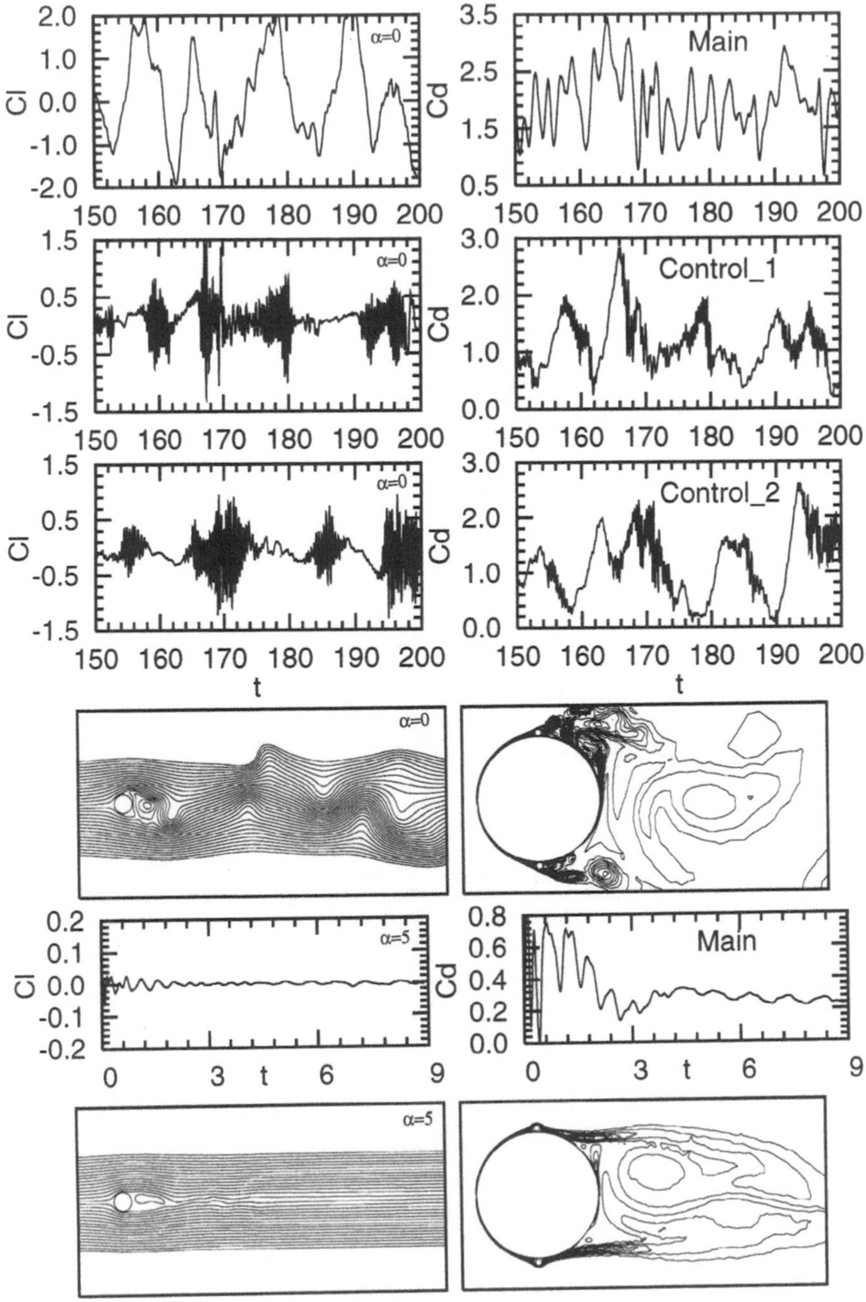

Figure 5. Time histories of the lift and drag coefficients for the main and control cylinders and stream-function and vorticity fields for $Re = 10^4$, $gap = .01D$ flow.

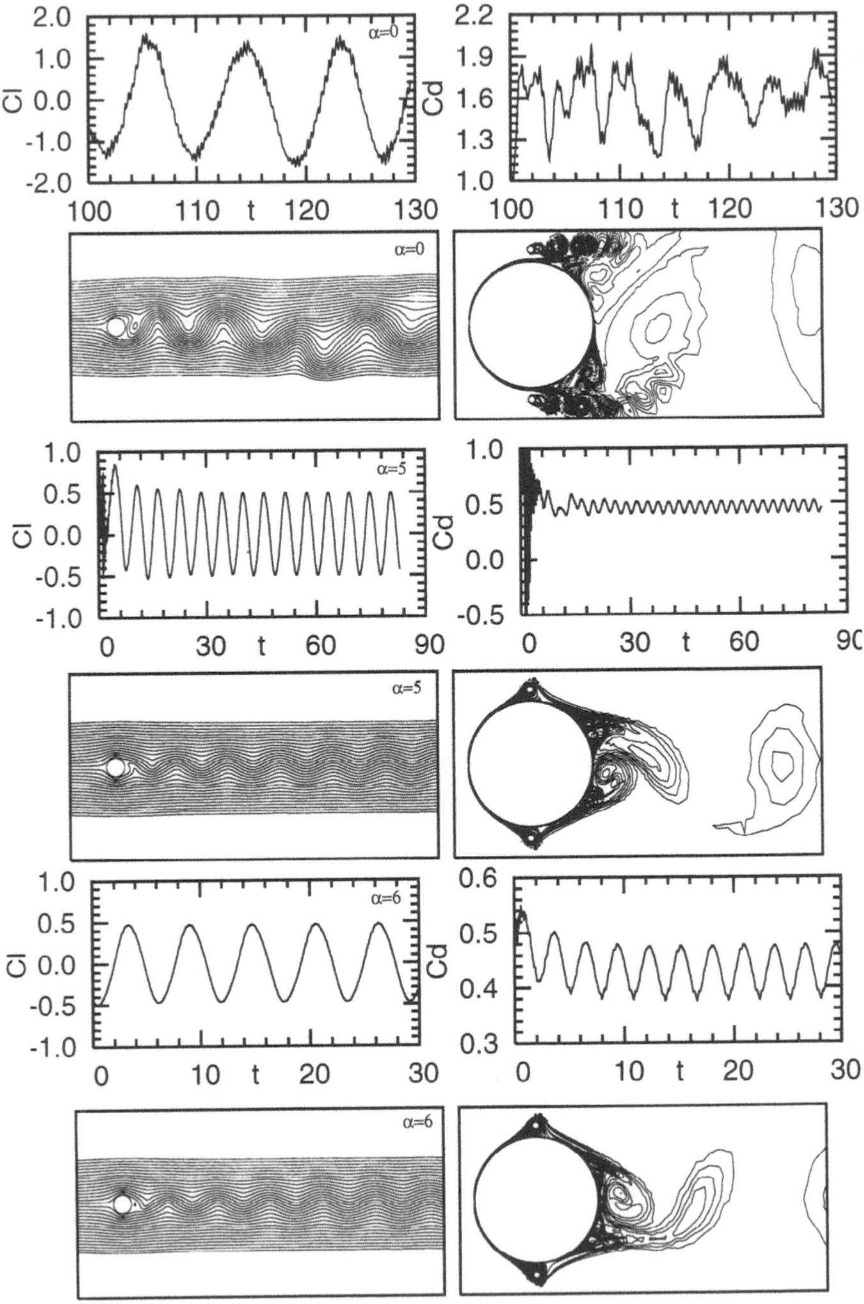

Figure 6. Time histories of the lift and drag coefficients for the main cylinder and stream-function and vorticity fields for $Re = 10^4$, $gap = .075D$ flow.

MOVING SURFACE BOUNDARY-LAYER CONTROL AS APPLIED TO SLENDER AND BLUFF BODIES

V. J. MODI[1], S. R. MUNSHI[2], T. YOKOMIZO[3]

1,2 *Department of Mechanical Engineering*
University of British Columbia
Vancouver, B.C., Canada V6T 1Z4

3 *Department of Mechanical Engineering*
Kanto Gakuin University
Mutsuura, Kanazawa, Yokohama, Japan 236

Abstract

Fluid dynamics and dynamics of two-dimensional airfoils as well as bluff bodies are studied in presence of the Moving Surface Boundary-layer Control (MSBC) using wind tunnel tests, numerical finite element analysis and flow visualization. Results suggest significant increase in the maximum lift coefficient and delay in the stall for airfoils, and decrease in the drag for bluff bodies. The concept proved effective in suppressing vortex resonance and galloping type of instabilities encountered by tall buildings, bridges, smokestacks and similar bluff structures.

1. Introduction

Application of moving wall for boundary-layer control has received relatively less attention (Modi, 1997). Using rotating cylinders as momentum injecting elements, the paper studies their effect on the flow fields associated with two-dimensional slender as well as bluff bodies (Figure 1). This is accomplished through an extensive wind tunnel test-program complemented by numerical analysis and flow visualization (Modi, Munshi, Bandyopadhyay and Yokomizo, 1996; Munshi, Modi and Yokomizo, 1996).

2. Results and Discussion

2.1. AIRFOILS

Figure 2 shows a family of two-dimensional symmetric airfoils, 15% thickness ratio, with different positions of the rotating elements used to inject momentum. Wind tunnel tests were carried out in the subcritical range of the Reynolds number ($R_e = 3\times10^4$ - 5×10^5) with a systematic variation of the angle of attack (α) and momentum injection

G.E.A. Meier and P.R. Viswanath (eds.),
IUTAM Symposium on Mechanics of Passive and Active Flow Control, 285–290.

parameter U_C/U (U_C = cylinder surface speed, U = free stream velocity). Results with airfoils showed an increase in the $C_{L, max}$ by as much as 210 %, and delay in stall up to 48°. Figure 3 summarizes the lift data for various locations of the rotating elements.

2.2. BLUFF BODIES

Next the attention was directed towards bluff geometries where a reduction in the drag is of more importance than the increasing in lift. Results showed a decrease in drag of the flat plate at $\alpha = 90°$ by as much as 75% for $U_C/U = 3$ (Figure 4), while the corresponding drag reduction for the D-section was observed to be ≈ 40% and that for the square prism was around 53% (Figure 5). Furthermore, the wind tunnel data indicated a significant increase in the Strouhal number suggesting narrowing of the wake-width and thus rendering the body effectively slender. It also modified variation of the excitation force with α favorably. These effects were used to advantage in suppressing vortex resonance and galloping type of instabilities (Figure 6), often encountered in industrial aerodynamic and offshore structures.

2.3. NUMERICAL SIMUALTIONS AND FLOW VISUALIZATION

Numerical simulation study was carried out using stream function-vorticity form of the Navier-Stokes equations in conjunction with the variable grid-size finite element analysis. It gave detailed information about the flow including the separation condition and time dependent wake. Finally, flow visualization in a closed circuit water channel, using slit lighting and polyvinyl chloride particles as tracers, showed rather dramatically effectiveness of the MSBC (Figure 7).

3. Concluding Remarks

A rather comprehensive study shows the MSBC to be remarkably effective in increasing the lift of the airfoil, delay in the onset of its stall, and a decrease in the drag of the bluff object. It is also successful in suppressing vortex resonance and galloping form of instabilities. Numerical simulations and flow visualization substantiated wind tunnel results rather well.

4. References

Modi, V. J. (1997) Moving surface boundary-layer control: a review, *Journal of fluids and structures* **11,** 627-663.

Modi, V. J., Munshi, S. R., Bandyopadhyay, G. and Yokomizo, T. (1996) A high performance airfoil with moving surface boundary-layer control, *Proceedings of the AIAA Atmospheric Flight Mechanics Conference,* AIAA Publisher, Washington D.C., pp. 688-698.

Munshi, S. R., Modi, V. J. and Yokomizo, T. (1996) Effect of momentum injection on the drag reduction and flow induced instabilities of a square prism, *International Journal of Offshore and Polar Engineering* **6,** 161-170.

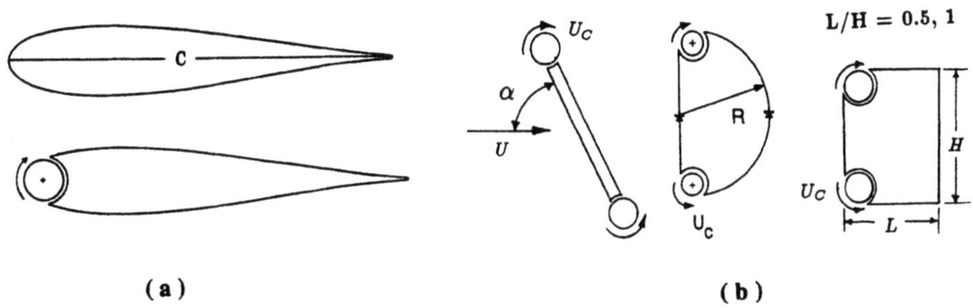

Figure 1. Schematic diagrams of the structural members with the Moving Surface Boundary-layer Control (MSBC) used in the study: (a) slender geometry represented by an airfoil; (b) bluff bodies.

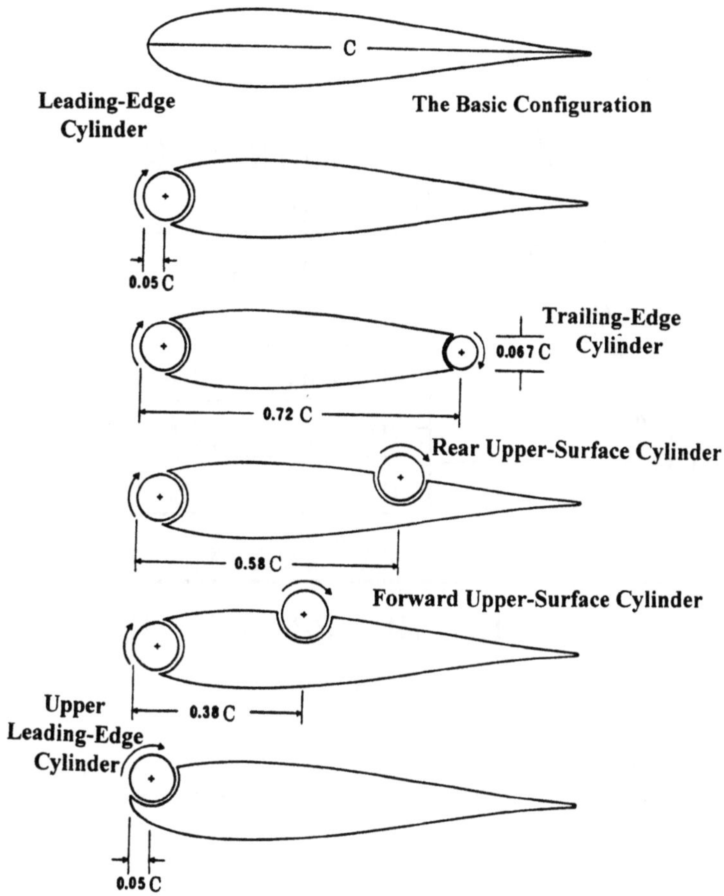

Figure 2. Various momentum injection configurations studied with the Joukowsky airfoil model.

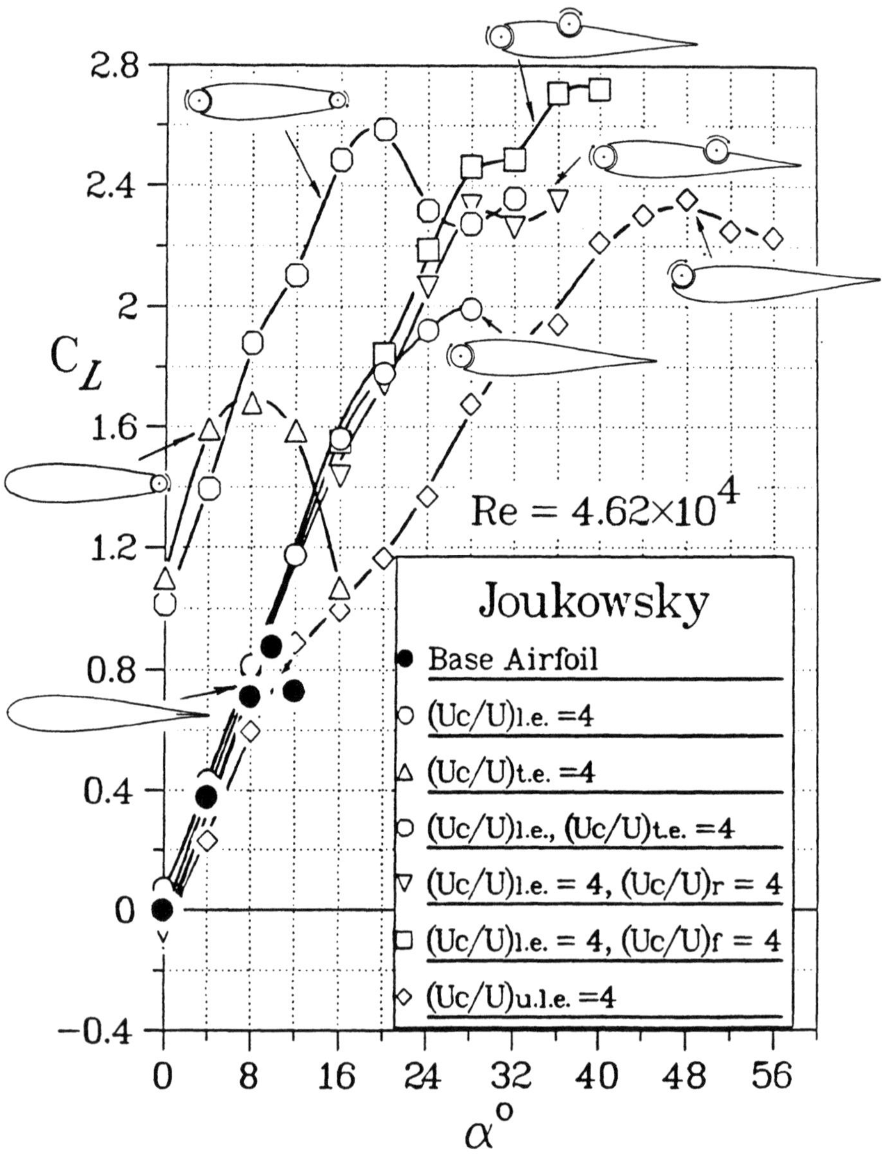

Figure 3. Plots showing effect of various configurations of momentum injection on the lift coefficient of a Joukowsky airfoil (l.e., leading edge; t.e., trailing edge; r, rear; f, forward).

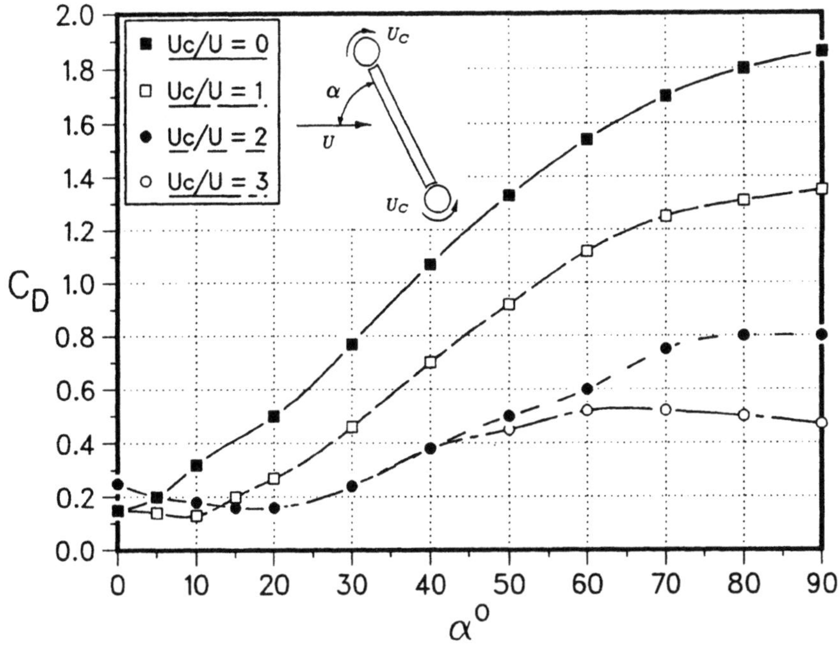

Figure 4. Effect of momentum injection on the drag of a flat plate. C_D, drag coefficient; α = angle of attack.

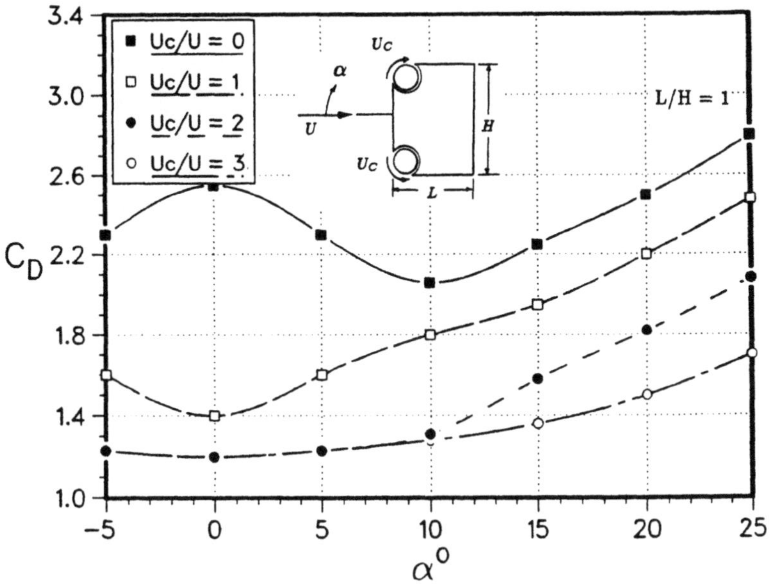

Figure 5. Variation of the drag coefficient with the angle of attack for a square prism as affected by the MSBC.

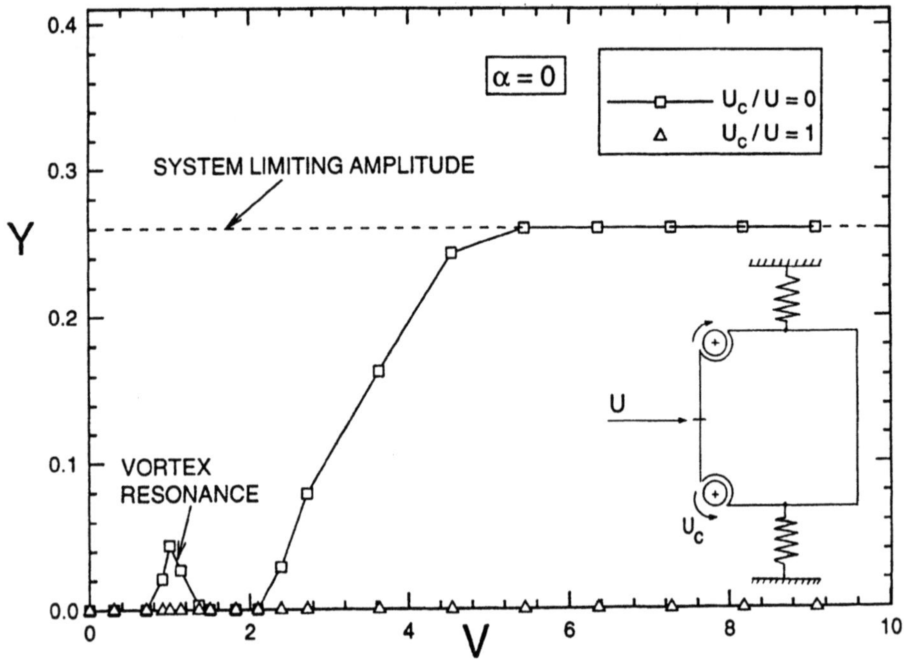

Figure 6. Dynamic response of a square prism without and with symmetric momentum injection. Note, both the vortex resonance and galloping are suppressed due to the MSBC. V, reduced speed; Y, dimensionless amplitude.

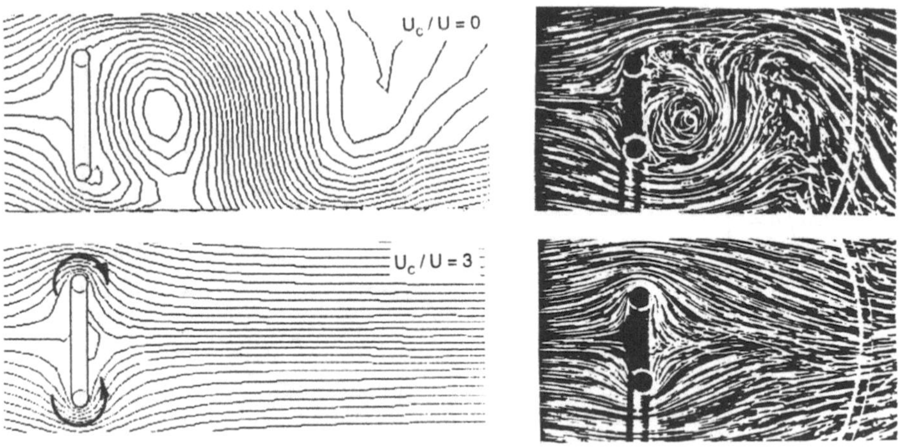

Figure 7. Finite element numerical simulation results and flow visualization photographs for a flat plate at $\alpha = 90^\circ$ showing effectiveness of the MSBC.

FLOW CONTROL BY PERIODIC FORCING: ACOUSTIC EXCITATION AND LOCAL INJECTION OF ARTIFICIAL DISTURBANCES

A.V. DOVGAL
Institute of Theoretical and Applied Mechanics,
Siberian Branch of the Russian Academy of Sciences,
630090 Novosibirsk, Russia

Abstract. Active flow control by excitation of artificial periodic disturbances is in focus. Two particular methods are considered, that is, external acoustic forcing and local injection of the perturbations by oscillatory suction at the body surface. Their application to boundary-layer transition and flow separation is discussed emphasizing basic control mechanisms.

1. Introduction

An approach to active flow control consists in manipulation of mean and nonstationary flow characteristics by some sort of artificially induced oscillations. Up to now, various control techniques were examined on that way, in particular, external acoustic excitation and local injection of periodic disturbances. Local forcing, in its turn, appeared in different forms, that is, oscillatory suction, periodic surface heating, vibrations, moving flaps. As a result, the above techniques are found as effective means of flow control in a number of aerodynamic problems. The present paper focuses application of two methods at subsonic flow velocities: external acoustics and periodic suction-blowing through slots, small holes or arrays of holes (some authors use the term "internal sound"). For other oscillatory control techniques see references in reviewing papers of Gad-el-Hak (1990), Gad-el-Hak & Bushnell (1991) and Fernholz (1993).

Published research data indicate a variety of fields where the methods can be successfully used including laminar-turbulent transition in boundary layers, flow separation, dynamics of vortices in free shear layers and bluff-body wakes, structure of turbulent flows. Of course, all these topics interfere with each other as far as flow control is concerned with. That is, controlling the transition one affects flow separation if it occurs, separation control often has an influence upon dynamics of vortices in the wake, and so on. Keeping this in mind, two particular problems are considered in what follows, that is, control of boundary layer transition and flow separation. First of all, basic aspects of application of the above control methods are in focus, mainly, to fix ideas on

G.E.A. Meier and P.R. Viswanath (eds.),
IUTAM Symposium on Mechanics of Passive and Active Flow Control, 291–298.

how do the methods work. Beyond the scope is oscillatory control just aimed at modification of vortex motion and turbulent structures. A great deal of research results on these points for free shear flows can be found in reviewing papers by Ho & Huerre (1984), Hussain & Husain (1988) and references therein. Acoustic control of regular vortices in bluff-body wakes (past cylinders) was explored by Blevins (1985) and Peterka & Richardson (1969). Results of Gu et al. (1993) and Gad-el-Hak & Blackwelder (1987) show how periodic suction-blowing modifies vortices occurring on delta wings.

The first point to be noticed when the oscillatory control methods are talking about is possibility to get a certain benefit of mean characteristics of the flow or to affect considerably its unsteady component by excitation of very small amplitudes. For example, Collins (1981) reported 2800 times difference between power expenditure for acoustic control of the leading-edge separation on an airfoil and aerodynamic savings which were obtained. Obviously, to get such a control effect the excitation should couple with flow instabilities and energize amplifying disturbances. Then, stimulation of an instability depends on excitation parameters. For a single-frequency forcing used in most of studies they are frequency, phase and amplitude of the oscillations which can be more or less crucial, it depends. Also important is where the flow to be excited, first of all, when it is perturbed locally through narrow slots or tiny holes. Reasonable choice of the control parameters is gained by available research data on the oscillatory control and its mechanisms as they could be realized.

2. Boundary-layer transition

In general, the control of laminar-turbulent transition in boundary layers may be aimed both to its acceleration or delay. However, the latter is much more interesting in most of aerodynamic problems and is usually implied when the transition control is spoken about. A number of research results indicates that external acoustics and periodic suction-blowing, as well as other oscillatory techniques, can be used to postpone the transition. Application of the methods to the transition control is based on the "waves cancellation" phenomenon, that is, suppression of a dominant Fourier component of the laminar flow disturbances by perturbations of the same frequency artificially excited in the boundary layer via receptivity mechanisms.

Experimentally this route of the transition control was modelled for different combinations of the boundary-layer excitation techniques. Gedney (1983) examined the flat-plate transition under external acoustic forcing and surface vibrations. It was shown that at appropriate phases and amplitudes of the oscillations the transition Reynolds number increased to a higher (about 1.5 times) value than it was when forcing the flow by acoustics or vibrations separately. Gilev & Kozlov (1985) initiated the flat-plate transition by a vibrating ribbon and controlled the disturbances by oscillatory suction through a transverse slot or array of small holes at the surface of the model. Similarly, they were able to reduce the boundary-layer disturbances and to delay the transition adjusting the phase difference between oscillations close to 180 degrees. Another possibility of the boundary-layer transition control was shown by Kosorygin & Polyakov (1990). They used a combination of external acoustics and a 2-D roughness element placed at the flat-plate surface. The excitation produced boundary-layer waves which were generated both at the roughness and in the upstream part of the flow. At appropriate streamwise position of the roughness cancellation of the disturbances was observed so that the amplitude of oscillations behind the roughness element became smaller than that in the flat-plate flow. Then, in (Kosorygin et al., 1995) similar result was obtained in another wind tunnel with recommendations on roughness dimensions and acoustic intensity to be used.

The technique of waves cancellation is still effective being applied not only to primary linear instability but to nonlinear and secondary boundary-layer disturbances, as well. The transition control in the plane channel flow by periodic suction-blowing was investigated numerically by Biringen (1984) for temporal problem and, later, by Danabasoglu et al. (1991) in spatial formulation. Navier-Stokes solutions obtained in those studies indicated that the control method enables to reduce the laminar flow disturbances: linear Tollmien-Schlichting waves, 2-D finite-amplitude disturbances and even 3-D nonlinear perturbations. An evidence that oscillatory suction can be used for control of secondary instabilities in a boundary layer was obtained recently in wind-tunnel tests by Boiko et al. (1998). They modelled stationary streamwise vortices of a swept wing flow unstable to secondary travelling waves. The latter were artificially excited by external acoustics, controlling oscillations were those introduced through a small hole on the surface underneath one of the vortices. Injection of the perturbations locally in spanwise direction made it possible to control just the primary vortex above the hole. Amplitude of the secondary waves sitting on that vortex was found to decrease provided that the phase of generated disturbances was properly chosen.

3. Separation bubbles, 2-D problem

While at the transition control discussed in the previous section external acoustics and oscillatory suction are aimed to diminish amplitudes of primary or, sometimes, secondary instabilities, separated flows are controlled by the methods through stimulation of flow disturbances. As perturbations increase the momentum transfer in the separated layer gets higher and the flow tends to reattach to the body surface. This makes a sense when improvement of flow characteristics needs reduction of a separation region or its total elimination.

A flow laminar separating and reattaching further downstream is, usually, unstable with disturbances amplifying behind the separation point. A result of the instability is the transition to turbulence occurring in the separation region or near by. Such flows which appear, for example, on airfoils at low angles of attack or small surface variations are known as "laminar" or "transitional" separation bubbles. As far as formation of a separation bubble is dominated by the transition process one can reduce the separated flow region stimulating the disturbances responsible for laminar flow breakdown. In experiments of Dovgal & Kozlov (1983) and Zaman & McKinzie (1991) it was done by external acoustics for mid-chord separation on airfoils. The studies showed that the bubbles got smaller at rather week (up to 90 - 100 dB) periodic forcing matching instability frequencies of the separating layer. Under the excitation disturbances growing in the separation region were observed which, according to Dovgal & Kozlov (1983), proved to be small-amplitude instability waves of the separated layer. Actually, in such case of separation control the excitation works as other environmental perturbations (for example, free-stream turbulence or surface roughness) which promote the shear-layer transition increasing initial amplitudes of the laminar flow disturbances.

Under certain conditions of separation (normally, at sharp edges, large steps, etc.) the unstable flow in separation bubbles behaves in quite different way. The transition to small-scale turbulence is shadowed by another feature of unsteady motion: origination of fairly well organized vortices growing in the separation region, interacting with each other and shedding from the bubble. In such cases separation bubbles can be controlled affecting the dynamics of quasi-periodic vortices. Then, the mean flow pattern does not tie in with the transition process and the control is applicable both for laminar and turbulent separating layers. Experimental data on this point were obtained by Bhattacharjee et al. (1986) for acoustic excitation of a back-step flow; Sigurdson & Roshko (1988), Kiya et al. (1991), Kiya et al. (1993) investigating separation bubbles on blunt circular cylinders under periodic blowing through narrow slots at the leading edge; Montividas et al. (1992) who examined a separation bubble behind a flap on a flat plate controlled by oscillatory jet at the surface. It comes out that the excitation in some frequency band reduces the separated flow region enhancing roll-up and amalgamation of the vortices. A vortex model explaining shortening of the separation bubble under the forcing by modification of vortex merging is proposed by Kiya et al. (1993). They found, also, that control efficiency can be increased exciting the flow at two frequencies: the most effective one and its harmonics in a proper phase of oscillations.

Realistic distinction between the above routes of separation control needs, at least, pretty different time scales of the shear-layer instability which provokes laminar-turbulent transition and of regular vortices rolling up in the separation region. However, it is not necessarily true for locally separated flows. The difference is more evident in flows which separate globally without reattaching back to the body surface.

4. Global separation, 2-D problem

A salient aerodynamic problem focused by many authors is control of separation on airfoils placed at high angles of attack. A lot of research data have been obtained on this point, mainly for low chord Reynolds numbers $Re_c = U_\infty c/\nu$ on the order of 10^4 to 10^5. The results show that the control methods considered here enable to reduce the region of globally separated flow and, thus, improve aerodynamic characteristics of an airfoil.

Collins & Zelenevitz (1975) excited the flow separating at the leading edge of an airfoil at high angles of attack by external sound and concluded that the flow could be reattached in some frequency range. "Optimum" frequencies were observed with the most profound control result. Similar effect of the external acoustics was found in a series of studies which were performed afterwards by Ahuja & Burrin (1984), Kozlov (1985), Marchman III et al. (1987), Moss (1979), Zaman et al. (1987), Zanin et al. (1992), Lushin (1992). In the last two studies the method was applied to separated flows both on unswept and swept wings. Results of those experiments made clear a mechanism of the acoustic control which consists in stimulation of laminar flow disturbances by coupling of the imposed oscillations with instability waves of the separated shear layer. That was the idea argued by Collins & Zelenevitz (1975), Moss (1979), Kozlov (1985), Zaman et al. (1987), Zanin et al. (1992). Zhigulev & Fedorov (1990) supported this point of view studying the leading-edge separation at Re $\sim 10^6$ under acoustic (ultrasonic) excitation and comparing measured data with results of linear stability calculations for the boundary layer at separation.

One expects that the above mechanism is meaningful when the separation can be controlled by rather week forcing, normally, at angles of attack close to the critical one. As incidence of the airfoil increases a stronger excitation is needed to get a pronounced aerodynamic effect and, then, the separation is controlled beyond the stimulation of linear shear layer instability. Nishioka et al. (1990) investigated the leading-edge separation affected by high-amplitude acoustic forcing. They observed reduction of the separated flow region caused by regular vortices generated close to the leading edge. It was noticed that the most effective frequency of excitation was lower than that of the linear separated-layer instability and the control result was explained by influence of the acoustics upon global dynamics of the vortex flow. This correlates to the data of Zaman (1992) who studied external acoustic effect on poststalled flows varying amplitude of the excitation. A finding was that optimum control frequencies turned lower as the intensity of excitation increased. At rather strong forcing they became too distant (an order of magnitude) from those of the linear separated-layer instability so that it could not be involved directly in the interaction between acoustics and the separated flow. The above observation let Zaman (1992) to explain by nonlinearity of the generated disturbances large difference in optimum frequencies which one can find comparing published data on separation control by external acoustic and local excitation techniques.

Periodic suction-blowing through slots and holes at the surface was tried for control of the separation on airfoils by Collins (1981), Huang et al. (1987), Hsiao et al. (1990), Chang et al. (1992), Hsiao et al. (1994), Dovgal (1993). The fact is that optimum frequencies reported by most of authors for external acoustics are in the range of St = 10 to 100 (for $St = fc/U_\infty$) but when the flow is excited locally they are on the order of unity. Low frequencies at St $\sim$ 1 are comparable with those of the vortices shedding into the wake of an airfoil. Thus, one expects that periodic disturbances injected locally into the separated flow affect the separation by energizing the large-scale vortex motion associated to global flow properties. Noteworthy is that the most effective frequencies observed at separation control by periodic suction-blowing are somewhat different. Those found in the studies of Huang et al. (1987), Hsiao et al. (1990), Chang et al. (1992), Dovgal (1993) are several times higher than the shedding frequency. On the other hand, Hsiao et al. (1994) observed the best result of control just matching the vortex shedding, the same was concluded by Huang et al. (1988) who controlled the trailing-edge separation on an airfoil. Deeper insight into processes of roll-up and interactions of the vortices initiated by separation may help to settle this point.

One more version of the mechanism by which the leading-edge separation is influenced by periodic suction-blowing at low levels of the excitation is given by Dovgal (1993). In that work the disturbances at St ~ 1 introduced into the flow were considered as "slow" modulation of the separated layer which stimulated the transition behind the point of separation. In such interpretation local excitation works similarly to small-amplitude external acoustics. The difference is that the transition is accelerated not by excitation of the instability waves but modification of the separated-layer stability characteristics.

Efficiency of the control by suction-blowing strongly depends on streamwise position of the slot (or holes). Exciting the leading-edge separation on airfoils through transverse slots Hsiao et al. (1990) and Dovgal & Grosche (1994) concluded that the flow was easier to control forcing it close to the separation point. The same was observed by Huang et al. (1988) for the trailing edge separation. Also they found that excitation before separation was more effective than behind the separation point.

Comparing external acoustics and oscillatory suction one can find that the latter technique needs smaller amplitudes of controlling oscillations. That was concluded by Collins (1981), Hsiao et al. (1990), Dovgal (1993) as sound pressure level was the measure of excitation intensity. In this respect, at alternative use of the methods the periodic suction-blowing is preferable.

5. Separated flows, 3-D aspects

Previous discussion was focused on research results obtained for nominally 2-D separation. However, three-dimensionality of separated flows can not be ignored under certain conditions. The mean flow may have a complex spatial structure even in high aspect ratio configurations as was observed in different geometries by Jaroch (1987), Papadopoulos et al. (1993), Winkelmann & Barlow (1980). Thus, further details on separation control by external acoustics and oscillatory suction are given by the studies taking into account spanwise flow variations.

External acoustic influence upon flow separation on a finite-span wing with emphasis on 3-D effects was examined in experiments of Lushin (1992). Gradually increasing sound pressure he found that reattachment did not occur instantly all over the span but progressively moved from tips of the model to its centreplane until the flow completely reattached at a high enough level of the excitation. Also, rearrangement of large-scale vortices spaced along the span was observed under poststall conditions when acoustics produced partial flow reattachment. One more evidence that 3-D phenomena are much important at separation control on finite-span wings is given by Zanin (1997). Exciting the high angle-of attack flow by external sound he observed fairly complex flow patterns associated to leading-edge separation bubbles appearing along the span and amalgamating at variation of the oscillations amplitude.

Control of the leading-edge separation on a finite-span wing by external sound focused to a beam by an acoustic mirror was dealt with in experiments of Kozlov et al. (1993). The flow pattern close to the surface of the model strongly depended on location of the sound spot at the wing. More effective was excitation fixed at the centreplane switching the flow to smaller region of turbulent separation further downstream. Otherwise, partial reattachment could be obtained on the forced side of the model. Dovgal & Grosche (1994) excited the leading-edge separation locally through transverse slots of different length at the wing surface. First, for a low aspect-ratio wing with end plates they found that global separation all over the model could be canceled forcing the flow in small regions much narrower than the span. The data indicated, again, that the most profound effect was produced by the disturbances generated in the central part of the wing.

The above results definitely show that 3-D effects may be of great importance at flow separation control. It appears that further development of the excitation techniques towards their optimization needs deeper insight into interaction between controlling disturbances with 3-D instability modes of separated flows which is far from being understood today.

6. Conclusion

External acoustic excitation and oscillatory suction proved to be valuable active control methods which enable to get beneficial effects in a number of aerodynamic problems. Flow control by the methods have been tackled by many authors. As a result, reasonable interpretation of the control mechanisms can be drawn from research data which are available today. Together with empirical correlations this makes it possible to predict fairly well optimum parameters of the excitation. Up until now, most of results on the topic have been obtained in the 2-D formulation usefully simplifying the problem. However, it does not take into account 3-D flow phenomena which are really meaningful for results and efficiency of the excitation in some cases. It seems, that further progress in fundamentals of flow control by the methods is largely related to that point.

Acknowledgment. Preparation of the paper was supported by the Russian Basic Sciences Foundation, grant no. 97-01-00821.

References

Ahuja, K.K. and Burrin, R.H. 1984 *AIAA-84-2298.*
Bhattacharjee, S., Scheelke, B., and Troutt, T.R. 1986 *AIAA J.* 24, 623-629.
Biringen, S. 1984 *Phys. Fluids* 27, 1345-1347.
Blevins, R.D. 1985 *J. Fluid Mech.* 161, 217-237.
Boiko, A.V., Kozlov, V.V., Syzrantsev, V.V., and Scherbakov, V.A. 1998 *Thermophysics and Aeromechanics* (Russia), to appear.
Chang, R.C., Hsiao, F.-B., and Shyu, R.-N. 1992 *J. Aircraft* 29, 823-829.
Collins, F.G. 1981 *AIAA J.* 19, 129-130.
Collins, F.G. and Zelenevitz, J. 1975 *AIAA J.* 13, 408-410.
Danabasoglu, G., Biringen, S., and Streett, C.L. 1991 *Phys. Fluids A* 3, 2138-2147.

Dovgal, A. 1993 *DLR-FB 93-16.*
Dovgal, A.V. and Grosche, F.-R. 1994 *DLR-IB 223-94 A 20.*
Dovgal, A.V. and Kozlov, V.V. 1983 *Fluid Dynamics* **18**, 205-209 (transl. from Russian).
Fernholz, H.H. 1993 *ZAMM* 73 (11), 287-300.
Gad-el-Hak, M. 1990 *AIAA J.* **28**, 1537-1552.
Gad-el-Hak, M. and Blackwelder, R.F. 1987 *AIAA J.* **25**, 1042-1049.
Gad-el-Hak, M. and Bushnell, D. 1991 *AIAA-91-0037.*
Gedney, Ch.J. 1983 *Phys. Fluids* **26**, 1158-1160.
Gilev, V.M. and Kozlov, V.V. 1985 *Preprint Inst. Theor. and Appl. Mech., Russian Acad. Sci. 1-85* (in Russian).
Gu, W., Robinson, O., and Rockwell, D. 1993 *AIAA J.* **31**, 1177-1186.
Ho, C.-M., and Huerre, P. 1984 *Ann. Rev. Fluid Mech.* **16**, 365-424.
Hsiao, F.-B., Liu, C.-F., and Shyu, J.-Y. 1990 *AIAA J.* **28**, 1440-1446.
Hsiao, F.-B., Shyu, R.-N., and Chang, R.C. 1994 *AIAA J.* **32**, 655-657.
Huang, L.S., Bryant, T.D., and Maestrello, L. 1988 *Proc. of AIAA / ASME / SIAM / APS 1st Nation. Fluid Dynam. Congr.*, 25-28 July 1988, Cincinnati, Ohio, 715-722.
Huang, L.S., Maestrello, L., and Bryant, T.D. 1987 *AIAA- 87-1261.*
Hussain, A.K.M.F. and Husain, H.S. 1988 *Turbulence Management and Relaminarization* (eds. H.W. Liepmann and R. Narasimha), Springer-Verlag, Berlin, 445-457.
Jaroch, M. 1987 *ZFW* **11**, 230-236.
Kiya, M., Mochizuki, O., Tanaka H., and Tsukasaki, T. 1991 *Separated Flows and Jets* (eds. V.V. Kozlov and A.V. Dovgal), Springer-Verlag, Berlin, 647-656.
Kiya, M., Shimizu, M., Mochizuki, O., Ido, Y., and Tezuka, H. 1993 *AIAA-93-3245.*
Kosorygin, V.S. and Polyakov, N.F. 1990 *Preprint Inst. Theor. and Appl. Mech., Russian Acad. Sci. 11-90* (in Russian).
Kosorygin, V.S., Radeztsky Jr., R.H., and Saric, W.S. 1995 *Laminar-Turbulent Transition* (ed. R. Kobayashi), Springer-Verlag, Berlin, 517-524.
Kozlov, V.V. 1985 *J. Appl. Mech. Tech. Phys.* (USSR) **26**, 254-257.
Kozlov, V.V., Grosche, F.-R., Dovgal, A.V., Bippes, H., Kuhn, A., and Stiewitt, H. 1993 *DLR-IB 222-93 A 10.*
Lushin, V.N. 1992 *Sib. Phisiko-Tehnic. Zh.* **4**, 64-68 (in Russian).
Marchman III, J.F., Sumantran, V., and Schaefer, C.G. 1987 *AIAA J.* **25**, 50-51.
Montividas, R.E., Acharya, M., and Metwally, M.H. 1992 *AIAA J.* **30**, 1133-1134.
Moss, N.J. 1979 *J. Sound and Vibration* **65**, 505-520.
Nishioka, M., Asai, M., and Yoshida, S. 1990 *AIAA J.* **28**, 1909-1915.
Papadopoulos, G., Otugen, M.V., and Vradis, G.C. 1993 *AIAA-93-0213.*
Peterka, J.A., and Richardson, P.D. 1969 *J. Fluid Mech.* **37**, 265-287.
Sigurdson, L.W. and Roshko, A. 1988 *Turbulence Management and Relaminarization* (eds. H.W. Liepmann and R. Narasimha), Springer-Verlag, Berlin, 497-514.
Winkelmann, A.E., and Barlow, J.B. 1980 *AIAA J.* **18**, 1006-1008.
Zaman, K.B.M.Q. 1992 *AIAA J.* **30**, 1492-1499.
Zaman, K.B.M.Q., Bar-Sever, A., and Mangalam, S.M. 1987 *J. Fluid Mech.* **182**, 127-148.
Zaman, K.B.M.Q. and McKinzie, D.J. 1991 *AIAA J.* **29**, 1075-1083.
Zanin, B.Yu. 1997 *Prikl. Mekh. i Tehn. Phisika* **38**(5), 80-84 (in Russian).
Zanin, B.Yu., Kozlov, V.V., and Lushin, V.N. 1992 *Sib. Phisiko-Tehnic. Zh.* **3**, 32-36 (in Russian).
Zhigulev, S.V. and Fedorov, A.V. 1990 *Uchenye Zapiski TSAGI* **21**(6), 58-66 (in Russian).

Control of leading-edge separation from an airfoil by internal acoustic excitation and self-induced resonances

Viktor Bader, Friedrich-Reinhard Grosche
DLR Institut für Strömungsmechanik
Bunsenstraße 10
37073 Göttingen, Germany

Introduction:

A significant problem of the operation of an airfoil is the leading-edge separation occurring at high angles of attack which results in formation of a large-scale separation region, so that the lift cannot increase further with the angle of attack. It is a task to find control mechanisms which are as effective as possible to attain a higher critical angle of incidence.
This paper demonstrates two methods of controlling the leading-edge separation: First, according to previous studies [1] internal acoustic excitation using horndrivers to create pressure pulsations in a cavity underneath a slot in the surface of the airfoil near the leading edge. The optimum frequencies to excite the flow were investigated. Similar investigations on leading edge separation have been reported by [6-8]. The second method shows the possibility of attaining an attached flow by utilizing self-excited resonance-oscillations of the slot-flow without any kind of external energy source. Such self-induced resonances could be obtained with suitable configurations of the slot-cavity system. For both methods the separated flow was excited by periodic suction and blowing through narrow slots in the surface near the leading edge.

Experimental Set-Up:

A sketch of the test set-up is shown in fig.1a. The NACA 0012-airfoil is mounted between two end-plates to obtain optimal two-dimensional flow conditions. The chord length of the airfoil is c = 30 cm and the span is also 30 cm. Pressure holes on the upper side of the model in the symmetry plane are used to measure the pressure distribution.
A cylindric cavity is placed inside the model near the leading edge (fig. 1b). It has a diameter of 1 cm and a length of 30 cm and was oriented in spanwise direction. Pressure oscillations inside the cavity cause periodic suction and blowing through a slot. This can excite flow instabilities. The pressure fluctuations within the cavity can be generated by a acoustic horndriver connected to both sides of the cavity. The slot has a length of 12 cm and is 0,5 mm wide. It is placed at 5 % of the chord length and also positioned symmetrically in spanwise direction. We have found that pressure oscillations can also be generated by self-excited resonances, if the cavity was closed except for the slot and if the flow velocity was high enough. For the experiments with self-excited resonances a second slot at 1,3 % of the chord length is added. It is connected to the same cavity.
The sound-pressure level A was measured always without flow by a microphone placed 2 mm above the slot in the plane of symmetry. A single hot-wire probe mounted on a traversing mechanism was used to measure the flow perturbations.

G.E.A. Meier and P.R. Viswanath (eds.),
IUTAM Symposium on Mechanics of Passive and Active Flow Control, 299–304.

The coordinate x' is the distance from the leading edge along the chord of the model, c is the chord length.

Results and Discussion:

Fig. 2 shows the variation of the pressure distribution on the airfoil with increasing sound pressure level A. The excitation frequency is 892 Hz and the angle of attack = 15° at $Re = 1{,}4 \times 10^5$. At this angle of attack and Re-number, the flow is separated completely from the airfoil under natural conditions as shown by the curve without excitation with A = 0 dB. The curve with the sound pressure level of 95,4 dB exhibits the flow reattached again to the model. This curve shows a suction peak near the leading edge instead of a flat distribution. The curve also indicates, with a high plateau of -Cp near the leading edge, the appearance of a separation bubble. With further increase of the sound pressure level A, only a small increase of the negative pressure coefficient takes place. An interesting result is a strong hysteresis with increasing and decreasing sound pressure level. Fig. 3 shows the dependence of the pressure at the chord position x'/c = 0,013 on increasing and decreasing sound pressure level. The highest value of -Cp in the case of attached flow was always found at this pressure hole (compare fig. 2 or fig. 5). In the range $83\ dB \leq A \leq 93\ dB$, it depends on the history of the flow whether it is separated or attached. If one comes from a high value of A, the flow remains attached until about 83 dB, although it becomes attached first at about 93 dB if one comes from a low value.
Fig. 4 shows, for three different angles of attack, the minimum sound pressure level A required to obtain attached flow as a function of the excitation frequency. In all cases the flow was separated under natural conditions. The slot was again at 5 % of the chord length. It is obvious that there are three frequency bands where only low sound pressure levels are necessary to obtain an attached flow. The first one is about 800 Hz, the second one about half of that and the third one below 1/10 of the first. The highest frequency band is determined by a coupling of the forcing with the linear instability of the separated free shear layer (Kelvin-Helmholtz-Instability). A peak at about 750 Hz is found in the spectra of disturbances in the unexcited separated free shear layer by means of hot-wire measurements. So the second effective frequency band in fig. 4 is centered at about the subharmonic of the natural free shear layer instability. A reason for the two first effective frequency bands can be the development of streamwise vortices in the free shear layer. In experiments in free shear layers behind splitter plates, structures in form of spatially stationary counter-rotating vortices in addition to the spanwise coherent structures were found [3]. It was also found that the vortex merging process of the primary vortices suppresses the activity of the streamwise vortices [4]. However, the further vortex merging process can be delayed by exciting the free shear layer with the natural instability frequency or the first subharmonic. In our case, the streamwise vortices could be responsible for the entrainment of the separated flow by the outer flow and they could deliver a higher momentum flux to the surface than under unexcited conditions. They could thus explain the first two effective frequency ranges.
Below the excitation frequency of about 150 Hz, no hysteresis loop like the one in fig.3 was found anymore. In this range the sound pressure level with the maximum gradient of -Cp is plotted in fig. 4. The third effective, lowest, frequency range could be explained by the development of large-scale vortical structures. In experiments on excited free shear layers behind splitter plates it was found that periodic excitations of the flow with frequencies below 1/10 of the natural instability and high enough excitation amplitude can develop the formation of large-scale vortical structures. Such an excitation phenomenon was called „collective interaction" [5]. These large-scale vortical structures could deliver a higher momentum flux to the surface and could thus explain the lowest effective frequency range.

Until now results were presented which were obtained at a Reynolds-Number of $Re = 1{,}4 \times 10^5$ (which means a flow velocity of $U = 7{,}3$ m/s). Above the flow velocity of about $U = 8$ m/s, self-excited resonances appeared in the configuration of open slot and cavity being closed at both sides (compare fig. 1b). It was found that the amplitudes of these resonances were high enough to excite the flow sufficiently to control separation. The critical angle of attack and therefore the maximum lift of the airfoil could be increased. Fig. 5 illustrates with the plot of pressure distributions at $Re = 2 \times 10^5$ in dependence of the angle of attack for open or closed slot at 5 % of the chord length. It is apparent that the flow remained attached at the angles of attack of 15° and 16° with the open slot, although it was separated under natural conditions, that is, with closed slot.
Another slot was now positioned at 1,3 % of the chord length, parallel to the first slot, and it was also connected to the same cavity. This slot was also 12 cm long and 0,5 mm wide. Fig. 6 shows the variation of the pressure coefficient at $x'/c = 0{,}013$ with increasing flow velocity for different slot configurations, and angle of attack = 15,5°. Under natural conditions the flow became attached at about 14 m/s, as indicated in the curve with the closed slot by a sudden increase of the negative pressure coefficient. The other curves with different slot configurations show a completely different behavior. The flow is already attached in each case at lower flow velocities. The most effective configuration seems to be the combination of both open slots. An explanation for this is that the configuration of one open slot at 5 % of the chord length develops much higher sound pressure levels (> 120 dB in the cavity) of the self-excited resonances (for attached flow) than in the case of one open slot at 1,3 % of the chord length (< 110 dB in the cavity). On the other hand the most effective position to excite the flow is near the separation line [2]. Therefore the configuration of both open slots should be the most favorable one here because the excitation happens very close to the separation position of the flow with a high sound pressure level. In all cases, several resonance frequency peaks were found in the range of 500 to 6000 Hz [2].

Summary:

Reattachment of a fully separated flow at a NACA 0012 airfoil at angles of incidence above the critical angle could be achieved by periodic suction and blowing through a narrow slot close to the leading edge of the airfoil. A small cavity below the slot was connected to a horn-driver providing periodic pressure fluctuations to drive the flow through the slot. The present research confirms the results of previous studies [1] that the mechanism for such a control consists of stimulation of momentum exchange across the separated free shear layer. Three optimum frequency bands were found where the periodic excitation was very effective (i.e. where low sound pressure levels were required to obtain an attached flow). The first frequency band was about the natural instability frequency of the separated free shear layer. The second frequency range was centered about the subharmonic of the natural shear layer instability frequency. It was assumed that excitation within these two frequency bands generate streamwise vortices in the free shear layer which promote entrainment of the outer flow. The third frequency range was below about 1/10 of the fundamental free shear layer instability. Large-scale vortical structures are believed to be responsible for this frequency band. Another major point of this paper is that at higher Reynolds numbers, self-excited resonance-oscillations of the slot-flow could be obtained with suitable configurations of the slot-cavity combination. These self-induced resonances can be used to suppress flow separation without any kind of external energy source.

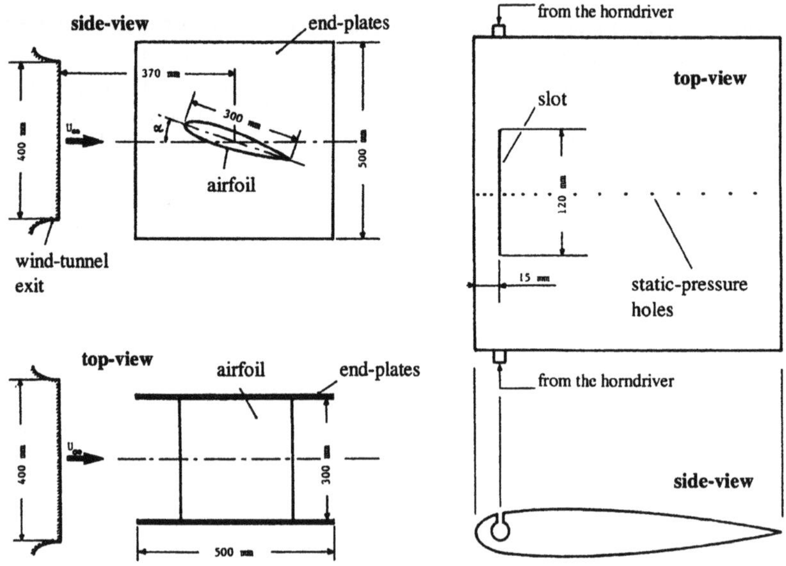

Fig. 1: a) Experimental sketch of the test set-up at the „Small Low Turbulence Wind Tunnel" with open test section of the DLR in Göttingen;
b) Schematic sketch of the NACA 0012-airfoil with a slot at 5 % of the chord length

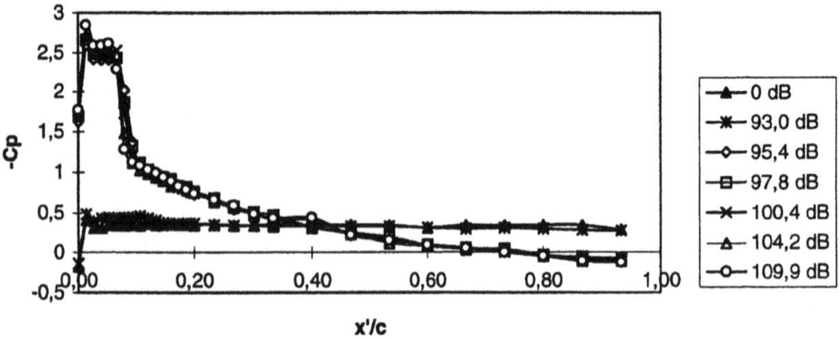

Fig. 2: Dependence of the pressure distribution on increasing sound pressure level A, with a slot at 5 % of the chord length. Excitation frequency = 892 Hz, angle of attack = 15°, Re = $1{,}4 \times 10^5$

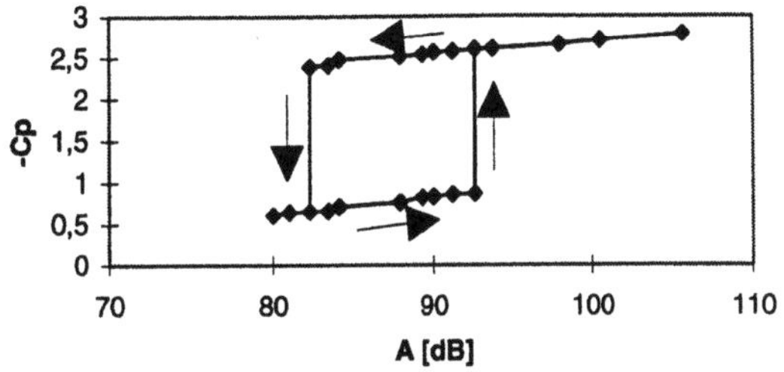

Fig 3: Dependence of the pressure at the chord position x'/c = 0,013 on increasing and decreasing sound pressure level, showing strong hysteresis.
Excitation frequency = 892 Hz, angle of attack = 15°, Re = $1{,}4 \times 10^5$

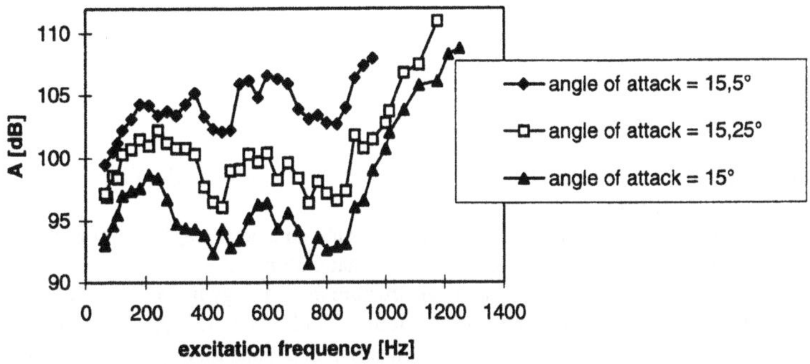

Fig 4: The minimum sound pressure level A to obtain attached flow, as function of excitation frequency, slot at 5 % of the chord length, Re = $1{,}4 \times 10^5$

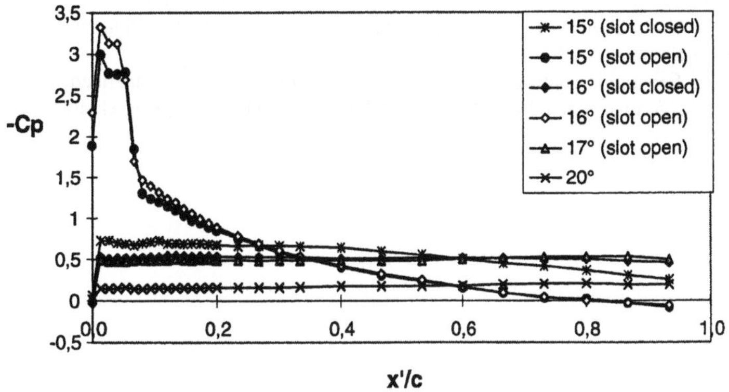

Fig. 5: Dependence of the pressure distribution on angle of attack for open or closed slot at 5 % of the chord length, Re = 2×10^5

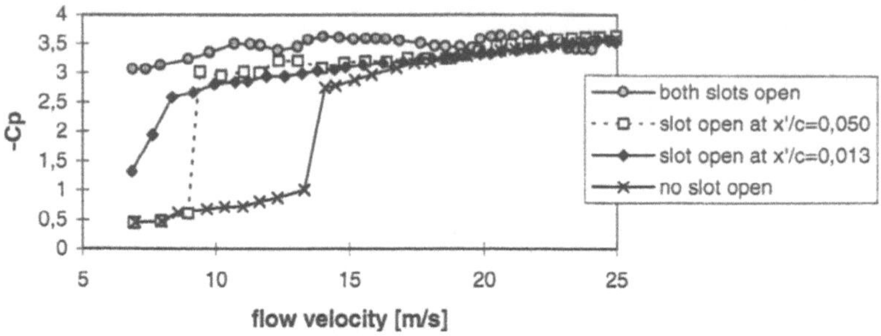

Fig. 6: Variation of the pressure at x'/c = 0,013 with increasing flow velocity for different slot configurations, angle of attack = 15,5°

References:

[1] Dovgal, A. (1993); „Control of leading-edge separation on an airfoil by localized excitation", DLR-FB 93-16

[2] Bader, V. (1997); „Steuerung der Vorderkantenablösung an einem Tragflügelprofil mittels interner Schallanregung", Diplomarbeit, Universität Göttingen

[3] Leboeuf, R.L. (1996); Mehta, R.D.; „Vortical structure morphology in the initial region of a forced mixing layer: roll-up and pairing", Journ. Fluid Mech. 315, pp. 175-221

[4] Huang, L.-S.; Ho, C.-M. (1990); „Small-scale transition in a plane mixing layer", Journ. Fluid Mech. 210, pp. 475-500

[5] Ho, C.-M.; Huerre, P. (1984); „Perturbed free shear layers"; Annual Review of Fluid Mechanics 16, pp. 365-424

[6] Huang, L.S.; Maestrello, L.; Bryant, T.D. (1987); „Separation control over an airfoil at high angles of attack by sound emanating from the surface"; AIAA-Paper 87-1261

[7] Hsiao, F.-B.; Liu, C.-F.; Shyu, J.-Y. (1990); „Control of wall-separated flow by internal acoustic excitation"; AIAA Journ. 28, pp. 1440-1446

[8] Erk, P.P. (1997); „Separation control on a Post-stall airfoil using acoustically generated perturbations", Number 328 in Fortschr.-Ber. VDI Reihe 7. VDI Verlag, Düsseldorf

CONTROL OF PIPE FLOW BY USE OF AN AEROACOUSTIC INSTABILITY

B. LANGE AND D. RONNEBERGER
Drittes Physikalisches Institut der Universität Göttingen
Bürgerstr. 42-44, 37073 Göttingen, Germany

Abstract. Ducts that are lined with a series of equally spaced cavities exhibit strong amplification of sound waves under specific flow conditions. Since the effect is coupled with a variation of the static pressure drop that depends on the amplitude of incident sound, the possibility is implied to control the flow velocity by sound. Based on these observations, an aeroacoustic valve has been designed combining a lined duct section as part of a flow channel and a conventional loudspeaker. As a feasibility test, the device is used as an actuator in an active noise control setup where low frequency fluctuations of the flow velocity in the duct are compensated for.

1. Introduction

In duct flow, fluctuations of the flow velocity are undesired in many cases. In pipes that are connected to relief valves of gas tanks, for example, flow fluctuations that occur while depressurizing the tank will cause considerable noise at the pipe outlet. The availability of fast and affordable digital signal processing units have given rise to efforts to actively stabilize the flow velocity by use of an auxiliary source — generally a dynamic loudspeaker. Conventional loudspeakers, however, tend to be rather inefficient at low frequencies. Some progress has been made recently in finding a device with high performance at low frequencies. Obier *et al.* [4] designed a loudspeaker that has been optimized for low frequency sound generation and that produces good results for frequencies above 20 Hz. A different concept was presented by Laumonier *et al.* [3], who used an oscillating valve for controlling the pressure drop in a duct. But until now, a completely satisfying solution has not been found.

G.E.A. Meier and P.R. Viswanath (eds.),
IUTAM Symposium on Mechanics of Passive and Active Flow Control, 305–310.

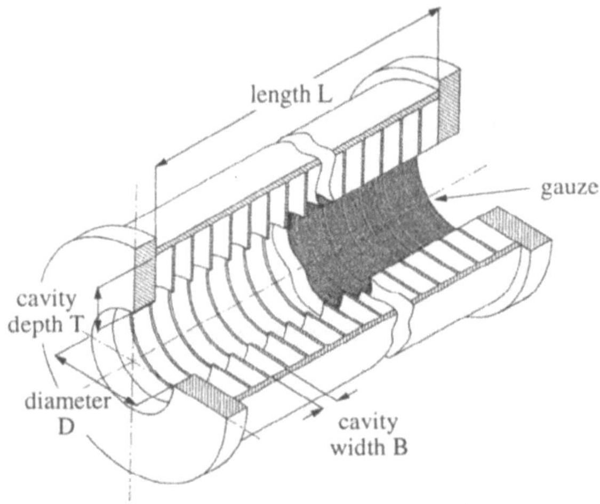

Figure 1. Duct lined with $\lambda/4$-resonators ($T = 75\,mm,\ B = 5\,mm$)

A novel approach to the problem was suggested by Brandes [1]. He studied the phenomenon of strong sound amplification in ducts that are lined with a sequence of radial symmetric $\lambda/4$-resonators as illustrated in figure 1. A power gain of up to 30 dB for sound waves travelling in the direction of the mean flow at frequencies somewhat above the resonance frequency of the resonators $f_{res} = 850\,\text{Hz}$ is observed. The bandwidth of the amplification range amounts to approx. 200 Hz. Appearantly, this amplification is caused by an aeroacoustic instability wave travelling in the boundary layer of the lining. The instability wave is excited by sound at the entrance of the lined duct section, it grows exponentially as it travels along the resonators and generates an amplified sound wave at the outlet of the lined section. The mean Reynolds shear stress and consequently also the mean flow through the lined duct are severely affected by the instability. In most cases, a drastic increase of the pressure drop is observed as a result of the sound/flow interaction. This allows control of the flow velocity by sound since the increase of the pressure drop is closely related to the amplitude of the incident sound wave.

2. Design of an aeroacoustic valve

The effect has been applied in designing an aeroacoustic valve using a lined duct section as part of a flow channel, including a conventional loudspeaker and a signal processor. The concept is illustrated on figure 2: Let the desired pressure drop be a low frequency sine wave. Some nonlinear function of this sine wave is taken as the envelope of a high frequency signal that is radiated

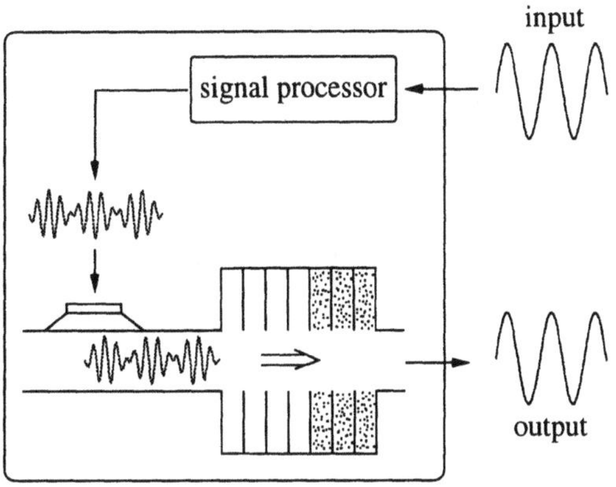

Figure 2. Scetch of the aeroacoustic valve

into the channel by the loudspeaker. The lined duct section then accounts for two effects: First, a variation of the pressure drop at the modulation frequency is generated. This component will precisely match the desired signal if the signal processing unit is capable of inverting the process by which the lined duct section affects the pressure drop due to high frequency sound incidence. Secondly, the modulated high frequency wave is amplified. This presumably unwanted effect can successfully be compensated for by use of a resonance type sound absorber, e. g. by filling the rear end cavities of the lining with some appropriate absorbing material.

Obviously, some attention has to be paid on modelling the envelope of the high frequency control signal with respect to the valve input. This requires a rather sophisticated knowledge of how the pressure drop along the lining is affected by the amplitude of incident sound.

2.1. LOW FREQUENCY SOUND GENERATION

The aeroacoustic valve is controlled by sound containing frequencies within the amplification range of the lining. An oscillating valve, on the other hand, will itself act as a dipole-type sound source under flow conditions. In order to understand the relationship between the valve input and output, a closer investigation of the pressure power spectra in the channel under flow conditions has been performed.

Figure 3 shows power spectra of signals picked up by condenser microphones up– and downstream of the aeroacoustic valve at Mach number $Ma = 0.22$, while random noise in the frequency range between 850 and 1300 Hz is radiated into the channel by the loudspeaker. The dotted-lined spectra are measured in the absence of the lined duct section. The high sound level at frequencies below 200 Hz is due to fan noise. In addition to

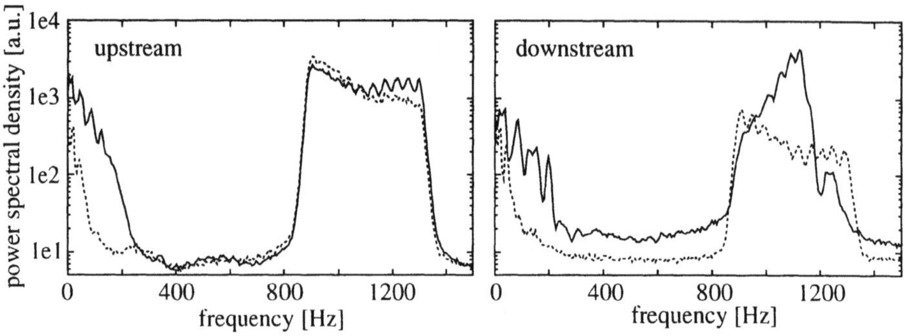

Figure 3. Pressure power spectrum up– and downstream of the valve location

that, noise that is generated by turbulent pressure fluctuations is present with a nearly constant power level at all frequencies. The spectra plottet solid are measured with the lined duct mounted as part of the flow channel. As expected, sound waves with frequencies between 900 and 1200 Hz that travel in the direction of the flow are amplified. This leads to an noticable increase of the power spectral density at the downstream microphone, while the spectrum measured upstream barely changes at frequencies within the amplification range. In addition to the sound amplification, a considerably higher sound level at frequencies below 200 Hz is observed. This effect is entirely due to sound generation by the aeroacoustic valve. It can be concluded from figure 3 that the frequency range of sound that can be generated by the valve is only limited by the bandwidth of the amplification of high frequency waves. The waviness of the spectrum at these low frequencies is accounted for by reflections at the outlet of the flow channel.

The function that describes the conversion of high frequency incident sound into low frequency output is obviously nonlinear. In order to specify this function, a coherence analysis in the frequency domain has been implemented. It has turned out that the questionend nonlinearity can primarily be regarded as quadratic. With this evidence in mind, it is promising to try and generate the envelope of the high frequency control sound by amplitude modulation of a carrier signal whose frequency lies within the amplification range of the lining.

2.2. PERFORMANCE OF THE AEROACOUSTIC VALVE USING AMPLITUDE MODULATION

After implementing an algorithm on the signal processing unit which modulates the amplitude of a carrier signal corresponding to the desired output, measurements have been made to give a performance estimate for the valve.

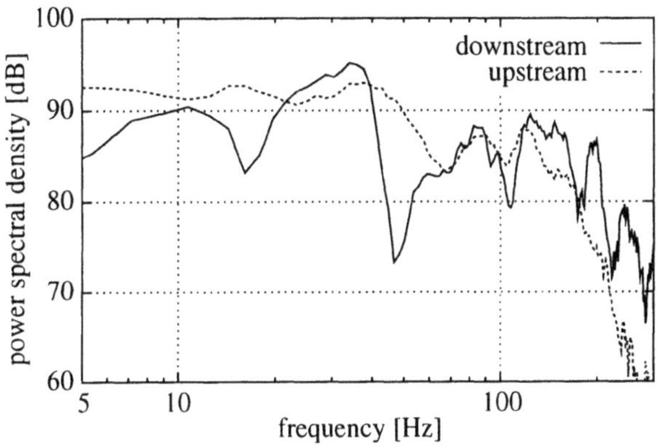

Figure 4. Power spectral density of the valve output

For that purpose, a low pass filtered random noise ($f < 300$ Hz) was taken as input for the device. Figure 4 shows a plot of the power spectral density measured up– and downstream of the valve at $Ma = 0.26$. The values are given in decibels with respect to 1 pW/Hz. The available power of approx. 85 dB downstream of the valve and even higher at the upstream microphone is indeed remarkable. Although frequencies below 5 Hz could not be measured on acount of hardware limitations, it is safe to assume high intensity sound generation even at extremely low frequencies. Integration over the frequency range from 5 to 300 Hz results in the overall output power of the valve which amounts to 114 and 117 db *re* 1 pW at the down– and upstream locations, respectively. Again, reflections at the outlet of the flow channel cause the spectra to be wavy.

3. Application of the aeroacoustic valve in an active noise control setup

As an application, the aeroacoustic valve is mounted as an actuator in a feed forward active noise control setup (see Kuo [2], for example). Low frequency noise is radiated into the flow channel at the inlet using a traditional loudspeaker. Simultaneously, the signal is digitized and used as reference for an adaptive LMS-algorithm. The compensation signal generated by the algorithm is then taken as input for the valve. Performance monitoring is done by observing the error signal that is measured by a microphone downstream of the valve.

In the flow channel used for the investigations, turbulent pressure fluctuations and fan noise which enters the flow channel from the rear end account for much of the error signal. With this observation in mind, it seems reasonable to consider the power spectral density only of that frac-

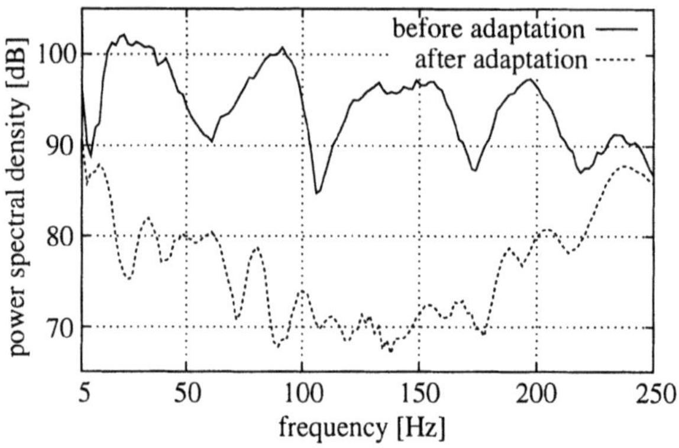

Figure 5. Power level before and after adaptation

tion of the error signal that is correlated with the reference noise. Figure 5 shows these spectra before and after adaptation, measured at $Ma = 0.26$. Attenuation of up to 25 dB could be accomplished for frequencies ranging between 5 and 250 Hz.

References

1. Brandes, M. (1995): Sound amplification in flow ducts lined with a periodic sequence of resonators, *Proc. 1st joint CEAS/AIAA Aeroac. Conf., Munich*, 893–901
2. Kuo, S. M. and Morgan, D. R. (1996): Active–Noise–Control Systems, Whiley–Interscience Publ.
3. Laumonier, J., Jean, P. and Hardouin, L. (1997): An active anechoic termination for low frequencies with mean flow, *Acta Acustica* **83**, 25–34
4. Obier, T. and Pfeifer, G. (1997): Ein neuartiger Lautsprecher für tiefe Frequenzen und hohe Schalldrücke, *Fortschritte der Akustik DAGA '97*

NEAR WALL TURBULENCE REACTION TO A LOCALIZED TIME-PERIODICAL BLOWING

S. TARDU
Laboratoire des Ecoulements Géophysiques et Industriels- LEGI
B.P. 53-X 38041 Grenoble-Cédex ; France

1. Introduction

The application of optimal control through suboptimal schemes (Bewley et al.,1993) or neural networks (Lee et al., 1997) and by the use of DNS resulted in drag reduction of about 20-30 % and has given quite attractive insight into the near wall physics somewhat indirectly. The major shortcoming of these strategies, however, is the need of a dense distribution of MEMS requiring intervention at the wall roughly every 4 to 10 wall units in the spanwise and streamwise directions. Another method could use a linear or non linear time-dependent forcing at the wall through a coarser distribution of intervention locations and find the optimal temporal variation of the disturbance to decrease or increase the drag. Note that this is similar but not identical to robust control problems. That is the reason why we look at the response of the near wall turbulence to a local time-periodical blowing/suction since a couple of years, the simpliest way of linear temporal disturbance beeing time-periodical (Tardu, 1998). Large-scale unsteady blowing/suction is already used in the control of separation, but the response and relaxation of the near wall turbulence to a localized intervention at the wall is not well understood even when the later is steady.

2. Definitions, Experimental set-up and Data reduction

An experimental model has been developed in the low-speed wind tunnel of our laboratory (Fig.1a) . The blowing and suction at the wall are done through spanwise slots of dimensions 0.6*100 mm corresponding to 10*1667 in wall units in the present working conditions. A special pulsating device has been designed for the present purpose. Quite satisfactory sinusoidal waveforms of the suction/blowing wall normal velocities have been obtained (Fig. 1b).

The wall shear stress measurements have been performed by means of a Cousteix-Houdeville wall hot-wire gauge (HWG) to avoid problems caused by the conduction into the substrate. Nice results have been obtained up to the statistics of order 4 and the details may be found in Tardu (1998).

In order to extract the deterministic and deduce the undeterministic part of the flow quantities the classical triple decomposition is used. A flow quantity $q(\vec{x}, t; T)$ is decomposed into a time mean $\overline{q}$ an oscillating $\tilde{q}$ and a fluctuating q' part, $q(\vec{x}, t; T) = \overline{q}(\vec{x}) + \tilde{q}(\vec{x}, t/T) + q'(\vec{x}, t)$, where T stands for the period of the oscillating blowing. The phase average is denoted by $\langle q \rangle = \overline{q} + \tilde{q}$. The modulation characteristics have been determined through a least square Fourier analysis.

In flows with local blowing/suction the severity parameter is defined as the ratio of the injection or suction flow to the *incoming* flow rate, i.e $\Theta = v_0 L_x / \int_0^\infty \overline{u}\, dy$ (Sano and Hirayama; 1985, Tardu ;1998). The injection velocity in steady blowing experiments

G.E.A. Meier and P.R. Viswanath (eds.),
IUTAM Symposium on Mechanics of Passive and Active Flow Control, 311–316.

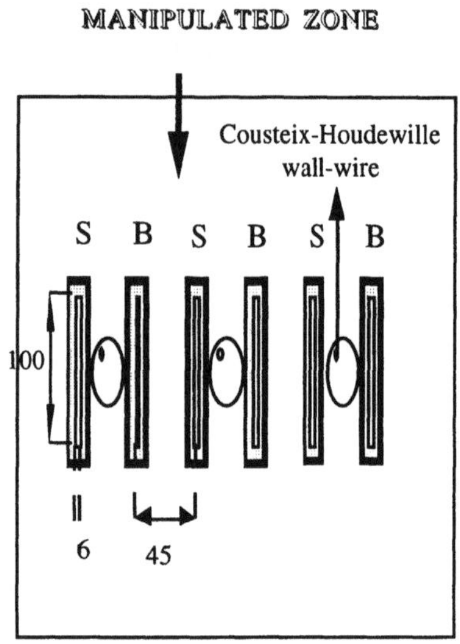

Figure 1a. Are shown the slots and their dimensions in mm . S and B refers respectively to suction and blowing.In this study only one blowing slit is used.

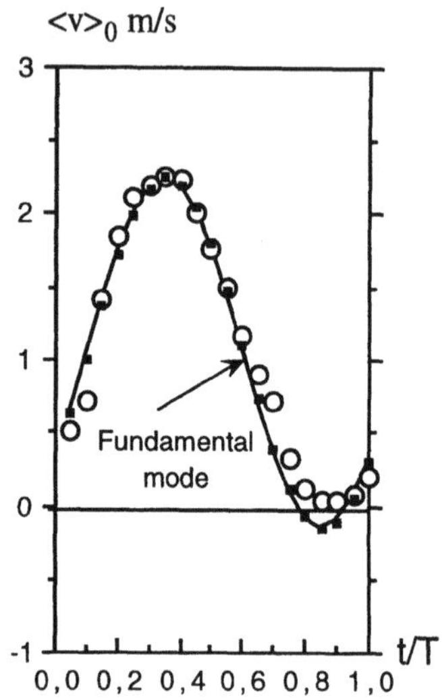

Figure 1b. Example of phase average of the injection frequency; the imposed frequency in wall units is $f^+= 0.017$.

investigated here is $\overline{v}_0= 1$m/s and the severity parameter is only $\Theta = 0.006$. The shape parameter just downstream of the slit at $x/\delta=0.1$ is H=1.4 under these circumstances. In unsteady blowing experiments, the injection velocity $<v_0>$ changes in a cyclic manner between 0 and 2 m/s. The maximum value of the severity parameter in the oscillation cycle is therefore $\Theta = 0.012$ and H=1.7 which is still below the limit of flows prone to separate. Note, by the way, that in DNS studies dealing with active control conducted so far, the severity parameter is zero, because of the pinpoint intervention. In practical situations this is impossible, and even a low Θ may affect profoundly the flow in the neighborhood of the injection/suction region. This may be an additional point to consider the DNS results with some caution.

The results reported in this study are obtained through a *local unsteady blowing* with an imposed frequency $f^+=0.017$ and an amplitude $\hat{v}_0^+ = 8$ ($^+$ denotes values scaled with inner variables). The free stream velocity is U_∞ =4m/s. Only part of the results concerning the space-time development of the wall shear stress could be reported here.

3. Results

3.1 *RELAMINARIZED* PHASE

Fig. 2a shows the cyclic modulation of the wall shear stress at $x^+=20$ and 40 downstream of the slit. The phase average $<\tau>$ is scaled with the time mean wall shear stress $\overline{\tau}_{SBL}$ of the unmanipulated standard boundary layer. The waveform of the injection velocity (to not scale) is also shown. The striking feature of the reaction of $<\tau>$ takes place during the acceleration phase of the injection velocity. The wall shear stress decreases rapidly during this phase until it reaches the laminar limit defined as the value that a laminar Blasius boundary layer would have at the same Reynolds number. The corresponding phase averages of the wall shear stress intensity $\langle\tau'\tau'\rangle/\sqrt{\tau'\tau'}_{SBL}$ are shown in Fig. 2b. The near wall turbulence activity is totally suppressed at $x^+=20$ during half of the oscillation cycle coinciding once more with the acceleration phase of $\langle v_0\rangle$. At $x^+=40$ there is a slight increase both in $\langle\tau\rangle$ and $\langle\tau'\tau'\rangle$ at t/T= 0.6. The close inspection of the data has shown that this corresponds to a transitional spot-like structure resulting from the set-up of a time space localized instability and that will be discussed hereafter.

The strong modification of the wall turbulence structure is better captured in Fig. 3 which shows the phase average of the skewness $\langle S\rangle_{du'/dt}$ of the time derivative of fluctuating streamwise velocity du'/dt and of the ejection frequency $\langle f_e^{+\,*}\rangle$ identified by VITA through the phase averaged thresholds at $y^+= 12$. Recall that $\langle S\rangle_{du'/dt}$ is related to the vorticity stretching and the non linearity in the inner layer. Fig. 3 shows that both the vorticity generation and production mechanisms are altered at high blowing frequency during half of the oscillation cycle. At the acceleration phase, the bursting activity together with the vortex stretching are largely suppressed.

The ensemble of ingredients showing relaminarization are present near the slot at $x^+< 40$ and during half of the cycle, namely:

*The wall shear stress decreases considerably until reaching the value that a laminar boundary layer would have at the same Reynolds number.

*Dissipation (estimated by using local isotropy and not shown here) dominates the near wall flow which is stabilized.

*The velocity fluctuations in the inner layer are not zero but their contribution to the dynamics of the flow becomes inconsequential.

*The frequency of active Reynolds stress producing events decrease considerably and a thin region near the wall extending to approximately 2-3 wall units grows from the wall being free of fluctuating streamwise vorticity. The thickness of this zone reaches almost 5 wall units during half of the oscillation cycle.

* The stretching of quasi-streamwise vorticity decreases strongly as indicated by even negative values of the skewness of du'/dt. This part of the oscillation cycle coincides also with large increases of the Taylor time scale (not shown here).

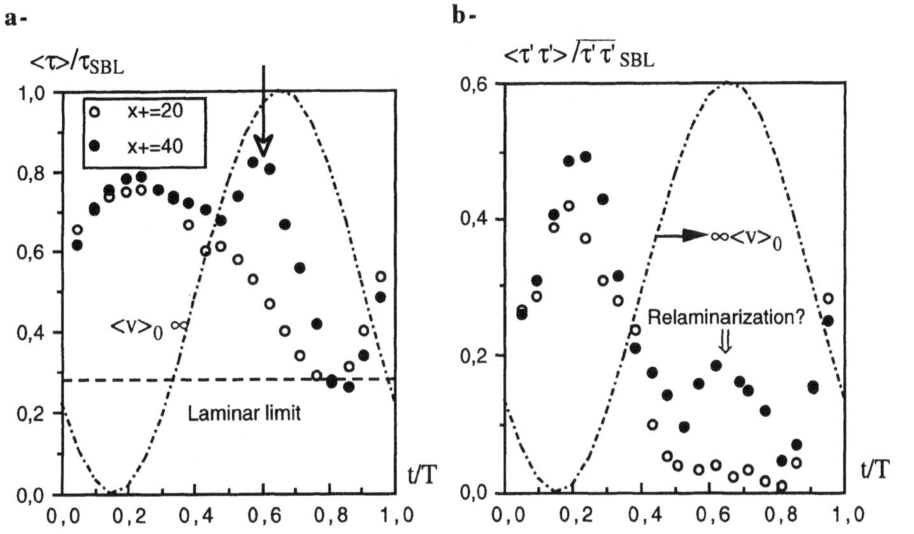

Figure 2. Phase averages of the wall shear stress (a) and wall shear stress intensity (b) at $x^+=20$ and 40 downstream of the injection slit. The imposed frequency is $f^+=0.017$.

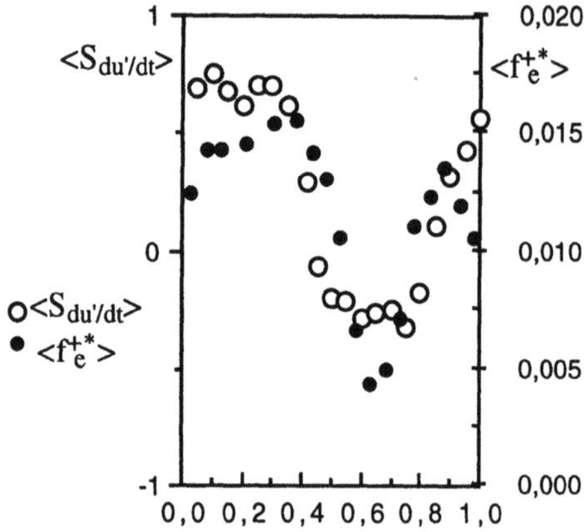

Figure 3. Phase averages of the skewness of time derivative of u' and the ejection frequency determined by VITA at $y^+=12$ and $x^+=30$; the imposed frequency is $f^+=0.017$.

3.2 BIRTH OF A COHERENT SPANWISE STRUCTURE EMERGING FROM THE *RELAMINARIZED* PHASE

The space time evolution of the near wall flow at further downstream locations is striking. First the velocity profiles become strongly inflexional at $x^+=40$ and approximately in the middle of the *deceleration* phase. Fig. 4a shows the phase averaged velocity profiles at t/T=0.4 (for reference) and 0.8 corresponding respectively to the middle of the acceleration and deceleration phases of the injection velocity. The presence of an inflexional point at t/T=0.8 where $<\tau>$ decreases to the laminar limit (Fig.2), is clear. It may, therefore, be argued that the *relaminarized* flow is possibly unstable to inviscid disturbances according to the Fjortoft's theorem. This behaviour is somewhat consistent with the general idea that time or space deceleration destabilizes the flow. It is emphasized here that the close inspectation of the phase averaged streamwise velocity profiles showed that the local gradient is never zero and that the vorticity does not change sign. Therefore, the observed behaviour is not due to a local unsteady separation according to Moore-Sears criteria. The unstabilized flow enters subsequently into *retransition* further downstream following the scheme reported by Narasimha and Sreenivasan (1973). This gives place to the accumulation and enhancement of a patch of spanwise vorticity *of the opposite sign* to the mean vorticity during the deceleration phase at t/T=0.8. The existence of this coherent patch may be clearly seen by the local maximum and minimum appearing in the $\sqrt{<u'u'>} / \bar{u}_{\tau\,SBL}$ profile shown in Fig.4b. The fact that it is positive may be understood through the sudden changes in the skewness of u' shown in Fig.4c and the mechanism suggested schematicaly in Fig 5. This patch rolls up into a coherent structure $\Omega'_{z\,COH} > 0$ approximately at $y^+= 12$. Fig. 6 gives a possible explanation of the effect of this coherent structure on the velocity profiles where it is suggested that $\Omega'_{z\,COH} > 0$ may accentuate the inflexion.The birth of this structure (its origin is approximately at $x^+=20$), in return, gives place downstream to a secondary spanwise vortical structure near the wall with opposite sign, i.e $\Omega'_{z\,SEC} < 0$, because of the non slip condition.The genesis of $\Omega'_{z\,SEC} < 0$ and its subsequent development are perfectly well localized both in time and space (the vortex-ring-like structure -similar to the streamwise Falco typical eddy- in Fig 6 is only suggestive for the moment). The whole structure is convected downstream with an advection velocity of $7\bar{u}_{\tau\,SBL}$, while $\Omega'_{z\,SEC} < 0$ is reinforced and $\Omega'_{z\,COH} > 0$ diffuses somewhat more rapidly. Consequently the wall shear stress increases almost in a Dirac function fashion at times and locations which are perfectly predictible as shown in Fig. 5. The whole phenomena relaxes further downstream at $x^+=300$ (not shown here). This mechanism occurs in the high frequency regime, for $f^+>0.007$ approximately. Therefore, it can nicely be used to *increase* the drag and prevent unsteady separation and/or to *decrease* it through distributed *blowing* locations and by *frequency modulation*. only with negligeable cost. On the other hand, the combination of phase shifted blowing/suction at $x^+>40$ may be efficient for the space-time control of $\Omega'_{z\ SEC}$.

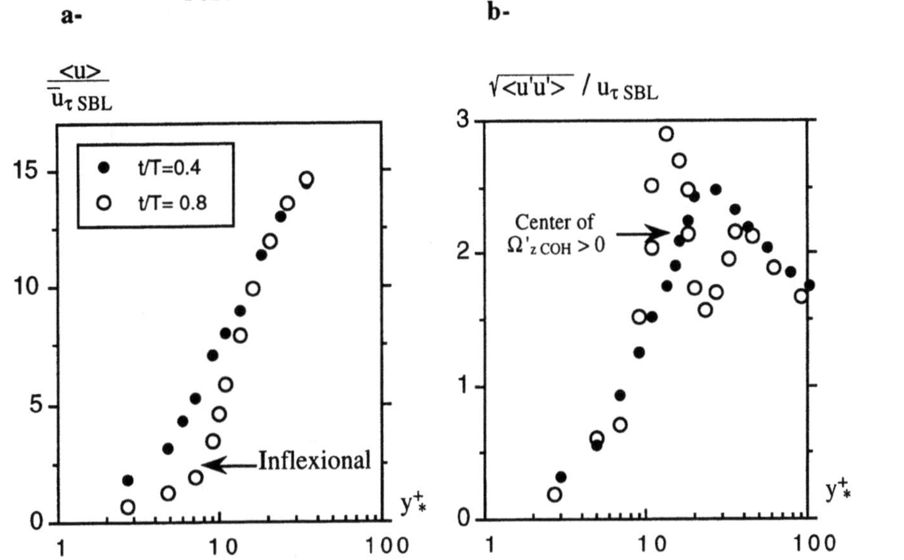

Figure 4. Distribution of (a) the streamwise velocity, (b) the streamwise turbulent intensity and (c) the skewness of u' at t/T=0.4-0.8 and $x^+=40$. The birth of $\Omega'_{z\ SEC} < 0$ is seen in (c) at t/T=0.6 .

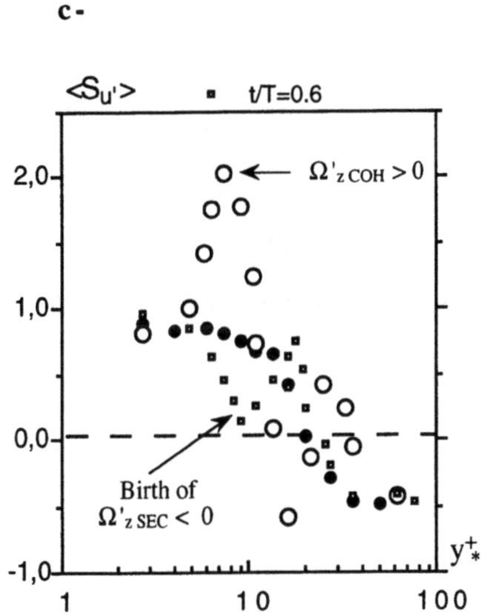

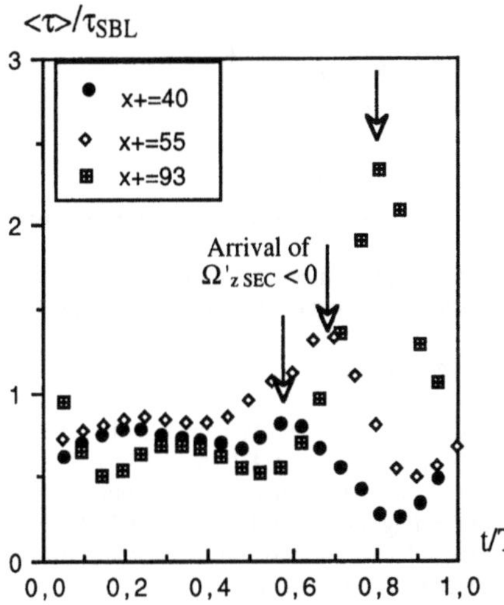

Figure 5. Phase average of the wall shear stress at further downstream positions

4. Discussion and Conclusion

In the case of blowing there is no substraction or addition of vorticity. The spanwise mean and fluctuating vorticity, together with *turbulent-drag* inducing quasi-streamwise vortical structures (QS) are displaced and pushed away from the wall by say $\delta\langle y_v\rangle \propto \langle v_0\rangle\, \delta\langle t_c\rangle$ where $\delta\langle t_c\rangle$ is the effective convection time of the structures as they are advected over the slit. Consider the spanwise fluctuating vorticity ω'_z which is *negative* in majority near the wall. The removing of ω'_z induces therefore a *global deficit* of $\delta\langle u'_w\rangle \propto \langle\omega'_{z0}\rangle\delta\langle y_v\rangle < 0$ in the instantaneous non slip velocity at the wall. This is subsequently corrected by the formation of a thin vortex sheet in front of the wall (and of its image) with vorticity of opposite sign to ω'_{z0} which is,by the way, also opposite to the mean vorticity. This sheet subsequently dilutes through diffusion. It turns out that, in the deceleration phase and when the imposed unsteadiness is sufficiently rapid, there is a net accumulation of this sheet leading to $\Omega'_{z\ COH} > 0$ after roll-up. This phase coincides with the uncoupling with the near wall flow of removed QS's, resulting in a temporal relaminarization. These are real unsteady effects occuring only when the blowing frequency is larger than a critical value. The ensemble of the results presented here shows how a time varying intervention at the wall may involve complex phenomena. It is hopped that they may provide new perspectives in the near wall control, through a combination of phase-shifted and eventualy frequency modulated local blowing/suction sites.

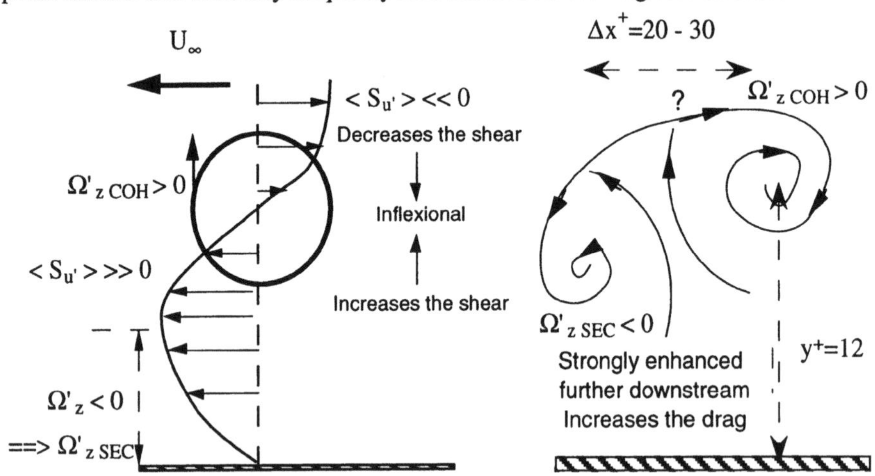

Figure 6. Possible mechanism of the genesis of the flow structure presumably associated with a shear layer instability during the *relaminarization* phase.

5. References

Bewley T., Choi H., Temam R., Moin P. " Optimal feedback control of turbulent channel flow" Annual Research Briefs, CTR; pp. 3-14 ; 1993

Lee C., Kim J., Babcock, D., Goodman, R., 1997 "Application of neural networks to turbulence control for drag reduction" Phys. Fluids 1, **9**, 1740-1747

Narasimha R., Sreenivasan K.R., 1973 "Relaminarization in highly accelerated boundary layers" J. Fluid Mech. ,**61**, 417-447.

Sano M., Hirayama N., 1985 " Turbulent boundary layers with injection and suction through a slit" Bulletin of JSME, **28**, 807-814

Tardu ,S., 1998 "Near wall turbulence control by local time-periodical blowing" Exp. Th. Fluid Science. **16/1-2**, 41-53.

Manipulation of Trailing Edge Separation on an Aerofoil Using a Periodic Cross-Flow

H.A. Siller (siller@pi.tu-berlin.de) and H.H. Fernholz
Hermann-Föttinger-Institut für Strömungsmechanik
Technische Universität Berlin,
Straße des 17. Juni 135, 10623 Berlin, FRG

Abstract. Experiments with an HQ-17 laminar aerofoil are described where flow separation in the recompression region of an aerofoil is reduced. A loudspeaker system is mounted inside the aerofoil. It generates an oscillating jet with a blowing and a suction phase in a spanwise slot in the aerofoil. Spanwise vortices are generated that enhance mixing processes, increasing the lift coefficient c_L. The lift gain depends mainly on the position of the separation line relative to the slot as well as the frequency and the amplitude of the excitation.

Keywords: high-lift aerodynamics, separation control, experiments

1. Introduction

During takeoff and landing, aircraft fly at high angles of attack where boundary layer separation starting at the trailing edge of the aerofoil reduces the lift forces. Passive methods, like vortex generators, stabilise the boundary layer very effectively, but increase the drag forces and reduce the performance of the aerofoil. A practical method of active boundary layer control that could be turned on at high angles of attack would reduce energy consumption and noise emission of aircraft. General applications of flow control are discussed in e.g. [4], recent examples in aerodynamics can be found in e.g. [1], [2] or [3].

The present study investigates the potential of manipulating the boundary layer in the recompression region of an aerofoil by a periodic disturbance. A loudspeaker system drives a periodic cross-flow in a spanwise slot in the upper surface of the aerofoil. Spanwise vortices are generated that stabilise the boundary layer by mixing fluid with high kinetic energy from the potential flow into the wall region.

2. Experimental setup

A schematic view of the experimental setup is given in figure 1. A section of a laminar glider profile of the type *HQ-17* is mounted in the test-section of a low-speed wind-tunnel. It is connected to a six-component force balance, the support being isolated from the flow by

G.E.A. Meier and P.R. Viswanath (eds.),
IUTAM Symposium on Mechanics of Passive and Active Flow Control, 317–322.

double walls. The chord length c of the aerofoil is 0.65 m, the span 1.55 m. The low-speed wind tunnel can be operated up to velocities of $U_\infty = 40$ m/s. The angle of attack can be adjusted by a stepper motor.

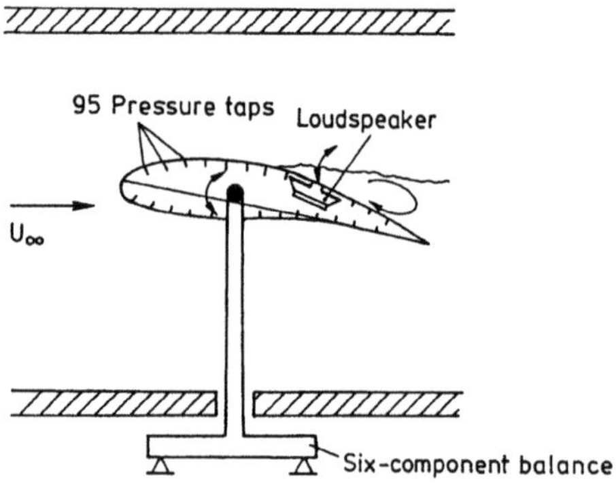

Figure 1. Experimental setup

A slot with a spanwise length of 1.05 m is cut into the upper surface of the aerofoil at $x/c = 0.66$. Underneath the slot, 8 loudspeakers are mounted, and a 1/4″ microphone is used to monitor the sound-pressure level inside the aerofoil. The aerofoil is equipped with 95 pressure taps. They are scanned using two Scanivalves (TM) which are mounted inside the aerofoil and connected with two differential pressure transducers located outside the wind tunnel. Data acquisition is controlled by a personal computer with an IEEE-bus card. No wind-tunnel corrections are applied to the data, because we are mainly concerned with the relative changes that are caused by the acoustic excitation.

3. Parameters for the acoustic excitation of the flow

To increase the lift on the aerofoil by delaying or suppressing flow separation, vortices have to be generated that mix the near-wall region with the potential flow efficiently. This mechanism consists of: i) the production of the disturbance in the slot and ii) the generation of vorticity by interaction of the disturbance with the boundary layer.

To optimise the method, the frequencies and amplitudes that most effectively delay or suppress boundary layer separation in a specified situation have to be determined first. Then, the system consisting of the loudspeakers, the pressure chamber and the slot has to be laid out so that it generates the required frequency and amplitude.

3.1. THE SOURCE OF THE EXCITATION

The loudspeakers act as pumps: during the upward stroke, a jet is produced in the slot, and during the downward stroke, the slot acts as a sink, drawing fluid from all sides. This generates a jet above the slot that is sustained by entrainment. The mean velocity of this jet measured at a fixed height above the slot can be used as a relative measure of the amplitude of the excitation.

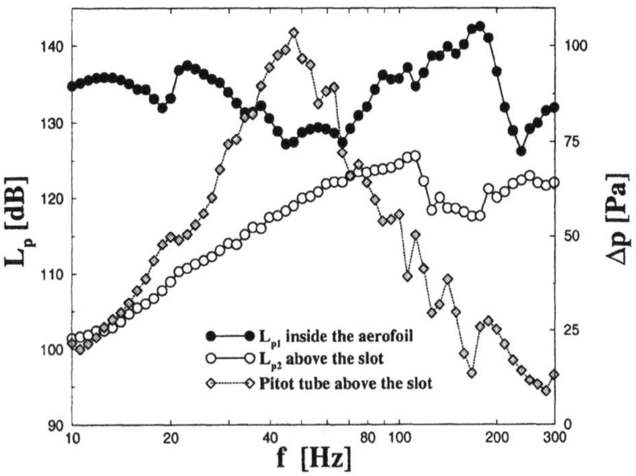

Figure 2. Frequency response of the excitation source

A simple approach was used to determine the frequency response of the system: A pitot tube was fixed 5 mm above the centre of the slot and the pressure difference against the ambient pressure was measured as a function of the frequency together with the sound-pressure levels L_{p1} inside the aerofoil and L_{p2} above the slot (see figure 2). There appears to be no correlation between L_{p1}, L_{p2} and the Pitot pressure. The frequency response of L_{p1} reflects the nonlinear behaviour of the loudspeakers and the acoustic properties of the volume they are mounted in. L_{p2} is a measure for the efficiency of sound transmission through the slot, it rises continuously with increasing frequency until it drops by 8 dB at 112 Hz. The Pitot pressure has an absolute maximum at approximately 48 Hz, the frequency where the highest flow velocities in the slot can be achieved.

3.2. THE EFFECT OF THE EXCITATION ON THE LIFT FORCES

The lift of the aerofoil depends on the flow speed U_∞, the ambient conditions ρ and ν, the angle of attack α, the position of the slot x_s, and the frequency f and the amplitude or energy of the excitation. In nondimensional terms, this relationship can be expressed as

$$c_L = f(Re_c, \alpha, x_s/c, St, Eu)$$

where Eu is the Euler number, the ratio of the energy of the excitation to the kinetic energy of the main flow. The flow speed will have little effect on c_L once a certain Reynolds number is exceeded. The Strouhal number will influence the efficiency of the excitation, since the wavelength of the vortices generated will have to match with lengthscales of the separation region. With the setup used here, the optimal frequencies were $40 < f < 80$ Hz. This corresponds to Strouhal numbers $0.4 < St < 0.8$, if the distance from the slot to the trailing edge of the aerofoil is used as the characteristic length scale.

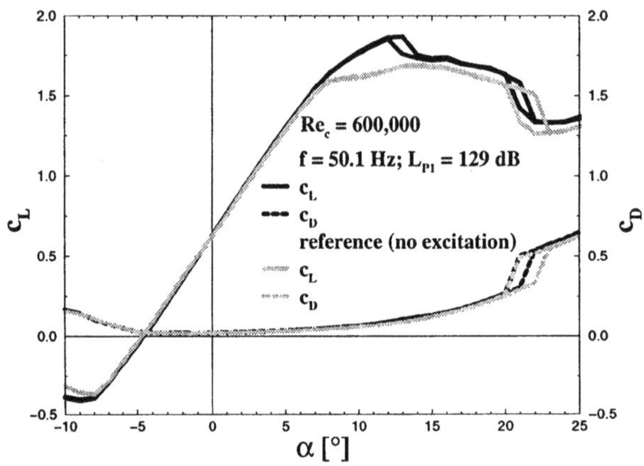

Figure 3. Lift and drag coefficients for a Reynolds number of 6.0 × 10^5

Figure 3 shows polars of the lift and drag coefficients of the aerofoil at a Reynolds number of 6.0×10^5. c_L rises linearly with α until the onset of separation from the trailing edge at $\alpha \approx 7°$. When the flow is excited at $f = 50.1$ Hz with a sound-pressure level of $L_{p1} = 129$ dB, c_L is increased for $\alpha > 7°$. The maximum of c_L at $\alpha = 13°$ is increased by 11.1%, but a hysteresis effect occurs: For increasing α, the excitation delays separation up to $\alpha \approx 13°$. When α is decreased, the flow is already separated and the excitation cannot force reattachment until $\alpha \approx 12°$.

If separation occurs downstream of the slot, the vortices generated at the slot are weakened by dissipation before they reach the separation region. If the slot is already inside the separation region, the disturbance cannot mix fluid from above efficiently. In this case, the slot is located at $x_s/c = 0.66$. The flow separates at this position for $\alpha \approx 13°$. Figure

3 shows a rapid decline of c_L for higher angles in the excited case, but increased lift for $7 \leq \alpha \leq 13$.

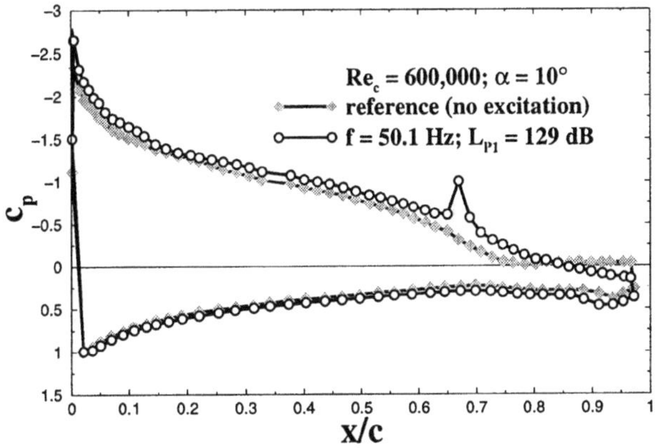

Figure 4. Pressure distribution around the aerofoil

Pressure distributions measured around the aerofoil at a fixed angle of attack of $\alpha = 10°$ are shown in figure 4. At the pressure tap downstream of the slot position (at $x_s/c = 0.66$), the disturbance has a strong local effect. While the flow separates at $x/c \approx 0.76$ in the reference case, it remains fully attached with the excitation turned on.

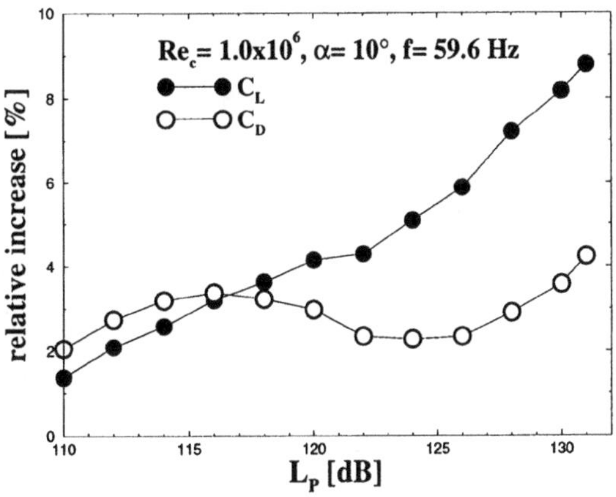

Figure 5. Lift and drag increase with rising sound-pressure level

Measurements of c_L and c_D at a fixed $\alpha = 10°$ and varying sound-pressure level L_{p1} are shown in figure 5. For sound-pressure levels below 116 dB, both lift and drag increase. For levels between 116 and 126 dB, c_L increases while c_D is reduced. For even higher sound-pressure levels, there is a steeper lift increase, coupled with an increase of the drag. Due to the limited power of the loudspeakers, a saturation point was not reached.

4. Conclusions

The lift forces on an aerofoil at high angles of attack can be increased by acoustic excitation that delays boundary layer separation. The lift increase depends on the Strouhal number, the Euler number and the position of the slot. The main problem is to match the properties of the source of the excitation with the frequencies and amplitudes required to control the separation most efficiently.

Further experiments are planned to study the influence of the different parameters. Especially questions regarding the optimisation of the source of the excitation, the system consisting of the slot, the loudspeakers and the chamber in between, demand further study.

Acknowledgements

The project was financed by the German ministry for education, science and technology. The authors would like to thank the Italian loudspeaker manufacturers *AUDYON* for their cooperation.

References

1. Erk, P.: 1997, *Separation control on a post-stall airfoil using acoustically generated pertubations*, No. 328 in Fortschrittberichte, Reihe 7: Strömungstechnik. VDI-Verlag.
2. Nishri, B. and I. Wygnanski: 1998, 'Effects of periodic excitatation on turbulent flow separation from a flap'. *AIAA Journal* **36**(4), 547–556.
3. Seifert, A., D. Eliahu, and D. Greenblatt: 1998, 'Use of piezoelectric actuators for airfoil separation control'. *AIAA Journal* **36**(8), 1535–1537.
4. Sigurdson, L. W.: 1995, 'The structure and control of a turbulent reattaching flow'. *Journal of Fluid Mechanics* **298**, 139–165.

TRANSITION CONTROL OF SUPERSONIC BOUNDARY LAYER ON FLAT PLATE

N.V. Semionov, A.D. Kosinov, A.A. Maslov
Institute of Theoretical and Applied Mechanics SB RAS, Novosibirsk, Russia

In the paper two methods of transition control of supersonic boundary layer on flat plate are considered. First one is a technique of an artificial turbulization of the supersonic boundary layer on the basis of the periodic electrical discharge, allowing smoothly to change a transition Reynolds number Re_{tr}. A dependency of value of Re_{tr} from artificial disturbance frequency is found. Second method is inhomogeneous heating of the model surface. Influence of the heating of a leading edge region on development of disturbances in the supersonic boundary layer is investigated. A gradient of temperatures on the leading part of the model allowed as to stabilize as to destabilize the supersonic boundary layer.

1. Introduction

In recent years considerable interest has been focused on the problem of transition control in shear layer [1-4]. Transition control mean both stabilization and destabilization of flow. The cooling of a model surface and suction is considered as perspective methods of a flow laminarization [2,5-7]. Destabilization methods of a supersonic boundary layer are well known. Homogeneous heating of the model surface always resulted in decreasing numbers Re_{tr} [8,9]. However in [10,11] it was obtained that favorable distribution of heating may lead to stabilization of a plate boundary layer at subsonic speeds. In the present paper a results of experiments on transition control by using artificial disturbances [12, 13] and stabilization/destabilization of supersonic boundary layer by inhomogeneous heating of leading edge region are presented.

2. Experimental Conditions

Experiments were carried out in T-325 supersonic wind tunnel of ITAM SB RAS at Mach number M=2. The hot-wire measurements were made in the boundary layer of the flat steel plates with a sharp leading edge which was 450 mm length, 200 mm wide and 10 mm thick. The plates was installed in the central plane of the test section under zero angle of attack. The glow discharge was used as the source of controllable disturbances for transition control. The disturbance generator design was based surface glow discharge [12] (coordinates of the source were x=18 mm, where x is the distance

G.E.A. Meier and P.R. Viswanath (eds.),
IUTAM Symposium on Mechanics of Passive and Active Flow Control, 323–328.

from the leading edge) or the glow discharge in camera [13] (coordinates of the source were x=35 mm).

The flow fluctuations were registered using of constant current and temperature hot-wire anemometer. Tungsten wire probes 5 micron in diameter and about 1 mm long were in use. The overheat ratio of hot-wire was about 0.8 that means the measured perturbations have corresponded to mass flux pulsations. The transition point was determined from the downstream total amplitude evolution of fluctuations in boundary layer. As the transition end we consider a position corresponds to the total amplitude maximum in this distributions.

To study the influence of surface irregular heating on disturbance development electroheating elements from a nichrome wire were mounted in the plate on a distance of 5 mm from the leading edge. Width of a heated up segment was 13 mm. The change of surface temperature distribution was implemented by change a current through the heated up elements. The surface temperature of the model was measured by using four «rustless steel - constantan» thermocouples that mounted flush with the model surface on distance 2.5, 12.5, 20, 25.9 mm from the leading edge. In the experiment the hot-wire was mounted at x=120 mm. The unit Reynolds number got out so that the heated up segment was in the region of the eigen oscillations amplification, and made $Re_1=6.5\times10$ m^{-1}.

3. Results

3.1. ACTIVE TRANSITION CONTROL

We are going to present here two different experiments on active transition control of supersonic boundary layer. First let us consider the data obtained about ten years ago. The experiments were carried out on flat plate with surface glow discharge. Using the technique it was possible to decrease transition Reynolds number changing a frequency and intensity of controllable disturbances in experiment [14]. At minimum amplitudes of controlled oscillations the Re_{tr} was the same like natural case. At the increasing of the controlled disturbance amplitude a inmonotonous decreasing of Re_{tr} versus frequency was observed. An example of the position of the transition end over

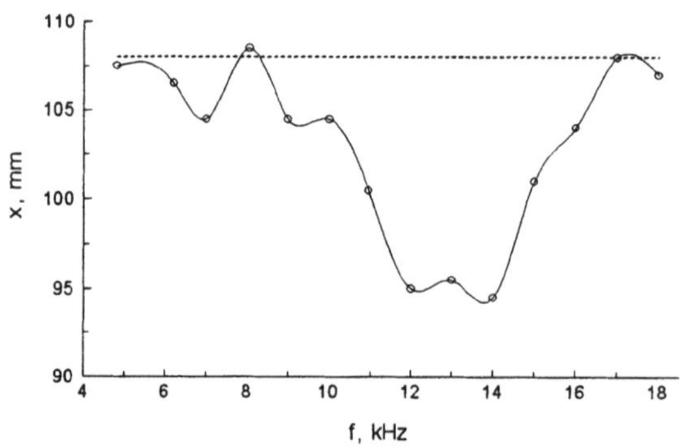

Figure 1.

frequency of controlled disturbances is shown in fig.1 ($Re_1 = 14\times10^6$ m^{-1}). Here a dotted line corresponds to the position of natural transition. The voltage on electrodes was fixed during experiment. It was obtained decreasing of Re_{tr} for $f\approx13\div14$ kHz and $f\approx7$ kHz. We can note that decreasing of Re_{tr} occur on divisible frequencies.

Decreasing Re_{tr} at $f\approx13\div14$ kHz is well agreed with results [15] where was obtained for various unit Reynolds numbers that the greatest growth these disturbances in flat plate boundary layer at M=2. Both results were obtained at the noisy conditions of T-325.

At maximum amplitude of artificial disturbances for all frequencies the transition began closed to the source ($x_{tr}\approx30$ mm). The controllable disturbances in the boundary layer are three-dimensional for the all initial amplitudes.

Second kind of experiments were performed by using glow discharge in chamber [16]. Under the given experimental conditions with controlled disturbance excitation laminar-turbulent transition on model occurred on 20 % closer to the leading edge than for natural pulsations (see fig.2). However in this case periodical initial disturbances (like «spikes») of large amplitude were quasi two-dimensional.

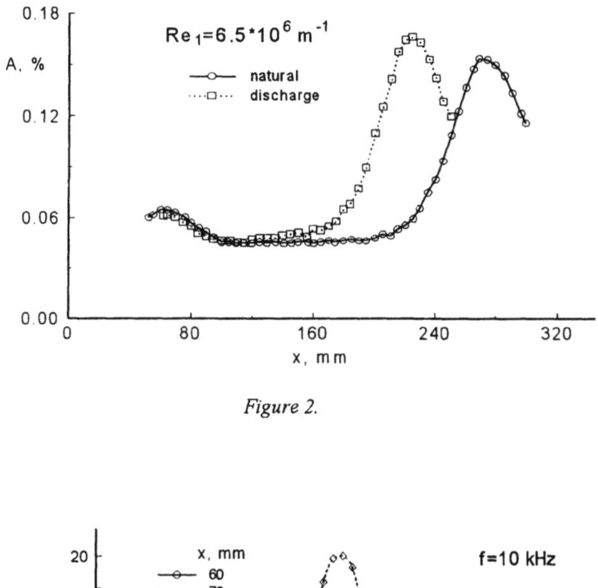

Figure 2.

In fig.3 evolution of amplitude β-spectra at frequency f=10 kHz, which had maximum initial amplitude and biggest growth are shown. The β-spectra have the maximum at β=0 for all frequencies at x=60 mm. We have observed the strong disturbances growth downstream (nearly 10 times from x=60 mm to x=110 mm) at f=10 kHz,

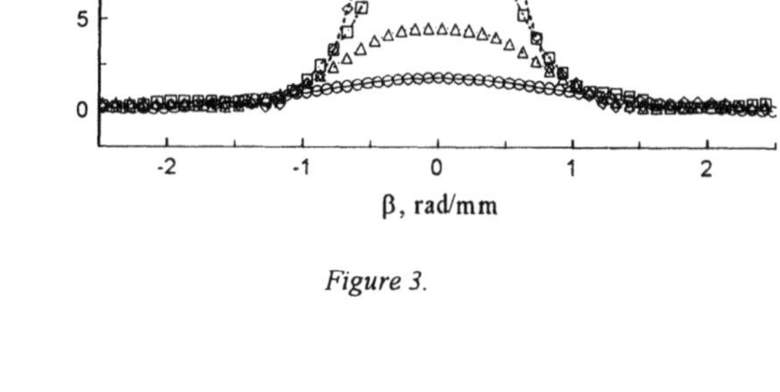

Figure 3.

while two-dimensional nature of wave spectra did not change. The results may be reason to reconsider a role of two-dimensional waves in supersonic boundary layer transition.

3.2. INHOMOGENEOUS HEATING

The experiments have been conducted for natural disturbances. For noisy conditions of T-325 it was obtained in [17], that disturbances with dimensionless frequency parameter $F=(0.1\div0.5)\cdot10^{-4}$ are more unstable in boundary layer at M=2, where $F=2\pi f/(Re_1\cdot U)$. The heated up segment of plate surface was in stable region of eigen waves for the mentioned above frequencies. The examples of distribution of integral fluctuations <*e*> versus of mean voltage *E* are shown in fig. 4. The curve «a» is obtained for insulated model, curve «b» and «c» - for heated up. In fig. 5 the appropriate downstream distributions of temperature are presented. If for case «b» decreasing amplitude of mass flow fluctuations is observed, i.e. increase of the boundary layer stability, for case «c» the amplitude of fluctuations has grown and the boundary layer is destabilized. The temperature distributions «b» and «c» differ by that for case «b» the maximum heating occurred in a vicinity of the first thermocouple, and for case «c» - in a vicinity of the second thermocouple. Hence, the distribution of temperature on the model surface influences on the supersonic boundary layer stability.

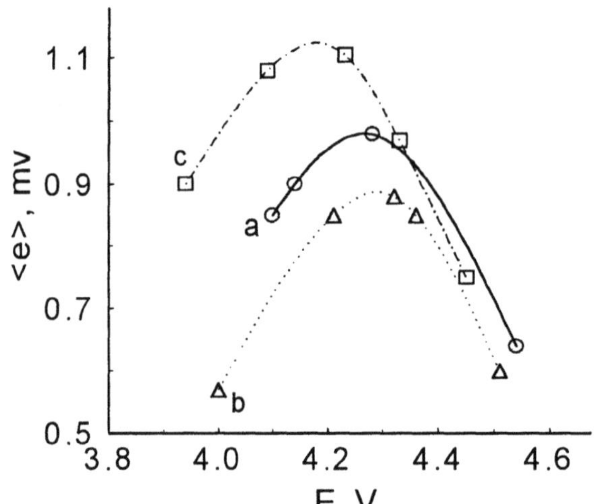

Figure 4.

For check of this assumption at various distributions of temperature were measured <*e*> and <e_f> in critical layer at $Y/\delta\approx0.5$, where δ - boundary layer thickness. Heating of the plate surface with maximum temperature in the vicinity of the second thermocouple always resulted in increase <*e*> and <e_f> on all frequency band rendering the greatest stability influence. So model heating up to 70°C resulted in increase of fluctuations amplitude <*e*> on 20 %. More interesting is the case of temperatures distributions with the

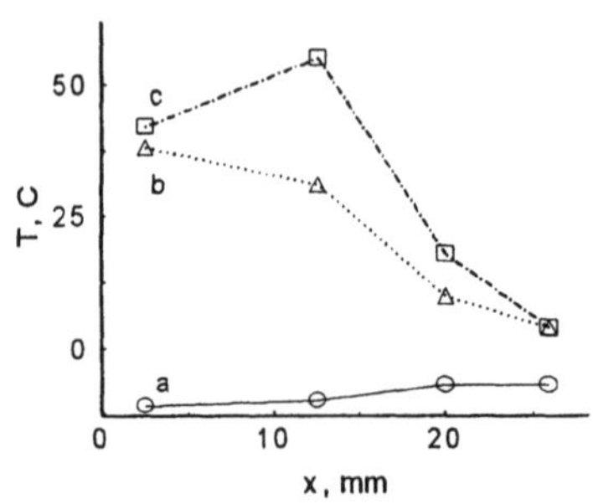

Figure 5.

maximum in the vicinity of the first thermocouple causing to stabilization. A small heating up to 20°C results in significant stabilization of supersonic boundary layer. Heating of model with the maximum on the first thermocouple up to 100°C leads to decreasing of mass flow fluctuations amplitude on 20 %.

Note that these temperature distributions on the model surface lead to completely different results, though the heating occurs in the stability region eigen oscillations. It was obtained an increase of boundary layer stability above the inhomogeneous heated up surface at subsonic speeds in [10]. There is the instability region in the plate leading edge zone at supersonic flow, caused by amplification of external acoustic disturbances in boundary layer [18, 19]. The possible mechanism of the boundary layer stabilization is following: if the maximum in distribution of temperature is in a vicinity of the first thermocouple, the maximum heating of the boundary layer occurs in middle of an instability region, caused by an external acoustic field. On an initial heated segment the heat flow is directed to the boundary layer - there is the heating of gas and amplification of disturbances. The surface becomes rather cold downstream and the propagation direction of a heat changes near to the model surface, as results in attenuation of disturbances [10]. Apparently, attenuation of eigen oscillations is caused by inhomogeneous heating in the disturbances generation region by a sound. If the maximum in temperature distribution is in the vicinity of the second thermocouple, the maximum heating of the boundary layer occurs near to the acoustic neutral curve for disturbances, rendering the greatest influence on transition. It leads to the more intensive growth of disturbances, induced by a sound. The combination of processes of inhomogeneous model heating and disturbance generation by an external acoustic field, on all visibility, is the reason of flow destabilization/stabilization in supersonic boundary layer.

4. Conclusions

Use of artificial disturbances created by the surface electrical discharge, provides an artificial turbulization of supersonic laminar boundary layer. Such method allows smoothly to reduce transition Reynolds number from values, appropriate to a undisturbed boundary layer, down to values, appropriate to the source location. Is shown, that the most intensive effect artificial periodic disturbances of small intensity render on the boundary layer on frequencies $f \approx 7$ and 14 kHz at experiments in the wind tunnel T-325.

Through inhomogeneous heating of the model surface in the leading edge zone it is possible to control development of disturbances in the supersonic boundary layer. Changing distribution of temperature along the model surface, it is possible how to stabilize, and to destabilize the boundary layer. Such influence, apparently, is caused by effect of inhomogeneous heating in the acoustic instability region on disturbances development in supersonic boundary layer, induced by an external sound.

5. References

1. Morkovin, M.V. (1990) On roughness-induced transition: facts, views, and speculations, in Hussaini M.Y. and Voigt R.G. (eds.), *Instability and Transition*, Springer-Verlag, Berlin et al., **1**, pp.281-295.
2. Bushnell, D.M. and Malic, M.R. (1985) Applications of stability theory to laminar flow control. In Dwoyer D.L. and Hussaini M.Y. (eds.), *Proc. Symp. On the Stability of Time Dependent and Spatially Varying Flows*, Springer-Verlag, NY, pp.1-17.
3. Reshotko, E. (1994) Boundary layer instability, transition and control, AIAA Paper N 94-0001.
4. Crouch, J. D. (1997) Transition prediction and control for airplane applications, AIAA Paper N 97-1907.
5. Mack, L.M. (1993) Effect of cooling on boundary layer stability at Mach number 3. In Ashpis D.E., Gatski T.B. and Hirsh R. (eds.), *Instability and Turbulence in Engineering Flows*, Kluwer Academic Publishers, Norwell, MA, pp.175-188.
6. Arnal, D., Reneaux, J., and Casalis, G. Numerical and experimental studies related to skin friction drag decreasing problem, *Transitional Boundary Layers in Aeronautics*, Noth-Holland, Amsterdam, 177-190.
7. Lysenko, V.I. and Maslov, A.A. (1984) The effect of cooling on the supersonic boundary layer stability and transition. In Kozlov V.V. (ed.), *Proc. IUTAM Symposium on Laminar-Turbulent Transition*, Springer-Verlag, Novosibirsk ,, pp.495-502.
8. Mack, L.M. Linear stability theory and the problem of supersonic boundary layer transition, *AIAA J.* **13** (1975), 423-448.
9. Lysenko, V.I., Maslov, A.A. and Semionov, N.V. Experimental study of heating influence on transition and stability of supersonic boundary layer, *J. Izv. SO AN USSR. Seria Tech. Nauk.* **13** (1981), vyp.3, 42-49 (in Russian).
10. Struminskiy, V.V. et al., Control of disturbances development in boundary layer by irregular surface heating, *J. Izv. SO AN USSR. Seria Tech. Nauk*, **11** (1988), vyp.3, 38-42 (in Russian).
11. Kazakov, A.V., Kogan, M.N., and Kuparev, B.A. About increase of stability of subsonic boundary layer at heated surface near to leading edge, *Doklady Ac. Sci. USSR.* **283** (1985), vyp.2, 333-335 (in Russian).
12. Kosinov, A.D., Maslov, A.A. and.Semionov, N.V. (1983) Techniques for introduction of artificial disturbances in a supersonic flow, Novosibirsk, Inst. Theor. Appl. Mech., Sib. Branch USSR Acad. Sci., *Preprint*, No.34-83 (in Russian).
13. Kosinov, A.D., Semionov, N.V. and Shevelkov, S.G. (1994) Investigation of supersonic boundary layer stability and transition using controlled disturbances, *Proc. 7 Int. Conf. on the Methods of Aerophys. Research*, Novosibirsk, **2**, pp.159-166.
14. Zinin, O.I.,Maslov, A.A. and Semionov, N.V. Artificial turbulization of supersonic boundary layer. *Modeling in Mechanics.* **6** (1992), №2, 71-76 (in Russian).
15. Lysenko, V.I., Maslov, A.A. and Semionov, N.V. (1980) Experimental study of nonlinear stage of disturbances development in supersonic boundary layer. *Applied aerogazdynamic and heating processes*, Novosibirsk,. pp.81-86 (in Russian).
16. Kosinov, A.D., Ermolaev, Yu.G. and Semionov, N.V. (1998) Experimental study of anomalous wave processes in supersonic boundary layer *Proc. 9 Int. Conf. on the Methods of Aerophys. Research*, Novosibirsk, **2**, pp.106-111.
17. Kosinov, A.D., Maslov, A.A. and Shevelkov, S.G.: Experiments on stability of supersonic boundary layers, *J. Fluid Mech.* **219** (1990), 621-633.
18. Lebiga, V.A., Maslov, A.A. and Pridanov, V.G. Experimental investigation of the stability of supersonic boundary layer on a flat insulated plate, *J. Archives Mech.* **31** (1979), N 3, 397-505.
19. Semionov, N.V., Kosinov, A.D. and Maslov, A.A. (1998) Experiments on supersonic leading edge receptivity to acoustic waves, In *Proc. 3rd Int. Conf. on Fluid Mechanics*, Beijing Institute of technology Press, Beijing, pp.225-230.

ACTIVE CONTROL OF TURBULENT BOUNDARY LAYERS FOR DRAG REDUCTION

John Kim
Mechanical and Aerospace Engineering Department
University of California, Los Angeles, CA 90095-1597

I. Introduction

This is a written version of my presentation at an IUTAM Symposium on Mechanics of Passive and Active Flow Control, September 7-11, 1998, Göttingen, Germany. I had given a similar talk at an AIAA meeting on the same subject, and a somewhat less condensed summary can be found in Kim (1997). Progress made since the AIAA meeting has been added to the present summary.

The ability to control turbulent flows is of significant economic interest. Successful control of turbulent boundary layers by reducing drag, for example, can result in a substantial reduction in operational costs for commercial aircrafts and ships. We have been exploring new approaches for boundary-layer control through numerical experiments. All approaches are based on our premise that the most effective way to control turbulent boundary layers is through proper manipulation of the organized turbulence structure in the wall region. After a short introduction to the turbulence structure and its relevance to the boundary-layer control, summaries resulting from applications of several different procedures are presented in this paper. The interested reader should refer to the original papers mentioned in each section for further details.

In this paper, I shall use (x, y, z) for the streamwise, wall-normal, and spanwise coordinates, respectively, and (u, v, w) for the corresponding velocity components.

II. Near-Wall Turbulence Structures

Kim (1992) has proposed the near-wall streamwise vortex as the most relevant turbulence structure from the perspective of drag reduction in turbulent boundary layers. This point of view is supported by the observation that streamwise vortices have been found to be responsible for both "ejection" and "sweep" events of the bursting process (Robinson 1991). Recent studies have also shown that the high skin-friction regions in turbulent boundary layers are closely related to the near-wall streamwise vortices (Kravchnko *et al.* 1993, Choi *et al.* 1994). Choi *et al.* (1994) showed that a significant drag reduction is possible when the surface boundary condition is modified to suppress the near-wall streamwise vortices. They also relate the drag-reducing mechanism on a riblet surface to the restricted interaction between the riblet and streamwise vortices (Choi *et al.* 1993a).

Our work on boundary-layer control, therefore, is based on the premise that the most effective way to control turbulent boundary layers for drag reduction is through proper manipulation of the near-wall streamwise vortices. All results discussed in this paper

G.E.A. Meier and P.R. Viswanath (eds.),
IUTAM Symposium on Mechanics of Passive and Active Flow Control, 329–336.

have been obtained by direct numerical simulations of turbulent and transitional channel flows using the numerical methods reported in Kim *et al.* (1987).

III. Application of Neural Network

The objective of the work reported in Lee *et al.* (1997) was to seek wall actuation in the form of blowing and suction at the wall, which were dependent only on the wall-shear stress to achieve a substantial drag reduction. This approach requires knowledge of how the wall-shear stresses respond to wall actuations (*i.e.*, the correlation between the wall-shear stresses and the wall actuations). Because of the complexity of the solutions to the Navier-Stokes equations, however, it is not possible to find such a correlation in closed form or to approximate it in simple form. Instead, we used a neural network to approximate the correlation that then predicts the optimal wall actuation to achieve the minimum of the skin-friction drag. Lee *et al.* (1997) described how neural networks with a small number of shared weights were constructed and trained off-line. An on-line control scheme for drag reduction based on that neural network was then implemented. A standard two-layer feed-forward network with hyperbolic tangent hidden units and a linear output unit was used. The functional form of the final neural network was:

$$v_{jk} = W_a \tanh\left(\sum_{i=-(N-1)/2}^{(N-1)/2} W_i \left.\frac{\partial w}{\partial y}\right|_{j,k+i} - W_b \right) - W_c \,,$$

$$1 \le j \le N_x \text{ and } 1 \le k \le N_z \,, \tag{1}$$

where W denotes weight, N is the total number of input weights, and the subscripts j and k denote the numerical grid point at the wall in the streamwise and spanwise directions, respectively. N_x and N_z are the number of computational grid points in each direction. The summation is done over the spanwise direction. Seven neighboring points ($N = 7$), including the point of interest, in the spanwise direction (corresponding to approximately 90 wall units) were found to provide enough information to adequately train and control the near-wall structures responsible for the high skin friction. Note that the input to the neural network is $\partial w/\partial y$ at the wall, not $\partial u/\partial y$. Initially, $\partial u/\partial y$ and $\partial w/\partial y$ at the wall at several instances in time were used as input data fields and the actuation at the wall was used for the output data of the network. Experimentally, we found that only $\partial w/\partial y$ at the wall from the current time was necessary for sufficient network performance. It should be noted that a brute force application of a neural network, whose architecture and input parameters were designed without consideration for relevant flow physics, led to either no convergence or excessively long training time.

Applying this control scheme to direct numerical simulations of a turbulent channel flow at low Reynolds number resulted in about 20% drag reduction (Figure 1). The computed flow fields were examined to determine the mechanism by which the drag reduction was achieved. The most salient feature of the controlled case was that the strength of the near-wall streamwise vortices was substantially reduced. This result further substantiates the notion that a successful suppression of the near-wall streamwise vortices leads to a significant reduction in drag.

An examination of the weight distribution from the on-line neural network led to a very simple control scheme that worked equally well while being computationally more

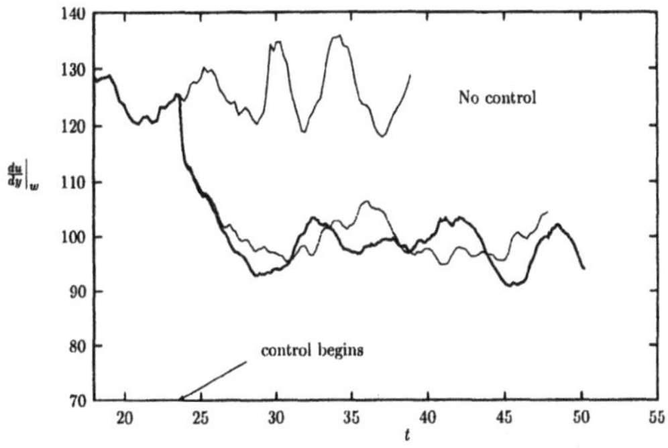

Figure 1: Mean wall-shear stress histories for various control schemes compared to the no-control case: ——— , no control; ——— , on-line control with neural network with 7x1 template; ········ , control with 7 fixed weights.

efficient. This simple control scheme indicates that the optimum blowing and suction at the wall should be in the form,

$$v_w \sim \overline{\frac{\partial}{\partial z}\frac{\partial w}{\partial y}\bigg|_w}, \tag{2}$$

where the overbar represents a local spatial average with high wavenumber components reduced. The converged weight distribution can be expressed analytically, thus making the implementation of this control scheme relatively easy. This control scheme produces a distribution of wall actuations that are very similar to those produced by the v-control of Choi *et al.* (1994).

IV. Application of Suboptimal Control Theory

Most previous control work has been rather ad hoc in that it was mainly based on the investigator's intuition. More systematic approaches using an optimal control theory have recently appeared. Choi *et al.* (1993b) proposed a "suboptimal" control procedure, in which the iterations required for a global optimal control were avoided by seeking an optimal condition over a short time period. The suboptimal control procedure was successfully applied to control of the Burgers equation. Bewley and Moin (1994) were the first to apply the suboptimal control procedure to a turbulent flow and reported a drag reduction of about 17%. The procedure requires full velocity information throughout the flow in order to solve an adjoint problem, from which a feedback control input was then derived. In spite of this obvious drawback, the fact that a control theory applied to a turbulent flow resulted in a substantial drag reduction is encouraging, especially since their control procedure was derived rigorously from a control theory in which a pre-determined cost functional was minimized.

In Lee *et al.* (1998), we demonstrated that a wise choice of the cost functional, coupled with a variation of the formulation, can lead to a more practical control law. We showed how to choose a cost functional and how to minimize it to yield simple feedback control laws that only require quantities at the wall as input. One of the laws requires spatial information on the wall pressure over the entire wall, and the other requires information, also over the entire wall, on one component of the wall shear stress. We then derived

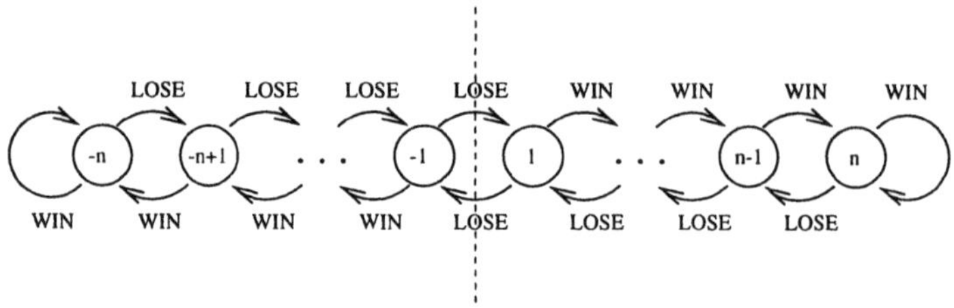

Figure 2: Automaton of the Goore scheme

more practical control schemes that only require local wall pressure or local surface-shear stress information, and showed that they work equally well.

Applying the suboptimal control theory to a turbulent channel flow produced, about 20% drag reduction (Lee *et al.* 1998). Turbulence characteristics of the controlled flow field are very similar to those obtained through the neural-network control.

V. Application of Goore Scheme

The objective of the paper by Lee and Kim (1998) was to explore whether a new control paradigm known as the Goore scheme could be applied to the boundary layer control. Details of the original Goore scheme can be found in Tung and Kleinrock (1993), who developed an algorithm for distributed control using simple finite state automata. Examples of their applications include the problems associated with multi-access communication and mobile robot coordination. In their original algorithm, there are a number of players who vote yes or no, and a referee who counts the number of yes votes. Each player (automaton) gets rewarded or penalized, independent of his vote, according to a given probability function, which is a function of the percentage of yes votes only. This reward probability function has a maximum value at a specific percentage. The goal is to construct automata such that, after enough number of trials, the percentage of yes votes approaches the optimum value that gives the highest reward.

Tung and Kleinrock (1993) presented an algorithm by which such automata could be built. In Tung and Kleinrock's algorithm, each player has a memory of size $2n$, ranging from $-n$ to n, skipping 0. If one's memory is positive (negative), his vote is yes (no). Whenever one gets rewarded, he moves his memory by one unit in the direction away from zero (i.e., he sticks to his previous vote); whenever one is penalized, his memory moves by one unit toward zero (i.e., he wants to flip his previous vote, but with relaxation). This process is illustrated in Figure 2. It was proven that this scheme works when the number of memory (n) is larger than the number of players (N). As the ratio n/N goes to infinity, the percentage of yes votes approaches the optimum one that produces the maximum probability of reward.

As a first step, we explored whether we could use a variation of the Goore scheme (instead of a neural network, for instance) to determine the optimal weights in Equation (1). The reward probability function was a function of the total drag. We used the following procedure. The weights in Equation (1) were varied according to a given probability function, which rewarded (penalized) players when the resulting total drag was smaller (larger) than that for an uncontrolled flow. The choice of the probability function is arbitrary. A simple linear function was used as a starting point. A preliminary result indicated that the Goore scheme was indeed able to find a set of weights close to the optimum set

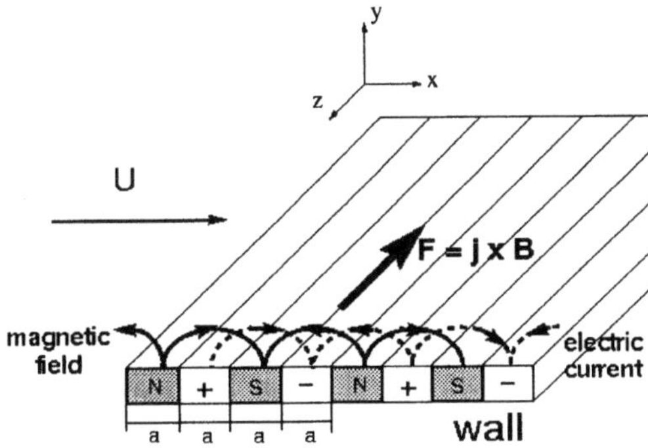

Figure 3: Schematic of an arrangement for the creation of a Lorentz force along the spanwise direction.

found by the neural network, resulting in a similar drag reduction to that shown in Figure 1.

Implementation of the Goore scheme is rather straightforward and involved computations are minimal, which makes it very attractive for many applications. In many real applications, it is desirable to have an integrated control system (i.e., sensors, actuators, control circuits) built into a miniature chip, a large number of which are to be deployed over control surfaces. In such a situation, the control logic should be as simple as possible and required computation must be as minimized as possible. Otherwise, required circuitry becomes too complicated to be implemented on a chip level.

VI. Application of Lorentz Force

The objective of the paper by Kim *et al.* (1997) was to investigate the possibility of utilizing the Lorentz force for flow control in turbulent boundary layers. In contrast to other control schemes discussed thus far, which employ a closed-loop feedback control, an open-loop control (i.e., no feedback, hence, no measurements required) was first tried with this approach. We started with the premise that the most effective way to affect the near-wall streamwise vortices, without first detecting them, is to apply perturbations perpendicular to the streamwise vortices. Perturbations perpendicular to the near-wall streamwise vortices in a turbulent boundary layer can be introduced easily through oscillating electro-magnetic forces generated from the arrangement shown in Figure 3. The spanwise Lorentz force, which decays in the wall-normal direction, can be made to oscillate in time by alternating the polarity of the electrodes shown in the figure.

Analyses of the above configuration for seawater have been carried out. Because seawater has very low conductivity, the magnetic Reynolds number that appears in the Maxwell equation is very small. The Navier-Stokes equations are then decoupled from the Maxwell equation, which allows the Lorentz force in the Navier-Stokes equations to be treated as an external body force. For a given magnetic field strength and electric potential, we used the following form of the body force, normalized by the wall-shear velocity (u_τ) and the channel half-height (δ):

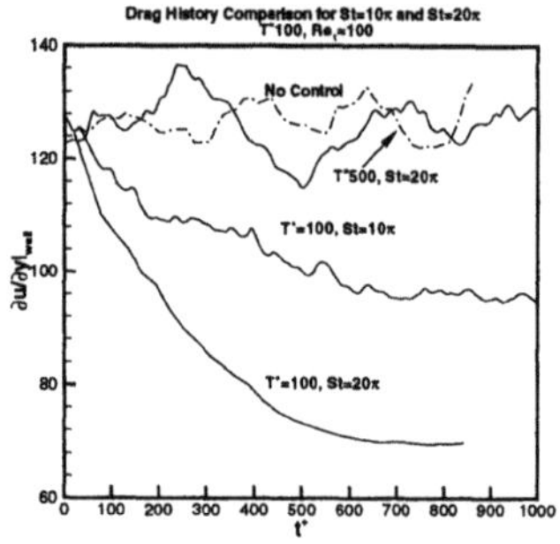

Figure 4: Drag history comparison between no control and open-loop control methods.

$$f_z = \frac{\pi \sigma V B_0}{4 a \rho u_\tau^2 / \delta} \, e^{-\pi y/a} \, \sin\left(\frac{2\pi}{T^+} t^+\right) \tag{3}$$

where σ, V, B_0, and a are the conductivity of seawater, applied voltage difference, magnetic field strength, and the streamwise width of the magnets, respectively. T^+ and t^+ are nondimensionalized period and time respectively.

A fully developed turbulent channel flow was subject to the above body force. The amplitude of the nondimensionalized body force was about the same as the spanwise pressure gradient in Akhavan's (1993) numerical experiment, in which a non-decaying oscillatory pressure gradient was imposed. Such magnitude can be obtained with magnets of B_0=500 gauss and applied potential V=100 volts for low Reynolds number flows. The spanwise width of the magnets, a, was chosen such that the body force decayed to 5% of the wall value at $y^+ = 30$. With $T^+ = 100$, computed results showed an immediate drop of the skin-friction drag by 50% (Figure 4), comparable to Akhavan's result, indicating that the spanwise oscillating Lorentz force limited to the wall region was as effective as the spanwise pressure gradient imposed across the whole channel in Akhavan's work. Investigation of near-wall turbulence structures showed that most turbulence structures in the wall region were suppressed, which is consistent with those results discussed in Sections III. and IV. A test with $T^+ = 500$ and the same amplitude indicated that the skin friction was increasing, again consistent with Akhavan's result, suggesting the existence of an optimal period of oscillation.

The efficiency of the control schemes by the oscillating Lorentz force was considered by estimating the power required to generate each type of force and comparing that with the amount saved from the reduction in skin-friction drag. We found that, even with reasonably strong magnets, the efficiency was very poor, although the amount of power put into the flow was less than the energy saved. The total efficiency was poor due to the excessive amount of power required in generating the Lorentz force in a weakly conducting fluid.

VII. Application of Systems Control Theory

In Joshi *et al.* (1997), a system theory framework is presented for the linear stabilization of a laminar channel flow. The linearized Navier-Stokes equations are converted into the state variable representation of the system using a Galerkin method. This representation relies on the fact that the motion of any finite-dimensional dynamic system can be expressed as a set of first-order ordinary differential equations in matrix form:

$$\frac{d\mathbf{x}}{dt} = A\mathbf{x} + B\mathbf{q} \tag{4}$$

$$\mathbf{z} = C\mathbf{x}\,. \tag{5}$$

The vector $\mathbf{x}$ is the state vector of the system, $\mathbf{q}$ is the input variable (*i.e.*, control input), and $\mathbf{z}$ is the output variable (*i.e.*, sensor output). The matrix A is the dynamic matrix of the system, and B and C are the input and output matrices, respectively.

Joshi *et al.* (1997) show that in addition to the well-studied system eigenvalues, the location of system zeros are important in linear stability control. The location of system zeros determines the effect of feedback control on both stable and unstable eigenvalues. In addition, system zeros can be used to determine sensor locations that lead to simple feedback control schemes. Joshi *et al.* showed that the channel flow can be stabilized using a simple, constant-gain feedback, integral-compensator controller. It was also shown that a controller designed with linear theory also has a strong stabilizing effect on two-dimensional finite-amplitude disturbances. As a result, the secondary instabilities resulting from infinitesimal three-dimensional disturbances in the presence of a finite-amplitude two-dimensional disturbance cease to exist (Joshi *et al.* 1997).

In our subsequent work (Joshi *et al.* 1995, Cortelezzi *et al.* 1997), linear quadratic Gaussian (LQG) optimal control theory is exploited to give faster perturbation energy dissipation. Controllers built using LQG methods for infinite-dimensional systems must, in theory, be infinite dimensional. Since it is impossible to implement infinite-dimensional controllers, we employed reduced-order controllers for the infinite order plant. It was shown that a reduced system, 8 states per wave numbers (256 total reduced states for 32 wave numbers) instead of 250 states per wave number (8000 total full states) performed equally well in suppressing finite-amplitude disturbances in a channel flow. We are currently applying the same approach to a turbulent boundary layer for drag reduction. It remains to be seen whether such a controller design based on the linearized Navier-Stokes system can still be effective in the fully nonlinear regime.

VIII. Concluding Remarks

Successful applications of a variety of control schemes to a turbulent channel flow for drag reduction have been described. Although the control schemes presented in this paper are significant improvements over earlier approaches that require velocity information throughout the flow field, there are many technical issues that must be resolved before these control schemes can be implemented in real practice. Precise blowing and suction distribution over a surface, for example, is difficult to implement. Other approaches, such as flaps and flexible surface, may prove to be more practical. We are currently working along this line. Another issue worth mentioning is the time delay between sensing and actuation, which was not included in any of our numerical experiments. In a real situation, there will be a finite time between sensing and actuation, which must be accounted for.

Acknowledgments

This paper describes the work being carried out by the author's students and colleagues at UCLA. I thank all of them. The financial support from Air Force Office of Scientific Research (F49620-95-0263, Dr. James McMichael; F49620-97-1-0514, Dr. Mark Glauser;

F49620-97-10276, Dr. Mark Jacob) and Office of Naval Research (N00014-95-1-0352, Dr. Patrick Purtell) during the course of this work is gratefully acknowledged. The computer time has been provided by the NAS Program of NASA Ames Research Center, San Diego Supercomputer Center, and DoD High Performance Computer Center.

References

Akhavan, R., Jung, W. J., Mangiavacchi, N. (1993) "Turbulence control in wall-bounded flows by spanwise oscillations," In F.T.M. Nieuwstadt, editor, *Applied Scientific Research*, **51**, *Advances in Turbulence IV*, Kluwer Academic Publishers.

Bewley, T. R. & Moin, P. (1994) "Optimal control of turbulent channel flows," *Active Control of Vibration and Noise*, ASME DE-Vol. 75.

Choi, H., Moin, P. & Kim, J. (1993a) "Direct numerical simulation of turbulent flow over riblets," *J. Fluid Mech.* **255**, 503.

Choi, H., Temam, R., Moin, P. & Kim, J. (1993b) "Feedback control for unsteady flow and its application to the stochastic Burgers equation," *J. Fluid Mech.* **253**, 509.

Choi, H., Moin, P. & Kim, J. (1994) "Active turbulence control for drag reduction in wall-bounded flows," *J. Fluid Mech.* **262**, 75.

Cortelezzi, L., Speyer, J. L., Lee, K. H. & Kim, J. (1997) "Robust reduced-order boundary layer controller," Bulletin of the American Physical Society, **42**, No.11, Nov. 1997.

Joshi, S. S., Speyer, J. L. & Kim, J. (1995) "Modelling and control of two dimensional Poiseuille flow," Proceedings of the 34th IEEE Conference on Decision and Control (CDC), New Orleans, LA, December 13-15, 1995.

Joshi, S. S., Speyer, J. L. & Kim, J. (1997) "A systems theory approach to the feedback stabilization of infinitesimal and finite-amplitude disturbances in plane Poiseuille flow," *J. Fluid Mech.* **332**, 157.

Kim, J. (1992) "Study of turbulence structure through numerical simulations: the perspective of drag reduction," in AGARD Report (R-786), AGARD FDP/VKI Special Course on "Skin Friction Drag Reduction," March 2-6, 1992, VKI, Brussels, Belgium.

Kim, J. (1997) "Taming Turbulence," AIAA Paper 97-1791, 28th AIAA Fluid Dynamics Conference, Snowmass Village, Colorado, June 29-July 2, 1997.

Kim, J. Moin, P. & Moser, R. D. (1987) "Turbulence statistics in fully developed channel flow at low Reynolds number," *J. Fluid Mech.* **177**, 133.

Kim, J., Berger, T., Lee, C., Lim, J., & Choi, H. (1997) "Turbulent boundary layer control utilizing the Lorenz force," in preparation; see also Bulletin of the American Physical Society, **42**, No.11, Nov. 1997.

Kravchenko, A. G., Choi, H. & Moin, P. (1993) "On the relation of near-wall streamwise vortices to wall skin friction in turbulent boundary layers," *Phys. Fluids A* **5**, 3307.

Lee, C., Kim, J. Babcock, D. & Goodman, R. (1997) "Application of neural networks to turbulence control for drag reduction," *Phys. Fluids* **9**, No. 6, 1740.

Lee, C., Kim, J. & Choi, H. (1998) "Suboptimal control of turbulent channel flow for drag reduction," *J. Fluid Mech.* **358**, 245.

Lee, C. & Kim, J (1998) "Application of the Goore Scheme to turbulence control for drag reduction," The Thirteenth U.S. National Congress of Applied Mechanics, Gainesville, Florida, June 21-26, 1998.

Robinson, S. K. (1991) "Coherent motions in the turbulent boundary layer," *Ann. Rev. Fluid Mech.* **23**, 601.

Tung, B. & Kleinrock, L. (1993) "Distributed control methods," *Proc. of the 2nd International Symposium on High Performance Distributed Computing*, Spokane, Washington, July 20-23, 1993.

THE EFFECTS OF FLOW INSTABILITIES ON THE ACTIVE CONTROL OF SEPARATION

A. DARABI, O. LIKHACHEV, R.NEUENDORF, I. WYGNANSKI
Department of Aerospace and Mechanical Engineering,
The University of Arizona,
Department of Fluid Mechanics and Heat Transfer,
Tel-Aviv University

1. Introduction

Steady-flow assumption provides a convenient criterion for flow separation linking the pressure gradient to the local characteristics of the boundary layer. However the boundary layer on the verge of separation is seldom (if at all) steady, it contains large eddies that are generated by instabilities. It was recently shown [1-3] that periodic addition of momentum may prevent separation with the same degree of control authority that is achieved by steady blowing, with one important difference: the momentum input required may be orders of magnitude smaller. This is particularly true when the input perturbations are amplified due to the instability of the mean flow [4]. It was also observed that the instability of a shear layer governs the minimum input required for a separated flow to reattach back to the surface. This determines the optimum frequency and amplitude of the harmonic perturbation necessary to force reattachment [4]. However the reattached flow encloses a bubble that may easily burst when the pressure gradient is further increased. An increase in the excitation frequency reduces the size of the bubble even though the amplitude of the excitation remains unchanged. This is so because the initial rate of amplification of the perturbations is higher and it increases the entraining capacity of the shear layer bounding the bubble. These concepts rely on the large span-wise coherence (i.e. the two-dimensionality) of the dominant turbulent structures that are triggered and enhanced by the periodic excitation. A separated mixing layer or one that bounds a bubble are not likely to be dominated by stream-wise vortices, however a boundary layer on a concave surface (such that exists on most of Liebeck's airfoils [5]) or a wall jet on a convex surface may indeed be. Two cases are discussed in the paper, one that contains large streamwise vortices and one that does not.

2. The Control of Separation Over a Flap

Mean vorticity contours plotted in fig. 1a, represent the flow over a flap that is deflected upward at the angle shown relative to the x-axis ($\alpha=26^{\circ}$). The negative vorticity region indicates the existence of a large bubble whose dimensions are commensurate with the length of the flap. A very similar bubble may exist just prior to complete separation. The

G.E.A. Meier and P.R. Viswanath (eds.),
IUTAM Symposium on Mechanics of Passive and Active Flow Control, 337–342.

time-averaged pressure distribution shown above the vorticity contours confirms the existence of a large bubble. There is an almost constant pressure region extending up to 30% of the flap length to be followed by an adverse pressure gradient that relaxes slowly

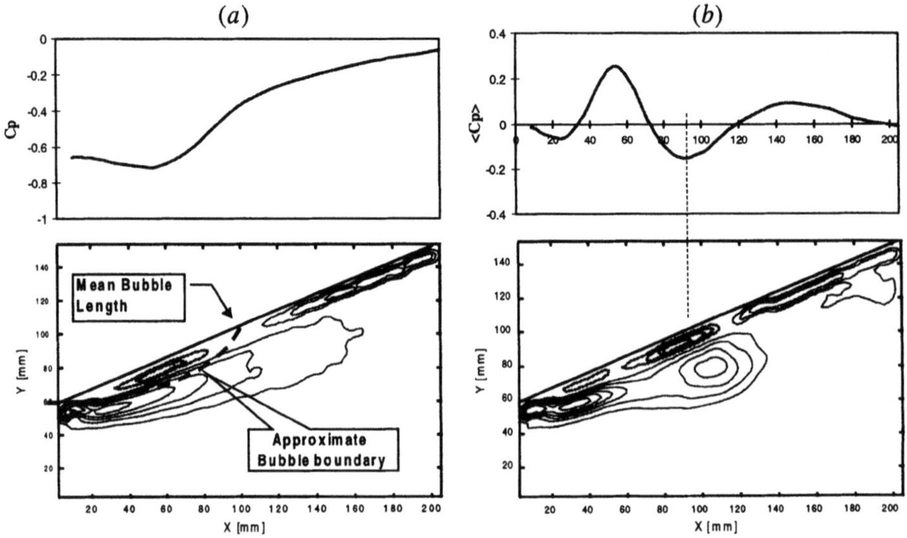

Figure 1. (a) Mean vorticity and wall pressure, (b) phase-locked vorticity and wall pressure.

as the trailing edge is approached. The mean length of the bubble can be approximately deduced from the vorticity but it was established from streamline pattern calculated from the measured velocity profiles. The bubble extends some 105*mm* from the flap-hinge when $\langle c_\mu \rangle = 0.01\%$. The vortex sheet shed from the upstream plate, flaps periodically up and down as a consequence of the imposed excitation. Thus the length of the bubble trapped near the flap-hinge varies throughout the excitation cycle. An example of the distribution of the phase-locked vorticity and pressure oscillations is shown in fig. 1b. It indicates that the bubble is much shorter because a large vortex has been detached from the original vortex-sheet and is being convected downstream. The passage of the vortex is characterized by a negative oscillation in the pressure coefficient resulting from the curvature of the local streamlines.

The computed streak lines (fig. 2a) illustrate the rollup of the vortex-sheet into discrete vortices with increasing distance from the flap hinge. One may recognize from this pattern that the flow bears a striking similarity to the classical mixing layer that forms the boundary between two parallel streams of different velocity. Nishri *et al.* [4] had shown that the most effective way to induce reattachment of a separated mixing layer is to excite it at a reduced frequency of $F^+ = fL_f / U_\infty \approx 1$. Their measurements revealed that the mixing layer was thickest opposite the trailing edge at that reduced frequency. Furthermore, the product indicating the state of stability of the layer, $\theta f / U_\infty \approx 0.035$, corresponded to the location at which the mixing layer became neutrally stable to the imposed harmonic perturbations. Excitation at either lower or higher frequencies proved to be less effective, the lower frequencies did not complete their amplification cycle by the time the perturbations arrived to the trailing edge, while the higher ones started to decay much too early. Power spectra indicated [6] that the amplitude of all the frequencies whose $F < 1$ increased with increasing x as they do in the canonical mixing layer. The amplitude distributions of the phase-locked streamwise velocity fluctuations, filtered at the forcing frequency resemble the amplitudes measured by

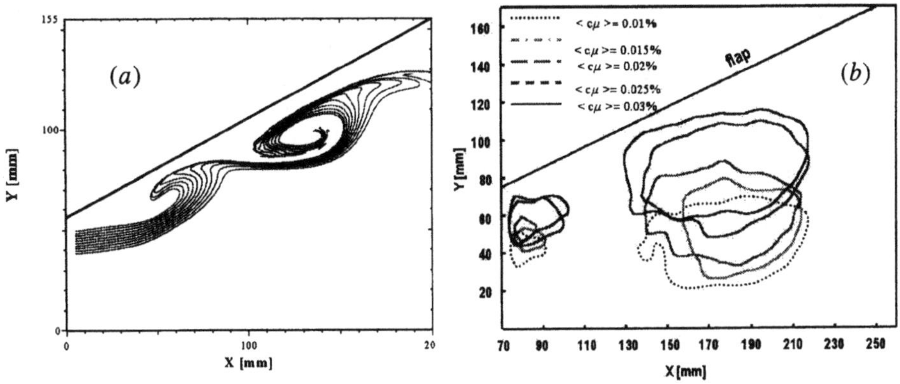

Figure 2. (a) Rollup of the vortex-sheet into discrete vortices, (b) Reattachment process resulting from increasing the input amplitudes.

Gaster *et al.* [6] in the forced mixing layer. When the forcing frequency corresponded to $F^+ = 1$ the amplitude distribution measured, agreed with the calculated neutrally stable mode. All this evidence provided a criterion for the optimum periodic forcing necessary to cause reattachment.

Phase locked vorticity measurements, using a PIV, confirmed the criteria for reattachment proposed in reference 4, they also indicated that the large structures in this flow do not undergo a major deformation during the reattachment process resulting from increasing the input amplitudes (fig. 2b). Since the flow bounding the bubble shown in fig. 1a also resembled the familiar mixing layer, an attempt was made to calculate the amplitudes and phase distributions at the excitation frequency using the familiar computational techniques [6]. Although these calculations predicted the general behavior of the large coherent structures surrounding the bubble the agreement between the simple model and experiment was only qualitative in nature. One probably has to account for pressure gradient, curvature and the large finite amplitudes in order to improve on this prediction.

Ensemble-averaged pressure distributions that were synchronized with the excitation signal enabled us to educe the phase of the surface pressure fluctuations (fig. 1b) and from them to calculate the propagation velocity of the pressure wave in the direction of streaming. The pressure wave at the frequency of excitation was coherent over the entire surface provided the excitation frequency was low ($F^+ < 3.5$). The phase velocity of the pressure wave increased with increasing distance from the flap-hinge in contrast to the free stream velocity that decelerated. This was confirmed by the PIV measurements and lead us initially to believe that the flow bounding the bubble near the flap hinge might have been absolutely unstable. More detailed assessment of the reverse flow in the bubble indicated that it was too weak to cause a global instability. Nevertheless the detrimental effects of weak steady blowing on the control of flow separation are under intensive scrutiny

3. Streamwise Vortices Resulting From Surface Curvature

The shape of the surface may also initiate the generation of streamwise vortices through a centrifugal instability of the turbulent flow. This instability may compete with the two-dimensional instability and reduce the spanwise coherence of the eddies associated with a uniform excitation along the span. Consequently, concave curvature may be detrimental to the control of separation of a boundary layer by periodic excitation. In order to assess the effects of curvature on the effectiveness of boundary layer control by periodic excitation, the evolution of a wall jet flowing over the exterior of a circular cylinder was investigated. This configuration was selected because the flow was found to spread outward much more rapidly than over a corresponding flat surface (at some locations the difference in the rate of spread of the two flows compared approached an order of magnitude) [7].

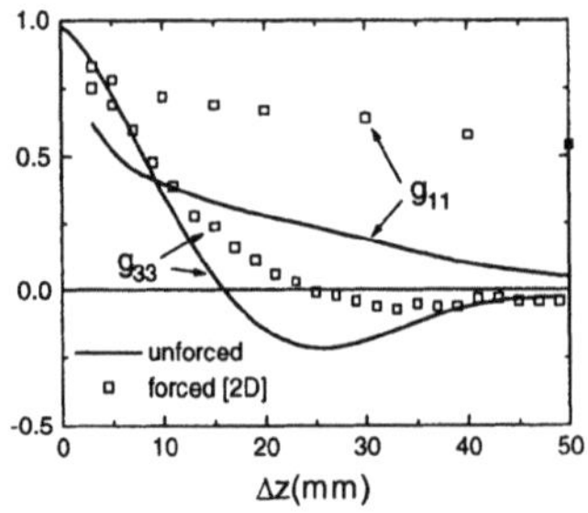

Figure 3. Cross correlation in the presence and the absence of plane excitation.

Cross correlation measurements in the presence and the absence of plane excitation revealed the competition between these two modes of instability (fig.3). The measurements were made using two hot wire probes that were separated by a spanwise distance, Δz at azimuthal angle $\theta - \theta_0 = 100^0$ (that in this case represents the distance from the virtual origin of the jet). The g_{11} correlation shown in the figure refers to a normalized product of the streamwise velocity component, u'_θ while the g_{33} correlation refers to the product of the spanwise velocity component. Streamwise vortices would generate alternating directions in the spanwise velocity component near the surface. The u_z-component provides the strongest indication for the existence of streamwise large-scale vortices since the normal component u_r, vanishes near the wall and the u_θ-component represents a secondary effect (i.e. a downward motion brings in high-momentum fluid to the vicinity of the surface that increases the local velocity and vice versa.). In the example shown, the natural g_{33} correlation attained a minimum value of -0.21, when the spanwise separation between the probes was approximately $25mm$. It implies that the averaged spanwise dimension of a counter rotating pair of the streamwise vortices is $\approx 50mm$ at that downstream location. This correlation was invariant to the spanwise location of the stationery probe. A simple traverse measuring the u_z-component of velocity (using an LDA) across the span indicated a periodicity that agreed well with the correlation measurements that used hot wire anemometers. The g_{11} correlation decayed exponentially with increasing probe separation and it only provided a measure to the scale of the naturally occurring spanwise vortices. Two-dimensional excitation emanating from the nozzle had a tremendous impact on the three-dimensionality of the flow. The g_{11} correlation at $\Delta z = 50mm$ increased 10 fold (from approximately 0.05 to 0.5). The negative g_{33} decreased by a factor of 4 and its spanwise location increased to $\Delta z = 30mm$. This suggests a diminution in the strength and coherence of the streamwise vortices and an increase in their scale.

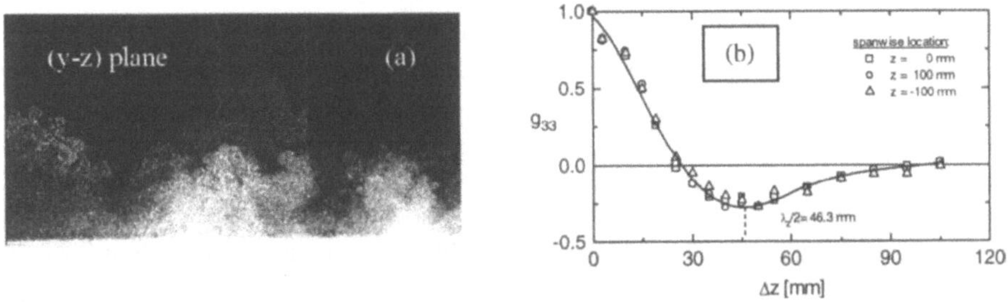

Figure 4. (a) Flow visualization and, (b) cross correlation at $(\theta - \theta_o) = 160^o$

A comprehensive investigation of the streamwise vortices occurring naturally in the flow was undertaken. Flow visualization (fig.4a), that used smoke as tracer particles and a laser-sheet that illuminated the plane normal to the flow direction, revealed clearly the

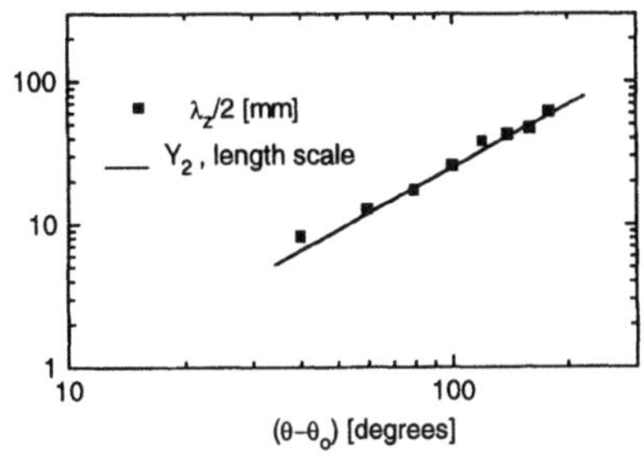

Figure 5. Wave length of the streamwise vortices versus downstream location.

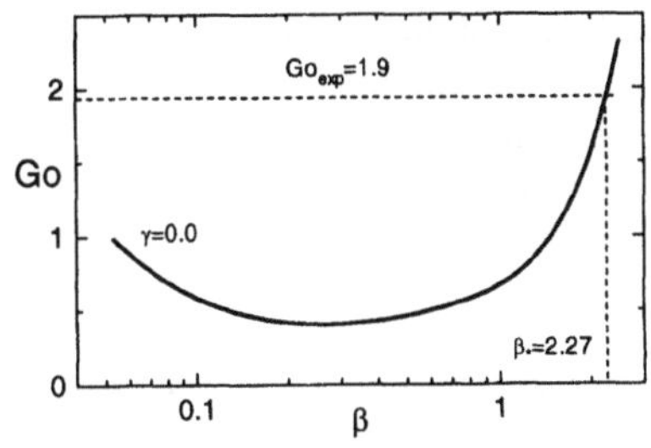

Figure 6. Neutral stability diagram for the longitudinal vortices in the turbulent wall jet around a circular cylinder.

existence of streamwise eddies. The scale of these eddies was in agreement with the scales educed from the g_{33} correlations (fig.4b). Figure 4b also illustrates how the spanwise wavelength of the longitudinal vortices was determined. When the procedure was repeated at various azimuthal angles along the cylinder (for $40^o<\theta<180^o$) [7] it became clear that the wave length of the streamwise vortices, λ_z increased with θ (fig.5). It turned out that λ_z scaled with the local width of the flow over the entire range of parameters considered, providing the following relationship: $\lambda_z \approx 2 \cdot Y_2$, (the solid line represents the mean length scale Y_2 at various downstream locations).

Calculations have been carried out to evaluate stability of the turbulent wall jet flowing over a circular cylinder with respect to the Gortler-type perturbations. One assumes that the longitudinal vortices generated are a manifestation of a linear centrifugal instability of the mean velocity profile. It is assumed that the effect of the random (smaller scale) turbulence on these vortices is the same as on the mean motion. The appropriate control parameter is $Go = \varepsilon * \mathrm{Re}_t$ since it is constant for the flow under consideration ($\varepsilon = Y_2/R$ is the curvature parameter and $\mathrm{Re}_t = Y_2 U_m / \nu_t$ is the turbulent Reynolds number). For the present experimental conditions $Go_{\mathbf{exp}} = 1.9$. The Gortler type stability calculations are shown in fig.6 assuming that the mean flow is locally parallel. In this figure, $\beta = 2\pi Y_2 / \lambda_z$ is the spanwise wavenumber of a vortex pair, rendered dimensionless with respect to Y_2. One can expect that the neutral perturbations will provide the dominant spanwise wavelength of the longitudinal vortex pairs observed in the experiment because these perturbations had undergone through the highest possible amplification process. The justification for this assumption is shown in fig.5 where the spanwise wavelength associated with the streamwise vortices is proportional to the local length scale. The calculation gives the critical wavenumber $\beta_* = 2.27$ for $Go_{\mathbf{exp}} = 1.9$ and $\varepsilon = 0.1$, while $\beta_{*\mathbf{exp}} \approx 3$. The good agreement between the theoretical model and the measured wavelength suggests that the centrifugal instability introduced by the curvature contributes to the transport of momentum across the flow.

4. Conclusions

The large coherent structures observed in turbulent shear flows such as the mixing layer, the wall jet, or the turbulent boundary layer in an adverse pressure gradient can be viewed as predominant instability modes. This point of view provides a basis for the enhancement of mixing, and control of separation by periodic excitation. The present paper reveals that stationary, streamwise vortices generated by surface curvature can also be predicted by stability considerations; in this case a centrifugal instability.

Referances

1. Seifert, A., Bachar, T., Koss, D., Shepshelovich, M., and Wygnanski, I.: Oscillatory blowing, a tool to delay boundary layer separation, *AIAA J.* **31** (1994), 2052-2060.
2. Seifert, A., Darabi, A., and Wygnanski, I.: Delay of airfoil stall by periodic excitation, *J. of Aircraft* **33**, No.3, (1996), 691-698.
3. Seifert, A. and Pack, L. G.: Oscillatory control of separation at high Reynolds numbers, *AIAA Paper 98-0214*, (1998).
4. Nishri, B. and Wynanski, I.: The effect of periodic excitation on turbulent flow separation from a flap, *AIAA J.* **36**, (1998), 547-556.
5. Liebck, R.H. and Camancho, P.P.: Airfoil design at low Reynolds numbers with constrained pitching moment, *UNDAS CP-77B123*.
6. Gaster, M., Kit, E., and Wygnanski, I.: Large scale structures in a forced turbulent mixing layer, *J. Fluid Mech.* **150**, (1985), 23-39.
7. Neuendorf, R. and Wygnanski, I.: On a turbulent wall jet flowing over a circular cylinder, accepted for publication in *J. Fluid Mech.*, (1998).

ACTIVE CONTROL OF KELVIN-HELMHOLTZ WAVES

J. E. FFOWCS WILLIAMS AND W. MÖHRING
Department of Engineering, University of Cambridge,
Cambridge CB2 1PZ, U.K.
Max-Planck-Institut für Strömungsforschung[1]
Postfach 2853, 37018 Göttingen, Germany

1. Introduction

One of the greatest nuisances resulting from the turbulence of high Reynolds number jets is their noise, a noise so loud as to threaten the viability of commercial supersonic aircraft on environmental grounds. Few with knowledge of turbulence's chaotic character can hold much expectation that turbulence will ever be controlled and made quiet. Avoiding turbulence altogether might be a simpler thing to do. A steady laminar jet would be silent, but so different from a jet's natural state that should it exist at all it would only do so under highly artificial conditions. The steady, silent jet conforms with all the laws of motion but is ruled out on stability grounds. Without extraneous interference, infinitesimal disturbances would grow and gain strength to become unmanageable and chaotic. Indeed it is this fact, that the mean flow so readily passes on its energy to parasitic eddies, that accounts for the robustness of shear layer turbulence.

Is it conceivable that the normal instability of the shear layer, the root cause of the jet's turbulence and noise, might possibly be avoided by active control? The benefits of control would surely be enormous and could well be sought if only one knew how to approach the problem [1, 2, 3, 4]. Control is bound to be difficult and may well turn out to be impractical but, from the results of the model considered below, we see no grounds for thinking the concept impossible. What we consider is the simplest idealisation of a shear flow, two-dimensional disturbances to an unbounded vortex sheet at the plane interface of a uniformly moving fluid and a fluid at rest. The flow

[1]Most of this research was done while W.M. was the quatercentenary visiting research fellow at Emmanuel College, Cambridge

G.E.A. Meier and P.R. Viswanath (eds.),
IUTAM Symposium on Mechanics of Passive and Active Flow Control, 343–348.

is inviscid and incompressible and could not persist because of the classical Kelvin-Helmholtz instability.

Weak enough disturbances can be regarded as a linear superposition of harmonic elements, each Fourier component having a growth rate equal to its frequency; all are unstable. A dispersion equation connects the spatial and temporal scales from which, at every frequency, one can see that the instability is a wave that travels downstream at half the flow speed, growing exponentially with downstream distance. If one were to monitor the strength of instability waves then one could consider using the information for the creation of exactly opposite waves, and hence by linear superposition, destructively interfering with and canceling disturbances downstream of the controller. This kind of thing works very well in silencing noise inside ducts by active methods [5]. Were this to prove feasible one could anticipate that conditions upstream of the controller would not be important because, providing of course that the disturbances were weak enough to be linear, the controller would be made to clean up the flow and pass on downstream an instability-free layer whose fate would then be determined by downstream conditions. It would be completely different from those with which we are familiar, which come from upstream already containing the seeds of chaos. It would instead be planted with more desirable seed and grow into something better than chaos.

We show below that it is possible to analyse the model problem exactly. Through the model we determine that a control scheme is possible, that it obeys all the constraints on physical reality and that the controller is causal. Of course the planar vortex sheet is not a jet, real disturbances are not two-dimensional and the viscosity is not zero even in the very high Reynolds Number cases that stimulate our interest. But our demonstration that it is definitely possible to devise a control scheme, in what is possibly the only example simple enough to work out exactly, does serve as encouragement to consider seriously the possibility that artificially maintained silent propulsive jets might one day exist.

2. Basic Relations

We consider small perturbations of an infinite or semi-infinite incompressible shear layer. We assume, that the unperturbed shear layer is at $y = 0$ and that there is a flow of uniform velocity U for $y > 0$ and no velocity for $y < 0$. Let $y = \eta(x, t)$ be the position of the perturbed shear layer. We require it to be small, such that higher order contributions may be neglected and that the boundary conditions can be fulfilled at $y = 0$. The flow in both half-planes $y < 0$ and $y > 0$ can be described by potentials Φ_i, by stream functions Ψ_i or by complex potentials $F_n = \Phi_n + i\,\Psi_n$. The index $n = 1$ refers to the upper halfplane with flow and the index $n = 2$ to the lower half-plane.

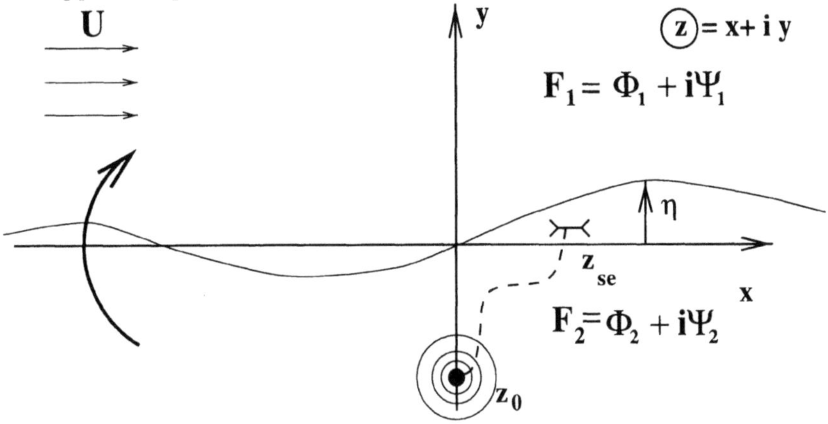

Fig. 1: Sketch of the flow with shear layer, sensor position z_{se} and point source.

To derive the boundary conditions at the shear layer, we use the method described in [6] and consider the mirror-reflection of the complex potential F_2 into the upper half-plane, which we also denote by F_2. In [6] it is shown that F_1 and F_2 fulfill the partial differential equations

$$\frac{\partial F_1}{\partial t} + U\frac{\partial F_1}{\partial z} - \frac{\partial F_2}{\partial t} = f(z,t) \quad \text{and} \quad \frac{\partial F_2}{\partial t} + U\frac{\partial F_2}{\partial z} + \frac{\partial F_1}{\partial t} = g(z,t) \quad (1)$$

where $f(z,t)$ and $g(z,t)$ are analytic functions in the upper halfplane. For an alternative approach, see Bechert [7]. From the continuity of pressure and particle displacement one finds that $f(z,t)$ and $g(z,t)$ have vanishing real and imaginary parts at $y = 0$ respectively. Both functions then admit an analytic continuation into the whole complex plane. For our application, we are interested in a flow with a shear layer and a point source of strength $Q(t)$ in the no-flow region. If the mirror image of the point source in the upper halfplane is at z_0, we obtain for $f(z,t)$ and $g(z,t)$

$$\begin{aligned} f &= Q(t)\frac{\partial}{\partial t}\left(\log(z-\overline{z_0}) - \log(z-z_0)\right), \\ g &= Q(t)\left(\frac{\partial}{\partial t} + U\frac{\partial}{\partial z}\right)\left(\log(z-\overline{z_0}) + \log(z-z_0)\right) \end{aligned} \quad (2)$$

where $\overline{z_0}$ denotes the complex conjugate of z_0.

For a stability analysis, one assumes that all functions are proportional to $e^{-j\omega t}$ with a second imaginary unit j. The algebraic and analytic complications which occur because of the use of two imaginary units i and j are described e.g. in [6]. Here it may be sufficient to mention, that functions depending on i and j can be considered to be a pair of ordinary complex functions containing either the imaginary units i or j. Then the eqs.(1) with the right hand sides taken from the eqs.(2) become ordinary differential equations with constant coefficients which can be solved using the method of variation of parameters. One then obtains for the complex potentials

$$\begin{aligned} \mathbf{F} &= \begin{pmatrix} F_1 \\ F_2 \end{pmatrix} = A(1-ij)\begin{pmatrix} e^{\frac{(1+j)\omega z}{U}} \\ 0 \end{pmatrix} + Q\begin{pmatrix} \log(z-\overline{z_0}) \\ \log(z-z_0) \end{pmatrix} \\ &+Q\begin{pmatrix} 1 & -1 \\ j & j \end{pmatrix}\begin{pmatrix} -\frac{1+j}{2}e^{\frac{(1+j)\omega z}{U}} \oint_{-j\infty}^{z} \frac{e^{-\frac{(1+j)\omega\zeta}{U}}}{\zeta-\overline{z_0}}\,d\zeta \\ \frac{1-j}{2}e^{\frac{(-1+j)\omega z}{U}} \int_{-\infty}^{z} \frac{e^{\frac{(-1+j)\omega\zeta}{U}}}{\zeta-\overline{z_0}}\,d\zeta \end{pmatrix} \\ &= A\mathbf{F}_{KH} + Q\mathbf{F}_Q. \end{aligned} \tag{3}$$

Eq.(3) shows, that the complex potentials $\mathbf{F}$ are superpositions of a Kelvin-Helmholtz wave of amplitude A, which is an arbitrary integration constant, and of the source solution $\mathbf{F}_Q$ of amplitude Q. The complex line integrals in $\mathbf{F}_Q$ contain a singularity at $\overline{z_0}$ in the lower halfplane and the value of the integral depends on the integration path. Both integrals contain i and j and correspond therefore to two ordinary line integrals. For z in the upper halfplane and positive real ω, three of the integrals can be evaluated in the upper halfplane and are therefore well defined. The mark at the first integral is to indicate, that the integration path of the remaining fourth integral should stay to the left of the singularity at $\overline{z_0}$. One then finds, that $\mathbf{F}$ grows in the upper halfplane at most logarithmically with the only exception of a sector near the positive x-axis. There one finds a dominating Kelvin-Helmholtz wave of amplitude $f_{KH}Q$ such that one has asymptotically

$$\mathbf{F} \sim (A + f_{KH}Q)\,\mathbf{F}_{KH} \quad \text{with} \quad f_{KH} = \pi\frac{1-j}{2}e^{\frac{(1+j)\omega}{U}z_0} \tag{4}$$

3. The Controlled Shear Layer

Eq.(4) shows that there is for every incident Kelvin-Helmholtz wave of amplitude A a source solution with a suitable source strength $Q = -A/f_{KH}$ such that there is no outgoing Kelvin-Helmholtz wave. From eq.(3) one finds that this controlled shear layer $\mathbf{F}_{CS}$ is given by

$$\mathbf{F} = Q\mathbf{F}_{CS} \quad \text{with} \quad \mathbf{F}_{CS} = (\mathbf{F}_Q - \frac{1}{f_{KH}}\mathbf{F}_{KH}). \tag{5}$$

The shear layer displacements of this controlled shear layer are shown in Fig. 2.

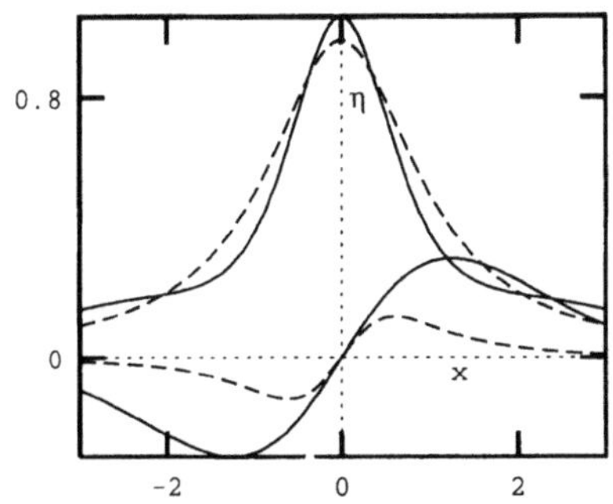

Fig. 2: Real and imaginary parts of the displacement of the controlled shear layer for $\omega = 1$ (full) and $\omega = 5$ (dashed).

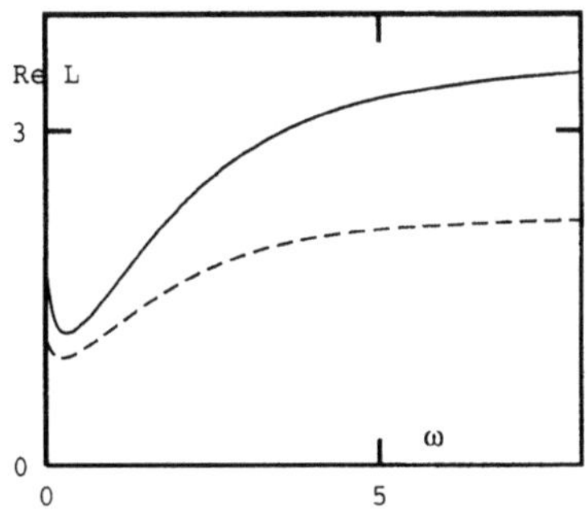

Fig. 3: The real part of the controll impedance (Sensor at 0.3 (full) and 0.6 (dashed).

To obtain the behaviour of the controlled shear layer, the source strength has to be chosen apropriately. A possible way to achieve this, is to relate the source strength Q with some transfer function L to the velocity u at a suitable sensor position z_{se} of the controlled shear layer. With u from eq.(5) one obtains $Q = Lu(z_{se}) = LQu_{CS}(z_{se})$ or

$$1 = L\, u_{CS}(z_{se}). \tag{6}$$

The transfer function is just the reciprocal value of the velocity at the sensor position. The velocity $u_{CS}(z_{se})$ depends on the frequency. A control for all frequencies is obtained, if eq.(6) is fulfilled for all frequencies. As an example, we consider a sensor position close to the shear layer in the region with mean flow. The real and imaginary parts of L are shown in Fig.(3) and Fig.(4) for two different sensor positions $z_{se} = 0.3$ and $z_{se} = 0.6$. All numerical examples were computed with mean flow velocity $U = 1$ and a distance of 1 from the actuator to the shear layer.

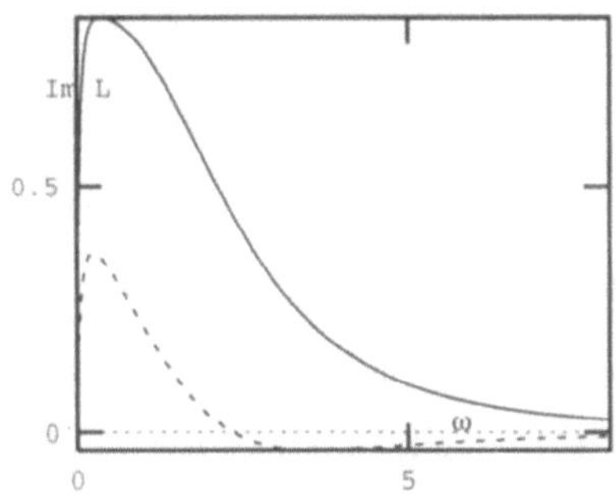

Fig. 4: The imaginary part of the controll impedance (Sensor at 0.3 (full) and 0.6 (dashed).

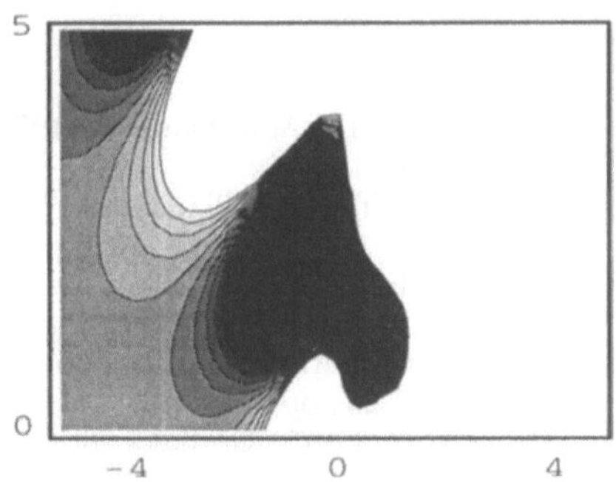

Fig. 5: Point sources in the shaded regions are effectively controlled. ($\omega = 1$, $z_{se} = 0.3$.)

Transfer functions are Fourier transforms of real functions. They should fulfill the physical requirements of

- reality, i.e. $L(\omega) = \overline{L(-\overline{\omega})}$
- causality, i.e. $L(\omega)$ is an analytic function in the upper complex ω-halfplane,
- eventual positive "dissipation", $\mathrm{Im}\,\omega L(\omega) > 0$ for real ω, implies stability,

see e.g. [8]. Reality can be used to determine $L(\omega)$ for negative frequencies. Causality is a very restrictive requirement and it is actually not fulfilled for $L(\omega)$ determined from eq.(6), $L(\omega)$ is not analytic. One observes however that the source solution $\mathbf{F}_Q$ in eq.(5) is analytic, the non-analyticity comes from the limit $x \to \infty$ involved in the determination of the Kelvin-Helmholtz wave. This shows that $L(\omega)$ can be approximated by causal impedances, e.g. by generating the incident Kelvin-Helmholtz wave by an upstream point source and modifying the control condition to ensure a zero potential at a finite distance downstream. These transfer functions are more complicated and differ from the L from eq.(6) only for small values of ω. For simplicity we keep the simpler L of eq.(6). Fig. 4 shows that a positive imaginary part of L can be achieved for a suitable sensor position, e.g. $z_{se} = 0.3$.

The impedance L has been designed to controll incident Kelvin-Helmholtz waves. To study its effectivity for other pertubations, the motion of the controlled shear layer with a point source has been determined. The results are shown in Fig.5. Kelvin-Helmholtz waves generated by point sources located in the shaded region are suppressed by more than 95%. The control scheme therefore suppresses effectively perturbations generated upstream of the sensor position.

References

1. Moin, P., Bewley T. (1994) Feedback control of turbulence. *Appl. Mech. Rev.* **47**, pp. S3–S13.
2. Ronneberger D., Ackermann V. (1979) Sound from stability waves in jets. *J. Sound Vib.* **62**, pp. 121–129.
3. Arbey H., Ffowcs Williams J.E. (1984) Active Cancellation of Pure Tones in an excited Jet. *J. Fluid Mech.* **149**, pp. 445–454.
4. Fiedler H.E. (1998) Control of free turbulent shear flow, in M. Gad-el-Hak, A. Pollard, J.-P. Bonnet, *Flow Control*. pp. 335–429. Springer.
5. Swinbanks M. A. (1973) The active control of sound propagation in long ducts. *J. Sound Vib.* **27**, pp. 411–436.
6. Möhring W. (1975) On flows with vortex sheets and solid plates. *J. Sound Vib.* **38** pp. 403–412.
7. Bechert, D.W. (1982) Excited waves in shear layers. *DFVLR-FB 82-23*
8. Landau L.D., Lifschitz E.M. (1966) *Lehrbuch der theoretischen Physik. Bd. V. Statistische Physik*.§125. Akademie Verlag.

DYNAMIC CONTROL OF THE JET-EDGE-FLOW

H. PRECKEL AND D. RONNEBERGER
3. Physikalisches Institut der Universität Göttingen
Bürgerstr. 42-44, 37073 Göttingen

1. Introduction

An important difference between the control of fluid mechanical systems and the control of electrical or mechanical systems is that even laminar flow systems with a low number of active degrees of freedom have an inherently high number of degrees of freedom which can be easily excited by an external control force. Thus the flow may change from a controlled into an uncontrolled state by activation of further degrees of freedom; in the extreme case the flow becomes turbulent. The objective of the present study is the applicability of different control strategies for the plane, laminar jet-edge-flow as shown in figure 1. The first step is to stabilize this flow with (adaptive) linear feedback control, and to determine the limits of this strategy. The next step is to extend the control to model and compensate for nonlinearities. A further extension will be to build a mathematical model of the whole system for control.

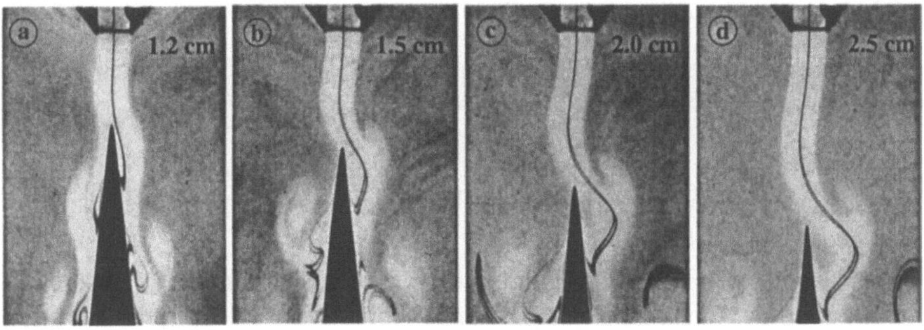

Figure 1. The plane, laminar jet-edge-flow for different edge distances (mean flow velocity 5.36 cm/s, nozzle width d=4mm, Re_d=214)

G.E.A. Meier and P.R. Viswanath (eds.),
IUTAM Symposium on Mechanics of Passive and Active Flow Control, 349–354.

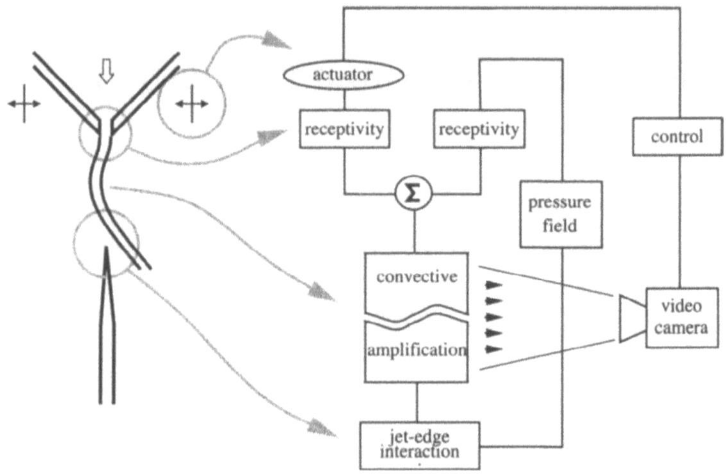

Figure 2. Block diagram of the system

2. Experimental setup

The system consists of a laminar plane jet leaving a nozzle and impinging on a plane wedge. The system performs self excited oscillations due to the convective amplification of the jet and the pressure feedback from the edge to the nozzle. The position of the jet is measured by means of dye which is injected in the center of the jet and monitored by a video camera connected to a real time video image processing system. The jet can be controlled by an external pressure gradient at the nozzle. The pressure gradient is caused by membranes which are mounted in the channel walls on the left and right side of the nozzle (fig. 2). The membranes are moved by a stepping motor connected to a microcontroller.

3. Physical description of the equivalent lumped elements of the system

The feedback loop can be divided into different elements as shown in figure 2. The receptivity of the jet to the pressure gradient at the nozzle is assumed to be linear.

In general the convective amplification of disturbances on the jet is an exponential function of the distance from the nozzle (figure 3a) in a small frequency range. Other frequencies are damped. This can easily be measured. In addition the amplification is dependent on the amplitude of the jet in a complicated way. The general trend is that the amplification becomes lower for higher amplitudes of the jet. An example is shown in

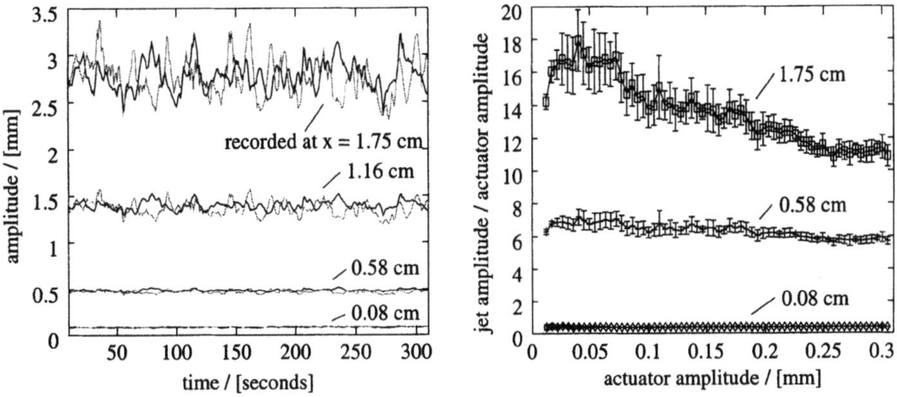

Figure 3. (a) Comparison of self-excited oscillations (——) and actuator-excited oscillations (--------) (without edge) at the same frequency and amplitude (nozzle-edge distance 2 cm, mean flow velocity 5.36 cm/s, nozzle width d=4mm, f=1.17 Hz, Re_d=214, Sr=0.55, actuator amplitude 0.22 mm) (b) Amplitude dependency of the transfer-factor from the actuator amplitude to the jet amplitude at 1.17 Hz

figure 3b.

The jet-edge interaction cannot be measured directly, because the resulting pressure is too low to be measured by a hydrophone. It is supposed to depend only slightly on the frequency and to be nonlinear with a saturation at high amplitudes.

The propagation of the pressure dipole field from the edge is known analytically to be linear.

4. Special problems modelling the jet-edge-system

Measuring the transfer functions of isolated elements of the feedback loop is difficult because the loop cannot be opened without changing the properties of the elements.

Due to the limited size of the flow channel, there are compensation flows in the channel which flow slowly upwards left and right of the jet. They supply the medium sucked in and transported down by the jet. If the jet-state changes, e.g. the amplitude becomes smaller, the compensation flows have to adjust accordingly. This happens on a timescale of minutes. It leads to variable boundary conditions, which are dependent on the jet-motion.

Another problem is that the oscillation of the system varies. In figure 3a the momentary jet-amplitude is shown for some typical parameters. The amplitude changes by ± 15% at a distance of 1.75 cm from the nozzle. This is true for both self-excited oscillations and actuator-excited oscillations at the same frequency and amplitude. External perturbations cause jet amplitudes below 0.1 mm, so they cannot be the reason for these fluctuations.

5. Simple coefficient feedback control

The easiest way to suppress the feedback from the edge to the nozzle is to measure the lateral jet displacement at some location between the nozzle and the edge and to feed the actuator by a delayed and inverted version of this signal. Delay and gain are subsequently adjusted so that the overall closed loop gain is reduced below one. This works for small flow velocities and small distances between the edge and the nozzle. This is equivalent to low gain of the convective instability and only a few possible degrees of freedom. Successful suppression of the motion was achieved with the parameters according to figure 1a. For longer nozzle-edge distances this simple technique fails as the system becomes unstable for higher modes (e.g. with parameters according to figures 1b-1d).

6. Feedback control with FIR-filter

To control the jet at larger nozzle-edge distances the feedback from the edge has to be compensated for in a broader range of frequencies. This can be achieved by using a digital FIR-filter in the control path. We used three methods for choosing the filter which all work for larger nozzle-edge distances than the simple coefficient compensation scheme. All methods have been tested with the parameters of figure 1c.

6.1. LMS ADAPTIVE ALGORITHM

A LMS adaption algorithm with reference and error signal in the middle between the edge and the nozzle is used. The filter is adapted to cancel the correlations between the reference signal (the disturbances on the jet flowing towards the edge) and the error (the disturbances that are fed back from the edge to the nozzle). The adaption process is very sensitive to external disturbances because the control loop easily becomes unstable during adaption. Then the system performs undamped oscillations of high amplitude and the jet becomes corrupted. If the adaption is successful a complete suppression of the motion is achieved in the situation of figure 1c while in 1d the adaption never succeeds.

6.2. GENETIC ALGORITHM

Further on a genetic algorithm minimizing the rms-value of the jet motion is used. A pool of 20 random coefficient filters is taken and the performance of each filter is evaluated for 30 seconds. Then two filters (parents) are selected with a probability proportional to the performance. A new filter is constructed by taking each coefficient with equal probability from one

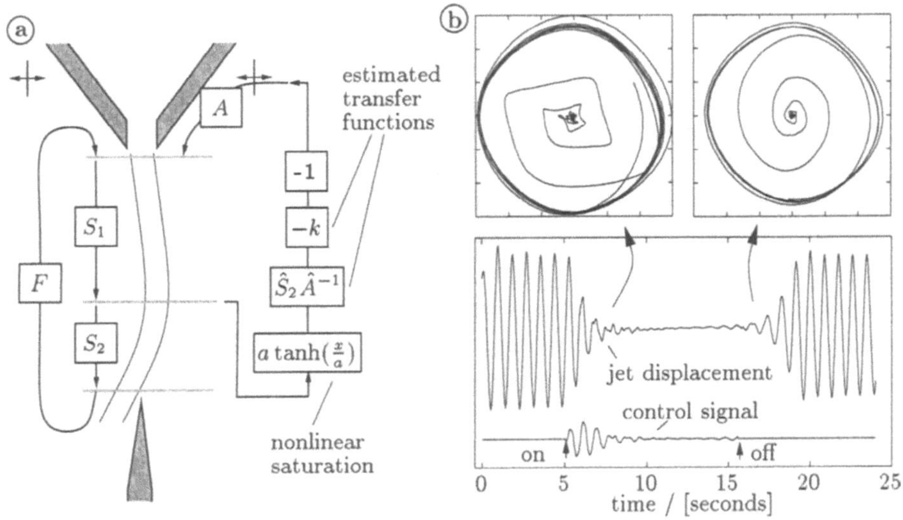

Figure 4. Example of simple feedback control with an FIR filter. a) Setup b) Time signal of jet-position with control switched on/off and state space projections

or the other parent. Additionally some random mutation is applied to the coefficients. All filters in the pool are replaced in this way. After 10 to 30 generations convergence is achieved.

The filters don't suppress the motion totally but they damp it better than 20 dB. To get further convergence the performance must be measured much longer. This is not possible in experimental studies because the adaption time becomes too long. For the parameters of figure 1c many short filters were found which lead to similar damping but all have totally different coefficients. For 1d no filter was found which damped the motion significantly.

6.3. TRANSFER FUNCTION MEASURED WITHOUT WEDGE

Another approach is shown in figure 4a. The transfer functions needed for control are measured without the wedge. First the transfer function of the jet S_2 is estimated by a FIR filter $\hat{S}_2$ using a LMS system identification algorithm. Then S_2A^{-1} is estimated by a FIR filter $\hat{S}_2\hat{A}^{-1}$ using $\hat{S}_2$ with a LMS system identification, too (A^{-1} is not causal because the actuator transfer function A has a delay; if S_2 has a longer delay than A, S_2A^{-1} is causal and can be used for realtime control). The transfer function F from the jet displacement at the wedge to the jet displacement at the nozzle is estimated by a negative constant $-k$. This can be confirmed qualitatively by measurements. Exact measurements are not possible due to the small jet displacement at the nozzle. The assumed nonlinear saturation distributed over S_2 and F is modeled by a tanh-function in front of S_2A^{-1}. Perhaps between S_2 and A^{-1} is a better place for the application of the nonlinear tanh-function but S_2 and A^{-1} cannot be separated because A^{-1} alone is not causal.

This scheme gives the best results out of the three tested ones. An example is shown in figure 4b. The first part of the time signal shows the uncontrolled state. Then the control is turned on for about ten seconds and turned off again. On top of the figure projections of the state space are shown for the start and end of control (time delay coordinates with a delay of a quarter of the uncontrolled oscillation period are used).

Due to the nonlinear saturation term this control scheme can be activated in any state of the system. In the other schemes the control must be turned on gently by increasing a gain factor from zero to one over some time to avoid high jet displacements which cannot be described by the simple linear model.

7. Summary and conclusions

Various different simple feedback control schemes were used to suppress the self-exited oscillations in the jet-edge-system. For low flow velocities and small nozzle-edge distances (e.g. figure 1a) a simple coefficient feedback scheme is sufficient.

To achieve control for higher nozzle-edge distances a FIR-filter is inserted in the feedback loop. Three different methods to determine a proper filter were shown. Despite of the simple model which neglects the nonlinearity of the jet instability at higher amplitudes and the complicated jet-edge-interaction, control is achieved for moderate edge-distances (figure 1b-c). This means that the pressure field produced by the jet impinging on the wedge is predicted from the video image of the jet with sufficient accuracy for suppressing self-excited oscillations in this case.

For higher edge-distances or higher flow velocities the prediction becomes worse and the jet-oscillations are not completely suppressed or the control loop performs self-excited oscillations (parameters of figure 1d). In this case a better model than the FIR filter must be used, which regards the nonlinearities. A first step in this direction is the nonlinear saturation term in the last-mentioned control scheme which leads to global stability of the controlled system.

An advanced control scheme should be based on a model of the system dynamics in a state space. For this, a design of the control signal should be constructed.

A LOW-ORDER MODEL-FOLLOWING STRATEGY FOR ACTIVE FLOW CONTROL

S. PRUDHOMME, L. LE LETTY
ONERA, Systems Control and Flight Dynamics Department
Centre d'Etudes et de Recherches de Toulouse,
2, avenue Edouard Belin, 31055 TOULOUSE cedex, FRANCE

Abstract

In the perspective of active flow control, an indirect approach is proposed, using a low-order model-following strategy. The objective of the control is to make the flow dynamics follow the dynamics of a given target *reference flow*. Both dynamics are written under the form of low order ordinary differential equations which approximate the partial differential equations. The first step is the derivation of such low order models, using proper orthogonal decomposition techniques. The second step is the design of the controller to achieve convenient model following. The equations for the derivation of the controller are given. A *distributed parameterization* is proposed for the *state feedback* case.

1. Motivation

1.1. DIRECT AND INDIRECT APPROACHES

Direct approaches for active flow control generally are based on the *optimal control* of the PDE that governs the flow dynamics. Most are of a theoretical nature [9, 7], need time consuming computations and lead to infinite dimensional controllers, so that they rarely are used for practical real time control in experiments.

Indirect approaches consist in designing simple low order models which are then used for the synthesis of the control laws with standard finite dimensional design procedures, in the same way as for controlling other kinds of system.

The approach proposed in this article is indirect. It is based on the *explicit model-following* control design technique. This technique is well known for finite dimensional systems. The key point is that it is adapted here for flow control, avoiding the manipulation of the infinite dimension of the flow dynamics for control law computation. More generally, it can be extended to distributed parameter systems control.

G.E.A. Meier and P.R. Viswanath (eds.),
IUTAM Symposium on Mechanics of Passive and Active Flow Control, 355–366.

1.2. AN INDIRECT APPROACH

The proposed *low-order model-following* strategy is characterized by two main items:

- For the system to be controlled, a model-following strategy consists in making its dynamics follow the dynamics of a reference system. Many applications are available in the literature, for instance for aircraft flight control [16]. The standard approach is to use a linear quadratic (LQ) formulation [8] which is adapted here for flow control. As illustrated figure 1, an extra signal is added to the reference input, in order to minimize the error between the *reference outputs* and the actual outputs. As the *reference dynamics* is explicitly needed for the implementation of the closed loop, such a control scheme is called explicit model-following. The gains of the controller are computed off-line, using the description of both dynamics.

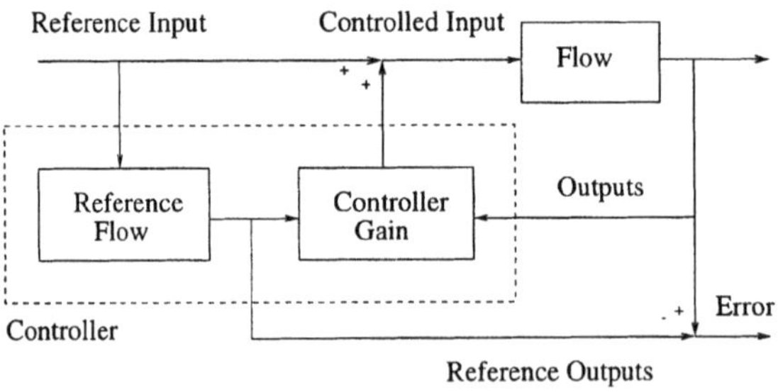

Figure 1. Explicit model-following control scheme

- The dynamics of the flow to be controlled and of the *reference flow* are governed by infinite dimensional PDE's. To design a controller with finite dimensional techniques, finite dimensional models must be constructed for both flows, under the form of ODE. This is why the strategy is called *low-order.* This is performed using proper orthogonal decomposition (POD) techniques [14, 1].

Section 2 presents the principle of reduced order models construction, using POD. Some difficulties mentioned in [11, 12, 17, 2] are partly solved to introduce the *reference input* in the equations. Model design is specialized for simultaneous flow dynamics modeling in section 3. Section 4 presents the derivation of the *model-following* controller. The equations are formulated under a special form, introducing distributed Lyapunov and Ricatti equations. Section 5 proposes a *distributed parameterization* for the controller and discusses the use of such a control strategy.

2. Low order model construction

2.1. PRINCIPLE

The PDE of the flow to be controlled in a domain $x \in \Omega$ with boundary Γ is denoted

$$\partial_t u = -\mathcal{L}_\Omega(u) \tag{1}$$

where $\mathcal{L}_\Omega$ is a space derivation operator which can be specialized for any kind of flow. There is no theoretical restriction for the form of $\mathcal{L}_\Omega$. Variables $u(x,t)$ can be multi-dimensional. For the sake of simplicity, the controlled boundary conditions are supposed to be written

$$u(x_c, t) = u_c(t) \tag{2}$$

with $x_c \in \Gamma$. This form may be a limitation, but can be easily extended to more complex boundary conditions as explained in 2.3.
For standard control of finite dimensional dynamical systems, ODE (also called *state space representation*) are used. They describe the dynamical relation between the inputs ν to the actuators, the internal variables (or *states*) a, and the outputs y from the sensors, of the form:

$$\begin{aligned} \dot{a} &= \mathcal{A}(a, \nu) \\ y &= \mathcal{C}(a, \nu) \end{aligned} \tag{3}$$

The derivation of such a model from the PDE is difficult. The challenge is to find finite dimensional states a to approximate the infinite dimensional variables u of the flow. Moreover, the actuators are located on the boundary, so that the flow is only partly controllable, and only a few sensors are available, generally located on the boundary, so that only a part of the flow is observable.

2.2. BASIC POD

POD techniques are not expected to solve the PDE. They use existing computational or experimental solutions to design simple models of flow dynamics. First studied from the mathematical point of view, they have been formulated within the aerodynamical context by Sirovich [14], and developed by Aubry [6, 1]. The first attempts to use POD for control purpose are very recent [17, 2, 11].
POD techniques are based on the Galerkin projection. Let $[\Phi_i(x)]_{i=1,2,\ldots}$ be an infinite dimensional normed orthogonal basis, with:

$$\int_\Omega \Phi_i(x)\Phi_j^T(x)dx = \delta_{ij} I$$

and I the identity matrix of same dimension as u.
Expanding the solution

$$u(x,t) = \sum_{i=1}^{\infty} a_i(t)\Phi_i(x)$$

and using the Galerkin projection

$$\int_{\Omega} \Phi_i(x)(\partial_t u + \mathcal{L}_{\Omega}(u))dx = 0$$

with orthogonal properties, it yields an infinite dimensional ODE:

$$\frac{da_i}{dt} = \mathcal{A}_i(a_1, a_2, ...)$$

where functions $\mathcal{A}_i$ have the same (non)linearity as PDE (1). With matrix notation:

$$\begin{aligned} \Phi(x) &= [\Phi_1(x)\ \Phi_2(x)\ ...]^T \\ a(t) &= [a_1(t)\ a_2(t)\ ...]^T \\ \dot{a} &= \mathcal{A}(a) \\ \mathcal{A}(a) &= -\int_{\Omega} \Phi \mathcal{L}_{\Omega}(\Phi^T a))dx \\ u(x,t) &= \Phi^T(x)a(t) \end{aligned}$$

This is a first step towards state space representation, but order is still infinite and controlled boundary conditions u_c of equation (2) do not explicitly appear as inputs.

To reduce the order, the idea is that only a finite set of Φ_i's may be enough to approximate the dynamics of the flow. To choose these functions, let $R(x, x')$ the autocorrelation tensor, computed from a known solution of equation (1) with suitable time dependent controlled inputs $u_c(t)$:

$$R(x, x') = \int_0^{\infty} u(x,t)u(x',t)dt$$

Let $[\lambda_i]_{i=1,2,...}$ its eigenvalues and $[\Phi_i(x)]_{i=1,2,...}$ its eigenfunctions or *modes*. The global energy

$$e = trace(R) = \sum_{i=1}^{\infty} \lambda_i \approx \sum_{i=1}^{r} \lambda_i$$

is the sum of the contribution of all modes. Only the few most significant modes are retained. The system is consequently truncated and is now finite dimensional. It has been shown in many examples, that a very small number of modes (generally less than 5) are enough for a good modeling of the flow dynamics ([5, 4, 3]).

2.3. INTRODUCTION OF A CONTROL FUNCTION

PDE (1) has no term at the right hand side, and has non homogeneous boundary conditions. The idea suggested in [17, 2] is to substitute an equivalent problem with homogeneous boundary conditions and with an extra term at the right hand side.

Let $\Phi_c(x)$ the stationary solution of PDE (1) with boundary condition $\Phi_c(x_c) = 1$, verifying $\mathcal{L}_\Omega(\Phi_c) = 0$. The new variable $\tilde{u} = u - \Phi_c^T u_c$ satisfies the following PDE with a right hand term:

$$\begin{aligned} \partial_t \tilde{u} &= -\mathcal{L}_\Omega(\tilde{u} + \Phi_c^T u_c) - \Phi_c^T \dot{u}_c \\ \tilde{u}(x_c, t) &= 0 \end{aligned} \tag{4}$$

with homogeneous boundary conditions. The controlled boundary condition $u_c(t)$ now appears as input. Actually, the most sensitive input seems to be $\dot{u}_c$ rather than u_c, so that in equation (3), the control which will be used for the design is $\nu = \dot{u}_c$. Then u_c becomes a state of the model which now includes an integrator.
The application of POD techniques on the new variable $\tilde{u}(x,t) = \Phi^T(x)a(t)$ then leads to homogeneous eigenfunctions $\Phi^T(x_c) = 0$, to the reconstructed solution

$$\begin{aligned} \dot{a} &= \mathcal{A}_0(a, u_c) + \dot{B}_0 \dot{u}_c \\ \mathcal{A}_0(a, u_c) &= -\int_\Omega \Phi \mathcal{L}_\Omega(\Phi^T a + \Phi_c^T u_c) dx \\ \dot{B}_0 &= -\int_\Omega \Phi \Phi_c^T dx \\ u &= \Phi^T a + \Phi_c^T u_c \end{aligned} \tag{5}$$

and to the following state space representation:

$$\begin{aligned} \begin{bmatrix} \dot{u}_c \\ \dot{a} \end{bmatrix} &= \begin{bmatrix} 0 \\ \mathcal{A}_0(u_c, a) \end{bmatrix} + \begin{bmatrix} I \\ \dot{B}_0 \end{bmatrix} \nu \\ u &= \begin{bmatrix} \Phi_c^T & \Phi^T \end{bmatrix} \begin{bmatrix} u_c \\ a \end{bmatrix} \end{aligned}$$

The PDE simplifies in the linear case:

$$\partial_t \tilde{u} = -\mathcal{L}_\Omega(\tilde{u}) - \mathcal{L}_\Omega(\Phi_c) u_c - \Phi_c^T \dot{u}_c$$

It yields:

$$\begin{aligned} \begin{bmatrix} \dot{u}_c \\ \dot{a} \end{bmatrix} &= \begin{bmatrix} 0 & 0 \\ B_0 & A_0 \end{bmatrix} \begin{bmatrix} u_c \\ a \end{bmatrix} + \begin{bmatrix} I \\ \dot{B}_0 \end{bmatrix} \nu \\ u &= \begin{bmatrix} \Phi_c^T & \Phi^T \end{bmatrix} \begin{bmatrix} u_c \\ a \end{bmatrix} \end{aligned} \tag{6}$$

with

$$\begin{aligned} A_0 &= -\int_\Omega \Phi \mathcal{L}_\Omega(\Phi^T) dx \\ \dot{B}_0 &= -\int_\Omega \Phi \Phi_c^T dx \\ B_0 &= -\int_\Omega \Phi \mathcal{L}_\Omega(\Phi_c^T) dx \end{aligned} \tag{7}$$

and $B_0 = 0$ if the control function can be chosen to verify $\mathcal{L}_\Omega(\Phi_c) = 0$.
In the nonlinear case, the linear combination $u - \Phi_c^T u_c$ remains motivated by the derivation of homogeneous conditions. This approach remains valid as long as the boundary conditions can be written as $\mathcal{L}_\Gamma(u(x_c,t)) = u_c(t)$ with $\mathcal{L}_\Gamma$ a linear space derivation operator. For more complex ones, it may not remain valid.

3. Application to Model-Following

3.1. SIMULTANEOUS *REFERENCE MODEL* DESIGN

Let the dynamics of the flow as in equations (1, 2), and the dynamics of the *reference flow*:

$$\begin{aligned} \partial_t u_r &= -\mathcal{L}_{\Omega r}(u_r) \\ u_r(x_c,t) &= u_c(t) \end{aligned}$$

where $\mathcal{L}_{\Omega r}$ a space derivation operator which may be different from $\mathcal{L}_\Omega$. A specific control function Φ_{rc} leads to the following manipulated solution:

$$\begin{aligned} \tilde{u}_r &= u_r - \Phi_{rc}^T u_c \\ \partial_t \tilde{u}_r &= -\mathcal{L}_{\Omega r}(\tilde{u}_r + \Phi_{rc}^T u_c) - \Phi_{rc}^T \dot{u}_c \\ \tilde{u}_r(x_c,t) &= 0 \end{aligned} \tag{8}$$

For finite dimensional model-following control design, low order models for both flows can be constructed within different procedures:

- Applying POD technique separately to the PDE of the flow to be controlled then to the PDE of the *reference flow* would lead to two sets of modes, respectively Φ, Φ_r with no reason to be identical.
- Applying POD technique to the manipulated solution $\tilde{u}$ of PDE (4) and using the obtained set of modes Φ for the Galerkin projection of the manipulated *reference solution* $\tilde{u}_r$ of PDE (8) yields:

$$\begin{aligned} \dot{a}_r &= \mathcal{A}_r(a_r,u_c) + \dot{B}_r \dot{u}_c \\ \mathcal{A}_r(a_r,u_c) &= -\int_\Omega \Phi \mathcal{L}_{\Omega r}(\Phi^T a_r + \Phi_{rc}^T u_c) dx \\ \dot{B}_r &= -\int_\Omega \Phi \Phi_{rc}^T dx \\ u_r &= \Phi^T a_r + \Phi_{rc}^T u_c \end{aligned} \tag{9}$$

  This leads to more suitable results. However, the modes Φ may not be complete enough for an accurate Galerkin projection of the *reference solution*, especially if the *reference dynamics* is significantly different from the exact one.
- The best solution consists in simultaneously applying POD technique on both PDE's, in order to get a unique set of modes Φ

$$\partial_t \begin{bmatrix} \tilde{u} \\ \tilde{u}_r \end{bmatrix} = -\begin{bmatrix} \mathcal{L}_\Omega(\tilde{u} + \Phi_c^T u_c) \\ \mathcal{L}_{\Omega r}(\tilde{u}_r + \Phi_{rc}^T u_c) \end{bmatrix} - \begin{bmatrix} \Phi_c^T \\ \Phi_{rc}^T \end{bmatrix} \dot{u}_c$$

$$\begin{bmatrix} \tilde{u}(x_c,t) \\ \tilde{u}_r(x_c,t) \end{bmatrix} = 0 \tag{10}$$

$$\begin{aligned}
\dot{\begin{bmatrix} a \\ a_r \end{bmatrix}} &= \mathcal{A}(\begin{bmatrix} a \\ a_r \end{bmatrix}, u_c) + \dot{B}\dot{u}_c \\
\mathcal{A}([\,], u_c) &= -\int_\Omega \begin{bmatrix} \Phi\mathcal{L}_\Omega(\Phi^T a + \Phi_c^T u_c) \\ \Phi\mathcal{L}_{\Omega r}(\Phi^T a_r + \Phi_{rc}^T u_c) \end{bmatrix} dx \\
\dot{B} &= -\int_\Omega \begin{bmatrix} \Phi\Phi_c^T \\ \Phi\Phi_{rc}^T \end{bmatrix} dx \\
\begin{bmatrix} u \\ u_r \end{bmatrix} &= \begin{bmatrix} \Phi^T a \\ \Phi^T a_r \end{bmatrix} + \begin{bmatrix} \Phi_c^T \\ \Phi_{rc}^T \end{bmatrix} u_c
\end{aligned}$$

which allows the simultaneous reconstruction of both models.

3.2. MEASUREMENTS

In order to use outputs for real time control, the following measurement equation must be added to equations (10):

$$\begin{aligned}
u_m &= \begin{bmatrix} \Phi_{cm}^T & \Phi_m^T \end{bmatrix} \begin{bmatrix} u_c \\ a \end{bmatrix} \\
u_{rm} &= \begin{bmatrix} \Phi_{rcm}^T & \Phi_m^T \end{bmatrix} \begin{bmatrix} u_c \\ a_r \end{bmatrix}
\end{aligned} \tag{11}$$

with $\Phi_m, \Phi_{cm}, \Phi_{rcm}$ the restriction of Φ, Φ_c, Φ_{rc} to the measurement locations in the exact and *reference* flow, or more generally, any space derivation of Φ, Φ_c, Φ_{rc}.

4. Design of the controller

4.1. CONSTRUCTION OF THE CLOSED LOOP

For the sake of simplicity, the equations are derivated for the linear case. With *reference input* ν_r, equations (9) and (11) give the following state space representation:

$$\begin{aligned}
\dot{\begin{bmatrix} u_c \\ a_r \end{bmatrix}} &= \begin{bmatrix} 0 & 0 \\ B_r & A_r \end{bmatrix} \begin{bmatrix} u_c \\ a_r \end{bmatrix} + \begin{bmatrix} I \\ \dot{B}_r \end{bmatrix} \nu_r \\
u_{rm} &= \begin{bmatrix} \Phi_{rcm}^T & \Phi_m^T \end{bmatrix} \begin{bmatrix} u_c \\ a_r \end{bmatrix}
\end{aligned}$$

Similarly, adding a control signal ν to the actuator on the real flow, according to figure 1, equations (6) and (11) write:

$$\begin{aligned}
\dot{\begin{bmatrix} u_c \\ a \end{bmatrix}} &= \begin{bmatrix} 0 & 0 \\ B_0 & A_0 \end{bmatrix} \begin{bmatrix} u_c \\ a \end{bmatrix} + \begin{bmatrix} I \\ \dot{B}_0 \end{bmatrix} (\nu + \nu_r) \\
u_m &= \begin{bmatrix} \Phi_{cm}^T & \Phi_m^T \end{bmatrix} \begin{bmatrix} u_c \\ a \end{bmatrix}
\end{aligned}$$

A controller must be designed for the realization of the control signal ν. A static structure is chosen:

$$\nu = \begin{bmatrix} K_0 & K_r \end{bmatrix} \begin{bmatrix} u_m \\ u_{rm} \end{bmatrix} \tag{12}$$

This leads to the following closed loop equations

$$\begin{aligned} \dot{\tilde{a}} &= F\tilde{a} + G\nu_r \\ y &= \begin{bmatrix} u_m \\ u_{rm} \end{bmatrix} = H\tilde{a} \end{aligned} \tag{13}$$

with

$$\begin{aligned} \tilde{a} &= \begin{bmatrix} u_c & a & u_{cr} & a_r \end{bmatrix}^T \\ F &= F_0 + G_0 K H \\ F_0 &= \begin{bmatrix} 0 & 0 & 0 & 0 \\ B_0 & A_0 & 0 & 0 \\ 0 & 0 & 0 & 0 \\ 0 & 0 & B_r & A_r \end{bmatrix} \\ K &= \begin{bmatrix} K_0 & K_r \end{bmatrix} \\ G_0 &= \begin{bmatrix} I & \dot{B}_0 & 0 & 0 \end{bmatrix}^T \\ G &= \begin{bmatrix} I & \dot{B}_0 & I & \dot{B}_r \end{bmatrix}^T \\ H &= \begin{bmatrix} \Phi_{cm}^T & \Phi_m^T & 0 & 0 \\ 0 & 0 & \Phi_{rcm}^T & \Phi_m^T \end{bmatrix} \end{aligned}$$

Remarks:

- Vector y includes all outputs from the real flow and from the *reference flow*. Though the objective is to minimize the output error $u_m - u_{rm}$, the controller uses u_m and u_{rm} through different gains, as shown in equation (12) and illustrated in figure 1;
- Two different integrators must be included in the augmented model, to distinguish the dynamics of both models;
- The structure of the augmented systems shows that the dynamics of the *reference model* (lower right part of matrix F) remains unchanged in closed loop, as it is not expected to be controlled. On the contrary, the closed loop dynamics of the real flow (upper left part of F) directly depends on the control gain K.

4.2. STRUCTURE OF THE CONTROLLER

Because of the dynamics of the *reference flow*, the controller has its own dynamics. Its inputs are the *reference input* and the actual measurements. Gains K_r and K_0 can be respectively interpreted as a *feedforward* gain (from the command) and a *feedback* gain (from the outputs). The controller state space representation is

$$
\begin{aligned}
\begin{bmatrix} \dot{u}_c \\ a_r \end{bmatrix} &= \begin{bmatrix} 0 & 0 \\ B_r & A_r \end{bmatrix} \begin{bmatrix} u_c \\ a_r \end{bmatrix} + \begin{bmatrix} I & 0 \\ \dot{B}_r & 0 \end{bmatrix} \begin{bmatrix} \nu_r \\ u_m \end{bmatrix} \\
\nu + \nu_r &= K_r \begin{bmatrix} \Phi_{rcm}^T & \Phi_m^T \end{bmatrix} \begin{bmatrix} u_c \\ a_r \end{bmatrix} + \begin{bmatrix} I & K_0 \end{bmatrix} \begin{bmatrix} \nu_r \\ u_m \end{bmatrix}
\end{aligned}
$$

4.3. DERIVATION OF THE CONTROLLER EQUATIONS

Considering the augmented system of equation (13) and the objective of achieving a good *reference following*, the controller will be computed to minimize the closed loop quadratic error criterion

$$J = \int_0^\infty (u_m - u_{rm})^T (u_m - u_{rm}) + \nu^T R \nu dt \tag{14}$$

with R a weighting matrix which penalizes the control effort. Considering the closed loop constraint

$$FV + VF + GG^T = 0$$

with V the closed loop impulse gramian [15], introducing a Lagrange parameter Λ and

$$Q = \begin{bmatrix} 1 & -1 \end{bmatrix}^T \begin{bmatrix} 1 & -1 \end{bmatrix}$$

the matrix transcription for $(u_m - u_{rm})^T(u_m - u_{rm})$, a new augmented criterion is used

$$J = trace(H^T(Q + K^T RK)HV + \Lambda(FV + VF + GG^T))$$

The optimality equations are:

$$
\begin{aligned}
\frac{\partial J}{\partial \Lambda} &= FV + VF^T + GG^T = 0 \\
\frac{\partial J}{\partial V} &= F^T \Lambda + \Lambda F + H^T (Q + K^T RK) H = 0 \\
\frac{1}{2} \frac{\partial J}{\partial K} &= RKHVH^T + G_0 \Lambda V H^T = 0
\end{aligned} \tag{15}
$$

and must be solved using iterative algorithms [18, 10]. Note that:

- The second equation is a kind of Ricatti equation, so that matrix Λ can also be interpreted as a closed loop gramian.
- The last equation writes

$$K = -R^{-1} G^T \Lambda V H^T (HVH^T)^{-1}$$

showing that the feedforward (K_r) and feedback (K_0) gains are computed simultaneously.

- All simplifies to the well know solution of the Ricatti equation for state feedback, or more generally when the output matrix is invertible (all states observable).

$$\begin{aligned} 0 &= F_0^T\Lambda + \Lambda F_0 + Q - H^T K^T R K H \\ KH &= -R^{-1} G_0^T \Lambda \end{aligned} \quad (16)$$

The gain K does not depend any more on gramian V and the first equation can then be removed.

5. Interpretation and illustration

5.1. INTRODUCTION OF DISTRIBUTED GRAMIANS

Due to the expressions of matrices in equations (7), and to the structure of F, G, H in equations (13), the following parameterization can be introduced

$$\begin{aligned} F &= \int_\Omega \Phi(x) f(x) dx \\ G &= \int_\Omega \Phi(x) g(x) dx \\ H &= \int_\Omega H\Phi(x)\Phi^T(x) dx \end{aligned}$$

The symmetric positive gramians V, Λ can also be parameterized as:

$$\begin{aligned} V &= \int_\Omega \Phi(x) v(x) \Phi^T(x) dx \\ \Lambda &= \int_\Omega \Phi(x) \lambda(x) \Phi^T(x) dx \end{aligned}$$

where $v(x), \lambda(x)$ are solution of distributed Lyapunov and Ricatti equations

$$\begin{aligned} 0 &= f\Phi v + v^T \Phi^T f^T + g g^T \\ 0 &= f^T \Phi \lambda + \lambda^T \Phi^T f + ... \\ ...+ & \; \Phi^T H^T (Q + K^T R K) H \Phi \end{aligned}$$

Symmetric positive matrixes $v(x), \lambda(x)$ can be interpreted, in the linear case only, as the distributed gramians of the infinite dimensional flow submitted to impulse input $\dot{u}_c$. The structure of the augmented system leads to the following structure for the gramian:

$$V = \begin{bmatrix} V_0 & V_{0r} \\ V_{0r}^T & V_r \end{bmatrix}$$

where V_r is the closed loop gramian of the reference flow, constant versus K, V_0 the closed loop gramian for the exact flow, function of the control gain K, and V_{0r} the correlation between the states of both flows. If V_{0r} is small, this correlation is poor and the controlled flow is different from the reference flow.

The objective is not reached. If the real flow is forced to follow the *reference flow*, then $V_0 \approx V_r \approx V_{0r}$, which can be related to controllability gramian assignment [19].

5.2. EVALUATION OF THE CONTROLLER GAIN

The parameterization of gain K as a distributed gain is not straightforward in the general case. It is presented here for state feedback, or more generally when the output matrix is invertible. The computation of the controller can be performed with simplified equations (16). Using the parameterization of Λ, and introducing a similar parameterization

$$\begin{aligned} HQH^T &= \int_\Omega \Phi(x)q(x)\Phi^T(x)dx \\ F_0 &= \int_\Omega \Phi(x)f_0(x)dx \end{aligned}$$

the gain writes

$$\begin{aligned} KH &= \int_\Omega k(x)H\Phi^T(x)dx \\ k(x)H &= -R^{-1}G_0^T\Phi(x)\lambda(x) \\ 0 &= f_0^T\Phi\lambda + \lambda\Phi^T f_0 + \dots \\ \dots + &\quad q - \lambda\Phi^T G_0 R G_0^T \Phi\lambda \end{aligned}$$

This distributed formulation shows that all distributed variables of the flow are implicitly used for the computation of the gain, and $q(x)$ is a distributed weighting for the contribution of the states to the outputs..

5.3. ILLUSTRATION

This strategy can be applied to flows with the following characteristics:
• The flow must be modeled with a PDE, and numerical solutions must be computable (or experimental results available);
• The dynamics must be simple enough to be contained in a limited set of *modes*, observable through the measured outputs;
• The reference flow must have a convenient dynamics, not too much different from the natural dynamics;
• PDE must have a stationary solution $\Phi_c(x)$ for $u_c(x_c) = 1$.
It is currently applied to the 2-D boundary layer transition location control with suction [13].

6. Conclusion

An indirect approach is proposed for active flow control. The objective is to make the flow dynamics follow a target flow called *reference flow*, using a closed loop architecture. A static controller is proposed, with two gains, one feeding forward into the actuators the outputs of the reference flow, one feeding back the measurements from the real flow. The design procedure includes the following steps:

- A low order model is constructed for the flow dynamics, using proper orthogonal decomposition (POD) techniques
- The modes for the decomposition are chosen for a simultaneous decomposition of the *reference flow* dynamics and construction of a low order *reference model.*
- An output feedback controller is then designed to minimize the error between both simplified models.

The equations for the computation of the controller are derived. Particularly, the controller can be interpreted within a distributed framework.

References

1. N. Aubry. On the hidden beauty of POD. *Theoretical and Computational Fluid Dynamics*, 2, 91.
2. J.A. Burns and Y.R. Ou. Active control of vortex shedding. *AIAA paper 94-0182*, 94.
3. L. Cordier, R. Manceau, J. Delville, and J.P. Bonnet. Low dimensional description of the dynamical behavior of a plane turbulent mixing layer. *Wokshop on Flow Control: fundamentals and practices*, 96.
4. A.E. Deane et al. Low dimensional models for complex geometry flows. *Physics of Fluids A*, 3(10), 91.
5. L. Cordier et al. Etude de la couche limite de mélange plane turbulente incompressible à l'aide d'un système dynamique d'ordre réduit. *32ème Congrès AAAF:Optimisation et Contrôle des écoulements en aérodynamique*, 96.
6. N. Aubry et al. The dynamics of coherent structures in the wall region of a turbulent boundary layer. *Journal of Fluid Mechanics*, 192, 88.
7. R. Glovinski and J.L. Lions. Exact and approximate controllability for distributed parameter systems. *Acta numérica*, 94.
8. H. Kwakernaak and R. Sivan. Linear optimal control systems. *J. Wiley*, 72.
9. J.L. Lions. Optimal control of systems governed by PDEs. *Springer-Verlag, New York*, 71.
10. D.D. Moerder and A.J. Calise. Convergence of a numerical algorithm for calculating optimal output feedback gains. *IEEE Trans. Aut. Contr.*, sept. 85.
11. S. Prudhomme and L. Le Letty. Flow variables estimation for active control, using POD and Kalman filtering. *European Fluid Mechanics Conference*, 97.
12. S. Prudhomme and L. Le Letty. Finite dimensional Kalman filtering for distributed parameter systems. *Submitted to Automatica*, 98.
13. S. Prudhomme, P. Mouyon, and P. Villedieu. On the use of low order models for active flow control. *International Workshop on Control in Fluid Mechanics and Combustion*, 98.
14. L. Sirovich. Turbulence and the dynamics of coherent structures. *Quarterly of Applied Mathematics*, XLV, 87.
15. V. Sreeman and P. Goddard. Model reduction using impulse response gramians: a frequency domain approach. *Int J. Systems Sci.*, 23(10), 92.
16. B.L. Stevens and F.L. Lewis. Aircraft control and simulation. *J. Wiley*, 92.
17. K.Y. Tang, W.R. Graham, and J. Peraire. Active flow control using a reduced order model and optimum control. *AIAA Fluid Dynamics Conference*, 96.
18. H.T. Toivonen. A globally convergent algorithm for the optimal constant output feedback problem. *Int. J. Contr.*, 85.
19. K. Yasuda and R.E. Skelton. Assigning controllability and observability gramians in feedback control. *Proc. AIAA Guid. Navig. and Contr. Conf.*, 91.

List of Participants

ASHILL, PATRICK
Defence Evaluation & Research Agency (DERA)
High Speed & Weapon Aerodynamics Dept., Bldg. 17
Bedford, MK41 6AE / UK
Fax: +44-1234 225848

ASMOLOV, EVGENY S.
Central Aero-Hydrodynamic Institute (TsAGI)
Dept. of Fundamental Researches
Zhukovsky 3
Moscow Region 140160 / RUSSIA
Fax: +7 095 556 4337
e-mail: aes@an.aerocentr.msk.su

AUGENSTEIN, EDGAR
French-German Research Institute of Saint-Louis (ISL)
Aerodynamics
P.O. Box 34
68301 Saint-Louis Cedex / FRANCE
Fax: +33 3 8969 5048
e-mail: ISL.D2@cedocar.fr

BADER, VIKTOR
DLR
Institut für Strömungsmechanik
Bunsenstraße 10
37073 Göttingen / GERMANY
Fax: +49-551 709-2829
e-mail: viktor.bader@dlr.de

BAKCHINOV, ANDREY A.
Chalmers University of Technology
Dept. of Thermo- and Fluiddynamics
Gothenburg / SWEDEN
Fax: +46-8-790 7654
e-mail: andrey@tfd.chalmers.se

BAO, FENG
DLR
Institut für Strömungsmechanik
Bunsenstraße 10
37073 Göttingen / GERMANY
Fax: +49-551 709-2829
e-mail: Feng.Bao@dlr.de

BARBERIS, DIDIER
ONERA
8 Rue des Vertugadins
92190 Meudon / FRANCE
Fax: +33-1 46 23 51 58
e-mail: barberis@onera.fr

BARDAKHANOV, SERGEY
Russian Academy of Sciences, Sib. Branch
Institute of Theoretical and Applied Mechanics
Institutskaya St. 4/1
630090 Novosibirsk / RUSSIA
Fax: +7-3832 35 2268
e-mail: bard@itam.nsc.ru

BECHERT, DIETRICH W.
DLR
Turbulenzforschung, Institut für Antriebstechnik
Müller-Breslau-Str. 8
10623 Berlin / GERMANY
Fax: +49 30-310 006-39
e-mail: Dietrich.Bechert@dlr.de

BETTS, CAROLINE J.
Defence Evaluation and Research Agency (DERA)
HWA Department
Building 17
Bedford MK41 6AE / UK
Fax: +44 1234 225848
e-mail: jfulker@dera.gov.uk

G.E.A. Meier and P.R. Viswanath (eds.),
IUTAM Symposium on Mechanics of Passive and Active Flow Control, 367.

BIELER, HERIBERT
Daimler-Benz Aerospace Airbus GmbH
Abt. EFV
28183 Bremen / GERMANY
Fax: +49-421 538 4486
e-mail: h.bieler.@airbus.dasa.de

BIPPES, HANS
DLR
Institut für Strömungsmechanik
Bunsenstraße 10
37073 Göttingen / GERMANY
Fax: +49-551 709-2829
e-mail: Hans.Bippes@dlr.de

BOHNING, RAINER
Universität Karlsruhe
Institut für Strömungslehre
Kaiserstraße 12
76128 Karlsruhe / GERMANY
Fax: +49 721 696 727
e-mail: Bohning@islhp7.mach.uni-karlsruhe.de

BOIKO, ANDREY V.
DLR
Institut für Strömungsmechanik, c/o Dr. Dallmann
Bunsenstraße 10
37073 Göttingen / GERMANY
Fax: +49 551 709-2404

BONNET, JEAN PAUL
Université de Poitiers, CEAT
Laboratoire d'Etudes Aerodynamiques (LEA)
43 route de l'Aérodrome
86036 Poitiers Cedex / FRANCE
Fax: +33-5 49 53 70 01
e-mail: bonnet@univ-poitiers.fr

BRUSE, MARTIN
DLR
Turbulenzforschung, Institut für Antriebstechnik
Müller-Breslau-Str. 8
10623 Berlin / GERMANY
Fax: +49-30-310006-39
e-mail: Martin.Bruse@dlr.de

CHANDRASEKHARA, MUGURU S.
NASA Ames Research Center
Navy-NASA Joint Institute of Aeronautics
Mail-Stop 260-1
Moffett Field, CA 94035-1000 / USA
Fax: +1 650 604-4511/4269
e-mail: mchandra@mail.arc.nasa.gov

COLLIN, ERWAN
Université de Poitiers, CEAT
Laboratoire d'Etudes Aerodynamiques (LEA)
43 route de l'Aérodrome
86036 Poitiers Cedex / FRANCE
Fax: +33-5 49 53 70 01
e-mail: collin@univ-poitiers.fr

CROUCH, JEFFREY
Boeing Commercial Airplane Group
Aerodynamic Research
P.O. Box 3707, MC 67-LF
Seattle, WA 98124-2207 / USA
Fax: +1 425-237-8281
e-mail: jeffrey.d.crouch@boeing.com

DALLMANN, UWE
DLR
Institut für Strömungsmechanik
Bunsenstrasse 10
37073 Göttingen / GERMANY
Fax: +49 551 709-2404
e-mail: uwe.dallmann@dlr.de

DANKERT, CARL
DLR
Institut für Strömungsmechanik
Bunsenstrasse 10
37073 Göttingen / GERMANY
Fax: +49 551 709-2800
e-mail: carl.dankert@dlr.de

DARGEL, GEORG
Daimler-Benz Aerospace Airbus GmbH
Abt. EFV
28183 Bremen / GERMANY
Fax: +49-421 538 4486
e-mail: g.dargel.@airbus.dasa.de

DELERY, JEAN
ONERA
Fundamental/Experimental
Aerodynamics Department
8 Rue des Vertugadins
92190 Meudon / FRANCE
Fax: +33 146 23 5158
e-mail: delery@onera.fr

DILLMANN, ANDREAS
Technische Universität Berlin
Hermann-Föttinger-Institut für
Strömungsmechanik
Müller-Breslau-Str. 8
10623 Berlin / GERMANY
Fax: +49 30 3142-1101
e-mail: Dillmann@pi.tu-berlin.de

DIMA, CORNELIU
Centro Italiano Ricerche Aerospaziali
(CIRA)
Via Maiorise
81043 Capua (CE) / ITALY
Fax: +39 823 62 3335
e-mail: c.dima@cira.it

DINKELACKER, ALBRECHT
Schöneberger Str. 38
37085 Göttingen / GERMANY
Fax: +49-551-5176-704

DOERFFER, PIOTR
Polish Academy of Sciences, IMP PAN
Institute of Fluid Flow Machinery
ul. Fiszera 14
80-952 Gdansk / POLAND
Fax: +48-58 341-61-44
e-mail:
doerffer@imppan.imp.pg.gda.pl

DOVGAL, ALEXANDER V.
Russian Academy of Sciences, Sib.
Branch
Inst. of Theoretical and Applied
Mechanics
Institutskaya St. 4/1
630090 Novosibirsk / RUSSIA
Fax: +7-3832 35 2268
e-mail: dovgal@itam.nsc.ru

EGAMI, YASUHIRO
Tohoku University
Institute of Fluid Science
2-1-1 Katahira, Aoba-ku
Sendai 980-8577 / JAPAN
Fax: +81-22 217 5280
e-mail: egami@ifs.tohoku.ac.jp

FFOWCS WILLIAMS, JOHN E.
University of Cambridge
Dept. of Engineering
Trumpington Street
Cambridge, CB2 1PZ / UK
Fax: +44-1223 464 815
e-mail: jefw@eng.cam.ac.uk

FULKER, JOHN L.
Defence Evaluation and Research
Agency (DERA)
HWA Department
Building 17
Bedford MK41 6AE / UK
Fax: +44 1234 225848
e-mail: jfulker@dera.gov.uk

GEISSLER, WOLFGANG
DLR
Institut für Strömungsmechanik
Bunsenstraße 10
37073 Göttingen / GERMANY
Fax: +49-551 709-2811
e-mail: wolfgang.geissler@dlr.de

GILLIES, ERIC A.
University of Glasgow
Dept. of Aerospace Engineering
James Watt Building
Glasgow G12 8QQ / UK
Fax: +44 141 330 5560
e-mail: ericg@aero.gla.ac.uk

GLEZER, ARI
School of Mechanical Engineering
Georgia Institute of Technology
Atlanta, GA 30332-0405 / USA
Fax: +1-404 894 8496
e-mail: ari.glezer@me.gatech.edu

GMELIN, CHRISTOPH
Universität Stuttgart
Institut für Aero- und Gasdynamik
Pfaffenwaldring 21
70550 Stuttgart / GERMANY
Fax: +49-711 6853438
e-mail: gmelin@iag.uni-stuttgart.de

GOBERT, JEAN-LOUIS
ONERA-CERT
Département Prospectives et Synthèses
2, Avenue Edouard Belin
31055 Toulouse Cedex / FRANCE
Fax: +33-5 62 25 25 93
e-mail: gobert@onecert.fr

GÖLLING, BURKHARD
DLR
Institut für Strömungsmechanik
Bunsenstraße 10
37073 Göttingen / GERMANY
Fax: +49 551 709-2829
e-mail: burkhard.goelling@dlr.de

GRÄSEL, JÜRGEN
RWTH Aachen
Stoßwellenlabor
Templergraben 55
52056 Aachen / GERMANY
Fax: +49-241 8888 175
e-mail: graesel@winston.swl.rwth-aachen.de

GROSCHE, FRITZ-REINHARD
DLR
Institut für Strömungsmechanik
Bunsenstraße 10
37073 Göttingen / GERMANY
Fax: +49-551 709-2829
e-mail: fr.grosche@dlr.de

GYR, ALBERT
ETH Zürich
Institut für Hydromechanik und Wasserwirtschaft
Hil-Gebäude G 37.1
8093 Zürich / SWITZERLAND
Fax: +41 1 633-1061
e-mail: gyr@ihw.baum.ethz.ch

HAGE, WOLFRAM
DLR
Institut für Antriebstechnik,
Müller-Breslau-Str. 8
10623 Berlin / GERMANY
Fax: +49-30 310006 39
e-mail: hage@pi.tu-berlin.de

HEIN, STEFAN
DLR
Institut für Strömungsmechanik
Bunsenstraße 10
37073 Göttingen / GERMANY
Fax: +49-551 709 2404
e-mail: Stefan.Hein@dlr.de

HO, CHIH-MING
University of California
Mechanical & Aerospace Engineering
Dept.
P.O. Box 951597
Los Angeles, CA 90095-1597 / USA
Fax: +1-310-206 2302
e-mail: chihming@ucla.edu

HÜBLER, ALFRED
University of Illinois at Urbana-Champaign
Department of Physics, Beckmann Institute
405 N. Mathew
Urbana, IL 61801 / USA
Fax: +1 217 244-8371
e-mail: a-huebler@uiuc.edu

HUSSAIN, FAZLE
University of Houston
Department of Mechanical Engineering
4800 Calhoun Road
Houston, Texas 77204-4792 / USA
Fax: +1-713 743 4503
e-mail: fhussain@UH.edu

IUSO, GAETANO
Politecnico di Torino
Dipartimento di Ingegneria Aeronautica e Spaziale (DIAS)
Corso Duca degli Abruzzi, 24
10129 Torino / ITALY
Fax: +39-11 564 6899
e-mail: iuso@polito.it

JANKE, ERIK
DLR
Institut für Strömungsmechanik
Bunsenstraße 10
37073 Göttingen / GERMANY
Fax: +49 551 709-2404
e-mail: erik.janke@dlr.de

KALTENBACH, HANS-JAKOB
Technische Universität Berlin
Hermann-Föttinger-Institut, Sekr. HF1
Straße des 17. Juni 135
10623 Berlin / GERMANY
Fax: +49-30 314 21110
e-mail: Kalten@math.tu-berlin.de

KATO, HIROHARU
University of Tokyo
Department of Environmental and Ocean Engineering
Hongo 7-3-1, Bunkyo
Tokyo 113 / JAPAN
Fax: +81-3 3815 8360
e-mail: kato@fluidlab.naoe.t.u-tokyo.ac.jp

KHAKIMOV, RUSTEM A.
Central Aerohydrodynamics Institute (TsAGI)
Power Unit Dept.
Zhukovsky Street, 1
140160 Moscow Region / RUSSIA
Fax: +7-095 556 4337
e-mail:
Khakimov@p19.f1101.n5020.z2.fidonet.org

KIBENS, VALDIS
The Boeing Company
P.O. Box 516, MC S 106-7126
St. Louis, MO 63166-0516 / USA
Fax: +1-314 777 1328
e-mail: valdis.kibens@boeing.com

KIM, JOHN
University of California, Los Angeles (UCLA)
Department of Mechanical and Aerospace Engineering
420 Westwood Plaza
Los Angeles, CA 90095-1597 / USA
Fax: +1 310-206-4830
e-mail: jkim@seas.ucla.edu

KOGAN, MIKHAIL N.
Central Aero-Hydrodynamic Institute
(TsAGI)
Dept. of Fundamental Researches
Zhukovsky 3
Moscow Region 140160 / RUSSIA
Fax: +7 095 556 4337
e-mail: mkogan@aerocentr.msk.su

KOHAMA, YASUAKI
Tohoku University
Institute of Fluid Science
2-1-1 Katahira, Aoba-ku
Sendai 980-8577 / JAPAN
Fax: +81-22 217 5280
e-mail: Kohama@ifs.tohoku.ac.jp

KOUMOUTSAKOS, PETROS
ETH Zürich
Institute of Fluid Dynamics
Sonneggstraße 3
8092 Zürich / SWITZERLAND
Fax: +41-1 632 1147
e-mail: petros@ifd.mavt.ethz.ch

KOZLOV, VIKTOR V.
Russian Academy of Sciences, Sib. Div.
Institute of Theroretical and Applied
Mechanics
Institutskaya Street 4/1
630090 Novosibirsk / RUSSIA
Fax: +7-3832 35 22 68
e-mail: kozlov@itam.nsc.ru

KREPLIN, HANS-PETER
DLR
Institut für Strömungsmechanik
Bunsenstraße 10
37073 Göttingen / GERMANY
Fax: +49 551 709-2829
e-mail: HP.Kreplin@dlr.de

KUMAR, ANAND
CSIR Centre for Mathematical
Modelling and Computer Simulation
Bangalore 560037 / INDIA
Fax: +91 80 526 0392
e-mail: kumar@cmmacs.ernet.in

LANGE, BJÖRN
Universität Göttingen
III. Physikalisches Institut
Bürgerstr. 42-44
37073 Göttingen / GERMANY
Fax: +49-551 39-7720
e-mail: blange@physik3.gwdg.de

LIN, JOHN C.
NASA Langley Research Center
Flow Modeling and Control Branch
Mail Stop 170
Hampton, VA 23681-2199 / USA
Fax: +1 757 864-7897
e-mail: J.C.Lin@LaRC.NASA.GOV

LU, MEI
DLR
Institut für Strömungsmechanik
Bunsenstraße 10
37073 Göttingen / GERMANY
Fax: +49-551 709-2829
e-mail: mei.lu@dlr.de

LUNDELL, FREDRIK
Royal Institute of Technology, KTH
Dept. of Mechanics
10044 Stockholm / SWEDEN
Fax: +46-8-790 7654
e-mail: fredrik@mech.kth.se

MANUILOVICH, SERGEI V.
Central Aero-Hydrodynamic Institute
(TsAGI)
Dept. of Fundamental Researches
Zhukosky 3
Moscow Region 140160 / RUSSIA
Fax: +7 95 556 43 37
e-mail: manu@recp.aerocentr.msk.su

MARDUEL, XAVIER
University of Graz
Institute of Mathematics
Heinrichstrasse 36
8010 Graz / AUSTRIA
Fax: +43-316 380 9815
e-mail:
xavier.marduel@kfunigraz.ac.at

MAY, CAROL
Cortana Corporation
520 N. Washington St., No. 200
Falls Church, VA 22046 / USA
Fax: +1-703 534 8005
e-mail: cmay@cortana.com

MEIER, GERD E.A.
DLR
Institut für Strömungsmechanik
Bunsenstraße 10
37073 Göttingen / GERMANY
Fax: +49 551 709-2889
e-mail: G.E.A.Meier@dlr.de

MEYER, DANIEL
Universität Stuttgart
Institut für Aero- und Gasdynamik
(IAG)
Pfaffenwaldring 21
70550 Stuttgart / GERMANY
Fax: +49-711 685 3438
e-mail: Daniel.Meyer@iag.uni-stuttgart.de

MEYER, ROBERT
Technische Universität Berlin, FB 10
Hermann-Föttinger-Inst. für
Strömungsmechanik
Müller-Breslau-Str. 8
10623 Berlin / GERMANY
Fax: +49-30-31000 639
e-mail: meyer@pi.tu-berlin.de

MITTAL, SANJAY
Indian Institute of Technology I.I.T.
Department of Aerospace Engineering
P.O. - I.I.T.
Kanpur 208016 / INDIA
Fax: +91-512 590 007
e-mail: smittal@iitk.ac.in

MOCHIZUKI, OSAMU
Hokkaido University
Division of Mechanical Science
Kita 13, Nishi 8
Kita-ku, Sapporo 060-8628 / JAPAN
Fax: +81-11 706 7889
e-mail: osamu@eng.hokudai.ac.jp

MODI, VINOD J.
University of British Columbia
Department of Mechanical Engineering
2324 Main Mall
Vancouver, B. C., V6T 1Z4 /
CANADA
Fax: +1 604 822 2403

MÖHRING, WILLI
Max-Planck-Institut für
Strömungsforschung
Bunsenstraße 10
37073 Göttingen / GERMANY
Fax: +49-551 5176-595
e-mail: wmoehri@gwdg.de

MORRISON, JONATHAN F.
Imperial College
Department of Aeronautics
Prince Consort Road
London SW7 2BY / UK
Fax: +44 171 584 8120
e-mail: j.morrrison@ic.ac.uk

MORZYNSKI, MAREK
Poznan University of Technology
Division of Basics of Machine Design
ul. Piotrowo 3
60-965 Poznan / POLAND
Fax: +48 61 8782 618
e-mail:
morzynski@stanton.bmd.put.poznan.pl

NARASIMHA, RODDAM
National Institute of Advanced Studies
Indian Institute of Science
Bangalore 560 012 / INDIA
Fax: +91 80 334 6634
e-mail: roddam@caos.iisc.ernet.in

NIKISHOVA, OLGA D.
National Academy of Science of the Ukraine
Institute of Hydromechanics
8/4 Zheliabov Str.
252057 Kiev / UKRAINE
Fax: +380-44 446 42 29
e-mail: vin@ihm.kiev.ua

OLSEN, JOHN F.
The University of Newcastle
Dept. of Mechanical Engineering
University Drive
Callaghan NSW 2308 / AUSTRALIA
Fax: +61-2 4921 7405
e-mail: meturb6@cc.newcastle.edu.au

ORELLANO, ALEXANDER
Universität der Bundeswehr München
Institut für Strömungsmechanik u. Aerodynamik
85577 Neubiberg / GERMANY
Fax: +49-89 6004 4092
e-mail: alexander.orellano@unibw-muenchen.de

PRECKEL, HARTWIG
Universität Göttingen
III. Physikalisches Institut
Bürgerstr. 42-44
37073 Göttingen / GERMANY
Fax: +49-551 39-7720
e-mail: hep@physik3.gwdg.de

PROTAS, BARTOSZ
Warsaw University of Technology
Institute of Aeronautics and Applied Mechanics
ul. Nowowiejska 24
00-665 Warszawa / POLAND
Fax: +48-22 628 2587
e-mail: bprotas@meil.pw.edu.pl

RONNEBERGER, DIRK
Universität Göttingen
III. Physikalisches Institut
Bürgerstr. 42-44
37073 Göttingen / GERMANY
Fax: +49-551/39-7720
e-mail:
Dirk.Ronneberger@DPI.physik.uni-goettingen.de

ROSEMANN, HENNING
DLR
Institut für Strömungsmechanik
Bunsenstraße 10
37073 Göttingen / GERMANY
Fax: +49 551 709-2811
e-mail: Henning.Rosemann@dlr.de

SARIC, WILLIAM S.
Arizona State University
Mechanical and Aerospace Engineering
P.O.Box 876106
Tempe, Arizona 85287-6106 / USA
Fax: +1 602 965-1384
e-mail: saric@asu.edu

SCHEWE, GÜNTER
DLR
Institut für Aeroelastik
Bunsenstraße 10
37073 Göttingen / GERMANY
Fax: +49-551 709-2862
e-mail: guenter.schewe@dlr.de

SEMIONOV, NICKOLAY V.
Russian Academy of Sciences, Sib. Div.
Institute of Theoretical and Applied Mechanics
Institutskaya 4/1
630090 Novosibirsk / RUSSIA
Fax: +7 3832 352 268
e-mail: semion@itam.nsc.ru

SENGUPTA, TAPAN K.
Indian Institute of Technology (I.I.T.)
Department of Aerospace Engineering
P.O. - I.I.T.
Kanpur 208016 U.P. / INDIA
Fax: +91-512 590 007,
e-mail: tksen@iitk.ernet.in

SERAUDIE, ALAIN
ONERA-CERT
Dept. DMAE
2, Avenue Edouard Belin
31055 Toulouse Cedex 4 / FRANCE
Fax: +33-5 62 25 25 83
e-mail: Alain.Seraudie@onecert.fr

SHARMA, SHAILENDRA D.
Indian Institute of Technology (I.I.T.) Bombay
Department of Aerospace Engineering
Powai
Mumbai-400 076 / INDIA
Fax: +91-22 578 2602
e-mail: sds@aero.iitb.ernet.in

SILLER, HENRI
Technische Universität Berlin
Hermann-Föttinger-Institut für Strömungsmechanik
Straße des 17. Juni 135
10623 Berlin / GERMANY
Fax: +49-30 314-21101
e-mail: Siller@pi.tu-berlin.de

SOBIECZKY, HELMUT
DLR
Institut für Strömungsmechanik
Bunsenstraße 10
37073 Göttingen / GERMANY
Fax: +49-551-709 2446
e-mail: helmut.sobieczky@dlr.de

STANEWSKY, EGON
DLR
Institut für Strömungsmechanik
Bunsenstraße 10
37073 Göttingen / GERMANY
Fax: +49-551 709-2811
e-mail: egon.stanewsky@dlr.de

TARDU, SEDAT
Lab. de Ecoulements Géophysique et Industrials (LEGI)
BP. 53-X
38041 Grenoble Cédex / FRANCE
Fax: +33 4 7682-5231
e-mail: Sedat.Tardu@hmg.inpg.fr

TATSUMI, TOMOMASA
International Institute for Advanced Studies (IIAS)
26-6 Chikuzen-dai, Momoyama
Kyoto 612-8032 / JAPAN
Fax: +81 75 611 4846 (home)
e-mail: tatsumi@skyblue.ocn.ne.jp

THIEDE, PETER
Daimler-Benz Aerospace Airbus GmbH
Abt. EFV
28183 Bremen / GERMANY
Fax: +49-421 538 4486
e-mail: p.thiede.@airbus.dasa.de

TUMIN, ANATOLI
Tel-Aviv University
Faculty of Engineering
Tel-Aviv, 69978 / ISRAEL
Fax: +972-3-640 7334
e-mail: tumin@eng.tau.ac.il

URZYNICOK, FRANK
Technische Universität Berlin
Hermann-Föttinger-Institut für Strömungsmechanik
Straße des 17. Juni 135
10623 Berlin / GERMANY
Fax: +49-30 314-21101
e-mail: urzynico@obiwan.pi.tu-berlin.de

VASANTA RAM, VENKATESA I.
Ruhr-Universität Bochum
Institut für Thermo- und Fluiddynamik
44780 Bochum / GERMANY
Fax: +49-234-7094 162
e-mail: VVR@lstm.ruhr-uni-bochum.de

VISWANATH, P.R.
National Aerospace Laboratories (NAL)
Experimental Aerodynamics Division
Post Bag 1779
Bangalore 560 017 / INDIA
Fax: +91 80 527 3942
e-mail: prv@ead.cmmacs.ernet.in

WARSOP, CLYDE
British Aerospace PLC
Sowerby Research Centre
FPC 267, PO Box 5
Filton, Bristol BS12 7QW / UK
Fax: +44 -1179 36 3733
e-mail: clyde.warsop@src.bae.co.uk

WENGLE, HANS
Universität der Bundeswehr München
Institut für Strömungsmechanik u. Aerodynamik
85577 Neubiberg / GERMANY
Fax: +49-89 6004 4092
e-mail: hans.wengle@unibw-muenchen.de

WONG, PETER W.C.
Aircraft Research Association Limited (ARA)
Manton Lane
Bedford MK41 7PF / UK
Fax: +44 1234 328 584
e-mail: pwang@ara.co.uk

WYGNANSKI, ISRAEL
University of Arizona
Department of Aerospace and Mechanical Engineering
Aero Building 16
Tucson, Arizona 85721 / USA
Fax: +1-520 621-8191
e-mail: wygy@bigdog.engr.arizona.edu

YOKOMIZO, TOSHIO
Kanto Gakuin University
Department of Mechanical Engineering
Mutsuura, Kanazawa
Yokohama 236 / JAPAN
Fax: +81-468-41-3926

YU, YUNG H.
NASA Ames Research Center
Fluid Mechanics Division
Mail-Stop 207-1
Moffett Field, Ca 94035 / USA
Fax: +1-650 604-2003
e-mail: yyu@mail.arc.nasa.gov

YURCHENKO, NINA F.
National Academy of Sciences
Institute of Hydromechanics
8/4 Zheliabov St.
252057 Kiev / UKRAINE
Fax: +380-44 446 4229
e-mail: yur@rdfm.kiev.ua

ZAPRYAGAEV, VALERY I.
Russian Academy of Sciences, Sib. Div.
Institute of Theoretical and Applied Mechanics
Institutskaya 4/1
630090 Novosibirsk / RUSSIA
Fax: +7 3832 342 268
e-mail: v.zapryagaev@itam.nsc.ru

Author Index

Mechanics

FLUID MECHANICS AND ITS APPLICATIONS

Series Editor: R. Moreau

Aims and Scope of the Series

The purpose of this series is to focus on subjects in which fluid mechanics plays a fundamental role. As well as the more traditional applications of aeronautics, hydraulics, heat and mass transfer etc., books will be published dealing with topics which are currently in a state of rapid development, such as turbulence, suspensions and multiphase fluids, super and hypersonic flows and numerical modelling techniques. It is a widely held view that it is the interdisciplinary subjects that will receive intense scientific attention, bringing them to the forefront of technological advancement. Fluids have the ability to transport matter and its properties as well as transmit force, therefore fluid mechanics is a subject that is particularly open to cross fertilisation with other sciences and disciplines of engineering. The subject of fluid mechanics will be highly relevant in domains such as chemical, metallurgical, biological and ecological engineering. This series is particularly open to such new multidisciplinary domains.

1. M. Lesieur: *Turbulence in Fluids.* 2nd rev. ed., 1990 ISBN 0-7923-0645-7
2. O. Métais and M. Lesieur (eds.): *Turbulence and Coherent Structures.* 1991 ISBN 0-7923-0646-5
3. R. Moreau: *Magnetohydrodynamics.* 1990 ISBN 0-7923-0937-5
4. E. Coustols (ed.): *Turbulence Control by Passive Means.* 1990 ISBN 0-7923-1020-9
5. A.A. Borissov (ed.): *Dynamic Structure of Detonation in Gaseous and Dispersed Media.* 1991 ISBN 0-7923-1340-2
6. K.-S. Choi (ed.): *Recent Developments in Turbulence Management.* 1991 ISBN 0-7923-1477-8
7. E.P. Evans and B. Coulbeck (eds.): *Pipeline Systems.* 1992 ISBN 0-7923-1668-1
8. B. Nau (ed.): *Fluid Sealing.* 1992 ISBN 0-7923-1669-X
9. T.K.S. Murthy (ed.): *Computational Methods in Hypersonic Aerodynamics.* 1992 ISBN 0-7923-1673-8
10. R. King (ed.): *Fluid Mechanics of Mixing.* Modelling, Operations and Experimental Techniques. 1992 ISBN 0-7923-1720-3
11. Z. Han and X. Yin: *Shock Dynamics.* 1993 ISBN 0-7923-1746-7
12. L. Svarovsky and M.T. Thew (eds.): *Hydroclones.* Analysis and Applications. 1992 ISBN 0-7923-1876-5
13. A. Lichtarowicz (ed.): *Jet Cutting Technology.* 1992 ISBN 0-7923-1979-6
14. F.T.M. Nieuwstadt (ed.): *Flow Visualization and Image Analysis.* 1993 ISBN 0-7923-1994-X
15. A.J. Saul (ed.): *Floods and Flood Management.* 1992 ISBN 0-7923-2078-6
16. D.E. Ashpis, T.B. Gatski and R. Hirsh (eds.): *Instabilities and Turbulence in Engineering Flows.* 1993 ISBN 0-7923-2161-8
17. R.S. Azad: *The Atmospheric Boundary Layer for Engineers.* 1993 ISBN 0-7923-2187-1
18. F.T.M. Nieuwstadt (ed.): *Advances in Turbulence IV.* 1993 ISBN 0-7923-2282-7
19. K.K. Prasad (ed.): *Further Developments in Turbulence Management.* 1993 ISBN 0-7923-2291-6
20. Y.A. Tatarchenko: *Shaped Crystal Growth.* 1993 ISBN 0-7923-2419-6

Kluwer Academic Publishers – Dordrecht / Boston / London

Mechanics

FLUID MECHANICS AND ITS APPLICATIONS

Series Editor: R. Moreau

21. J.P. Bonnet and M.N. Glauser (eds.): *Eddy Structure Identification in Free Turbulent Shear Flows.* 1993 ISBN 0-7923-2449-8
22. R.S. Srivastava: *Interaction of Shock Waves.* 1994 ISBN 0-7923-2920-1
23. J.R. Blake, J.M. Boulton-Stone and N.H. Thomas (eds.): *Bubble Dynamics and Interface Phenomena.* 1994 ISBN 0-7923-3008-0
24. R. Benzi (ed.): *Advances in Turbulence V.* 1995 ISBN 0-7923-3032-3
25. B.I. Rabinovich, V.G. Lebedev and A.I. Mytarev: *Vortex Processes and Solid Body Dynamics.* The Dynamic Problems of Spacecrafts and Magnetic Levitation Systems. 1994 ISBN 0-7923-3092-7
26. P.R. Voke, L. Kleiser and J.-P. Chollet (eds.): *Direct and Large-Eddy Simulation I.* Selected papers from the First ERCOFTAC Workshop on Direct and Large-Eddy Simulation. 1994 ISBN 0-7923-3106-0
27. J.A. Sparenberg: *Hydrodynamic Propulsion and its Optimization.* Analytic Theory. 1995 ISBN 0-7923-3201-6
28. J.F. Dijksman and G.D.C. Kuiken (eds.): *IUTAM Symposium on Numerical Simulation of Non-Isothermal Flow of Viscoelastic Liquids.* Proceedings of an IUTAM Symposium held in Kerkrade, The Netherlands. 1995 ISBN 0-7923-3262-8
29. B.M. Boubnov and G.S. Golitsyn: *Convection in Rotating Fluids.* 1995 ISBN 0-7923-3371-3
30. S.I. Green (ed.): *Fluid Vortices.* 1995 ISBN 0-7923-3376-4
31. S. Morioka and L. van Wijngaarden (eds.): *IUTAM Symposium on Waves in Liquid/Gas and Liquid/Vapour Two-Phase Systems.* 1995 ISBN 0-7923-3424-8
32. A. Gyr and H.-W. Bewersdorff: *Drag Reduction of Turbulent Flows by Additives.* 1995 ISBN 0-7923-3485-X
33. Y.P. Golovachov: *Numerical Simulation of Viscous Shock Layer Flows.* 1995 ISBN 0-7923-3626-7
34. J. Grue, B. Gjevik and J.E. Weber (eds.): *Waves and Nonlinear Processes in Hydrodynamics.* 1996 ISBN 0-7923-4031-0
35. P.W. Duck and P. Hall (eds.): *IUTAM Symposium on Nonlinear Instability and Transition in Three-Dimensional Boundary Layers.* 1996 ISBN 0-7923-4079-5
36. S. Gavrilakis, L. Machiels and P.A. Monkewitz (eds.): *Advances in Turbulence VI.* Proceedings of the 6th European Turbulence Conference. 1996 ISBN 0-7923-4132-5
37. K. Gersten (ed.): *IUTAM Symposium on Asymptotic Methods for Turbulent Shear Flows at High Reynolds Numbers.* Proceedings of the IUTAM Symposium held in Bochum, Germany. 1996 ISBN 0-7923-4138-4
38. J. Verhás: *Thermodynamics and Rheology.* 1997 ISBN 0-7923-4251-8
39. M. Champion and B. Deshaies (eds.): *IUTAM Symposium on Combustion in Supersonic Flows.* Proceedings of the IUTAM Symposium held in Poitiers, France. 1997 ISBN 0-7923-4313-1
40. M. Lesieur: *Turbulence in Fluids.* Third Revised and Enlarged Edition. 1997 ISBN 0-7923-4415-4; Pb: 0-7923-4416-2

Kluwer Academic Publishers – Dordrecht / Boston / London

Mechanics

SOLID MECHANICS AND ITS APPLICATIONS

Series Editor: G.M.L. Gladwell

Aims and Scope of the Series

The fundamental questions arising in mechanics are: *Why?, How?,* and *How much?* The aim of this series is to provide lucid accounts written by authoritative researchers giving vision and insight in answering these questions on the subject of mechanics as it relates to solids. The scope of the series covers the entire spectrum of solid mechanics. Thus it includes the foundation of mechanics; variational formulations; computational mechanics; statics, kinematics and dynamics of rigid and elastic bodies; vibrations of solids and structures; dynamical systems and chaos; the theories of elasticity, plasticity and viscoelasticity; composite materials; rods, beams, shells and membranes; structural control and stability; soils, rocks and geomechanics; fracture; tribology; experimental mechanics; biomechanics and machine design.

1. R.T. Haftka, Z. Gürdal and M.P. Kamat: *Elements of Structural Optimization.* 2nd rev.ed., 1990 ISBN 0-7923-0608-2
2. J.J. Kalker: *Three-Dimensional Elastic Bodies in Rolling Contact.* 1990 ISBN 0-7923-0712-7
3. P. Karasudhi: *Foundations of Solid Mechanics.* 1991 ISBN 0-7923-0772-0
4. *Not published*
5. *Not published.*
6. J.F. Doyle: *Static and Dynamic Analysis of Structures.* With an Emphasis on Mechanics and Computer Matrix Methods. 1991 ISBN 0-7923-1124-8; Pb 0-7923-1208-2
7. O.O. Ochoa and J.N. Reddy: *Finite Element Analysis of Composite Laminates.* ISBN 0-7923-1125-6
8. M.H. Aliabadi and D.P. Rooke: *Numerical Fracture Mechanics.* ISBN 0-7923-1175-2
9. J. Angeles and C.S. López-Cajún: *Optimization of Cam Mechanisms.* 1991 ISBN 0-7923-1355-0
10. D.E. Grierson, A. Franchi and P. Riva (eds.): *Progress in Structural Engineering.* 1991 ISBN 0-7923-1396-8
11. R.T. Haftka and Z. Gürdal: *Elements of Structural Optimization.* 3rd rev. and exp. ed. 1992 ISBN 0-7923-1504-9; Pb 0-7923-1505-7
12. J.R. Barber: *Elasticity.* 1992 ISBN 0-7923-1609-6; Pb 0-7923-1610-X
13. H.S. Tzou and G.L. Anderson (eds.): *Intelligent Structural Systems.* 1992 ISBN 0-7923-1920-6
14. E.E. Gdoutos: *Fracture Mechanics.* An Introduction. 1993 ISBN 0-7923-1932-X
15. J.P. Ward: *Solid Mechanics.* An Introduction. 1992 ISBN 0-7923-1949-4
16. M. Farshad: *Design and Analysis of Shell Structures.* 1992 ISBN 0-7923-1950-8
17. H.S. Tzou and T. Fukuda (eds.): *Precision Sensors, Actuators and Systems.* 1992 ISBN 0-7923-2015-8
18. J.R. Vinson: *The Behavior of Shells Composed of Isotropic and Composite Materials.* 1993 ISBN 0-7923-2113-8
19. H.S. Tzou: *Piezoelectric Shells.* Distributed Sensing and Control of Continua. 1993 ISBN 0-7923-2186-3

Kluwer Academic Publishers – Dordrecht / Boston / London

Mechanics

SOLID MECHANICS AND ITS APPLICATIONS

Series Editor: G.M.L. Gladwell

20. W. Schiehlen (ed.): *Advanced Multibody System Dynamics.* Simulation and Software Tools. 1993 ISBN 0-7923-2192-8
21. C.-W. Lee: *Vibration Analysis of Rotors.* 1993 ISBN 0-7923-2300-9
22. D.R. Smith: *An Introduction to Continuum Mechanics.* 1993 ISBN 0-7923-2454-4
23. G.M.L. Gladwell: *Inverse Problems in Scattering.* An Introduction. 1993 ISBN 0-7923-2478-1
24. G. Prathap: *The Finite Element Method in Structural Mechanics.* 1993 ISBN 0-7923-2492-7
25. J. Herskovits (ed.): *Advances in Structural Optimization.* 1995 ISBN 0-7923-2510-9
26. M.A. González-Palacios and J. Angeles: *Cam Synthesis.* 1993 ISBN 0-7923-2536-2
27. W.S. Hall: *The Boundary Element Method.* 1993 ISBN 0-7923-2580-X
28. J. Angeles, G. Hommel and P. Kovács (eds.): *Computational Kinematics.* 1993 ISBN 0-7923-2585-0
29. A. Curnier: *Computational Methods in Solid Mechanics.* 1994 ISBN 0-7923-2761-6
30. D.A. Hills and D. Nowell: *Mechanics of Fretting Fatigue.* 1994 ISBN 0-7923-2866-3
31. B. Tabarrok and F.P.J. Rimrott: *Variational Methods and Complementary Formulations in Dynamics.* 1994 ISBN 0-7923-2923-6
32. E.H. Dowell (ed.), E.F. Crawley, H.C. Curtiss Jr., D.A. Peters, R. H. Scanlan and F. Sisto: *A Modern Course in Aeroelasticity.* Third Revised and Enlarged Edition. 1995 ISBN 0-7923-2788-8; Pb: 0-7923-2789-6
33. A. Preumont: *Random Vibration and Spectral Analysis.* 1994 ISBN 0-7923-3036-6
34. J.N. Reddy (ed.): *Mechanics of Composite Materials.* Selected works of Nicholas J. Pagano. 1994 ISBN 0-7923-3041-2
35. A.P.S. Selvadurai (ed.): *Mechanics of Poroelastic Media.* 1996 ISBN 0-7923-3329-2
36. Z. Mróz, D. Weichert, S. Dorosz (eds.): *Inelastic Behaviour of Structures under Variable Loads.* 1995 ISBN 0-7923-3397-7
37. R. Pyrz (ed.): *IUTAM Symposium on Microstructure-Property Interactions in Composite Materials.* Proceedings of the IUTAM Symposium held in Aalborg, Denmark. 1995 ISBN 0-7923-3427-2
38. M.I. Friswell and J.E. Mottershead: *Finite Element Model Updating in Structural Dynamics.* 1995 ISBN 0-7923-3431-0
39. D.F. Parker and A.H. England (eds.): *IUTAM Symposium on Anisotropy, Inhomogeneity and Nonlinearity in Solid Mechanics.* Proceedings of the IUTAM Symposium held in Nottingham, U.K. 1995 ISBN 0-7923-3594-5
40. J.-P. Merlet and B. Ravani (eds.): *Computational Kinematics '95.* 1995 ISBN 0-7923-3673-9
41. L.P. Lebedev, I.I. Vorovich and G.M.L. Gladwell: *Functional Analysis.* Applications in Mechanics and Inverse Problems. 1996 ISBN 0-7923-3849-9
42. J. Menčik: *Mechanics of Components with Treated or Coated Surfaces.* 1996 ISBN 0-7923-3700-X
43. D. Bestle and W. Schiehlen (eds.): *IUTAM Symposium on Optimization of Mechanical Systems.* Proceedings of the IUTAM Symposium held in Stuttgart, Germany. 1996 ISBN 0-7923-3830-8

Kluwer Academic Publishers – Dordrecht / Boston / London

Mechanics

SOLID MECHANICS AND ITS APPLICATIONS

Series Editor: G.M.L. Gladwell

44. D.A. Hills, P.A. Kelly, D.N. Dai and A.M. Korsunsky: *Solution of Crack Problems.* The Distributed Dislocation Technique. 1996 ISBN 0-7923-3848-0
45. V.A. Squire, R.J. Hosking, A.D. Kerr and P.J. Langhorne: *Moving Loads on Ice Plates.* 1996 ISBN 0-7923-3953-3
46. A. Pineau and A. Zaoui (eds.): *IUTAM Symposium on Micromechanics of Plasticity and Damage of Multiphase Materials.* Proceedings of the IUTAM Symposium held in Sèvres, Paris, France. 1996 ISBN 0-7923-4188-0
47. A. Naess and S. Krenk (eds.): *IUTAM Symposium on Advances in Nonlinear Stochastic Mechanics.* Proceedings of the IUTAM Symposium held in Trondheim, Norway. 1996 ISBN 0-7923-4193-7
48. D. Ieşan and A. Scalia: *Thermoelastic Deformations.* 1996 ISBN 0-7923-4230-5
49. J. R. Willis (ed.): *IUTAM Symposium on Nonlinear Analysis of Fracture.* Proceedings of the IUTAM Symposium held in Cambridge, U.K. 1997 ISBN 0-7923-4378-6
50. A. Preumont: *Vibration Control of Active Structures.* An Introduction. 1997 ISBN 0-7923-4392-1
51. G.P. Cherepanov: *Methods of Fracture Mechanics: Solid Matter Physics.* 1997 ISBN 0-7923-4408-1
52. D.H. van Campen (ed.): *IUTAM Symposium on Interaction between Dynamics and Control in Advanced Mechanical Systems.* Proceedings of the IUTAM Symposium held in Eindhoven, The Netherlands. 1997 ISBN 0-7923-4429-4
53. N.A. Fleck and A.C.F. Cocks (eds.): *IUTAM Symposium on Mechanics of Granular and Porous Materials.* Proceedings of the IUTAM Symposium held in Cambridge, U.K. 1997 ISBN 0-7923-4553-3
54. J. Roorda and N.K. Srivastava (eds.): *Trends in Structural Mechanics.* Theory, Practice, Education. 1997 ISBN 0-7923-4603-3
55. Yu. A. Mitropolskii and N. Van Dao: *Applied Asymptotic Methods in Nonlinear Oscillations.* 1997 ISBN 0-7923-4605-X
56. C. Guedes Soares (ed.): *Probabilistic Methods for Structural Design.* 1997 ISBN 0-7923-4670-X
57. D. François, A. Pineau and A. Zaoui: *Mechanical Behaviour of Materials.* Volume I: Elasticity and Plasticity. 1998 ISBN 0-7923-4894-X
58. D. François, A. Pineau and A. Zaoui: *Mechanical Behaviour of Materials.* Volume II: Viscoplasticity, Damage, Fracture and Contact Mechanics. 1998 ISBN 0-7923-4895-8
59. L. T. Tenek and J. Argyris: *Finite Element Analysis for Composite Structures.* 1998 ISBN 0-7923-4899-0
60. Y.A. Bahei-El-Din and G.J. Dvorak (eds.): *IUTAM Symposium on Transformation Problems in Composite and Active Materials.* Proceedings of the IUTAM Symposium held in Cairo, Egypt. 1998 ISBN 0-7923-5122-3
61. I.G. Goryacheva: *Contact Mechanics in Tribology.* 1998 ISBN 0-7923-5257-2
62. O.T. Bruhns and E. Stein (eds.): *IUTAM Symposium on Micro- and Macrostructural Aspects of Thermoplasticity.* Proceedings of the IUTAM Symposium held in Bochum, Germany. 1999 ISBN 0-7923-5265-3

Kluwer Academic Publishers – Dordrecht / Boston / London

Mechanics

SOLID MECHANICS AND ITS APPLICATIONS

Series Editor: G.M.L. Gladwell

63. F.C. Moon: *IUTAM Symposium on New Applications of Nonlinear and Chaotic Dynamics in Mechanics.* Proceedings of the IUTAM Symposium held in Ithaca, NY, USA. 1998 ISBN 0-7923-5276-9
64. R. Wang: *IUTAM Symposium on Rheology of Bodies with Defects.* Proceedings of the IUTAM Symposium held in Beijing, China. 1999 ISBN 0-7923-5297-1
65. Yu. I. Dimitrienko: *Thermomechanics of Composites under High Temperatures.* 1999 ISBN 0-7923-4899-0
66. P. Argoul, M. Frémond and Q.S. Nguyen (eds.): *IUTAM Symposium on Variations of Domains and Free-Boundary Problems in Solid Mechanics.* Proceedings of the IUTAM Symposium held in Paris, France. 1999 ISBN 0-7923-5450-8
67. F.J. Fahy and W.G. Price (eds.): *IUTAM Symposium on Statistical Energy Analysis.* Proceedings of the IUTAM Symposium held in Southampton, U.K. 1999 ISBN 0-7923-5457-5

Kluwer Academic Publishers – Dordrecht / Boston / London

MIX
Papier aus verantwortungsvollen Quellen
Paper from responsible sources
FSC® C105338

If you have any concerns about our products,
you can contact us on
ProductSafety@springernature.com

In case Publisher is established outside the EU,
the EU authorized representative is:
Springer Nature Customer Service Center GmbH
Europaplatz 3, 69115 Heidelberg, Germany

Printed by Libri Plureos GmbH
in Hamburg, Germany